VOLUME ONE HUNDRED AND TWENTY ONE

VITAMINS AND HORMONES

Antioxidants

EDITORIAL BOARD

VOLUME ONE HUNDRED AND TWENTY ONE

VITAMINS AND HORMONES

Antioxidants

Series Editor

GERALD LITWACK, PhD
Toluca Lake, North Hollywood, California

ELSEVIER

ACADEMIC PRESS
An imprint of Elsevier

Academic Press is an imprint of Elsevier
50 Hampshire Street, 5th Floor, Cambridge, MA 02139, United States
525 B Street, Suite 1650, San Diego, CA 92101, United States
The Boulevard, Langford Lane, Kidlington, Oxford OX5 1GB, United Kingdom
125 London Wall, London, EC2Y 5AS, United Kingdom

First edition 2023

ISBN: 978-0-443-15768-4
ISSN: 0083-6729

For information on all Academic Press publications
visit our website at https://www.elsevier.com/books-and-journals

Publisher: Zoe Kruze
Acquisitions Editor: Leticia M. Lima
Developmental Editor: Federico Paulo S. Mendoza
Production Project Manager: Vijayaraj Purushothaman
Cover Designer: Greg Harris

Typeset by STRAIVE, India

Contents

Contributors

Azizeh Abdolmaleki
Department of Chemistry, Tuyserkan Branch, Islamic Azad University, Tuyserkan, Iran

Orinhamhe G. Agbadua
Institute of Pharmacognosy, Interdisciplinary Excellence Centre, University of Szeged, Szeged, Hungary

Shahin Ahmadi
Department of Chemistry, Faculty of Pharmaceutical Chemistry, Tehran Medical Sciences, Islamic Azad University, Tehran, Iran

Pooja Prakash Atpadkar
Laboratory of Bioactive heterocycles and Catalysis (BHC Lab), Department of Medicinal Chemistry, National Institute of Pharmaceutical Education and Research, Lucknow, UP, India

Diana A. Averill-Bates
Département des sciences biologiques (Center of Excellence in Orphan Diseases Research - Courtois Foundation (CERMO-FC), Research Group in Environmental Toxicology (TOXEN)), Université du Québec à Montréal, Montréal, QC, Canada

Gyorgy T. Balogh
Department of Chemical and Environmental Process Engineering, Budapest University of Technology and Economics, Budapest; Department of Pharmacodynamics and Biopharmacy, University of Szeged, Szeged, Hungary

Yerramathi Babu Bhagath
Food Technology Division, College of Science, Sri Venkateswara University, Tirupati, Andhra Pradesh, India

Gagan B.N. Chainy
Department of Biotechnology, Utkal University, Bhubaneswar, Odisha, India

Sandeep Chaudhary
Laboratory of Bioactive heterocycles and Catalysis (BHC Lab), Department of Medicinal Chemistry, National Institute of Pharmaceutical Education and Research, Lucknow, UP; Laboratory of Organic and Medicinal Chemistry (OMC Lab), Department of Chemistry, Malaviya National Institute of Technology, Jaipur, India

Jugun Prakash Chinta
Department of Chemistry, National Institute of Technology-Warangal, Telangana, India

Monica P. Colaiácovo
Department of Genetics, Blavatnik Institute, Harvard Medical School, Boston, MA, United States

John A. Corbett
Department of Biochemistry, Medical College of Wisconsin, Milwaukee, WI, United States

Prekshi Garg
Amity Institute of Biotechnology, Amity University, Lucknow, Uttar Pradesh, India

Agnieszka Gęgotek
Department of Inorganic and Analytical Chemistry, Medical University of Bialystok, Bialystok, Poland

Shraboni Ghosh
National Institute of Plant Genome Research, New Delhi, India

David Goltzman
McGill University Health Centre and McGill University, Montreal, QC, Canada

Tania Gómez-Sierra
Antioxidant Biochemistry Laboratory, Department of Biology, Faculty of Chemistry, National Autonomous University of Mexico (UNAM), Mexico City, Mexico

Sumanth Gopavaram
Laboratory of Bioactive heterocycles and Catalysis (BHC Lab), Department of Medicinal Chemistry, National Institute of Pharmaceutical Education and Research, Lucknow, UP, India

Maria Fernanda Hornos Carneiro
Department of Pharmacy, Faculty of Chemistry and Pharmacy, Pontificia Universidad Católica de Chile, Santiago, Chile

Attila Hunyadi
Institute of Pharmacognosy, Interdisciplinary Excellence Centre; Interdisciplinary Centre for Natural Products, University of Szeged, Szeged, Hungary

Marjan Jebeli Javan
Department of Chemistry, Faculty of Pharmaceutical Chemistry, Tehran Medical Sciences, Islamic Azad University, Tehran, Iran

Alexis Paulina Jiménez-Uribe
Antioxidant Biochemistry Laboratory, Department of Biology, Faculty of Chemistry, National Autonomous University of Mexico (UNAM), Mexico City, Mexico

Manoj Majee
National Institute of Plant Genome Research, New Delhi, India

Kola Manjula
Food Technology Division, College of Science, Sri Venkateswara University, Tirupati, Andhra Pradesh, India

Dengshun Miao
Nanjing Medical University, Nanjing, China

Pablo Muriel
Laboratory of Experimental Hepatology, Department of Pharmacology, Cinvestav-IPN, Mexico City, Mexico

Ariadna Jazmín Ortega-Lozano
Antioxidant Biochemistry Laboratory, Department of Biology, Faculty of Chemistry, National Autonomous University of Mexico (UNAM), Mexico City, Mexico

José Pedraza-Chaverri
Antioxidant Biochemistry Laboratory, Department of Biology, Faculty of Chemistry, National Autonomous University of Mexico (UNAM), Mexico City, Mexico

Karla Jaqueline Ramírez-Magaña
Antioxidant Biochemistry Laboratory, Department of Biology, Faculty of Chemistry, National Autonomous University of Mexico (UNAM), Mexico City, Mexico

Aluru Rammohan
Ural Federal University, Yekaterinburg, Russian Federation

Erika Ramos-Tovar
Postgraduate Studies and Research Section, School of Higher Education in Medicine-IPN, Mexico City, Mexico

Anindya Roy
Department of Chemistry, National Institute of Technology-Warangal, Telangana, India

Dipak Kumar Sahoo
Department of Veterinary Clinical Sciences, College of Veterinary Medicine, Iowa States University, Ames, IA, United States

Anurag Singh
Department of Biochemistry, University of Lucknow, Lucknow, Uttar Pradesh, India

Elżbieta Skrzydlewska
Department of Inorganic and Analytical Chemistry, Medical University of Bialystok, Bialystok, Poland

Neha Srivastava
Excelra Knowledge Solution Pvt Ltd, NSL Arena, Uppal, Hyderabad, India

Prachi Srivastava
Amity Institute of Biotechnology, Amity University, Lucknow, Uttar Pradesh, India

Jennifer S. Stancill
Department of Biochemistry, Medical College of Wisconsin, Milwaukee, WI, United States

Gergely Takács
Department of Chemical and Environmental Process Engineering, Budapest University of Technology and Economics; Mcule.com Ltd., Budapest, Hungary

Grigory V. Zyryanov
Ural Federal University; I. Ya. Postovsky Institute of Organic Synthesis, Ural Division of the RAS, Yekaterinburg, Russian Federation

About the editor

Gerald (Gerry) Litwack was educated at Hobart College (BA) with graduate work at the University of Wisconsin, Madison (MS and PhD in biochemistry). After spending a summer at the University of Wisconsin as an instructor in a course on enzymes, he sailed to Paris (Liberté) where he spent a year as a postdoctoral fellow of the National Foundation for Infantile Paralysis at the Laboratoire de Chimie Biologique of the Sorbonne. He then became Assistant Professor of Biochemistry at Rutgers University, later becoming Associate Professor at the Graduate School of Medicine of the University of Pennsylvania. Future posts took him to the Fels Institute for Cancer Research and Molecular Biology at the Temple University School of Medicine where he was the Laura H. Carnell Professor of Biochemistry and the Deputy Director of the institute. Later, he became the Chairman of the Department of Biochemistry and Molecular Pharmacology, Professor of Biochemistry, Vice Dean for Research, and the Director of the Institute for Apoptosis at Thomas Jefferson University Kimmel Medical College with future appointments as Visiting Scholar at the Geffen School of Medicine at UCLA, the Founding Chair of the Department of Basic Science of the Geisinger Commonwealth School of Medicine, and, as his final position, Professor of Molecular and Cellular Medicine, and Associate Director of the Institute for Regenerative Medicine at Texas A&M University School of Medicine. Dr. Litwack was Sabbatical Visitor at the University of California, Berkeley; the University of California, San Francisco; the Courtauld Institute of Biochemistry (London); and the Wistar Institute. He was appointed Emeritus Professor and/or Chairman at Rutgers University, Thomsas Jefferson University, and Geisinger Commonwealth School of Medicine. He has published more than 350 papers in scientific journals, is named on more than 20 patents, and has held numerous editorial positions on biochemical and cancer journals. He has authored 3 textbooks in the areas of biochemistry or endocrinology and has edited more than

65 books. Following his retirement, he lives with his family in Los Angeles where he continues his work as an author and editor and paints in watercolor during his leisure time.

Preface

We live in a polluted environment. The air, water, soil, and food are polluted with chemicals, some of which are carcinogenic. Even some of our medicines are carcinogenic. The air we breathe does not appear to contain more than 18% oxygen and when it is inhaled, the oxygen is combined with pollutants. Theoretically, for optimal health, about 30% oxygen in unpolluted air would be required. In consequence, the reductive capacity in cells is insufficient to reduce the mass of reactive oxygen species (ROS). ROS take the form of hydroxyl ions, hydroxy radicals, superoxide anions, and hydrogen peroxide. ROS can lead to double-strand breaks in DNA through the conversion of guanine to 8-oxyguanine, as well as causing damage to proteins and lipids.

The endless volume of pollutants in our environment produces more ROS than can be handled. This leads to many disease states, including inflammation, cardiovascular disease, cancer, and aging. The cell depends on its level of glutathione as a major antioxidant. In addition to its antioxidant activity, glutathione is involved in glutathione peroxidase, which converts hydrogen peroxide to water. Another major function of glutathione involves ligandin and the three other glutathione S-transferase isozymes. This protective enzyme family converts xenobiotics and certain carcinogens into their water-soluble sulfate derivatives, allowing their excretion. Other important antioxidants are vitamin C, vitamin E, polyphenolic compounds, peroxiredoxin, and thioredoxin.

Some ROS can be produced in normal cellular functions (although these amounts are small and actually could be beneficial); some ROS can be produced in the mitochondria when electrons are passed from Complex I or II to Complex III involving ubiquinone and the electron transport chain, wherein some electrons may escape the pathway and react with oxygen to form superoxide. In the cellular membrane, catalysis by NADPH oxidase produces superoxide. Beta-oxidation of triglycerides in the mitochondria can also produce some hydrogen peroxide. The enzyme superoxide dismutase converts superoxide to hydrogen peroxide and the peroxisomal catalase converts hydrogen peroxide, harmlessly, to a molecule of water and an atom of oxygen.

This book, summarizing current knowledge, should be of interest to biochemists, chemists, researchers, and students as well as clinicians.

In this volume, the Editorial Board of *Vitamins and Hormones* has been formalized and is presented on a dedicated page, a page that will appear in future volumes of the series.

The illustration on the cover is the molecule of glutathione retrieved from the *Molecule of the Month* (212) that is a part of the Protein Data Bank contributed by Dr. David S. Goodsell.

Publication of this book has been mediated by the efforts of Paulo Mendoza and Leticia Lima of Elsevier and completion has been accomplished by the production group in Chennai, India.

GERALD LITWACK
North Hollywood, CA
October 26, 2022

CHAPTER ONE

In silico study of natural antioxidants

Shahin Ahmadi[a,]*, Azizeh Abdolmaleki[b], and Marjan Jebeli Javan[a]

[a]Department of Chemistry, Faculty of Pharmaceutical Chemistry, Tehran Medical Sciences, Islamic Azad University, Tehran, Iran
[b]Department of Chemistry, Tuyserkan Branch, Islamic Azad University, Tuyserkan, Iran
*Corresponding author: e-mail address: ahmadi.chemometrics@gmail.com

Contents

Abstract

Antioxidants are the body's defense system against the damage of reactive oxygen species, which are usually produced in the body through various physiological processes. There are various sources of these antioxidants such as endogenous antioxidants in the body and exogenous food sources. This chapter provides important information on methods used to investigate antioxidant activity and sources of plant antioxidants. Over the past two decades, numerous studies have demonstrated the importance of in silico research in the development of novel natural and synthesized antioxidants. In silico methods such as quantitative structure-activity relationships (QSAR), pharmacophore, docking, and virtual screenings are play critical roles in designing effective antioxidants that may be synthesized and tested later. This chapter introduces the available in silico approaches for different classes of antioxidants. Many successful applications of in silico methods in the development and design of novel antioxidants are thoroughly discussed. The QSAR, pharmacophore, molecular docking techniques, and virtual screenings process summarized here would help readers to find out the proper mechanism for the interaction between the free radicals and antioxidant compounds. Furthermore, this chapter focuses on introducing new QSAR models in

Vitamins and Hormones, Volume 121
ISSN 0083-6729
https://doi.org/10.1016/bs.vh.2022.09.001

combination with other in silico methods to predict antioxidants activity and design more active antioxidants. In silico studies are essential to explore largely unknown plant tissue, food sources for antioxidant synthesis, as well as saving time and money in such studies.

Abbreviations

ORAC assay	oxygen radical absorbance capacity assay
HAT mechanism	hydrogen atom transfer mechanism
PCL assay	photo chemiluminescence assay
FRAP assay	ferric reducing antioxidant power assay
TPTZ	iron [III]-2,4,6-tripyridyl-S-triazine
SET	single-electron transfer
CUPRAC assay	cupric reducing antioxidant capacity assay
NC	neo cuproine
TEAC assay	trolox equivalent antioxidant capacity assay
ABTS	2,2′-azinobis-(3-ethylbenzothiazoline-6-sulfonic acid)
DPPH assay	2,20-diphenyl-1-picrylhydrazyl radical assay
QSAR approach	quantitative structure-activity relationship approach
OECD principals	European Organization for Economic Co-operation and Development principals
ML algorithms	machine learning algorithms
SLR	simple linear regression
MLR	multiple linear regression
GA-MLR	genetic algorithm with multiple linear regression
PCR	principal component regression
PLSR	partial least squares regression
RF	random forest
ANN	artificial neural networks
SVM	support vector machines
GPR	Gaussian process regression
GBM	gradient boosting machine
XGBoost	extreme gradient boosting
CCC	concordance correlation coefficient
IIC	index of ideality of correlation
AD	applicability domain
TEAC activity	trolox equivalent antioxidant capacity activity
IC_{50}	50% inhibitory concentration
SETPT mechanism	electron transfer-proton transfer mechanism
SPLET mechanism	sequential proton loss-electron transfer mechanism
BDE	bond dissociation enthalpy
IP	ionization potential
PDE	proton dissociation enthalpy
PA	proton affinity
ETE	electron transfer enthalpy
SMILES	simplified molecular input-line entry system
VCEAC	vitamin C equivalent antioxidant capacity
GFA	genetic function approximation

G/PLS	genetic partial lease squares
RSA	radical scavenging activity
QM	quantum mechanics
MFA	molecular field analysis
G-QSAR	group based QSAR
HQSAR	hologram QSAR
CoMFA	comparative molecular field analysis
CoMSIA	comparative molecular similarity indices analysis
GRIND	GRid independent descriptors
ADME properties	absorption, distribution, metabolism and elimination properties
CADD	computer aided drug design
AI methods	artificial intelligence methods
RMSE	root mean square error
R^2	square correlation coefficient
Q2	leave-one-out cross-validated square correlation coefficient
DFT	density functional theory
CORAL software	correlation and logic

1. Antioxidants

The production of reactive oxygen species as a result of the elaborate oxidative processes that occur in the human body is a powerful precursor of systemic cellular and tissue damage.

Antioxidants eliminate these free-radical intermediates as an oxidation process inhibitor by oxidizing themselves (Brewer, 2011), even at utterly little concentration; therefore, to stop these oxidation reactions, they have arranged the physiological action in the body to protect the body from dangerous chain reactions (Miguel, 2010). Hence, many researchers reviewed antioxidants' structure as nature's response to environmental and physiological stress, cardiovascular disease, cancer, and aging (Valko et al., 2007). The body's internal protection system *contrarily* these free radicals play a principal part, which are aided by antioxidants added in the meals. Typically, antioxidants are split into two main classes: natural and synthetic.

Moreover, these antioxidants are categorized as nonenzymatic and enzymatic. Enzymatic antioxidants essentially comprise superoxide dismutase, catalase, and glutathione peroxidase. Furthermore, various enzymes are in the body that assists in the whole antioxidant span, which detects in the serum (Anwar, Rahman, Javed, & Muhammad, 2012). Nonenzymatic antioxidants

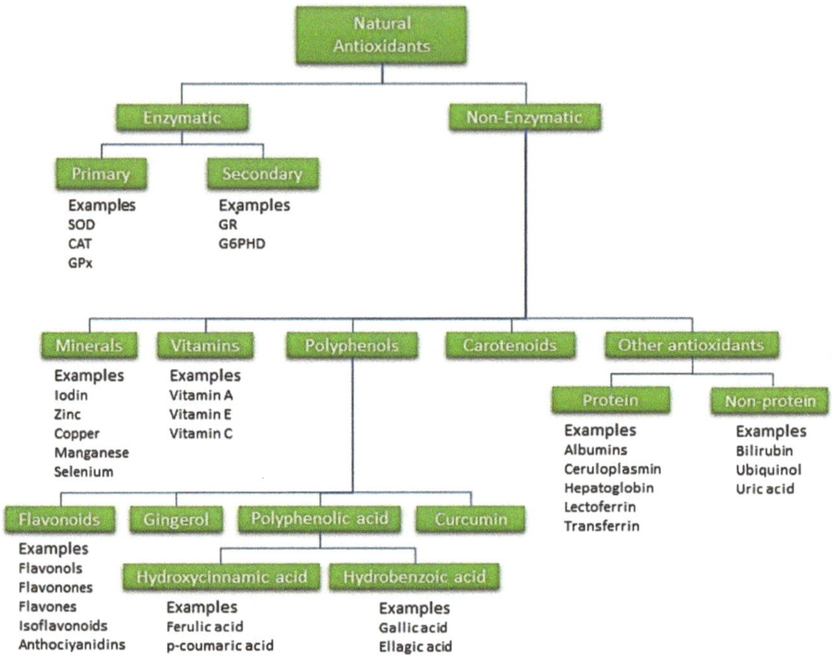

Fig. 1 Categorization and subcategorization of antioxidants discovered in natural sources.

comprise various segments primarily vitamins A, C, E, and to a minor amount enzyme cofactor (Q10), vitamin D, some minerals (Zn and Se) and peptides.

A detailed categorization and subcategorization are depicted in Fig. 1.

Recently, some toxicological researches concerning the usage of synthetic antioxidants have revealed their undesirable or poor impacts. These records have motivated the researchers to concentrate their research on searching the natural originals with sensible antioxidant capacity (Ramalakshmi, Kubra, & Rao, 2008). Moreover, in the context of using these natural antioxidants, the accessibility and providence are considerable concerns. The natural antioxidants can be classified into different subsets. However, two crucial groups are similar to antioxidants from consistently used or regular natural diets (e.g., beans, fruits, vegetables, cereals) and furthermore from herb or plant sources that have goodish antioxidant activity but are not the common food supply (e.g., wild herbs and pharmaceutical plants).

Today, the goal of researchers around the world is to find natural sources of antioxidants, provided that these compounds are affordable and environmentally friendly. This discovery will be the premier replacement of synthetic additives in the cosmetic, food, and pharmaceutical industries (Binic, Lazarevic, Ljubenovic, Mojsa, & Sokolovic, 2013). However, notable damaging impacts of synthetic additives have not been announced till now.

Meat and meat products are sensitive to oxidative reactions. The oxidative essence of muscle food is due to the excess amount of lipids and the attendance of oxidative metal ions (Cu, Fe). The prime reason for quality declines in meat and meat products is protein, and lipid oxidations (Fig. 2). These oxidative procedures mostly emerge to be connected and the oxidation of one of these arouses the creation of chemical compounds that can intensify the oxidation of the others (Faustman, Sun, Mancini, & Suman, 2010; Kumar, Yadav, Ahmad, & Narsaiah, 2015).

Oxidative failure in muscle foods exhibits an increases of flavor, discoloration, nutrient and drop fatalities, and also the generation of toxic structures (Falowo, Fayemi, & Muchenje, 2014). Several types of research on the utilization of natural antioxidants in meat production have described that they were applied to inhibit oxidation and stopping the formation of

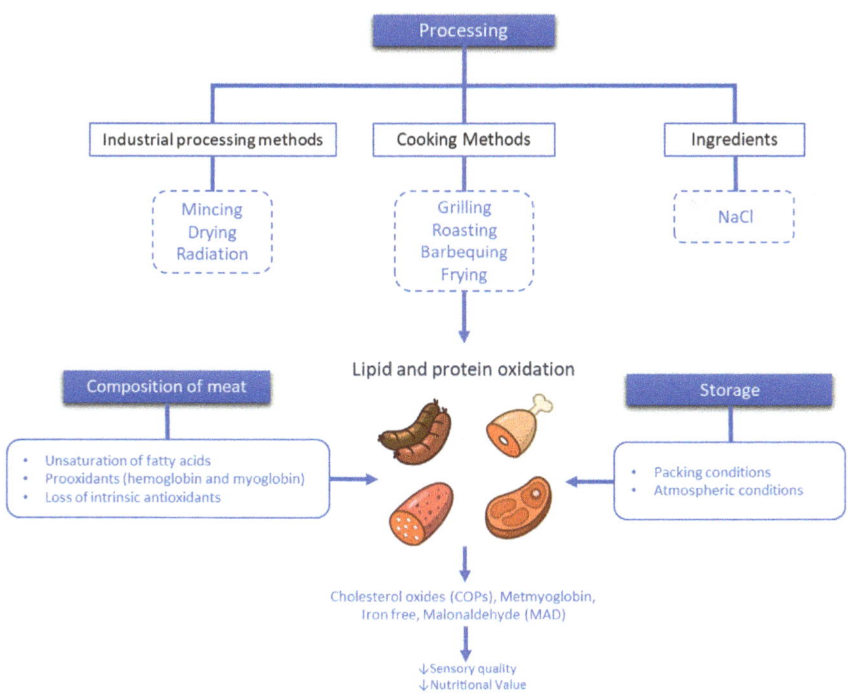

Fig. 2 Parameters that influence protein and oxidation of meat and meat products.

cytotoxic functions (Falowo et al., 2014; Kumar et al., 2015). Therefore, natural antioxidants could enhance the healthful profile of meat, which is damaged by its harmful aggregation of cytotoxic compounds. It is worth noting that the investigation of natural antioxidants will be the main focus of researchers in the future decade (Shebis, Iluz, Kinel-Tahan, Dubinsky, & Yehoshua, 2013).

The main concept of this chapter is to outline and sum up the natural origins with antioxidant capability. This brief information provides expert and amateur readers with valuable data that collects unbiased facts about nutrient and non-nutrient sources in nature, which can be part of a regular diet to improve the body's antioxidant capacity. In addition, it is necessary to discover the chemical structures of antioxidants and the analytical methods used in the discovery of those antioxidants (Javan & Javan, 2014).

From an empirical opinion, the outcomes of investigation on the effect of processing and storing of natural antioxidants are valorous in action.

2. Natural antioxidants

Mother nature is always a rich source of numerous components that work for human health.

Many of these natural sources usually include vegetables, used fruits, spices, herbs, and edible mushrooms that can be part of a natural diet. In addition, there is a huge list of medicinal plants that have significant health-improving abilities.

It is worth mentioning that, scientists have appended several marine origins like seagrass and algae plus in the index of these natural sources (Javan, Javan, & Tehrani, 2013).

Vegetables and fruits are highly essential in diet contents, broadly familiar for their health-improving impacts and nourishing worth. They got a fundamental role as everyday foods in history because of their exceeding quantity of electrolytes, minerals, and vitamins, especially vitamins E and C. However, several present pieces of research researches are disclosing their phytochemical add-up with antioxidant activities (Slavin & Lloyd, 2012). These antioxidants trap the free radicals or oxidants made as a consequence of various deterioration and illness procedures like cardiovascular disorders, cancers and diabetes. Therefore, systematic use of vegetables and fruits can decrease the chance of death related to these disorders (Nahak, Suar, & Sahu, 2014). A large number of natural antioxidants convert lipid radicals into more stable products by chain breakdown. As mentioned earlier, antioxidants obtained from fruits and vegetables mainly have a phenolic structure, which consists of vitamins, polyphenols, and minerals (Hurrell, 2003). Minerals, such as iron, copper, zinc, manganese, and selenium, take action as a cofactor of many enzymes in the antioxidant process, loss of which can assuredly frustrate their enzymatic scavenging activity (Sonia, Mini, & Geethalekshmi, 2016). This chapter will expand upon the several natural origins of antioxidants that optimistically assist in arranging the good corrections in diet combination.

2.1 Origins of natural antioxidants

The main sources of natural antioxidants include cereals, oilseeds, plants of the Lamiaceae family, coffee and tea, legumes, tree nuts, berries and fruits. The possible antioxidant inclusion of plant materials or the antioxidant capacity of extracts obtained from them depends on the number of phenolic compounds present in the extracts or plants (Fig. 3).

2.2 Use of natural antioxidants in the meat products

Fig. 4 shows the applications of natural antioxidants in the meat industry. In explaining the above diagram, it should be noted that natural antioxidants obtained from non-vegetable compounds are structures that show antioxidant ability as long as they are present in fresh and prepared meat. Honey, oregano, sage (Sampaio, Saldanha, Soares, & Torres, 2012), hydrolyzed proteins and peptides are sources of these antioxidants (Jiang & Xiong, 2016).

Moreover, several plant or animal-based agricultural by-products were suggested as natural antioxidants for producing meat products and preparations (Fig. 5).

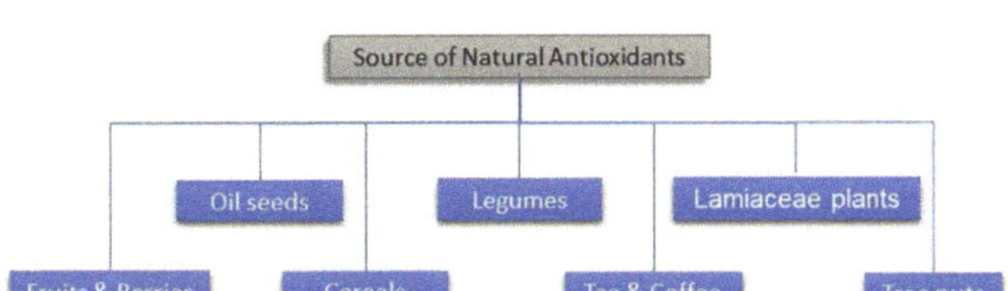

Fig. 3 Source of natural antioxidants.

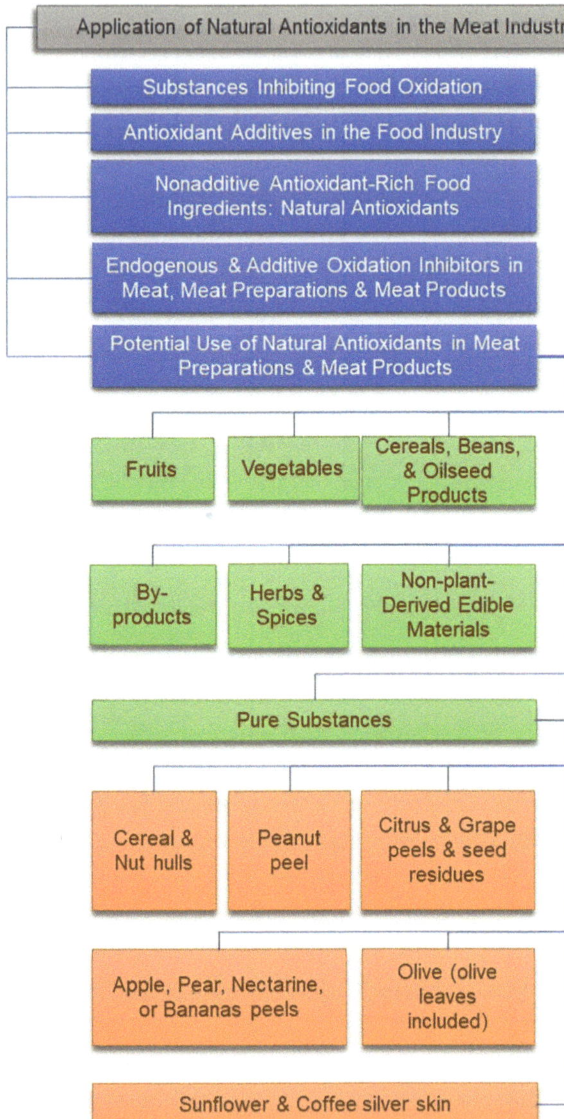

Fig. 4 The applications of natural antioxidants in the meat industry.

2.3 Methods employed for the measurement of antioxidant activity

2.3.1 ORAC (oxygen radical absorbance capacity) assay

In this method (Adom & Liu, 2005; Glazer, 1990; Huang, Ou, Hampsch-Woodill, Flanagan, & Deemer, 2002; Ou, Hampsch-Woodill, & Prior, 2001) RO_2 is created using thermal decay of (AAPH). The reaction of

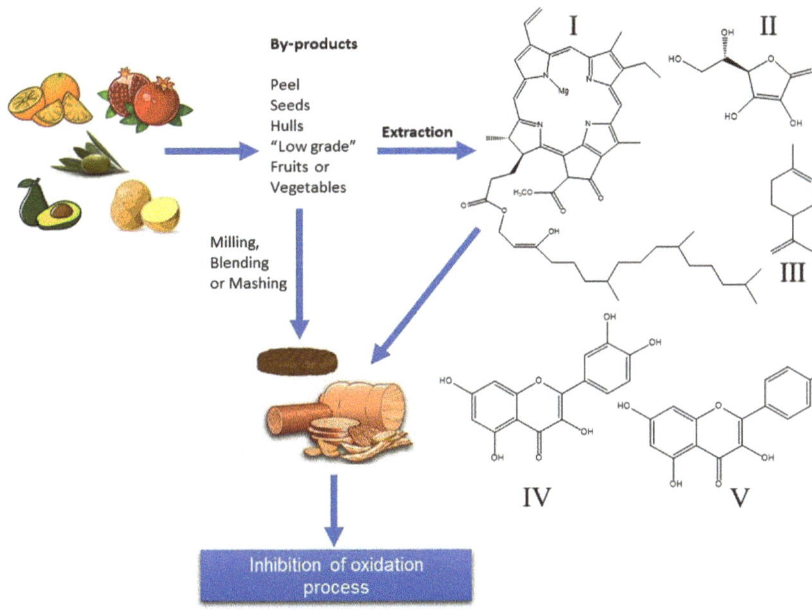

Fig. 5 Agro-industrial derivates impede the oxidation process. (I) Chlorophyll (II) Vitamin C (III) Limonene (IV) Quercetin (V) Kaempferol.

the obtained peroxyl radicals with fluorescein or its derivatives leads to a decrease in the fluorescence signal at an excitation/emission wavelength pair of 515/493 nm.

In peroxyl radicals' reactions using the HAT mechanism, the fadeaway of fluorescein in the sample is protected. The reaction is as follows:

$$2RO_2^{\bullet+}(FL)OH\ (\text{fluorescence } 515) \rightarrow 2ROOH + (FL)O^{\bullet}(HAT) \quad (1)$$

2.3.2 Photochemiluminescence (PCL) assay

In the earliest section of the PCL method (Popov & Lewin, 1994, 1996), luminol (5-amino-2,3-dihydro-1,4-phthalazinedione) is photo-degraded derived in the producing/suppressing of $O_2^{\bullet -}$:

$$\text{Luminol} + h\nu_1\ (UV) \rightarrow L^* + {}^3O_2 \rightarrow [L^*O_2] \rightarrow L^{\bullet +} \rightarrow L^{\bullet +} + O_2^{\bullet -} \quad (2)$$

When $O_2^{\bullet -}$ radicals and luminol are formed, they act through a sequence of reactions that evolve to produce blue luminescence.

$$L^{\bullet +} + O_2^{\bullet -} \rightarrow N_2 + AP^{*2-} \rightarrow AP^{2-} + h\nu_2\ (\text{blue at 360nm}) \quad (3)$$

In the reaction introduced over, AP^{*2-} is an excited amino phthalate anion, and AP^{2-} is the amino phthalate anion at the ground state. In the HAT reaction, the antioxidant types are available in the reaction will overtop the luminol radical. The blue luminescence is generated till the concentration rises.

2.3.3 FRAP (ferric reducing antioxidant power) assay

Benzie and Strain established the FRAP assay (1996) to estimate the ferric-reducing power of human plasma (Benzie & Strain, 1996). This procedure modified to evaluate the ferric reducing antioxidant power of plant syrups (Pulido, Bravo, & Saura-Calixto, 2000).

The FRAP assay was used in a microtiter plate reader in a 96-well format by Dragsted et al. (2004). In FRAP, Fe^{3+}–TPTZ (iron [III]-2,4,6-tripyridyl-S-triazine) reduce to Fe^{2+}–TPTZ through SET with an antioxidant structure. The outcome is a severe blue color:

$$Fe^{3+} - TPTZ + ArOH \rightarrow Fe^{2+} - TPTZ\ (\text{blue at } 595) + [ArOH]^{\bullet +}(SET) \quad (4)$$

2.3.4 CUPRAC (cupric reducing antioxidant capacity) assay

Regarding reaction speed, redox reactions with iron are often slower than copper ones. Similar iron, copper ions correlate with chelating agents containing nitrogen like 1,10-phenanthroline or 2,20-bipyridine. In the CUPRAC method (Apak, Güçlü, Özyürek, & Karademir, 2004; Özyürek et al., 2011), free copper (II) reduce to copper (I) in the presence of neocuproine (NC) (2,9-dimethyl-1,10-phenanthroline) and the coordinated complex Cu(I)–NC at a scale of 2:1 is produced. The mentioned reaction is as follows:

$$Cu^{2+} + ArOH - 2NC \rightarrow Cu^{+} - (NC)_2(\text{bue at } 450\text{nm}) + [ArOH]^{\bullet +} \quad (5)$$

2.3.5 TEAC (Trolox equivalent antioxidant capacity) assay

Miller, Rice-Evans, Davies, Gopinathan, and Milner (1993) developed the TEAC assay for the estimation of the antioxidant activity of human plasma upon the trapping of free-radical cation ($ABTS^{\cdot +}$) using antioxidants (Miller et al., 1993). The TEAC method is commonly admitted as a SET method. Although, an ABTS cation radical can be annihilated by HAT and SET procedures:

$$ABTS^{\bullet+}\text{(green at 734nm)} + ArOH \rightarrow ABTS\text{ (colorless)} + [ArOH]^{\bullet+}\text{ (SET)} \quad (6)$$

$$ABTS^{\bullet+}\text{(green at 734nm)} + ArOH \rightarrow ABTS(H)\text{ (colorless)} + ArO^{\bullet}\text{ (HAT)} \quad (7)$$

2.3.6 DPPH (2,20-diphenyl-1-picrylhydrazyl radical) assay

The DPPH˙ method is often operated because of its comparative cheapness (Blois, 1958; Bondet, Brand-Williams, & Berset, 1997; Brand-Williams, Cuvelier, & Berset, 1995). DPPH undertakes a HAT (Brand-Williams et al., 1995; Litwinienko & Ingold, 2003), SET (Foti, Daquino, & Geraci, 2004; Huang, Ou, & Prior, 2005) or mixed (Schaich, 2005) mechanisms based on the forwarding reaction:

$$DPPH^{\bullet}\text{(violet at 515nm)} + ArOH \rightarrow DPPH(H)\text{ (colorless)} + ArO^{\bullet}\text{ (HAT)} \quad (8)$$

$$DPPH^{\bullet}\text{(violet at 515nm)} + ArOH \rightarrow DPPH^{-}\text{ (colorless)} + ArO^{\bullet+}\text{ (SET)} \quad (9)$$

2.3.7 β-Carotene-linoleic acid (linoleate) assay

In this method (Miller, 1971) an emulsion is used, linoleic acid produces the pentadienyl free radical that attacks the β-carotene molecules and the distinctive orange color of the emulsion fades (Frankel, 2005). Phenolic antioxidants neutralize the linoleate free radical to protect beta-carotene from extinction. This method can be observed by estimating the absorbance of the sample at 470 nm (Harnly, 2017).

3. Computer assisted study of natural antioxidant activities

3.1 QSAR studies

We need efficient and robust computational methods to screen chemical databases and virtual chemical libraries against molecules with known activity or properties to identify and design novel structures with desirable activity or properties. For this goal, the quantitative structure-activity relationship (QSAR) approach is widely used. The QSAR modeling provides a powerful way to create and exploit the relationship between activity and structural features to develop new compounds. QSAR study is the application of mathematical and statistical techniques to develop the models to predict the chemical, physical, and biological properties of new compounds. A QSAR model can be summarized as follows:

$$\text{Biological activity} = f(\text{descriptors}) \tag{10}$$

A QSAR procedure tries to minimize the error of prediction, for example, in the form of the sum of squares between predicted and observed activities. A QSAR method tries to minimize the difference between predictions and observations. The QSAR model development process can be divided into several stages based on the European Organization for Economic Co-operation and Development (OECD) principles.

3.1.1 OECD principles

There are five OECD guidelines for the development of a reliable QSAR model:

- Principle 1. a defined endpoint

The first principle intends to ensure clarity in the endpoint predicted by a given model, since a given endpoint can be determined by different experimental protocols and under different conditions. It is therefore essential to use the homogeneous data determined according to harmonized protocols.

- Principle 2: an unambiguous algorithm

The purpose of an unambiguous algorithm is to ensure transparency in the model algorithm, which predicts an endpoint based on chemical structure and/or physicochemical properties. Obviously, in the case of commercially developed models, this information is not always publicly available. However, without this information, the performance of a model cannot be independently determined, which is likely to be a barrier to regulatory acceptance. Reproducibility of the predictions is included in this principle.

- Principle 3: a defined domain of applicability

The need to define an applicability domain reflects that QSARs are models which are inevitably associated with limitations in terms of the types of chemical structures, physicochemical properties, and mechanisms of action for which the models can generate reliable predictions.

- Principle 4: appropriate measures of goodness-of-fit, robustness and, predictivity

The appropriate revised criteria for goodness-of-fit, robustness, and predictability include the goal of the original Setubal Principles 5 and 6. The principal statement is intended to simplify the overall set of principles, but does not lose the distinction between the internal performance of a model and the predictability of a model.

- Principle 5: a mechanistic interpretation, if possible

It is recognized that from a scientific point of view, it is not always possible to provide a mechanical interpretation of a QSAR, or even there are multiple mechanical interpretations of a given model. The lack of a mechanical interpretation of a model does not mean that a model is potentially unreliable in the regulatory context. The absence of a mechanistic interpretation for a model does not mean that a model is not potentially useful in the regulatory context. The purpose of this principle is not to reject models with no apparent mechanistic basis, but to ensure that some consideration is given to the possibility of a mechanistic association between the descriptors used in a model and the endpoint being predicted, and to ensure that this association is documented.

3.1.2 QSAR steps

The main workflow of QSAR modeling is presented in Fig. 6. The details of each step are described in the steps below.

- Step 1. The curated chemical dataset

Usually, dataset for QSAR modeling is collected from databases and the literature. To build a reliable QSAR model, first, biological and chemical curation is required to confirm the accuracy, consistency, and reproducibility of the reported experiment. New workflows and various tools are available to manage small and large datasets (Ambure, Gajewicz-Skretna, Cordeiro, & Roy, 2019; Fourches, Muratov, & Tropsha, 2016).

- Step 2. Descriptor generation

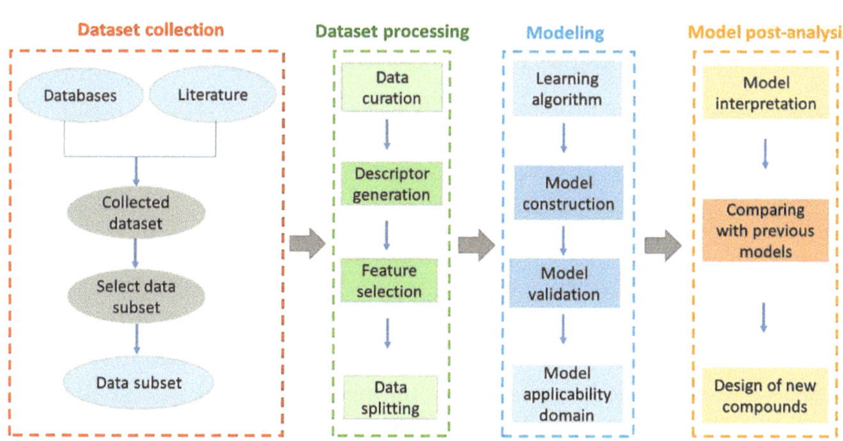

Fig. 6 Schematic diagram of the QSAR study workflow.

Descriptors as independent variables are a numerical description of the molecular structure used to develop QSAR models. Molecular descriptors play a crucial role in QSAR modeling. Information encoded by molecular descriptors usually depends on the type of chemical representation and the specific algorithm for its computation (Khan, 2016). The topological, geometrical, constitutional, and physicochemical descriptors are some significant classes of descriptors. Constitutional descriptors are the simplest and most commonly used descriptors, reflecting the molecular composition of a compound without any geometrical information. Dragon and CORAL software have many constitutional descriptors are utilized in various QSAR models (Ahmadi & Akbari, 2018; Ahmadi & Ganji, 2016; Habibpour & Ahmadi, 2017).

- Step 3. The data splitting

Prior to using a QSAR model for external predictions, its predictive ability must be indicated by statistical criteria. In the absence of a real external data set, the best way to validate the model's predictability is to perform its statistical external validation. In the external statistical validation, the entire dataset is divided into training and test sets. Usually, this division is done using random splitting (Ahmadi, 2020; Ahmadi, Ghanbari, Lotfi, & Azimi, 2021). The rational splitting approaches can intelligently divide dataset into training and test sets (Martin et al., 2012). The rational partitioning algorithms have been developed which attempt to "intelligently" split datasets into training and test sets to produce more predictive models. Some logical algorithms such as the Kohonen Self Organizing Map method (Ahmadi & Habibpour, 2017; Zupan, Novič, & Ruisánchez, 1997), the Kennard-Stone method, D-optimal design (de Aguiar, Bourguignon, Khots, Massart, & Phan-Than-Luu, 1995), D-optimal onion design (Olsson, Gottfries, & Wold, 2004), and sphere exclusion based methods (Golbraikh et al., 2003) is used for data splitting.

- Step 4. Building QSAR model

There are different machine learning (ML) algorithms to build several dozen QSAR models for revealing relationships between structures and activities of compounds.

Simple linear regression (SLR) and multiple linear regression (MLR) are two popular algorithms for QSAR construction. The main advantage of MLR is its simplistic form and easily interpretable mathematical QSAR model. However, MLR is extremely sensitive to regression outliers and lead to overfitting when too many redundant or highly correlated features are used in the equation. Genetic algorithm combined with MLR (GA-MLR) can overcome these shortcomings (Ahmadi & Habibpour, 2017; Ahmadi, Khazaei, & Abdolmaleki, 2014).

Principal component regression (PCR) and partial least squares regression (PLSR) are used to model an activity when there are a large number of highly correlated descriptors. These algorithms reduced the dimensions of the models by converting a large number of correlated descriptors into a small number of uncorrelated variables called principal components (Gedeck, Kramer, & Ertl, 2010; Ghasemi, Ahmadi, & Brown, 2011). However, the above linear algorithms can only create the linear relationship between activity and descriptors, and it is not safe to always assume a linear correlation between descriptors and activity in the real world. Therefore, they may be unable to detect the nonlinear relationships in the data sufficiently.

In the past two decades, a wide range of ML algorithms has been widely used to model the nonlinear relationship between molecular features and the biological activity of structures. The most crucial ML algorithms such as random forest (RF) (Svetnik et al., 2003), artificial neural networks (ANN) (Jain, Mao, & Mohiuddin, 1996), and support vector machines (SVM) (Byvatov & Schneider, 2003) have been widely used to develop QSAR models. Nowadays, with the development of automatic tuning analysis and deep learning-based approaches, people have again started to use ANN in QSAR modeling (Ghasemi, Mehridehnavi, Fassihi, & Pérez-Sánchez, 2018; Ghasemi, Mehridehnavi, Perez-Garrido, & Perez-Sanchez, 2018). Some other efficient ML algorithms, such as Gaussian process regression (GPR), gradient boosting machine (GBM), and extreme gradient boosting (XGBoost), also demonstrated excellent performance in QSAR studies (Schroeter et al., 2007; Sheridan, Wang, Liaw, Ma, & Gifford, 2016; Svetnik et al., 2005).

- Step 5. Model validation

External validation of QSAR models is a crucial task to ensure the reliability of the model to verify the actual prediction ability of models on compounds never used in model development.

For models built on small data sets, model validation is usually investigated by internal validation. The most famous internal validation criterion is leave-one-out cross-validated R^2 (Q^2). Thus, the high value of Q^2 appears to be the necessary but not the sufficient condition for the model to have a high predictive power (Golbraikh & Tropsha, 2002). However, other validation criteria have been proposed by authors, such as $Q_{F1}{}^2$ (Tropsha, Gramatica, & Gombar, 2003), $r_m{}^2$ (Roy, 2007), $Q_{F2}{}^2$ (Roy, 2007), $Q_{F3}{}^2$ (Roy, 2007), concordance correlation coefficient (CCC) (Chirico & Gramatica, 2011), and Index of Ideality of Correlation (IIC) (Toropova & Toropov, 2019).

- Step 6. Applicability domain

The need to make more reliable predictions has led to studies of the so-called Applicability Domain (AD), which indicates the chemical space from which a model is derived and where a reliable prediction is considered. Gadaleta et al. reviewed several approaches for examining the AD of a QSAR model (Gadaleta, Mangiatordi, Catto, Carotti, & Nicolotti, 2016).

- Step 7. Model interpretation

Although higher predictive models are often desirable, the interpretability of a QSAR model is also essential. Most researchers agree that the interpretability of a QSAR model is as important as its predictability. Identifying descriptors of increase or decrease should be helpful for researchers who intend to modify the structure (Polishchuk, 2017). Some packages like CORAL are beneficial for structure modification because the features of models are very simple and interpretable.

Step 8. Design of new compounds

Interpretable QSAR models usually help the researchers design new or modified molecules.

3.1.3 QSAR of antioxidants

Different classes of natural or synthesized compounds have been identified that exhibit antioxidant activity. The change in the substitution pattern of these antioxidant structures affects the change in their physicochemical properties, which in turn affects their reactivity with toxic free radicals. Over the past two decades, diverse classes of such chemicals have attracted considerable attention from various groups of pharmaceutical chemists because of their ability to scavenge free radicals and thus inhibit systemic damage such as lipid peroxidation. Reliable QSAR models developed by researchers worldwide based on various classes of antioxidants are discussed in the following and summarized in Table 1.

Prediction of antioxidant activity of different classes of antioxidants

- QSAR of phenolic antioxidants

QSAR modeling of antioxidants of wine polyphenols was modeled by Rastija and Medic-Saric using 2D and 3D dragon descriptors. The proposed models for the antioxidant activity of polyphenols showed that the zero-order connectivity index and molar refractivity are critical parameters for modeling the free radical scavenging (TEAC) activity of polyphenols belonging to different groups (phenolic acids and flavonoids, flavans, flavonols, and stilbene). Also, IC_{50} modeling of flavonoids showed that lipophilicity and van der Waals volume are the essential variables for the

Table 1 The comparison between some of the previous QSAR models for activity prediction of difference class of antioxidants.

Antioxidant type	End point	Software or descriptor type	Feature selection method	Machine learning method	Data set size		R^2		RMSE		Reference
					Training	Test	Training	Test	Training	Test	
Polyphenols	TEAC	Dragon	Best-MLR	MLR	10	–	0.833–0.845	–	0.180–0.186	–	Rastija and Medić-Šarić (2009)
Flavonoid derivatives	IC_{50}	Dragon	Best-MLR	MLR	8	–	0.858–0.950	–	7.87–13.4	–	Rastija and Medić-Šarić (2009)
NO donor phenols	pIC_{50}	Cerius 2 software	GFA and G/PLS	MLR and PLS	25	8	0.824–0.935	0.678–0.924	0.284–0.443	–	Farahani, Sohrabi, and Ghasemi (2018)
Natural phenolic acids	pIC_{50}	DFT-based	–	SLR and MLR	20	–	0.739–0.745	–	–	–	Chen, Xiao, Zheng, and Liang (2015)
Phenolic compounds	TEAC	CORAL and graph of atomic orbitals	Monte Carlo method	MLR	82	14	0.899–0.938	0.886–0.966	0.430–0.639	0.467–0.597	Ahmadi, Mehrabi, Rezaei, and Mardafkan (2019)
Flavonoid derivatives	TEAC	DFT approach with MOPAC	–	SLR and MLR	38		0.958–0.966	–	0.424–0.554	–	Amić and Lučić (2010)
Flavonoid derivatives	VCEAC	DFT approach with MOPAC	–	SLR and MLR	36	–	0.852–0.910	–	23.81–29.74	–	Amić and Lučić (2010)
Isoflavones, isoflavanes and biphenyl ketones	pIC_{50}	Cerius 2 software	GFA and G/PLS	GFA, PLS	64	22	0.673	0.759	0.259	–	Mitra, Saha, and Roy (2010)

Continued

Table 1 The comparison between some of the previous QSAR models for activity prediction of difference class of antioxidants.—cont'd

Antioxidant type	End point	Software or descriptor type	Feature selection method	Machine learning method	Data set size		R^2		RMSE		Reference
					Training	Test	Training	Test	Training	Test	
Flavonoid derivatives	RSA (%)	DFT-based	–	MLR	29	–	0.768	–	20.18	–	Sarkar, Middya, and Jana (2012)
Flavonoid derivatives	$LogIC_{50}$ (μM)	DFT-based	–	MLR	19	5	0.816	–	0.311	–	Djeradi, Rahmouni, and Cheriti (2014)
Flavonoid derivatives	TEAC	DFT-based	–	ANN							Žuvela, David, and Wong (2018)
Chromones derivatives	$LogEC_{50}$ (μM)	MFA	G/PLS	PLS	30	5	0.868	0.924	–	–	Samee, Nunthanavanit, and Ungwitayatorn (2008)
Chromones derivatives	pIC_{50}	SYBYLE	–	CoMFA	38	10	0.978	–	0.095	–	Phosrithong and Ungwitayatorn (2013)
Chromones derivatives	pIC_{50}	SYBYLE	–	CoMSIA	38	10	0.976	–	0.129	–	Phosrithong and Ungwitayatorn (2013)
Coumarin derivatives	pIC_{50}	Cerius 2 software	GFA	MLR	32	13	0.906–0.928	0.833–0.908	–	–	Mitra, Saha, and Roy (2013a)
Coumarin derivatives	pIC_{50}	Cerius 2 software	G/PLS	PLS	32	13	0.875–0.882	0.784–0.859	–	–	Mitra et al. (2013a)

Coumarin derivatives	pIC_{50}	Cerius 2 software	G/PLS	PLS	35	15	0.884	0.938	–	–	Mitra, Saha, and Roy (2013b)
Pulvinic acid and coumarine derivatives	Fenton reaction	CORAL	Monte Carlo method	MLR	101	9	0.79	0.86	–	–	Ahmadi et al. (2021)
Pulvinic acid and coumarine derivatives	Gama activity	CORAL	Monte Carlo method	MLR	79	12	0.86	0.86	–	–	Ahmadi et al. (2021)
Pulvinic acid and coumarine derivatives	UV activity	CORAL	Monte Carlo method	MLR	100	10	0.83	0.80	–	–	Ahmadi et al. (2021)
Tripeptides	Log(TEAC)	SYBYLE	–	MLR	50	5	0.668	–	0.476	–	Uno, Kodama, Yukawa, Shidara, and Akamatsu (2020)
Tripeptides	TEAC	SYBYLE	–	PLS	41	13	0.996–0.998	–	0.025–0.039	–	Yan, Lin, Zhang, Liang, and Wu (2020)
Tripeptides	FRAP and ABTS	SYBYLE	–	PLS	146	52	0.885–0.914	–	0.586–0.659	–	Guo et al. (2019)
Tripeptides	FRAP and ABTS	CORAL	Monte Carlo	MLR	81	26	0.6262–0.8271	0.9365	0.173–0.403	0.15	Toropova, Toropov, Roncaglioni, and Benfenati (2021)

predicting antioxidant activity. They also showed that the number and the arrangement of free hydroxyl groups on the flavonoid skeleton, or phenol ring of phenolic acids together with the shape, size, mass, and steric properties of the molecules, had significant effects on the activity of these compounds (Rastija & Medić-Šarić, 2009). Here, four OECD principles have been applied for QSAR modeling. The AD of the model (Principle 3) was not estimated.

Mitra, Saha, and Roy (2011) developed different QSAR models for 33 phenolic derivatives bearing NO donor groups. These models mainly inferred that presence of substituted aromatic carbons, long- chain branched substituents, and an oxadiazole-N-oxide ring with an electronegative atom-containing group substituted at the five position of the parent nucleus, increases the positively charged surface area and the volume of the molecules increase the antioxidant activity of molecules. The antioxidant activity of compounds increases with long-chain branched substituents lacking symmetry about the center of mass of the molecule. The authors suggested fifteen new antioxidants in this category and reported their activity based on the approved models (Mitra et al., 2011). QSAR has been developed based on the G/PLS technique, the DModX (Distance to Model in X-space) method (Wold, Sjöström, & Eriksson, 2001) implemented in the SIMCA software (UMETRICS SIMCA-P 10.0, info@umetrics.com: www.umetrics.com, Umea, Sweden, 2002) has been utilized for detecting the applicability domain of the developed model. In this research, all OECD principles have been used to validate the QSAR model.

Chen et al. studied the structure-thermodynamics-antioxidant relationships of 20 natural phenolic acids and derivatives based on $DPPH^{\cdot}$ scavenging assay, DFT calculations at the B3LYP/6–311++G(d,p) levels of theory, and QSAR modeling. The authors investigated three main working mechanisms including: hydrogen atom transfer (HAT), electron transfer-proton transfer (SETPT) and sequential proton loss-electron transfer (SPLET), in four micro-environments (gas-phase, benzene, water, and ethanol). Subsequently, computed thermodynamics parameters (BDE, IP, PDE, PA, and ETE) were compared with the experimental, radical scavenging activities against $DPPH^{\cdot}$. A combination of theoretical and experimental studies showed that the extended delocalization and intra-molecular hydrogen bonds have two main effects on the stability of the radicals. The C=O or C=C in COOH, COOR, C=CCOOH and C=CCOOR groups, and orthodiphenolic functionalities were shown to favorably stabilize the specific radical species to enhance the radical scavenging activities, while the presence of the single OH in the ortho position of the COOH group disfavors the activities. The authors concluded that HAT was the thermodynamically preferred mechanism in the gas phase and benzene, whereas SPLET in water and ethanol. Here, four OECD principles have been applied for QSAR modeling. The AD for the model (Principle 3) was not reported (Chen et al., 2015).

Ahmadi et al. developed the QSPR models of the radical scavenging activities for 96 natural phenolic antioxidants based on the graph of atomic orbitals by CORAL software. The QSAR models are developed with the representation of the molecular structure by the Simplified Molecular Input-Line Entry System (SMILES) and the Monte Carlo optimization was used for the calculation of the correlation weights of optimal SMILES-based descriptors. The data set was randomly divided three times into the training, invisible training, calibration and, validation sets. The reliable proposed models are constructed using SMILES information, without any information on physicochemical parameters, the quantum mechanics descriptors, or the 3D-structure of the antioxidant compounds (Ahmadi et al., 2019). The AD of QSAR models was defined by "defect" definition in CORAL software and the number of outliers has been estimated. All the OECD principles for QSAR model validation have been applied here.

- QSAR of flavonoids antioxidants

Amić and Lučić developed two QSAR models for experimental TEAC values of 38 flavonoids and two QSAR models for experimental vitamin C equivalent antioxidant capacity (VCEAC) of 36 flavonoids measured by the ABTS free radical. RM1 and PM6 semiempirical quantum mechanical calculations have been carried out in MOPAC 2009 software for modeling. The QSAR models have been developed based on the hydrogen atom transfer (HAT) mechanism of free radical scavenging of flavonoids encoded by minimal bond dissociation enthalpy values (BDEmin) and the number of hydroxyl groups (nOH). The presented models have been validated by cross-validated statistical parameters, and there are no external test sets (Amić & Lučić, 2010). The QSAR model validation was done by cross-validation without any external validation. In this paper, four OECD principles have been applied for QSAR modeling. The AD for model (Principle 3) was not reported.

Mitra et al. developed some QSAR models using genetic function approximation (GFA) and genetic partial lease squares(G/PLS) on 86 flavone derivatives (Fig. 7) which were modeled previously using the Fujita-Ban method. The total dataset was divided by the k-means clustering technique into 64 training and 22 test set. They used the additional descriptors viz.,

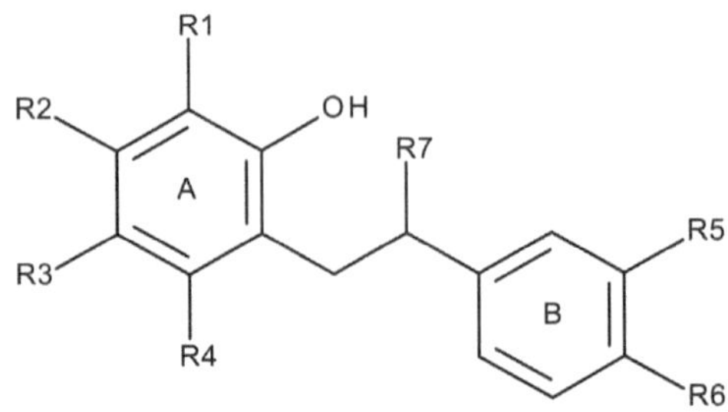

Fig. 7 The general structure of 86 molecules.

topological, structural, spatial and quantum chemical ones. They investigated the influence of different substituents on the free radical scavenging activity of the flavones. They suggested that the hydroxyl substituent at different positions of the A and B rings substantially influences the antioxidant activity of these molecules, in contrast, a methoxy substituent at R6 position of the B ring increases the activity. Increasing in the number of hydroxy substituents enhances the availability of hydrogen bond donor groups on the antioxidant molecules and consequently facilitates the neutralization of toxic free radicals. Moreover, molecules don't have the ene fragment of the pyran ring and molecules with an open pyran ring show an increased activity value. Here, four OECD principles for modeling have been applied. The AD for the model (Principle 3) was not presented (Mitra et al., 2010).

Sarkar et al. constructed a QSAR model for the radical scavenging activity (RSA) of a series of 29 flavonoids using density functional theory (DFT)-based quantum chemical descriptors. The QSAR model results show that hardness (η), group electrophilic frontier electron density (F_A^E) and group philicity (ω_b^+) of individual flavonoids are important for in vitro antioxidant activity. Here, four OECD principles for modeling have been applied. The AD for the model (Principle 3) was not presented (Sarkar et al., 2012).

Djeradi et al. (2014) studied a QSAR model for the antioxidant activity of 24 flavonoids based on electronic structure descriptors which are Fukui indices. The indices are calculated at the DFT/B3LYP level of chemical quantum theory. The dataset is divided into 19 train and 5 test compounds and the model is built based on the MLR method. The best QSAR equation with three parameters, i.e., f_k^+, f_k^-, and f_k^0, indicating a reliable regression model. Here, four OECD principles for modeling have been applied. The AD for the model (Principle 3) was not presented (Djeradi et al., 2014).

Žuvela et al. constructed QSAR models of TEAC of flavonoids using machine learning methods, such as ANN, in antioxidant activity prediction of flavonoids based on quantum mechanical parameters that are not usually interpretable. Interpreting the ANN models is often omitted or performed erroneously altogether. They compared six approaches (PaD, PaD2, weights, stepwise, perturbation, and profile) to study the contribution, importance descriptors and interpretation of ANN models. The sum of ranking differences was employed to rank the six methods according to the contributions of the calculated QM molecular descriptors toward the target (TEAC). The model results indicate that the PaD, PaD2, and profile techniques are the most stable and lead to a valid interpretation of the observed correlations. Here, the AD of the molecules was checked using the leverage approach. Thus, all the OECD principles for the QSAR model validation have been applied (Žuvela et al., 2018).

- QSAR of chromone derivatives

Samee et al. synthesized and tested the DPPH radical scavenging activities for 7-hydroxy, 8-hydroxy and 7, 8-dihydroxy synthetic chromone derivatives (Fig. 8). Furthermore, they developed the 3D-QSAR models of antioxidant activity of synthetic chromone derivatives using the genetic PLS (G/PLS) method for model construction. Dataset is divided into training (30 molecules) and test set (5 molecules). The molecular field analysis (MFA) model of antioxidants was done by field fit alignment. The MFA equation showed that the electronegative group on the benzoyl ring and the electropositive group on the phenyl ring are the main factors controlling the antioxidant activity of the mentioned compounds (Samee et al., 2008). Here, four OECD principles have been used for modeling. The AD for the model (Principle 3) was not presented.

Mitra et al. developed four QSAR models, namely 3D-pharmacophore mapping, CoMSIA, HQSAR, and group-based QSAR (G-QSAR) methods

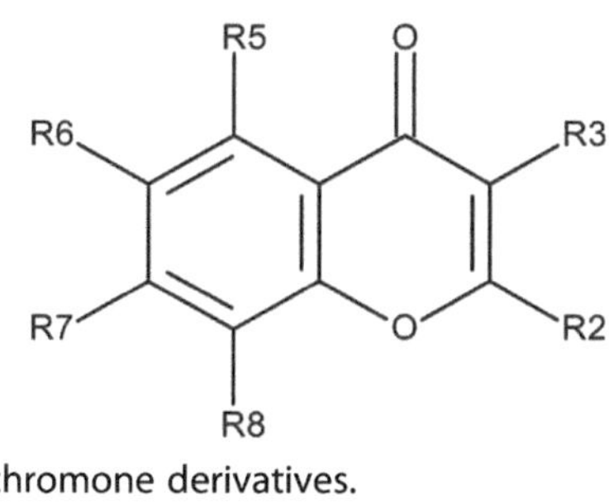

Fig. 8 The skeleton of chromone derivatives.

for 36 synthetic chromone derivatives (Mitra, Saha, & Roy, 2012). The hydrogen-bond acceptor descriptor was identified as an influence feature in all four models. Therefore, hydroxyl substituents at the R8 position, and benzoyl substitutes at the R3 position of the chromone nucleus were essential components for enhancing antioxidant activity. Moreover, the ketonic group at C4 increases the abilities of the compounds to interact with the toxic free radicals by a mechanism of electron transfer followed by deprotonation. The CoMSIA analysis showed that bulky substituents at the R2 position were unfavorable. In the three-dimensional pharmacophore model, the presence of the aromatic ring and the hydrophobic properties on the substituents at the positions R2 and R3, respectively, show that such groups separated by the specific distance of 5.890 Å were necessary to increase the activities of the compounds. Similar results were achieved from the CoMSIA analysis, in which these substituents map to the hydrophobically favored contours. In addition, the hologram QSAR (HQSAR) contour analysis also indicated the major impact of these fragments, with the green color for the substituent at R3 showing its maximum contribution. Finally, the G-QSAR models were utterly consistent with the remaining models, revealing the significant effect of the hydroxyl substitution at the R8 position on the antioxidant activity profiles of the chromone derivatives, moreover on the remaining essential properties, such as the presence of the substituted benzoyl fragment at the R3 position and the substituted aromatic fragment at the R2 position. In this work, the AD of the molecules was investigated using the leverage approach. So, all five OECD principles for QSAR model validation have been applied.

The 3D-QSAR approach based on comparative molecular field analysis (CoMFA) and comparative molecular similarity indices analysis (CoMSIA) has been carried out for DPPH antioxidant activity for a series of 48 chromone derivatives by Phosrithong and Ungwitayatorn (2013). The chromone compounds were geometry optimized by AM1 and HF/6- 31G* calculations. The results of CoMFA and CoMSIA were compared between different alignment strategies. The contour maps provide fruitful structure-radical scavenging activity relationships, which are helpful for designing new compounds with higher activity. In this study, the AD for the model (Principle 3) was not presented. So, four OECD principles have been used for modeling.

- QSAR of coumarins

Mitra et al. studied different QSAR models, including the descriptor-based QSAR model, 3D pharmacophore, and HQSAR models, to identify the critical structural features indicating the potential antioxidant activity profile

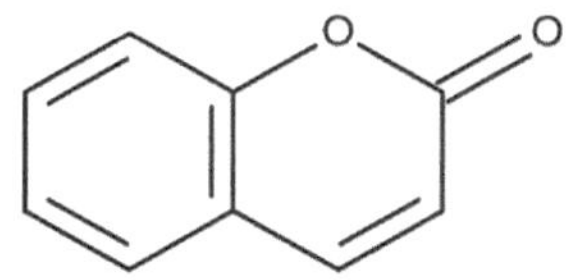

Fig. 9 The main structure of the coumarin core.

of a series of coumarin derivatives (Mitra et al., 2013a). The results of all models show the essential role of the central coumarin core (fused benzene ring and the oxygen atom of the pyran ring) for the optimal antioxidant activity of the compounds (Fig. 9). Here, the AD for the model (Principle 3) was not presented. Thus, four OECD principles have been used for modeling.

Mitra et al. developed the descriptor-based QSAR, 3D pharmacophore, HQSAR, and G-QSAR for 50 coumarin derivatives. The descriptor-based QSAR model demonstrated primarily the importance of the ketonic oxygen fragment, followed by the secondary amine group and the unsubstituted aromatic carbon fragments (Mitra et al., 2013b). The AD for the model (Principle 3) was not reported. So, four OECD principles have been assessed for modeling.

Ahmadi et al. developed different QSAR models through SMILES of molecules and based on the Monte Carlo optimization approach to predict the antioxidant activity of 79 derivatives of pulvinic acid, 23 of coumarin, as well as nine structurally nonrelated compounds against three radiation sources of Fenton, gamma, and UV. They used the index of ideality correlation as the target function to find the best models. The dataset was randomly divided into four sets, including training, invisible training, validation, and calibration; this division was repeated three times randomly. The optimal descriptors were picked up from a hybrid model by combining the hydrogen-suppressed graph and SMILES descriptors based on the objective function. The results of three randomized sets showed that simple, robust, reliable, and predictive models were achieved for training, invisible training, validation, and calibration sets of all three models. The mechanistic interpretation of the increasing and decreasing descriptors has been made. The presence of oxygen with branching from oxygen atom (O...(...) and presence of double covalent bonds (=...) as increasing promotors and the presence of oxygen connected to the aliphatic carbon. The presence of carbon with branching (C...(...) as decreasing promotors (Ahmadi et al., 2021).

The AD of QSAR models was given by the "defect" definition in CORAL and the number of outliers was estimated. So, all the OECD principles for QSAR model validation have been applied here.

- QSAR of peptides

Uno et al. synthesized 55 tripeptides using 20 natural amino acids, and their antioxidant activity was evaluated based on the TEAC Assay. The TEAC antioxidant activity of these tripeptides was modeled using QREG, version 2.05, with MLR analysis. The dataset was divided into training and validation sets of 50 and 5 tripeptides. The model results of the QSAR study indicate that the presence of a cysteine residue at any position, an aromatic amino acid at the C-terminus, higher hydrophobicity of the N-terminal residue, and smaller HOMO-LUMO energy gap of the middle residue are main parameters for increasing the antioxidant activity (Uno et al., 2020). Here, the AD of QSAR models was not given and consequently, only four OECD principles of QSAR modeling have been applied.

Yan et al. performed the 3D-QSAR models for the TEAC value of 54 tripeptides by SYBYL-X 2.0 software. The tripeptides were divided into training and test sets consisting of 41 and 13 tripeptides. The 3D-QSAR models were created by the PLS regression method. CoMFA and CoMSIA methods were used to study the relationship between structure and TEAC values. Based on the model finding, the small and hydrophilic group in the N-terminal of the tripeptide or bulky and hydrophobic group in the C-terminal enhance the antioxidant activity. Two new tripeptides, GWY and QWY, were designed based on the model, and their antioxidant activity was evaluated by the same assay of the database in vitro and exhibited good activity (Yan et al., 2020). In this research, the AD of QSAR models was not presented and so, only four OECD principles of QSAR modeling have been applied.

The stable and predictive 3D-QSAR study of CoMFA and CoMSIA for 198 antioxidant tripeptides was performed by Guo et al. to guide the design and virtual screening for new peptides. The dataset was ranked in descending order of antioxidant activities and randomly divided into the training set and the test sets consisting of 52 and 146 tripeptides. The contour maps show that steric, electrostatic, hydrophilic, and hydrogen bond acceptor force fields have a greatly contribute to the antioxidant activity that can provide the graphical information for rational design and modification of new peptides. Moreover, they designed 10 novel tripeptides and evaluated their relative antioxidant activity by FRAP and ABTA assay and most of them showed good antioxidant activity (Guo et al., 2019). Here, the AD of QSAR models was not investigated, and only four OECD principles of QSAR modeling have been applied.

Toropova et al. developed the reliable QSAR models for antioxidant activity of 107 tripeptides from frog skin using CORAL software based on the Monte Carlo optimization. They used only the structure of tripeptides represented by sequences of one-symbol abbreviations of the corresponding amino acids to construct the QSAR models (Toropova et al., 2021). The AD of the model was obtained according "defect" statistics criterion, and the QSAR models have been thoroughly validated according to five OECD principles.

3.2 Molecular docking

Molecular recognition shows a key role in helping scientists to understand the essential biomolecular experience such as enzyme-substrate, drug-protein, and drug-nucleic acid interactions. A comprehensive understanding of the common view that governs the nature of the interactions (hydrogen bonding, van der Waals, electrostatic) between the ligands and their corresponding targets (protein or nucleic acid) offer a conceptual support for designing the specific potential drug with the desired potency. Thus, these structural evidence for an interesting target and evaluation candidate ligands generate such knowledge for practical applications (Cummings, DesJarlais, Gibbs, Mohan, & Jaeger, 2005).

In this regard, *in silico* drug design as a helpful tool have used to find hits in the initial step to lead optimization at the final step. Among many reported investigations, various reports suggest molecular docking as a worthy strategy to predict 3D structures of protein-ligand complexes. Molecular docking can tackle experimental complications connecting the structure resolve of these complexes; it can provide deep structural comprehension of biomolecular interactions. However, the process of binding a drug candidate to its protein target is not simple; several factors can influence the interaction between two molecules. Accordingly, these methods can be modified as an answer to docking problems. In contrast to great advances and successes, the common application of docking procedures, and the rapid and accurate prediction of protein–ligand interactions is still a challenging field to search, and some deficiencies still exist. A number of these limitations that can confuse the quantitative explanation of the process are:

- The flexibility or mobility of both ligand and target
- The effect of the protein environment on the charge dispersal over the ligand

- Their interactions with the surrounding water molecules

Meanwhile, we know that molecular docking can be helpful to the different fields of drug discovery. It has many potential usages in finding possible leads through structure–activity, lead optimization, and virtual screening studies.

In other words, molecular docking is significant *in silico* technique commonly used for identifying drugs lead by 3D docking ligand to the 3D structure of a pre-choose target site. Then, it can be done with optimization of binding configurations and calculation of binding probability based on paired molecular interactions (Abdolmaleki, Shiri, & Ghasemi, 2018; Chaudhary & Mishra, 2016; Gozalbes et al., 2008; Kitchen, Decornez, Furr, & Bajorath, 2004; March-Vila et al., 2017). The computational process of docking is divided into two main stages:

- The correct situation of the ligand at the protein binding-site
- Calculation of scoring function to estimate the ligand affinity

The current field principally adapts to dock ligand molecules into macromolecule targets, mostly protein and its usage is growing year by year (Abdolmaleki et al., 2018; Ghosh, Chandar, Lo, & Ganguly, 2016; Pei, Yin, Ma, & Lai, 2014; Zhong et al., 2018). The goal of running a docking simulation of ligand–protein is calculating the primary binding mode(s) of a small molecule with a protein of known 3D structures of its components. Ligand or drug uses its action by binding to specific molecular parts of the cell, such as modulating biochemical processes in an illness-adapting manner. Non-covalent interactions are an essential aspect of structure-based drug discovery that has been recognized with diverse fields associated to materials, biological sciences, chemical, physical, and pharmacological studies.

In a typical interaction model between two molecules, the following energies are relevant:

- Electrostatics (non-bonding interactions)
- Covalent bond (Aljoundi, Bjij, El Rashedy, & Soliman, 2020)
- Hydrophobic
- H-bonding (H-bond donating and accepting)
- van der Waals (non-bonding interactions)
- Energy (De)solvation

There are some examples of 3D-QSAR for a single target in a combination with molecular docking studies, for GRid Independent Descriptors (GRIND) (Ahmadi & Ghasemi, 2014), CMFA and CoMSIA descriptors (Abedi, Ghasemi, & Ebrahimzadeh, 2013; Ghasemi, Pirhadi, & Ayati, 2011), CoMSIA descriptors for 4D-QSAR (Ghasemi, Safavi-Sohi, & Barbosa, 2012).

Multi-targeted drug discovery is an innovative field for inhibiting of multi-target activity and searching for new drugs (Abdolmaleki & Ghasemi, 2017). Table 2 shows several natural-based hybrid compounds.

An example of hybrid compound design is the synthesis and bioactivity evaluation of a series of the drug like compounds by Sangshetti et al. (2016). They investigated ADME properties, structural-activity relationship (SAR), and binding mode of derivatives by docking and other *in silico* methodologies. They reported a series of hybrid molecules contains of indole and coumarin centers joined with Chaconne like a chain. Fig. 10 shows the planning of these structures in such bioactive derivatives. The diversity of these moieties in different pharmaceuticals (without or by adding another molecule) as well as in nature is the reasoning behind the usage of these three fragments in their design. Marrapu, Mittal, Shivahare, Gupta, and Bhandari (2011) indicated that maximum antioxidant effects were achieved for the indole derivatives such as tryptamine and tryptophan with IC_{50}s of 6.00 ± 0.60 and $3.50 \pm 0.4\,\mu M$, respectively. Also, we know Butein, a naturally happening substance having a chalcone backbone, is the active component of *Dalbergia odorifera*. Besides, it is found in the heartwood of Acacia, and has been reported to have antioxidant properties (Sogawa et al., 1994; Wang, Weng, & Cheng, 2000).

3.3 Pharmacophore model

Pharmacophore is defined as the essential geometric arrangement of atoms or functional groups necessary to produce a given biological response. Pharmacophore methods are complemented for molecular docking by assisting the choice of drug leads with higher structural flexibility, whereas molecular docking offers a more accurate calculation of specific binding interactions. Pharmacophore is an abstract concept for describing electronic and steric features that is essential to understanding the optimal macromolecular/receptor interactions with a ligand. Several factors influence the quality of pharmacophore models, including:

- a general problem of estimating ligand binding affinities
- molecular overlay
- sensitivity to training datasets
- quality of conformational sampling
- selection of anchoring points
- selection of a few relevant chemical features among many possible choices
- consideration of tautomeric and protonation status of compounds

Table 2 Atypical natural-based hybrid molecules.

No.	Structures	Hybrid components	Reference
1	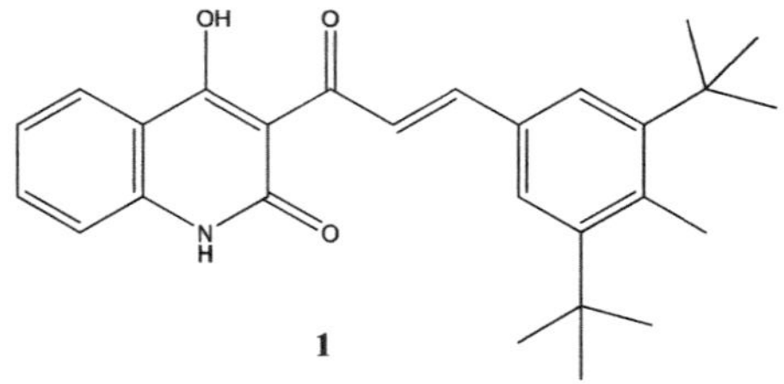 **1**	Quinolinone and chalcone	Roussaki, Hall, and Lima (2013)
2a, 2b	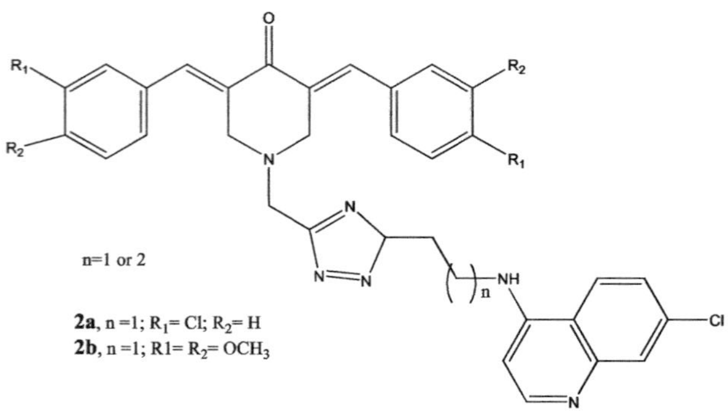 n=1 or 2 **2a**, n =1; R_1= Cl; R_2= H **2b**, n =1; R1= R_2= OCH_3	C5 -curcuminoid and 4 -aminoquinoline	Kandi, Manohar, and Vélez-Genera (2015)
3	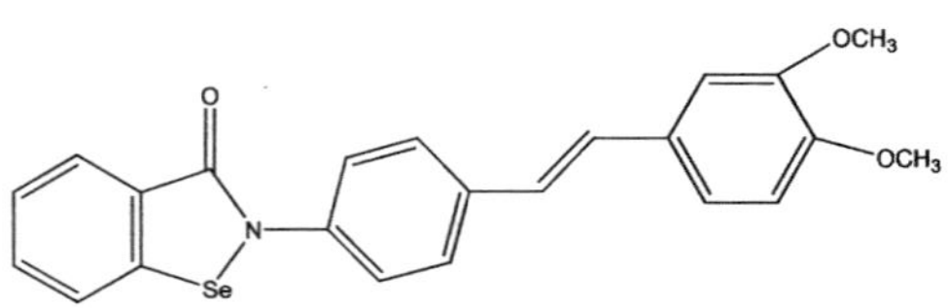 **3**	Resveratrol and ebselen	Yan, Guo, and Wang (2015)

4	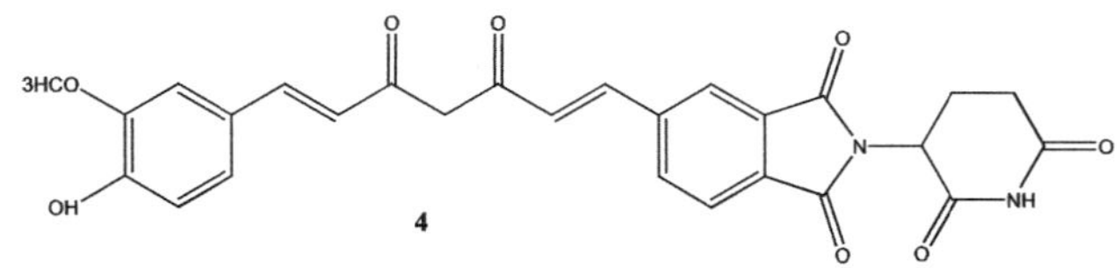	Curcumin and thalidomide	Lui, Zhang, and Chojnacki (2013)
5	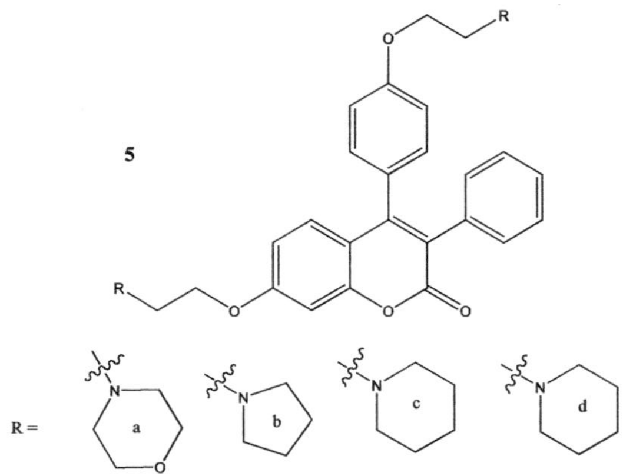	Triphenylethylene and coumarin	Chen, Li, and Yao (2013)
6	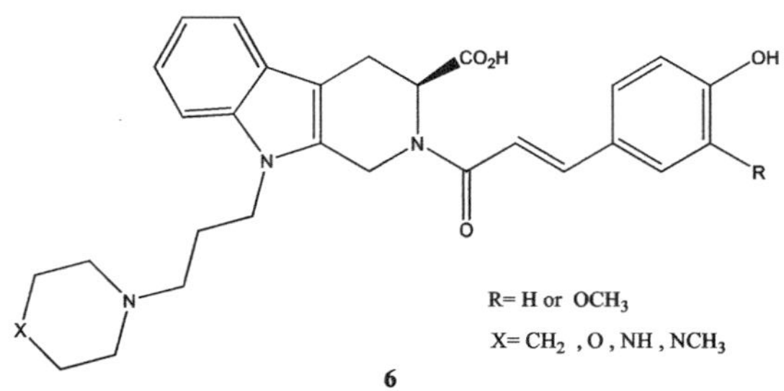	Tetrahydro-β- carboline and hydroxylcinnamic acid	Lin, Xia, and Yao (2014)

Continued

Table 2 Atypical natural-based hybrid molecules.—cont'd

No.	Structures	Hybrid components	Reference
7	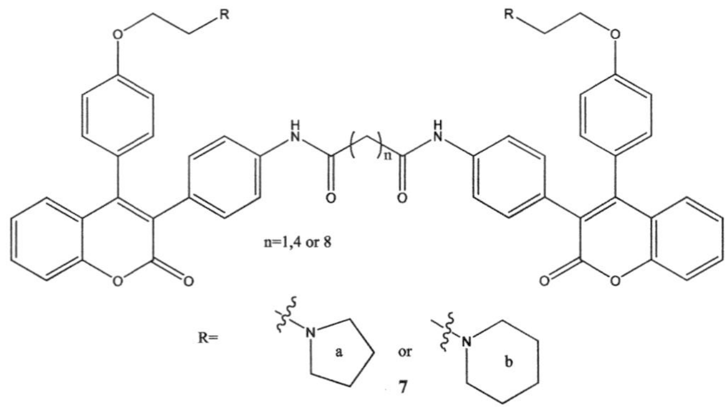 **7**	Triphenylethylene and coumarin	Tan, Yao, and Gu (2014)
8a, 8b	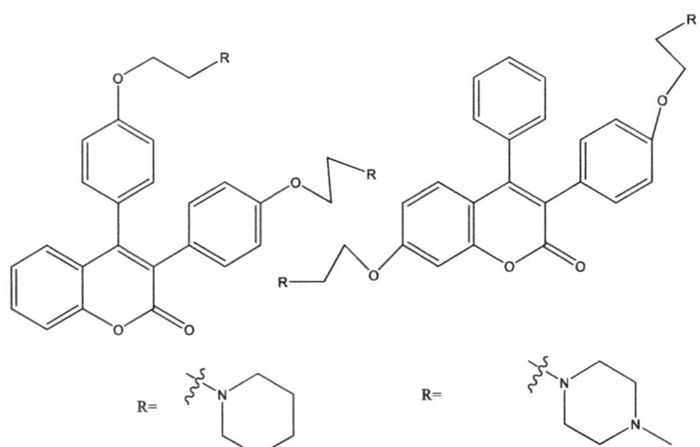 **8**	Triphenylethylene and coumarin	Zhao, Yao, and Li (2014)

9	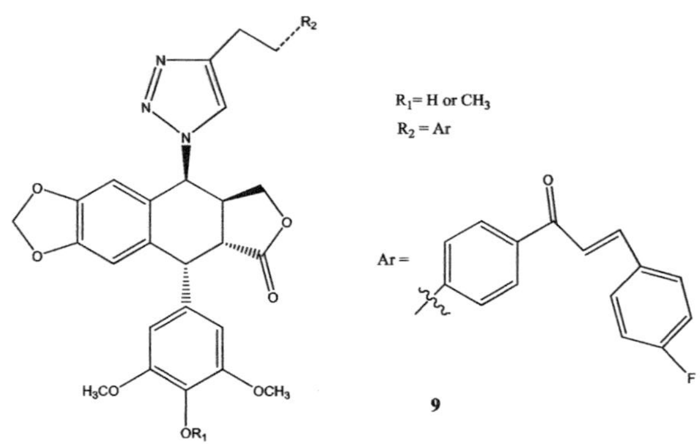 **9**	Epipodophyllotoxin and chalcone	Bandy, Kilkarni, and Hruby (2015)
10	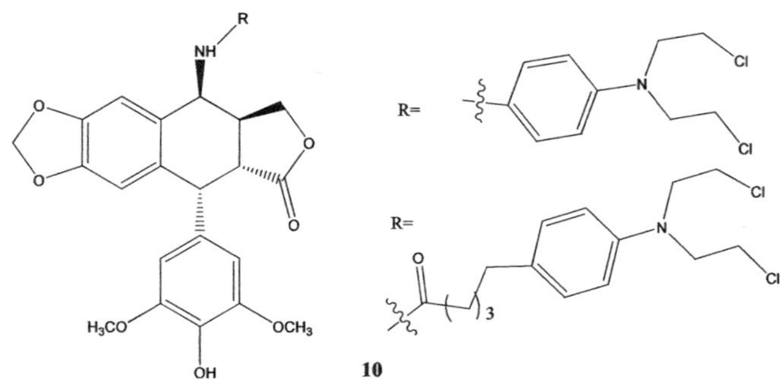 **10**	Epipodophyllotoxin and nitrogen-mustard	Yadav, Wu, Patel, Yalowich, and Hasinoff (2014)

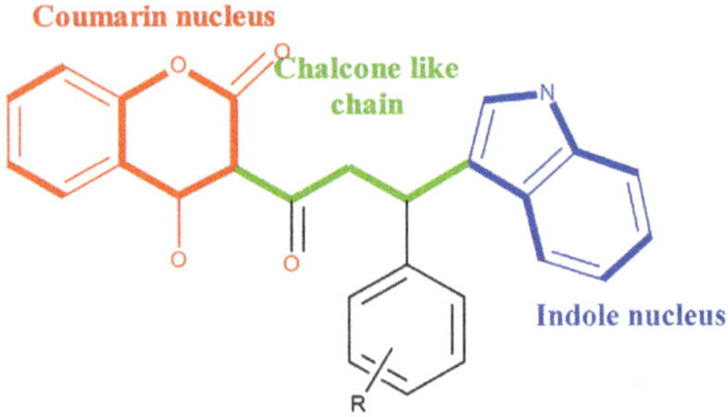

Fig. 10 Coumarin-Chalcone-Indol as a core for designing of hybrid compound derivatives.

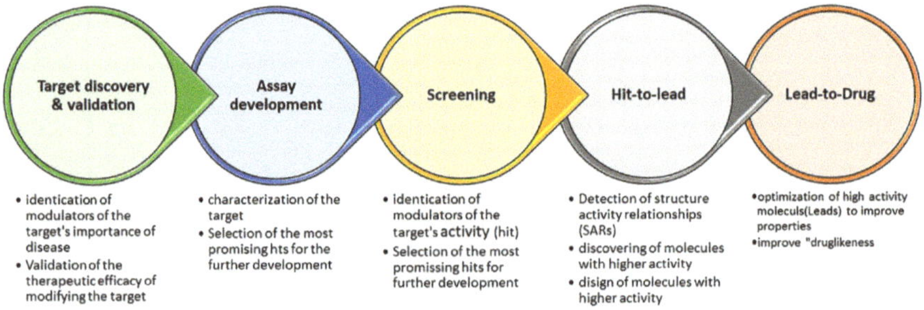

Fig. 11 General steps to the development of a new investigational drug.

For example, Queiroz, Gomes, Moraes, and Borges (2009) investigated the structure–activity relationship to resolve antioxidant pharmacophore for resveratrol through Functional of Density Theory and quantum chemistry computations. They showed that 4-hydroxystilbene is the antioxidant pharmacophore of resveratrol. The antioxidant activity of resveratrol connects to the stabilization energy of the 4-hydroxystilbene. Besides, the π-type electron system controls the radicals' stability, and the unpaired electrons are distributed mainly to the B-benzene ring, O-atom in the *para* position, and a double bond in resveratrol hydroxylated derivatives.

3.4 Integration method

Drug discovery is a challenging and highly complicated task. As we see in Fig. 11, it starts by picking a disease and detecting a target, discovering different leads or hits, and finally uses pre-clinical and clinical trials. During the drug finding process, some drug candidates have failed for improper ADMET properties. So, developing *in silico* methods is critical in the early stages of drug discovery to decrease the time and cost of this process (Ghaleb et al., 2020). Molecular docking is useful for problem solving of an optimization project or simulation of some medicinal processes. For that reason, it may be joined with other computer-aided drug design (CADD) methods such as pharmacophore modeling and molecular dynamics to find the best fit for drug-protein complexes.

Binding affinity prediction for protein-ligand is one of the critical challenges in drug discovery. In recent years, machine learning methodologies significantly developed in this task. Though, current procedures of model assessment are overly optimistic in calculating generality to new targets, and there does not exist a standard dataset of sufficient size to compare performance between models. Morrone, Weber, Huynh, Luo, and Cornell (2020) combined a standard docking procedure and deep learning to improve the prediction of binding mode(pose) for protein-ligand complexes. They applied structural data from protein-ligand interactions as input for a convolutional neural network to model the generation of activity and the prediction of binding mode.

4. Conclusion and future direction

The history of traditional therapies indicates medicinal plants and natural extracts (such as Resveratrol Curcumin, and Cinnamophilin) have been applied as a source for disease treatment. The therapeutic efficacy of these sources may be due to individual ingredients or a combination of individual compounds with multi-target activity. Curcumin, and resveratrol are well-known examples of multi-target compounds that can interact with large numbers of targets (Athar, Back, Kopelovich, Bickers, & Kim, 2009; Shehzad & Lee, 2010). Also, natural products have been shown multi-target activity more than many synthetic molecules. As a result, they are potentially valuable starting points for multi-target design (Park, Lee, Ahn, & Kim, 2009). These suitable properties can be studied deeply by intelligent data-driven methods. Today, various scientific approaches, powerful instrumental techniques, and fast data analysis strategies such as artificial intelligence (AI) methods bring the ability to achieve more knowledge of natural compounds. AI technology can learn and explore patterns in large volumes of data to find information quickly. Also, it is capable to quickly extracting insights in order to automate tasks for a variety of aims. Intelligence tools can have significant advantages over traditional methods for accurately analyzing chemical data and activity prediction.

According to Table 1, several articles on QSAR modeling of natural antioxidants have been published. To develop QSAR modeling of natural antioxidants, it is necessary to review natural antioxidant activity data, pre-process them and build new models using new software and methods. In many of the reported models, there is no applicability domain (AD) of models and their external validation. Also, as a result of robust and predictive QSAR models, some active antioxidants can be designed and synthesized. In addition, based on the results of predictive QSAR models for peptide-based antioxidants, new and more active antioxidants can be designed and developed as additives in functional foods and nutritional supplements.

References

Žuvela, P., David, J., & Wong, M. W. (2018). Interpretation of ANN-based QSAR models for prediction of antioxidant activity of flavonoids. *Journal of Computational Chemistry, 39*(16), 953–963.

Abdolmaleki, A., & Ghasemi, J. B. (2017). Dual-acting of hybrid compounds—A new dawn in the discovery of multi-target drugs: Lead generation approaches. *Current Topics in Medicinal Chemistry, 17*(9), 1096–1114.

Abdolmaleki, A., Shiri, F., & Ghasemi, J. B. (2018). Computational multi-target drug design. In *Multi-target drug design using chem-bioinformatic approaches* (pp. 51–90). Springer.

Abedi, H., Ghasemi, J. B., & Ebrahimzadeh, H. (2013). 3D-QSAR, CoMFA, and CoMSIA of new phenyloxazolidinones derivatives as potent HIV-1 protease inhibitors. *Structural Chemistry, 24*, 433–444. https://doi.org/10.1007/s11224-012-0092-1.

Adom, K. K., & Liu, R. H. (2005). Rapid peroxyl radical scavenging capacity (PSC) assay for assessing both hydrophilic and lipophilic antioxidants. *Journal of Agricultural and Food Chemistry, 53*(17), 6572–6580.

Ahmadi, S. (2020). Mathematical modeling of cytotoxicity of metal oxide nanoparticles using the index of ideality correlation criteria. *Chemosphere, 242*, 125192.

Ahmadi, S., & Akbari, A. (2018). Prediction of the adsorption coefficients of some aromatic compounds on multi-wall carbon nanotubes by the Monte Carlo method. *SAR and QSAR in Environmental Research, 29*(11), 895–909.

Ahmadi, S., & Ganji, S. (2016). Genetic algorithm and self-organizing maps for QSPR study of some N-aryl derivatives as butyrylcholinesterase inhibitors. *Current Drug Discovery Technologies, 13*(4), 232–253.

Ahmadi, P., & Ghasemi, J. B. (2014). 3D-QSAR and docking studies of the stability constantsof different guest molecules with beta-cyclodextrin. *Journal of Inclusion Phenomena and Macrocyclic Chemistry, 79*(3–4), 423–435.

Ahmadi, S., & Habibpour, E. (2017). Application of GA-MLR for QSAR modeling of the arylthioindole class of tubulin polymerization inhibitors as anticancer agents. *Anti-Cancer Agents in Medicinal Chemistry (Formerly Current Medicinal Chemistry-Anti-Cancer Agents), 17*(4), 552–565.

Ahmadi, S., Ghanbari, H., Lotfi, S., & Azimi, N. (2021). Predictive QSAR modeling for the antioxidant activity of natural compounds derivatives based on Monte Carlo method. *Molecular Diversity, 25*(1), 87–97.

Ahmadi, S., Khazaei, M. R., & Abdolmaleki, A. (2014). Quantitative structure–property relationship study on the intercalation of anticancer drugs with ct-DNA. *Medicinal Chemistry Research, 23*(3), 1148–1161.

Ahmadi, S., Mehrabi, M., Rezaei, S., & Mardafkan, N. (2019). Structure-activity relationship of the radical scavenging activities of some natural antioxidants based on the graph of atomic orbitals. *Journal of Molecular Structure*, *1191*, 165–174.

Aljoundi, A., Bjij, I., El Rashedy, A., & Soliman, M. E. (2020). Covalent versus non-covalent enzyme inhibition: Which route should we take? A justification of the good and bad from molecular modelling perspective. *The Protein Journal*, *39*, 97–105.

Ambure, P., Gajewicz-Skretna, A., Cordeiro, M. N. D., & Roy, K. (2019). New workflow for QSAR model development from small data sets: Small dataset curator and small dataset modeler. integration of data curation, exhaustive double cross-validation, and a set of optimal model selection techniques. *Journal of Chemical Information and Modeling*, *59*(10), 4070–4076.

Amić, D., & Lučić, B. (2010). Reliability of bond dissociation enthalpy calculated by the PM6 method and experimental TEAC values in antiradical QSAR of flavonoids. *Bioorganic & Medicinal Chemistry*, *18*(1), 28–35.

Anwar, H., Rahman, Z., Javed, I., & Muhammad, F. (2012). Effect of protein, probiotic, and symbiotic supplementation on serum biological health markers of molted layers. *Poultry Science*, *91*(10), 2606–2613.

Apak, R., Güçlü, K., Özyürek, M., & Karademir, S. E. (2004). Novel total antioxidant capacity index for dietary polyphenols and vitamins C and E, using their cupric ion reducing capability in the presence of neocuproine: CUPRAC method. *Journal of Agricultural and Food Chemistry*, *52*(26), 7970–7981.

Athar, M., Back, J. H., Kopelovich, L., Bickers, D. R., & Kim, A. L. (2009). Multiple molecular targets of resveratrol: Anti-carcinogenic mechanisms. *Archives of Biochemistry and Biophysics*, *486*(2), 95–102.

Bandy, A. H., Kilkarni, V. V., & Hruby, V. J. (2015). Design, synthesis, and biological and docking studies of novel epipodophyllotoxin–chalcone hybrids as potential anticancer agents. *Medicinal Chemistry Communications*, *6*, 94–104.

Benzie, I. F., & Strain, J. J. (1996). The ferric reducing ability of plasma (FRAP) as a measure of "antioxidant power": The FRAP assay. *Analytical Biochemistry*, *239*(1), 70–76.

Binic, I., Lazarevic, V., Ljubenovic, M., Mojsa, J., & Sokolovic, D. (2013). Skin ageing: Natural weapons and strategies. Evidence-Based Complementary and Alternative Medicine, 2013, 827248.

Blois, M. S. (1958). Antioxidant determinations by the use of a stable free radical. *Nature*, *181*(4617), 1199–1200.

Bondet, V., Brand-Williams, W., & Berset, C. (1997). Kinetics and mechanisms of antioxidant activity using the DPPH free radical method. *LWT-Food Science and Technology*, *30*(6), 609–615.

Brand-Williams, W., Cuvelier, M.-E., & Berset, C. (1995). Use of a free radical method to evaluate antioxidant activity. *LWT-Food Science and Technology*, *28*(1), 25–30.

Brewer, M. (2011). Natural antioxidants: Sources, compounds, mechanisms of action, and potential applications. *Comprehensive Reviews in Food Science and Food Safety*, *10*(4), 221–247.

Byvatov, E., & Schneider, G. (2003). Support vector machine applications in bioinformatics. *Applied Bioinformatics*, *2*(2), 67–77.

Chaudhary, K. K., & Mishra, N. (2016). A review on molecular docking: Novel tool for drug discovery. *Database*, *3*(4), 1029.

Chen, H., Li, S., & Yao, Y. (2013). Design, synthesis, and anti-tumor activities of novel triphenylethylene–coumarin hybrids, and their interactions with Ct-DNA. *Bioorganic & Medicinal Chemistry Letters*, *23*(17), 4785–4789.

Chen, Y., Xiao, H., Zheng, J., & Liang, G. (2015). Structure-thermodynamics-antioxidant activity relationships of selected natural phenolic acids and derivatives: An experimental and theoretical evaluation. *PLoS One*, *10*(3), e0121276.

Chirico, N., & Gramatica, P. (2011). Real external predictivity of QSAR models: How to evaluate it? Comparison of different validation criteria and proposal of using the concordance correlation coefficient. *Journal of Chemical Information and Modeling, 51*(9), 2320–2335.

Cummings, M. D., DesJarlais, R. L., Gibbs, A. C., Mohan, V., & Jaeger, E. P. (2005). Comparison of automated docking programs as virtual screening tools. *Journal of Medicinal Chemistry, 48*(4), 962–976.

de Aguiar, P. F., Bourguignon, B., Khots, M., Massart, D., & Phan-Than-Luu, R. (1995). D-optimal designs. *Chemometrics and Intelligent Laboratory Systems, 30*(2), 199–210.

Djeradi, H., Rahmouni, A., & Cheriti, A. (2014). Antioxidant activity of flavonoids: A QSAR modeling using Fukui indices descriptors. *Journal of Molecular Modeling, 20*(10), 1–9.

Dragsted, L. O., Pedersen, A., Hermetter, A., Basu, S., Hansen, M., Haren, G. R., et al. (2004). The 6-a-day study: Effects of fruit and vegetables on markers of oxidative stress and antioxidative defense in healthy nonsmokers. *The American Journal of Clinical Nutrition, 79*(6), 1060–1072.

Falowo, A. B., Fayemi, P. O., & Muchenje, V. (2014). Natural antioxidants against lipid–protein oxidative deterioration in meat and meat products: A review. *Food Research International, 64*, 171–181.

Farahani, S. R., Sohrabi, M. R., & Ghasemi, J. B. (2018). A detailed structural study of cytotoxicity effect of ionic liquids on the leukemia rat cell line IPC-81 by three dimensional quantitative structure toxicity relationship. *Ecotoxicology and Environmental Safety, 158*, 256–265.

Faustman, C., Sun, Q., Mancini, R., & Suman, S. P. (2010). Myoglobin and lipid oxidation interactions: Mechanistic bases and control. *Meat Science, 86*(1), 86–94.

Foti, M. C., Daquino, C., & Geraci, C. (2004). Electron-transfer reaction of cinnamic acids and their methyl esters with the DPPH• radical in alcoholic solutions. *The Journal of Organic Chemistry, 69*(7), 2309–2314.

Fourches, D., Muratov, E., & Tropsha, A. (2016). Trust, but verify II: A practical guide to chemogenomics data curation. *Journal of Chemical Information and Modeling, 56*(7), 1243–1252.

Frankel, E. (2005). *Lipid Oxidation* (2nd ed.). Bridgewater, UK: The Oily Press, PJ Barnes & Associates.

Gadaleta, D., Mangiatordi, G. F., Catto, M., Carotti, A., & Nicolotti, O. (2016). Applicability domain for QSAR models: Where theory meets reality. *International Journal of Quantitative Structure-Property Relationships (IJQSPR), 1*(1), 45–63.

Gedeck, P., Kramer, C., & Ertl, P. (2010). Computational analysis of structure–activity relationships. *Progress in Medicinal Chemistry, 49*, 113–160.

Ghaleb, A., Aouidate, A., Ayouchia, H. B. E., Aarjane, M., Anane, H., & Stiriba, S.-E. (2020). In silico molecular investigations of pyridine N-Oxide compounds as potential inhibitors of SARS-CoV-2: 3D QSAR, molecular docking modeling, and ADMET screening. *Journal of Biomolecular Structure and Dynamics, 40*(1), 143–153.

Ghasemi, J. B., Ahmadi, S., & Brown, S. (2011). A quantitative structure–retention relationship study for prediction of chromatographic relative retention time of chlorinated monoterpenes. *Environmental Chemistry Letters, 9*(1), 87–96.

Ghasemi, F., Mehridehnavi, A., Fassihi, A., & Pérez-Sánchez, H. (2018). Deep neural network in QSAR studies using deep belief network. *Applied Soft Computing, 62*, 251–258.

Ghasemi, F., Mehridehnavi, A., Perez-Garrido, A., & Perez-Sanchez, H. (2018). Neural network and deep-learning algorithms used in QSAR studies: Merits and drawbacks. *Drug Discovery Today, 23*(10), 1784–1790.

Ghasemi, J. B., Pirhadi, S., & Ayati, M. (2011). 3D-QSAR studies of 2-arylbenzoxazolesas novel cholesteryl ester transfer protein inhibitors. *Bulletin of the Korean Chemical Society, 32*(2), 645–650.

Ghasemi, J. B., Safavi-Sohi, R., & Barbosa, E. G. (2012). 4D-LQTA-QSAR and docking study on potent Gram-negative specific LpxC inhibitors: A comparison to CoMFA modeling. *Molecular Diversity*, *16*, 203–213.

Ghosh, S., Chandar, N. B., Lo, R., & Ganguly, B. (2016). Effective docking program for designing reactivator for treating organophosphorus inhibited AChE. *JSM Chemistry*, *4*, 1032.

Glazer, A. N. (1990). [14] Phycoerythrin fluorescence-based assay for reactive oxygen species. *Methods in Enzymology*, *186*, 161–168.

Golbraikh, A., & Tropsha, A. (2002). Beware of q2! *Journal of Molecular Graphics and Modelling*, *20*(4), 269–276.

Golbraikh, A., Shen, M., Xiao, Z., Xiao, Y.-D., Lee, K.-H., & Tropsha, A. (2003). Rational selection of training and test sets for the development of validated QSAR models. *Journal of Computer-Aided Molecular Design*, *17*(2), 241–253.

Gozalbes, R., Simon, L., Froloff, N., Sartori, E., Monteils, C., & Baudelle, R. (2008). Development and experimental validation of a docking strategy for the generation of kinase-targeted libraries. *Journal of Medicinal Chemistry*, *51*(11), 3124–3132.

Guo, H., Wang, Y., He, Q., Zhang, Y., Hu, Y., Wang, Y., et al. (2019). In silico rational design and virtual screening of antixoidant tripeptides based on 3D-QSAR modeling. *Journal of Molecular Structure*, *1193*, 223–230.

Habibpour, E., & Ahmadi, S. (2017). QSAR modeling of the arylthioindole class of colchicine polymerization inhibitors as anticancer agents. *Current Computer-Aided Drug Design*, *13*(2), 143–159.

Harnly, J. (2017). Antioxidant methods. *Journal of Food Composition & Analalysis*, *64*, 145–146.

Huang, D., Ou, B., & Prior, R. L. (2005). The chemistry behind antioxidant capacity assays. *Journal of Agricultural and Food Chemistry*, *53*(6), 1841–1856.

Huang, D., Ou, B., Hampsch-Woodill, M., Flanagan, J. A., & Deemer, E. K. (2002). Development and validation of oxygen radical absorbance capacity assay for lipophilic antioxidants using randomly methylated β-cyclodextrin as the solubility enhancer. *Journal of Agricultural and Food Chemistry*, *50*(7), 1815–1821.

Hurrell, R. F. (2003). Influence of vegetable protein sources on trace element and mineral bioavailability. *The Journal of Nutrition*, *133*(9), 2973S–2977S.

Jain, A. K., Mao, J., & Mohiuddin, K. M. (1996). Artificial neural networks: A tutorial. *Computer*, *29*(3), 31–44.

Javan, A. J., & Javan, M. J. (2014). Electronic structure of some thymol derivatives correlated with the radical scavenging activity: Theoretical study. *Food Chemistry*, *165*, 451–459.

Javan, A. J., Javan, M. J., & Tehrani, Z. A. (2013). Theoretical investigation on antioxidant activity of bromophenols from the marine red alga Rhodomela confervoides: H-atom vs electron transfer mechanism. *Journal of Agricultural and Food Chemistry*, *61*(7), 1534–1541.

Jiang, J., & Xiong, Y. L. (2016). Natural antioxidants as food and feed additives to promote health benefits and quality of meat products: A review. *Meat Science*, *120*, 107–117.

Kandi, S. K., Manohar, S., & Vélez- Genera, C. E. (2015). C5-curcuminoid-4-aminoquinoline based molecular hybrids: Design, synthesis and mechanistic investigation of anticancer activity. *New Journal of Chemistry*, *39*, 224–234.

Khan, A. U. (2016). Descriptors and their selection methods in QSAR analysis: Paradigm for drug design. *Drug Discovery Today*, *21*(8), 1291–1302.

Kitchen, D. B., Decornez, H., Furr, J. R., & Bajorath, J. (2004). Docking and scoring in virtual screening for drug discovery: Methods and applications. *Nature Reviews. Drug Discovery*, *3*(11), 935–949.

Kumar, Y., Yadav, D. N., Ahmad, T., & Narsaiah, K. (2015). Recent trends in the use of natural antioxidants for meat and meat products. *Comprehensive Reviews in Food Science and Food Safety*, *14*(6), 796–812.

Lin, Y., Xia, X., & Yao, R. (2014). Synthesis and in vitro biological evaluation of hybrids from tetrahydro-β-carboline and hydroxylcinnamic acid as antitumor carcinoma agents. *Chemical & Pharmaceutical Bulletin, 62*(4), 343–349.

Litwinienko, G., & Ingold, K. (2003). Abnormal solvent effects on hydrogen atom abstractions. 1. The reactions of phenols with 2, 2-diphenyl-1-picrylhydrazyl (dpph•) in alcohols. *The Journal of Organic Chemistry, 68*(9), 3433–3438.

Lui, K., Zhang, D., & Chojnacki, J. (2013). Design and biological characterization of hybrid compounds of curcumin and thalidomide for multiple myeloma. *Organic & Biomolecular Chemisrty, 11*(29), 4757–4763.

March-Vila, E., Pinzi, L., Sturm, N., Tinivella, A., Engkvist, O., Chen, H., et al. (2017). On the integration of in silico drug design methods for drug repurposing. *Frontiers in Pharmacology, 8*, 298.

Marrapu, V. K., Mittal, M., Shivahare, R., Gupta, S., & Bhandari, K. (2011). Synthesis and evaluation of new furanyl and thiophenyl azoles as antileishmanial agents. *European Journal of Medicinal Chemistry, 46*(5), 1694–1700.

Martin, T. M., Harten, P., Young, D. M., Muratov, E. N., Golbraikh, A., Zhu, H., et al. (2012). Does rational selection of training and test sets improve the outcome of QSAR modeling? *Journal of Chemical Information and Modeling, 52*(10), 2570–2578.

Miguel, M. G. (2010). Antioxidant and anti-inflammatory activities of essential oils: A short review. *Molecules, 15*(12), 9252–9287.

Miller, H. (1971). A simplified method for the evaluation of antioxidants. *Journal of the American Oil Chemists' Society, 48*(2), 91.

Miller, N. J., Rice-Evans, C., Davies, M. J., Gopinathan, V., & Milner, A. (1993). A novel method for measuring antioxidant capacity and its application to monitoring the antioxidant status in premature neonates. *Clinical Science, 84*(4), 407–412.

Mitra, I., Saha, A., & Roy, K. (2010). Chemometric modeling of free radical scavenging activity of flavone derivatives. *European Journal of Medicinal Chemistry, 45*(11), 5071–5079.

Mitra, I., Saha, A., & Roy, K. (2011). Chemometric QSAR modeling and in silico design of antioxidant NO donor phenols. *Scientia Pharmaceutica, 79*(1), 31–58.

Mitra, I., Saha, A., & Roy, K. (2012). In silico development, validation and comparison of predictive QSAR models for lipid peroxidation inhibitory activity of cinnamic acid and caffeic acid derivatives using multiple chemometric and cheminformatics tools. *Journal of Molecular Modeling, 18*(8), 3951–3967.

Mitra, I., Saha, A., & Roy, K. (2013a). Predictive modeling of antioxidant coumarin derivatives using multiple approaches: Descriptor-based QSAR, 3D-pharmacophore mapping, and HQSAR. *Scientia Pharmaceutica, 81*(1), 57–80.

Mitra, I., Saha, A., & Roy, K. (2013b). Quantification of contributions of different molecular fragments for antioxidant activity of coumarin derivatives based on QSAR analyses. *Canadian Journal of Chemistry, 91*(6), 428–441.

Morrone, J. A., Weber, J. K., Huynh, T., Luo, H., & Cornell, W. D. (2020). Combining docking pose rank and structure with deep learning improves protein–ligand binding mode prediction over a baseline docking approach. *Journal of Chemical Information and Modeling, 60*(9), 4170–4179.

Nahak, G., Suar, M., & Sahu, R. (2014). Antioxidant potential and nutritional values of vegetables: A review. *Research Journal of Medicinal Plant, 8*(2), 50–81.

Olsson, M., Gottfries, J., & Wold, S. (2004). D-optimal onion designs in statistical molecular design. *Chemometrics and Intelligent Laboratory Systems, 73*(1), 37–46.

Ou, B., Hampsch-Woodill, M., & Prior, R. L. (2001). Development and validation of an improved oxygen radical absorbance capacity assay using fluorescein as the fluorescent probe. *Journal of Agricultural and Food Chemistry, 49*(10), 4619–4626.

Özyürek, M., Güçlü, K., Tütem, E., Başkan, K. S., Erçağ, E., Çelik, S. E., et al. (2011). A comprehensive review of CUPRAC methodology. *Analytical Methods, 3*(11), 2439–2453.

Park, K., Lee, S., Ahn, H.-S., & Kim, D. (2009). Predicting the multi-modal binding propensity of small molecules: Towards an understanding of drug promiscuity. *Molecular BioSystems*, *5*(8), 844–853.
Pei, J., Yin, N., Ma, X., & Lai, L. (2014). Systems biology brings new dimensions for structure-based drug design. *Journal of the American Chemical Society*, *136*(33), 11556–11565.
Phosrithong, N., & Ungwitayatorn, J. (2013). Ligand-based CoMFA and CoMSIA studies on chromone derivatives as radical scavengers. *Bioorganic Chemistry*, *49*, 9–15.
Polishchuk, P. (2017). Interpretation of quantitative structure–activity relationship models: Past, present, and future. *Journal of Chemical Information and Modeling*, *57*(11), 2618–2639.
Popov, I. N., & Lewin, G. (1994). Photochemiluminescent detection of antiradical activity: II. Testing of nonenzymic water-soluble antioxidants. *Free Radical Biology and Medicine*, *17*(3), 267–271.
Popov, I. N., & Lewin, G. (1996). Photochemiluminescent detection of antiradical activity; IV: Testing of lipid-soluble antioxidants. *Journal of Biochemical and Biophysical Methods*, *31*(1–2), 1–8.
Pulido, R., Bravo, L., & Saura-Calixto, F. (2000). Antioxidant activity of dietary polyphenols as determined by a modified ferric reducing/antioxidant power assay. *Journal of Agricultural and Food Chemistry*, *48*(8), 3396–3402.
Queiroz, A. N., Gomes, B. A., Moraes, W. M., Jr., & Borges, R. S. (2009). A theoretical antioxidant pharmacophore for resveratrol. *European Journal of Medicinal Chemistry*, *44*(4), 1644–1649.
Ramalakshmi, K., Kubra, I. R., & Rao, L. J. M. (2008). Antioxidant potential of low-grade coffee beans. *Food Research International*, *41*(1), 96–103.
Rastija, V., & Medić-Šarić, M. (2009). QSAR study of antioxidant activity of wine polyphenols. *European Journal of Medicinal Chemistry*, *44*(1), 400–408.
Roussaki, M., Hall, B., & Lima, S. C. (2013). Synthesis and anti-parasitic activity of a novel quinolinone–chalcone series. *Bioorganic & Medicinal Chemistry Letters*, *23*(23), 6436–6441.
Roy, K. (2007). On some aspects of validation of predictive quantitative structure–activity relationship models. *Expert Opinion on Drug Discovery*, *2*(12), 1567–1577.
Samee, W., Nunthanavanit, P., & Ungwitayatorn, J. (2008). 3D-QSAR investigation of synthetic antioxidant chromone derivatives by molecular field analysis. *International Journal of Molecular Sciences*, *9*(3), 235–246.
Sampaio, G., Saldanha, T., Soares, R., & Torres, E. (2012). Effect of natural antioxidant combinations on lipid oxidation in cooked chicken meat during refrigerated storage. *Food Chemistry*, *135*(3), 1383–1390.
Sangshetti, J. N., Khan, F. A. K., Kulkarni, A. A., Patil, R. H., Pachpinde, A. M., Lohar, K. S., et al. (2016). Antileishmanial activity of novel indolyl–coumarin hybrids: Design, synthesis, biological evaluation, molecular docking study and in silico ADME prediction. *Bioorganic & Medicinal Chemistry Letters*, *26*(3), 829–835.
Sarkar, A., Middya, T. R., & Jana, A. D. (2012). A QSAR study of radical scavenging antioxidant activity of a series of flavonoids using DFT based quantum chemical descriptors—The importance of group frontier electron density. *Journal of Molecular Modeling*, *18*(6), 2621–2631.
Schaich, K. (2005). Developing a rational basis for selection of antioxidant screening and testing methods. In *Paper presented at the I international symposium on natural preservatives in food systems* (p. 709).
Schroeter, T. S., Schwaighofer, A., Mika, S., Ter Laak, A., Suelzle, D., Ganzer, U., et al. (2007). Predicting lipophilicity of drug-discovery molecules using gaussian process models. *ChemMedChem: Chemistry Enabling Drug Discovery*, *2*(9), 1265–1267.
Shebis, Y., Iluz, D., Kinel-Tahan, Y., Dubinsky, Z., & Yehoshua, Y. (2013). Natural antioxidants: Function and sources. *Food & Nutrition Sciences*, *4*(6).

Shehzad, A., & Lee, Y. (2010). Curcumin: Multiple molecular targets mediate multiple pharmacological actions: A review. *Drugs Future*, *35*(2), 113.
Sheridan, R. P., Wang, W. M., Liaw, A., Ma, J., & Gifford, E. M. (2016). Extreme gradient boosting as a method for quantitative structure–activity relationships. *Journal of Chemical Information and Modeling*, *56*(12), 2353–2360.
Slavin, J. L., & Lloyd, B. (2012). Health benefits of fruits and vegetables. *Advances in Nutrition*, *3*(4), 506–516.
Sogawa, S., Nihro, Y., Ueda, H., Miki, T., Matsumoto, H., & Satoh, T. (1994). Protective effects of hydroxychalcones on free radical-induced cell damage. *Biological and Pharmaceutical Bulletin*, *17*(2), 251–256.
Sonia, N., Mini, C., & Geethalekshmi, P. (2016). Vegetable peels as natural antioxidants for processed foods—A review. *Agricultural Reviews*, *37*(1), 35–41.
Svetnik, V., Liaw, A., Tong, C., Culberson, J. C., Sheridan, R. P., & Feuston, B. P. (2003). Random forest: A classification and regression tool for compound classification and QSAR modeling. *Journal of Chemical Information and Computer Sciences*, *43*(6), 1947–1958.
Svetnik, V., Wang, T., Tong, C., Liaw, A., Sheridan, R. P., & Song, Q. (2005). Boosting: An ensemble learning tool for compound classification and QSAR modeling. *Journal of Chemical Information and Modeling*, *45*(3), 786–799.
Tan, G., Yao, Y., & Gu, Y. (2014). Cytotoxicity and DNA binding property of the dimers of triphenylethylene–coumarin hybrid with one amino side chain. *Bioorganic & Medicinal Chemistry Letters*, *24*(3), 2825–2830.
Toropova, A. P., & Toropov, A. A. (2019). Does the index of ideality of correlation detect the better model correctly? *Molecular Informatics*, *38*(8-9), 1800157.
Toropova, A. P., Toropov, A. A., Roncaglioni, A., & Benfenati, E. (2021). The index of ideality of correlation improves the predictive potential of models of the antioxidant activity of tripeptides from frog skin (Litoria rubella). *Computers in Biology and Medicine*, *133*, 104370.
Tropsha, A., Gramatica, P., & Gombar, V. K. (2003). The importance of being earnest: Validation is the absolute essential for successful application and interpretation of QSPR models. *QSAR & Combinatorial Science*, *22*(1), 69–77.
Uno, S., Kodama, D., Yukawa, H., Shidara, H., & Akamatsu, M. (2020). Quantitative analysis of the relationship between structure and antioxidant activity of tripeptides. *Journal of Peptide Science*, *26*(3), e3238.
Valko, M., Leibfritz, D., Moncol, J., Cronin, M. T., Mazur, M., & Telser, J. (2007). Free radicals and antioxidants in normal physiological functions and human disease. *The International Journal of Biochemistry & Cell Biology*, *39*(1), 44–84.
Wang, W., Weng, X., & Cheng, D. (2000). Antioxidant activities of natural phenolic components from Dalbergia odorifera T. Chen. *Food Chemistry*, *71*(1), 45–49.
Wold, S., Sjöström, M., & Eriksson, L. (2001). PLS-regression: A basic tool of chemometrics. *Chemometrics and Intelligent Laboratory Systems*, *58*(2), 109–130.
Yadav, A. A., Wu, X., Patel, D., Yalowich, J. C., & Hasinoff, B. B. (2014). Structure-based design, synthesis and biological testing of etoposide analog epipodophyllotoxin–N-mustard hybrid compounds designed to covalently bind to topoisomerase II and DNA. *Bioorganic & Medicinal Chemistry*, *22*(21), 5935–5949.
Yan, J., Guo, Y., & Wang, Y. (2015). Design, synthesis, and biological evaluation of benzoselenazolestilbene hybrids as multi-target-directed anti-cancer agents. *European Journal of Medicinal Chemisty*, *95*, 220–229.
Yan, W., Lin, G., Zhang, R., Liang, Z., & Wu, W. (2020). Studies on the bioactivities and molecular mechanism of antioxidant peptides by 3D-QSAR, in vitro evaluation and molecular dynamic simulations. *Food & Function*, *11*(4), 3043–3052.

Zhao, L., Yao, Y., & Li, S. (2014). Cytotoxicity and DNA binding property of triphenylethylene–coumarin hybrids with two amino side chains. *Bioorganic & Medicinal Chemistry Letters*, *24*(3), 900–904.
Zhong, F., Xing, J., Li, X., Liu, X., Fu, Z., Xiong, Z., et al. (2018). Artificial intelligence in drug design. *Science China Life Sciences*, *61*(10), 1191–1204.
Zupan, J., Novič, M., & Ruisánchez, I. (1997). Kohonen and counterpropagation artificial neural networks in analytical chemistry. *Chemometrics and Intelligent Laboratory Systems*, *38*(1), 1–23.

CHAPTER TWO

Hydrogen peroxide detoxification through the peroxiredoxin/ thioredoxin antioxidant system: A look at the pancreatic β-cell oxidant defense

Jennifer S. Stancill and John A. Corbett*
Department of Biochemistry, Medical College of Wisconsin, Milwaukee, WI, United States
*Corresponding author: e-mail address: jcorbett@mcw.edu

Contents

Abstract

Reactive oxygen species (ROS), such as hydrogen peroxide, are formed when molecular oxygen obtains additional electrons, increasing its reactivity. While low concentrations of hydrogen peroxide are necessary for regulation of normal cellular signaling events, high concentrations can be toxic. To maintain this balance between beneficial and deleterious concentrations of hydrogen peroxide, cells utilize antioxidants. Our recent work supports a primary role for peroxiredoxin, thioredoxin, and thioredoxin reductase as the oxidant defense pathway used by insulin-producing pancreatic β-cells. These three players work in an antioxidant cycle based on disulfide exchange, with oxidized targets ultimately being reduced using electrons provided by NADPH. Peroxiredoxins also participate in hydrogen peroxide-based signaling through disulfide exchange with redox-regulated target proteins. This chapter will describe the catalytic mechanisms of thioredoxin, thioredoxin reductase, and peroxiredoxin and provide an in-depth look

Vitamins and Hormones, Volume 121
ISSN 0083-6729
https://doi.org/10.1016/bs.vh.2022.11.001

at the roles these enzymes play in antioxidant defense pathways of insulin-secreting β-cells. Finally, we will evaluate the physiological relevance of peroxiredoxin-mediated hydrogen peroxide signaling as a regulator of β-cell function.

1. The many roles of ROS in mammalian cells

Reactive oxygen species (ROS) is a broad term used to represent molecules generated when molecular oxygen accepts electrons, thus making these molecules more prone to participate in reduction-oxidation (redox) reactions. Examples of ROS include superoxide ($O_2^{\bullet -}$) and hydrogen peroxide (H_2O_2). A major mechanism of superoxide generation is cellular respiration, in which free oxygen accepts electrons that "leak" out from the electron transport chain during mitochondrial oxidative phosphorylation (Zhao, Jiang, Zhang, & Yu, 2019). Under conditions in which oxidative phosphorylation and electron leak are increased, superoxide production may also increase (Maritim, Sanders, & Watkins, 2003; Robertson, Harmon, Tran, & Poitout, 2004). Superoxide is also produced via NADPH oxidases (NOX or DUOX), which catalyze the addition of an electron (using NADPH) to molecular oxygen (Lassegue & Griendling, 2010) and are primarily localized to the plasma membrane, endoplasmic reticulum membrane, and endosome membrane (Sies & Jones, 2020). Superoxide is a major source of hydrogen peroxide that is formed via superoxide dismutation, either spontaneously or through a reaction catalyzed by superoxide dismutase (SOD) (Sies, 2017).

One of the primary mechanisms by which hydrogen peroxide exerts its effects on target molecules is through oxidation of protein thiol residues (Auten & Davis, 2009; Poole, 2015). The peroxide reacts with a cysteine residue (S^-), forming a sulfenate (SO^-) that can be further modified to form an intermolecular disulfide (SS), potentially altering the structure and function of the target protein (Stadtman & Levine, 2000). Under normal conditions, the intracellular concentration of hydrogen peroxide is maintained in the low nanomolar range (Sies & Chance, 1970; Sies & Jones, 2020). Due to its relative stability and target selectivity, hydrogen peroxide can act as a signaling molecule (Sies & Jones, 2020). Indeed, low concentrations of hydrogen peroxide (in the range of 1–100 nM) are known to regulate a variety of cellular functions, including proliferation, survival, and angiogenesis (Murrell, Francis, & Bromley, 1990; Sies & Jones, 2020; Stone & Yang, 2006; Yasuda et al., 1999). However, higher concentrations of hydrogen peroxide

(10 μM or greater) are deleterious to the cell, causing uncontrolled modification of target molecules (Stone & Yang, 2006). Unchecked oxidation of proteins or lipids can negatively alter their function, and DNA oxidation can cause breaks, potentially leading to cell death (Auten & Davis, 2009; Cooke, Evans, Dizdaroglu, & Lunec, 2003; Stadtman & Levine, 2000). Further, hydrogen peroxide can react with free metal ions to produce the highly reactive hydroxyl radical ($^{\bullet}OH$) (Sies, 2017). To help keep hydrogen peroxide concentrations within the beneficial rather than deleterious range, several enzymes catalyze the reduction of hydrogen peroxide to water, including peroxiredoxins, glutathione peroxidases, and catalase. The actions of peroxiredoxins in detoxifying hydrogen peroxide will be the focus of this chapter, with a particular emphasis on the roles of these antioxidants in pancreatic β-cells. But first, we will describe the antioxidant partners of peroxiredoxins: thioredoxin and thioredoxin reductase.

2. Thioredoxin and thioredoxin reductase

Two thioredoxin isoforms exist in mammals: Trx1 (encoded by *TXN1*) is localized to the cytoplasm, and Trx2 (encoded by *TXN2*) is localized to the mitochondria. Thioredoxin (Trx) was first discovered in *E. coli* in 1964 by Laurent and colleagues as the electron donor for ribonucleotide reductase, the enzyme that catalyzes the conversion of nucleotides to deoxynucleotides (Laurent, Moore, & Reichard, 1964). Since then, thioredoxin has been found to be an electron donor to methionine sulfoxide reductases that reduce methionine sulfoxides (Brot & Weissbach, 1983), and to peroxiredoxins that reduce hydrogen peroxide. Peroxiredoxins are discussed in the subsequent section of this chapter. Thioredoxin's role as an electron donor makes it an important player in maintaining proper redox balance in the cell. Thioredoxin contains a cysteine residue at the N-terminus, and this thiol can react with an oxidized residue (disulfide) on a target protein (such as peroxiredoxin) to form a mixed disulfide intermediate (Fig. 1A). The thiol group on the C-terminal cysteine of thioredoxin can then reduce the mixed disulfide intermediate to result in a reduced target protein and an oxidized thioredoxin molecule (Fig. 1A).

In order to maintain its function as an electron donor, thioredoxin requires thioredoxin reductase (TrxR) to reduce the disulfide that is formed when it reacts with a target protein. Two TrxR isoforms are ubiquitously expressed in mammals: TrxR1 (*TXNRD1*) in the cytoplasm and TrxR2 (*TXNRD2*) in the mitochondria. A third isoform, TrxR3 (*TXNRD3*) is

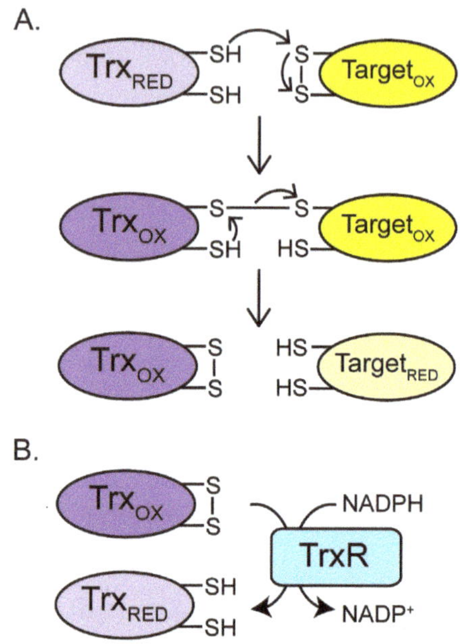

Fig. 1 Mechanisms of thioredoxin and thioredoxin reductase. (A) The N-terminal cysteine thiol of thioredoxin reacts with the disulfide of an oxidized target protein, such as peroxiredoxin, resulting in the formation of a mixed disulfide intermediate between the two proteins. The C-terminal cysteine thiol of thioredoxin reacts with the disulfide bond, resulting in an intramolecular disulfide in thioredoxin and a reduced target protein. Curved arrows represent the flow of electrons. (B) Thioredoxin reductase utilizes electrons from NADPH to reduce the intramolecular disulfide bond of thioredoxin, recycling the antioxidant to its reduced form. *Abbreviations* used: Trx (thioredoxin), TrxR (thioredoxin reductase), RED (reduced), and OX (oxidized).

specific to the testes. All mammalian TrxRs are selenoproteins and contain a C-terminal selenocysteine (Sec) residue. Given that Sec is a better nucleophile than Cys (higher propensity to donate electrons), the incorporation of a Sec residue into the active site of TrxR may grant the enzyme a higher catalytic efficiency and allow it to react with a broader spectrum of substrates than if a Cys residue was instead present in the active site (Arner, 2010; Gromer et al., 2003; Lee et al., 2000). The selenol contained in the selenocysteine residue of TrxR reacts with oxidized thioredoxin, reducing the disulfide in thioredoxin and oxidizing TrxR in the process (Fig. 1B). Electrons from NADPH reduce TrxR, allowing the cycle to continue (Fig. 1B) (Rhee & Woo, 2011). In many cell types, cytoplasmic glucose-6-phosphate dehydrogenase (G6PD) is used to generate NADPH through the conversion of glucose-6-phosphate to 6-phospho-gluconate. Mouse embryonic stem cells that are deficient of G6PD are sensitive to death triggered by hydrogen peroxide, suggesting a role for this enzyme in the mammalian cytoplasmic antioxidant defense (Pandolfi et al., 1995).

3. Catalytic mechanisms and isoforms of peroxiredoxins

The enzyme now called peroxiredoxin was originally discovered in *S. cerevisiae* by Kim and colleagues in 1988 as a protein that protected against oxidative damage when a thiol was used as an electron donor in the oxidation system but not when the electron source was replaced with ascorbate (Kim, Kim, Lee, Rhee, & Stadtman, 1988). They later determined that thioredoxin (Trx) and thioredoxin reductase (TrxR) were necessary for the reducing capabilities of their newly discovered protein, which they named thioredoxin peroxidase (Chae, Chung, & Rhee, 1994). This class of antioxidants was later re-named "peroxiredoxin" in mammalian systems because not all isoforms of this family require thioredoxin to reduce peroxide (Chae et al., 1994).

Oxidant detoxification by peroxiredoxins occurs in three steps: oxidation, resolution, and reduction (Fig. 2). The oxidation step occurs when peroxiredoxin reacts with an oxidant, like hydrogen peroxide, peroxynitrite, or lipid peroxides. A highly conserved cysteine residue called the "peroxidatic" Cys is oxidized to form a sulfenic acid (Fig. 2A). The second step of the process, called "resolution," occurs when another cysteine residue, termed the "resolving" Cys, reacts with the sulfenic acid produced in the oxidation step, forming a disulfide bond (Fig. 2B). Because the resolution step is slower than the oxidation step (Portillo-Ledesma et al., 2018), peroxiredoxins are susceptible to hyperoxidation at higher peroxide concentrations. Under these conditions, the peroxidatic Cys can be further oxidized to form a sulfinic or sulfonic acid. The sulfinic acid can be reduced by an enzyme called sulfiredoxin (Biteau, Labarre, & Toledano, 2003), but the sulphonic acid is thought to be irreversible (Rhee & Woo, 2011). The final step (reduction) occurs when the disulfide formed in the resolution step is reduced, typically by the cysteine thiol residue of thioredoxin, as described in the previous section (Fig. 2C).

There are six mammalian peroxiredoxins belonging to three different classes, each with slightly different catalytic mechanisms: typical 2-Cys, atypical 2-Cys, and 1-Cys. The catalytic mechanisms of typical and atypical 2-Cys isoforms differ in the nature of the disulfide bond formed in the "resolution" step. During oxidant detoxification by typical 2-Cys

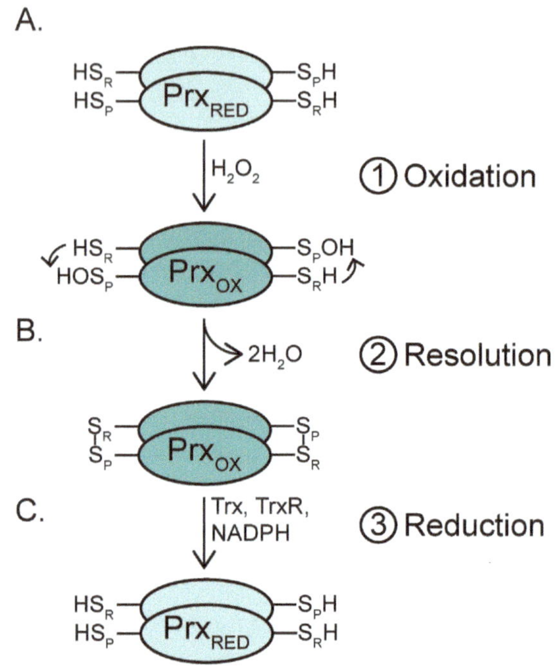

Fig. 2 Mechanism of typical 2-Cys peroxiredoxins. (A) The peroxidatic cysteine thiol (S_PH) of peroxiredoxin is oxidized to S_POH by hydrogen peroxide. This is known as the "Oxidation" step. (B) Reaction with the resolving cysteine thiol (S_RH) on a neighboring peroxiredoxin results in the formation of an intermolecular disulfide bond and the release of H_2O. This is called the "Resolution" step. (C) The disulfide formed in the Resolution step is reduced by interaction with thioredoxin and thioredoxin reductase, resulting in the recycled, reduced peroxiredoxin. This is the "Reduction" step. Abbreviations used: Prx (peroxiredoxin), Trx (thioredoxin), TrxR (thioredoxin reductase), RED (reduced), OX (oxidized), S_P (peroxidatic cysteine thiol), and S_R (resolving cysteine thiol). Curved arrows represent the flow of electrons.

peroxiredoxins, a disulfide bond is formed between two separate peroxiredoxin molecules (Rhee & Woo, 2011). In contrast, an atypical 2-Cys peroxiredoxin forms a disulfide bond between two cysteine residues contained in the same peroxiredoxin molecule, meaning that a second peroxiredoxin is not needed for oxidant detoxification (Rhee & Woo, 2011). Typical 2-Cys isoforms are the most abundant, with PRDX1, PRDX2, PRDX3, and PRDX4 falling into this category. There is only one atypical 2-cys peroxiredoxin known to exist in mammals, PRDX5 (Rhee & Woo, 2011). PRDX6 is the only known mammalian peroxiredoxin of the 1-Cys class and utilizes a completely different catalytic mechanism than any other isoform: no disulfide bond is formed because there is no other nearby cysteine reside available (Rhee & Woo, 2011). Because of this difference, PRDX6 does not rely on reduction by thioredoxin, thioredoxin reductase, or NADPH, but instead utilizes glutathione (Rhee & Woo, 2011).

The peroxiredoxin isoforms also differ in their subcellular localization. PRDX1 and PRDX2 are the most abundant and are cytoplasmic (Rhee & Woo, 2011). These two isoforms may be thought to be redundant, yet they display different rates of "resolution" and likely serve different functions (Dalla Rizza, Randall, Santos, Ferrer-Sueta, & Denicola, 2019). Indeed, we have shown that specific depletions of either isoform have divergent effects on mammalian cell viability (Stancill, Happ, Broniowska, Hogg, & Corbett, 2020). PRDX3 is mitochondrial, and PRDX4 is localized to the endoplasmic reticulum (Rhee & Woo, 2011; Tavender, Sheppard, & Bulleid, 2008). PRDX5 has broad localization, being found in the mitochondria, cytoplasm, and peroxisome (Knoops et al., 1999). Finally, PRDX6 is found in the cytoplasm and lysosome (Rhee, Woo, Kil, & Bae, 2012).

4. Relevance to human disease: Roles of thioredoxin/peroxiredoxin in protecting pancreatic β-cells from oxidative damage

4.1 Oxidative stress in pancreatic β-cells

In the United States alone, diabetes mellitus, a group of diseases characterized by chronically high blood glucose, affects 29.1 million people (National Diabetes Statistics Report, 2014). Although the causes of disease development are varied, all involve failure of pancreatic β-cells to secrete sufficient amounts of insulin to meet body demands. The most common, type 2 diabetes (T2D), accounts for 90–95% of all diabetes cases and develops due to progressive worsening β-cell function amplified by age, inactivity, obesity, and/or genetic risk factors (National Diabetes Statistics Report, 2014). The less common type 1 diabetes (T1D), accounts for 5% of cases and is classically referred to as juvenile diabetes because it commonly presents in childhood (National Diabetes Statistics Report, 2014). β-Cells are selectively destroyed by an autoimmune process in this form of the disease, necessitating daily insulin injections to control blood glucose. Globally, the incidence of diabetes has significantly increased over the last 20 years and is projected to continue this trajectory (Lin et al., 2020).

Due to the prevalence of diabetes in the United States and to the medical burden caused by the disease (Lin et al., 2020), there has been a major research effort to determine factors contributing to β-cell dysfunction and death. Pancreatic β-cells, residing in the islets of Langerhans along with other endocrine cell types, respond to an increase in blood glucose concentration with an increase in mitochondrial oxidation of the hexose sugar, resulting in ATP accumulation and ATP-dependent membrane depolarization via closure of gated potassium channels, an influx of Ca^{2+} ions into the cell, and Ca^{2+}-dependent secretion of insulin granules (Henquin, 2011; Prentki, Matschinsky, & Madiraju, 2013). Unique to this cell type is the coupling of glycolysis to oxidative phosphorylation, resulting in metabolism of most of the carbons in glucose to CO_2 on supply of the hexose (Jitrapakdee, Wutthisathapornchai, Wallace, & MacDonald, 2010; MacDonald et al., 2005; Schuit et al., 1997). In this way, the rate of glucose sensing is proportional to the rate of oxidation and, thus, to insulin secretion (Matschinsky, 1996).

This metabolic feature has led to the hypothesis that chronically high blood glucose levels, such as the case in diabetes, may increase oxidative phosphorylation, electron leak, and superoxide production in β-cells (Maritim et al., 2003; Robertson et al., 2004). In agreement, ROS production in insulinoma cells (Neal et al., 2016; Pi et al., 2007) and rat pancreatic islets (Leloup et al., 2009) is elevated by stimulatory glucose concentrations, and islets from diabetic rodents and patients have increased markers of oxidative damage (Dominguez, Ruiz, Gussinye, & Carrascosa, 1998; Ghiselli, Laurenti, De Mattia, Maiani, & Ferro-Luzzi, 1992; Gopaul et al., 1995; Ihara et al., 1999; Nourooz-Zadeh, Tajaddini-Sarmadi, McCarthy, Betteridge, & Wolff, 1995; Rehman et al., 1999; Sakuraba et al., 2002; Shin et al., 2001; Tanaka, Gleason, Tran, Harmon, & Robertson, 1999). In addition to increased ROS generation by electron leak, NADPH oxidases may contribute to oxidative damage in β-cells exposed to proinflammatory cytokines (Kowluru, 2020; Subasinghe, Syed, & Kowluru, 2011). Further, β-cells were reported to have lower levels of antioxidants like catalase and glutathione peroxidase compared to the kidney and liver (Grankvist, Marklund, & Taljedal, 1981; Lenzen, 2008; Lenzen, Drinkgern, & Tiedge, 1996; Tiedge, Lortz, Munday, & Lenzen, 1998). These lines of evidence have fostered the widespread hypothesis that β-cells are vulnerable to oxidative damage and that oxidative stress contributes to development of T1D and T2D (Lenzen, 2017). However, more recent studies (discussed in detail in the subsequent sections) have identified roles for thioredoxin, thioredoxin reductase, and peroxiredoxin in the β-cell antioxidant defense, challenging the hypothesis that β-cells are vulnerable to oxidative damage.

4.2 Protective roles of thioredoxin and thioredoxin reductase in β-cells

Some of the earliest studies examining the potential role for thioredoxin in protecting β-cells from oxidative stress utilized overexpression models. For example, Hotta and colleagues showed that the onset of diabetes was delayed by the β-cell selective overexpression of human thioredoxin in a non-obese diabetic (NOD) mouse model for T1D or in a streptozotocin-based diabetes model (Hotta et al., 1998). Others have shown in a NOD mouse model of disease reoccurrence that disease frequency is reduced when mice are transplanted with islets over-expressing human thioredoxin as compared to the transplantation of islets isolated from wild type control mice (Chou & Sytwu, 2009). In vitro studies using rodent-derived insulinoma cells have also been used to examine whether thioredoxin plays a protective role in β-cells. For example, small molecule thioredoxin mimetics protected INS832/13 cells from damage caused by treatment with auranofin, a thioredoxin reductase inhibitor (Cohen-Kutner et al., 2013). Additionally, MIN6 cells were protected from hypoxia-induced cell death by addition of exogenous TRX1 (Hanschmann et al., 2020). Together, the studies summarized above suggest that β-cells are protected from a number of environmental stressors by addition of exogenous thioredoxin.

Overall, studies examining the effects of enhanced expression of thioredoxin in β-cells have been enlightening as they have provided a potential mechanism to attenuate β-cell damage in a disease setting. However, they have fallen short of providing information related to the endogenous antioxidant pathways used by β-cells as a defense from oxidative damage. Studies have shown that the expression of thioredoxin interacting protein (TXNIP), a genetically encoded thioredoxin inhibitor, is increased in human islets following high (20 mM) glucose exposure and is elevated in islets obtained from diabetic rodent models (Shalev, 2014; Shalev et al., 2002). These associations of TXNIP expression with hyperglycemia and diabetes suggests that TXNIP may play a role in β-cell damage. In agreement with this hypothesis, TXNIP overexpression causes β-cell death while TXNIP depletion promotes β-cell survival (Chen, Fontes, Saxena, Poitout, & Shalev, 2010; Chen et al., 2008; Minn, Hafele, & Shalev, 2005). Mechanistically, oxidized TXNIP inhibits reduced thioredoxin by binding to two cysteine residues and forming a stable, mixed disulfide, thus preventing thioredoxin from reducing oxidized target proteins (Hwang et al., 2014; Patwari, Higgins, Chutkow, Yoshioka, & Lee, 2006).

The interaction depends on TXNIP being oxidized and on thioredoxin being reduced, suggesting that TXNIP is a redox-sensitive protein and that reduction of TXNIP or oxidation of thioredoxin would prevent the interaction between these two players (Patwari et al., 2006). Damage of β-cells by TXNIP upregulation may be mediated by redox-dependent and redox-independent mechanisms (Spindel, World, & Berk, 2012), including high-glucose induced ROS and apoptosis (Chen, Hui, et al., 2008; Chen, Saxena, Mungrue, Lusis, & Shalev, 2008) and disrupted adipogenesis (Chutkow & Lee, 2011).

The observations that increased expression of a thioredoxin inhibitor promotes β-cell damage suggest that endogenous thioredoxin may be necessary for β-cell survival. In support of this idea, we have shown that INS832/13 insulinoma cells and rat islets are more susceptible to damage and death in response to hydrogen peroxide delivered continuously by glucose oxidase when thioredoxin reductase is inhibited by auranofin or depleted by siRNAs (Stancill, Broniowska, Oleson, Naatz, & Corbett, 2019). Interestingly, this susceptibility was not observed in response to hydrogen peroxide delivered as a bolus (Stancill et al., 2019). These divergent outcomes are dependent on the method of oxidant delivery and are related to the relative rate of the oxidation and resolution steps of the peroxiredoxin catalytic mechanism that result in increased susceptibility of peroxiredoxins to hyperoxidation at high oxidant concentrations (discussed earlier in this chapter (Rhee & Woo, 2011)). When delivered as a bolus, hydrogen peroxide likely overwhelms the thioredoxin/peroxiredoxin antioxidant system, rendering it useless for antioxidant defense due to hyperoxidation. Our findings support the hypothesis that thioredoxin reductase is necessary for the β-cell defense against oxidant damage as assessed under conditions in which the oxidant is continuously-delivered.

As previously mentioned, thioredoxin reductase requires electrons from NADPH to reduce thioredoxin, and glucose-6-phosphate dehydrogenase (G6PD) is hypothesized to be a main contributor to this reducing pool (Pandolfi et al., 1995; Rhee & Woo, 2011). In MIN6 insulinoma cells deficient of G6PD, basal levels of ROS are elevated, and this is associated with increased DNA damage and apoptosis, suggesting a defect in antioxidant defenses (Zhang et al., 2010). In mice with a global knockout of G6PD, islets are smaller than the size observed in wildtype control animals, suggesting a role for G6PD in β-cell development (Zhang et al., 2010). Finally, enhanced glucose tolerance is associated with increased β-cell function and reduced pancreatic oxidative damage in transgenic mice globally over-expressing

G6PD. These findings provide further evidence to support a role for NADPH generated by G6PD in β-cell ROS detoxification via thioredoxin reductase (De la Rosa et al., 2021).

4.3 Protective roles of peroxiredoxins in β-cells

In 2002, Bast and colleagues first reported that peroxiredoxins are endogenously expressed in β-cells (Bast, Wolf, Oberbaumer, & Walther, 2002), an observation that has since been confirmed by others (Jin et al., 2020; Stancill et al., 2019; Zhao & Wang, 2012). PRDX1 and PRDX2 were found in the cytoplasm of insulinoma cells and mouse islets, and were found to be more abundant in the islets compared to the surrounding pancreatic acinar tissue (Bast et al., 2002). They also observed that the expression of these peroxiredoxins was not stagnant, but increased in response to several oxidative stressors, including inflammatory cytokines, hydrogen peroxide, alloxan, and streptozotocin (Bast et al., 2002). This observation is somewhat controversial as others have observed either no change in expression or decreased expression of PRDX1 or PRDX2 in response to similar stimuli (Jin et al., 2020; Stancill, Kasmani, Khatun, Cui, & Corbett, 2021; Zhao & Wang, 2012). Independent of changes in peroxiredoxin expression in response to stressors, the findings of Bast et al. (Bast et al., 2002) revealed that peroxiredoxins are indeed expressed in pancreatic islets, and, thus, may play a role in promoting β-cell survival.

To determine the specific roles of peroxiredoxins in the β-cell antioxidant defense, more recent studies have utilized depletion or overexpression models. We found that INS832/13 cell death is increased in response to continuously-delivered superoxide or hydrogen peroxide when peroxiredoxins were globally inhibited by conoidin A (Stancill et al., 2020). Specific depletion of the cytoplasmic isoform PRDX1 using targeted siRNAs phenocopied the effect of the pharmacological inhibition, while depletion of the other main cytoplasmic isoform, PRDX2, had no effect on hydrogen peroxide-mediated β-cell death (Stancill et al., 2020). These findings suggest that PRDX1 plays a primary protective role in the response of β-cells to continuously-delivered ROS. Others have shown that depletion of PRDX2 in a different insulinoma cell line, MIN6, sensitized these cells to death in response to palmitic acid, inflammatory cytokines, or streptozotocin, and that overexpression PRDX2 had the converse effect (Zhao & Wang, 2012). The differences in these studies may be attributed to differences in the species (mouse vs. rat) of the insulinoma cell model, differences in the stimuli used to cause damage, or differences in the methods of oxidant addition (bolus vs. continuous delivery).

In addition to PRDX1 and PRDX2 in the cytoplasm, the mitochondrial isoform, PRDX3, and the ER-localized isoform, PRDX4, are expressed in β-cells (Mehmeti, Lortz, Elsner, & Lenzen, 2014; Stancill et al., 2019; Wolf et al., 2010). However, likely due to the dominant action of the cytoplasmic 2-Cys isoforms, PRDX3 and PRDX4 were found to be dispensable for the β-cell defense against hydrogen peroxide (Mehmeti et al., 2014; Stancill et al., 2020). Others have shown that overexpression of either isoform protects insulinoma cells against hydrogen peroxide-mediated damage (bolus addition), and mice with a global overexpression of PRDX4 are protected against streptozotocin-induced diabetes and have reduced islet damage (Ding et al., 2010; Mehmeti et al., 2014; Wolf et al., 2010). These studies demonstrate that the dominant isoform in the endogenous antioxidant defense of β-cells is likely PRDX1, but that other isoforms, when overexpressed, are capable of providing additional protection from oxidant damage.

PRDX6, a unique 1-Cys PRDX isoform that does not utilize thioredoxin or thioredoxin reductase, is expressed in β-cells and has been implicated in diabetes pathogenesis (Pacifici et al., 2014; Paula, Ferreira, Boschero, & Souza, 2013). RINm5F insulinoma cells are more susceptible to death induced by inflammatory cytokines or hydrogen peroxide delivered as a bolus when PRDX6 is depleted (Paula et al., 2013). In addition, mice with a global knockout of PRDX6 are glucose intolerant with significantly reduced β-cell mass (Pacifici et al., 2014) These studies suggest that this unique peroxiredoxin isoform also functions in maintaining β-cell survival in response to oxidative stressors.

5. Peroxiredoxin-mediated hydrogen peroxide signaling

As discussed earlier in this chapter, basal cytosolic hydrogen peroxide levels are maintained in the range of 1–10 nM while hydrogen peroxide concentrations in the mid-nanomolar range are conducive to signaling events (Sies & Jones, 2020). Classically, it has been hypothesized that hydrogen peroxide directly reacts with redox-regulated cysteine residues to oxidize target proteins, modulating their function. In this view, peroxiredoxins are competitors for hydrogen peroxide-based signaling events because they scavenge the necessary oxidant. Therefore, peroxiredoxins must be temporarily inactivated or depleted, resulting in locally high hydrogen peroxide concentrations, for this hydrogen peroxide-mediated signaling to occur. However, the biochemical feasibility of direct oxidation of target proteins by hydrogen peroxide has been questioned (Stocker, Van Laer, Mijuskovic, & Dick, 2018). Cysteine thiol residues of peroxiredoxins have very high reactivity with hydrogen peroxide, with rate constants in the range of 10^5–$10^8 M^{-1} s^{-1}$ (Carvalho et al., 2017; Manta et al., 2009; Peskin et al., 2007; Portillo-Ledesma et al., 2018). Further, cytoplasmic peroxiredoxins are highly abundant, accounting for approximately 1% of the total soluble protein in several mammalian cell types (Chae, Kim, Kang, & Rhee, 1999). These properties of peroxiredoxins are in stark contrast to typical redox-regulated proteins, which usually have hydrogen peroxide reactivity orders of magnitude lower than peroxiredoxins (10^1–$10^2 M^{-1} s^{-1}$) and are less abundant (Marinho, Real, Cyrne, Soares, & Antunes, 2014). Given their drastically different characteristics, it is unclear how thiols on redox-regulated proteins could outcompete peroxiredoxins for interaction with hydrogen peroxide. In this view, peroxiredoxins are always more likely to interact with hydrogen peroxide than a redox-regulated protein, meaning that target proteins would have to receive oxidizing equivalents (in the form of disulfide exchange) from thiols of peroxiredoxin in order for hydrogen peroxide-mediated signaling to occur.

2-Cys peroxiredoxins have been shown to participate in hydrogen peroxide "redox relays" via the formation of mixed disulfide intermediates with other proteins (Fig. 3) (Sobotta et al., 2015). Here, peroxiredoxins are oxidized at the peroxidatic Cys by hydrogen peroxide (Fig. 3A), but instead of forming a disulfide bond by reacting with the resolving Cys of a neighboring peroxiredoxin, a mixed disulfide is formed between the peroxidatic Cys and the cysteine thiol of a redox-regulated target (Fig. 3B). This results in the oxidation of the target protein (Fig. 3C), potentially altering its function and allowing peroxiredoxins to play a critical signaling role.

6. Hydrogen peroxide signaling and β-cell function

While high concentrations of hydrogen peroxide are thought to be deleterious to β-cell survival, low concentrations of the oxidant are hypothesized to promote β-cell function. Indeed, basal insulin secretion from INS832/13 insulinoma cells is enhanced by addition of small concentrations of hydrogen peroxide as a bolus (Pi et al., 2007). Conversely, insulin secretion stimulated by high glucose or by membrane depolarization is blunted

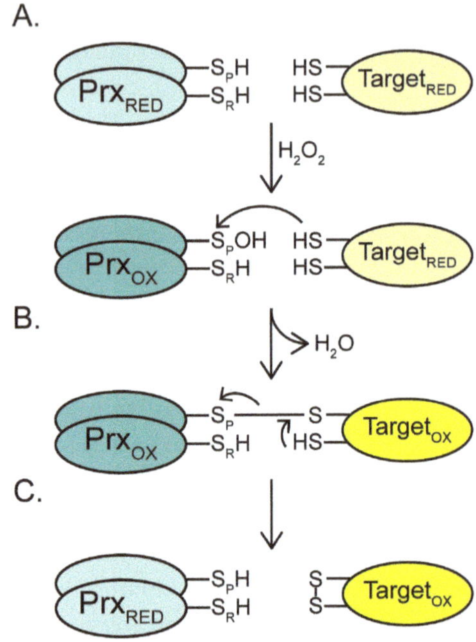

Fig. 3 Potential mechanism of peroxiredoxin-mediated redox relay. (A) The peroxidatic cysteine thiol of peroxiredoxin (S_PH) is oxidized to S_POH by hydrogen peroxide. (B) Reaction with a redox-sensitive cysteine thiol on a target protein results in the formation of a mixed disulfide intermediate between the two proteins and the release of H_2O. (C) The disulfide is transferred to the target protein, resulting in an oxidized target and reduced peroxiredoxin. Abbreviations used: Prx (peroxiredoxin), RED (reduced), OX (oxidized), S_P (peroxidatic cysteine thiol), and S_R (resolving cysteine thiol). Curved arrows represent the flow of electrons.

when antioxidants (catalase, *N*-acetyl-L-cysteine, or vitamin E analog) are added to insulinoma cells or rat islets (Leloup et al., 2009; Pi et al., 2007).

To our knowledge, no direct evidence for the existence of peroxiredoxin-mediated redox relays, such as identification of interactions between peroxiredoxin and target proteins or the visualization of mixed disulfide intermediates, has been found in β-cells. Nonetheless, indirect evidence can be found. Glucose-stimulated insulin secretion (GSIS) in insulinoma cells or mouse islets is blunted when thioredoxin reductase is inhibited (Cohen-Kutner et al., 2013; Stancill, Hansen, Mathison, Schmidt, & Corbett, 2022). PRDX2 deficiency in *C. elegans* (Olahova & Veal, 2015), PRDX3 deficiency in RINm5F insulinoma cells (Wolf et al., 2010), or G6PD deficiency in MIN6 insulinoma cells (Zhang et al., 2010) are associated with decreases in GSIS. Recently, β-cells harvested from mice with enhanced global transgenic expression of G6PD were shown to have increased electrical activity when exposed to stimulatory concentrations of glucose (De la Rosa et al., 2021). Logically, one would predict that depletion or inhibition of thioredoxin reductase or peroxiredoxins, thus increasing intracellular ROS, would enhance GSIS, if hydrogen peroxide directly modified target proteins to exert its positive effects on β-cell function. But, in fact, the opposite outcome is observed: inhibition of members of this antioxidant cycle negatively affects GSIS, suggesting that peroxiredoxins are necessary for the signaling to occur, potentially through a redox relay. Based on this evidence, we hypothesize that peroxiredoxins play at least two physiological roles in β-cells (Fig. 4)

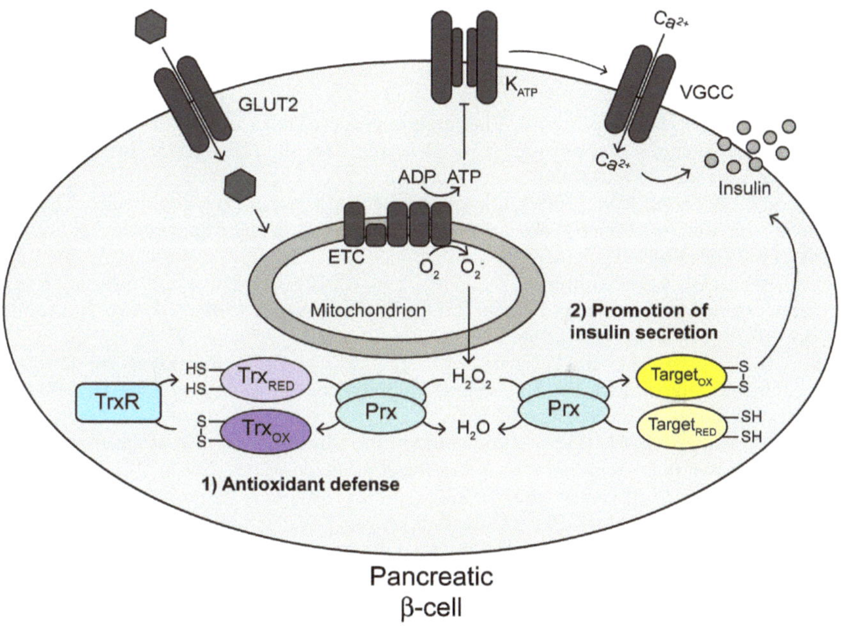

Fig. 4 A dual role for peroxiredoxins in the β-cell antioxidant defense and in promotion of insulin secretion. Increased blood glucose concentration stimulates glucose metabolism, increased ATP:ADP ratio, inhibition of ATP-sensitive potassium channels, membrane depolarization, and calcium-mediated insulin release in β-cells. During this process, mitochondrial oxidative metabolism promotes the formation of superoxide and hydrogen peroxide via electron leak. (1) Peroxiredoxins in the cytoplasm are necessary for the β-cell defense against elevated levels of hydrogen peroxide, as they reduce this ROS to water, utilizing thioredoxin and thioredoxin reductase. (2) Peroxiredoxin may also participate in redox relays, transferring oxidizing equivalents obtained from hydrogen peroxide to oxidize target proteins (that have not yet been identified) to promote glucose-stimulated insulin secretion. Abbreviations used: GLUT2 (glucose transporter 2), ETC (electron transport chain), ADP (adenosine diphosphate), ATP (adenosine triphosphate), K_{ATP} (ATP-sensitive potassium channel), VGCC (voltage-gated calcium channel), Prx (peroxiredoxin), Trx (thioredoxin), TrxR (thioredoxin reductase), RED (reduced), and OX (oxidized).

(Stancill & Corbett, 2021). First, they protect β-cells from oxidative damage by detoxifying hydrogen peroxide when it reaches cytotoxic levels. Second, they play a fundamental role in GSIS by mediating hydrogen peroxide-based signaling relays, ultimately making them indispensable for β-cell function and survival.

Funding

The writing of this chapter was supported by the National Institutes of Health grants K99DK129709 (to J.S.S.), and R01DK052194 and R01AI044458 (to J.A.C.), by a grant from the Medical College of Wisconsin Cancer Center (to J.A.C.), and by gifts from the Forest County Potawatomi Foundation and the Scott Tilton Foundation (to J.A.C.).

References

Arner, E. S. (2010). Selenoproteins-What unique properties can arise with selenocysteine in place of cysteine? *Experimental Cell Research*, *316*(8), 1296–1303. https://doi.org/10.1016/j.yexcr.2010.02.032.

Auten, R. L., & Davis, J. M. (2009). Oxygen toxicity and reactive oxygen species: The devil is in the details. *Pediatric Research*, *66*(2), 121–127. https://doi.org/10.1203/PDR.0b013e3181a9eafb.

Bast, A., Wolf, G., Oberbaumer, I., & Walther, R. (2002). Oxidative and nitrosative stress induces peroxiredoxins in pancreatic beta cells. *Diabetologia*, *45*(6), 867–876. https://doi.org/10.1007/s00125-002-0846-1.

Biteau, B., Labarre, J., & Toledano, M. B. (2003). ATP-dependent reduction of cysteine-sulphinic acid by *S. cerevisiae* sulphiredoxin. *Nature*, *425*(6961), 980–984. https://doi.org/10.1038/nature02075.

Brot, N., & Weissbach, H. (1983). Biochemistry and physiological role of methionine sulfoxide residues in proteins. *Archives of Biochemistry and Biophysics*, *223*(1), 271–281. https://doi.org/10.1016/0003-9861(83)90592-1.

Carvalho, L. A. C., Truzzi, D. R., Fallani, T. S., Alves, S. V., Toledo, J. C., Jr., Augusto, O., et al. (2017). Urate hydroperoxide oxidizes human peroxiredoxin 1 and peroxiredoxin 2. *The Journal of Biological Chemistry*, *292*(21), 8705–8715. https://doi.org/10.1074/jbc.M116.767657.

Chae, H. Z., Chung, S. J., & Rhee, S. G. (1994). Thioredoxin-dependent peroxide reductase from yeast. *The Journal of Biological Chemistry*, *269*(44), 27670–27678. (Retrieved from https://www.ncbi.nlm.nih.gov/pubmed/7961686).

Chae, H. Z., Kim, H. J., Kang, S. W., & Rhee, S. G. (1999). Characterization of three isoforms of mammalian peroxiredoxin that reduce peroxides in the presence of thioredoxin. *Diabetes Research and Clinical Practice*, *45*(2–3), 101–112. (Retrieved from https://www.ncbi.nlm.nih.gov/pubmed/10588361).

Chae, H. Z., Robison, K., Poole, L. B., Church, G., Storz, G., & Rhee, S. G. (1994). Cloning and sequencing of thiol-specific antioxidant from mammalian brain: Alkyl hydroperoxide reductase and thiol-specific antioxidant define a large family of antioxidant enzymes. *Proceedings of the National Academy of Sciences of the United States of America*, *91*(15), 7017–7021. https://doi.org/10.1073/pnas.91.15.7017.

Chen, J., Fontes, G., Saxena, G., Poitout, V., & Shalev, A. (2010). Lack of TXNIP protects against mitochondria-mediated apoptosis but not against fatty acid-induced ER stress-mediated beta-cell death. *Diabetes*, *59*(2), 440–447. https://doi.org/10.2337/db09-0949.

Chen, J., Hui, S. T., Couto, F. M., Mungrue, I. N., Davis, D. B., Attie, A. D., et al. (2008). Thioredoxin-interacting protein deficiency induces Akt/Bcl-xL signaling and pancreatic beta-cell mass and protects against diabetes. *The FASEB Journal*, *22*(10), 3581–3594. https://doi.org/10.1096/fj.08-111690.

Chen, J., Saxena, G., Mungrue, I. N., Lusis, A. J., & Shalev, A. (2008). Thioredoxin-interacting protein: A critical link between glucose toxicity and beta-cell apoptosis. *Diabetes*, *57*(4), 938–944. https://doi.org/10.2337/db07-0715.

Chou, F. C., & Sytwu, H. K. (2009). Overexpression of thioredoxin in islets transduced by a lentiviral vector prolongs graft survival in autoimmune diabetic NOD mice. *Journal of Biomedical Science*, *16*, 71. https://doi.org/10.1186/1423-0127-16-71.

Chutkow, W. A., & Lee, R. T. (2011). Thioredoxin regulates adipogenesis through thioredoxin-interacting protein (Txnip) protein stability. *The Journal of Biological Chemistry*, *286*(33), 29139–29145. https://doi.org/10.1074/jbc.M111.267666.

Cohen-Kutner, M., Khomsky, L., Trus, M., Aisner, Y., Niv, M. Y., Benhar, M., et al. (2013). Thioredoxin-mimetic peptides (TXM) reverse auranofin induced apoptosis and restore insulin secretion in insulinoma cells. *Biochemical Pharmacology*, *85*(7), 977–990. https://doi.org/10.1016/j.bcp.2013.01.003.

Cooke, M. S., Evans, M. D., Dizdaroglu, M., & Lunec, J. (2003). Oxidative DNA damage: Mechanisms, mutation, and disease. *The FASEB Journal*, *17*(10), 1195–1214. https://doi.org/10.1096/fj.02-0752rev.

Dalla Rizza, J., Randall, L. M., Santos, J., Ferrer-Sueta, G., & Denicola, A. (2019). Differential parameters between cytosolic 2-Cys peroxiredoxins, PRDX1 and PRDX2. *Protein Science*, *28*(1), 191–201. https://doi.org/10.1002/pro.3520.

De la Rosa, A., Gomez-Cabrera, M. C., Vinue, A., Gonzalez-Navarro, H., Sanchez-Andres, J. V., & Vina, J. (2021). Overexpression of glucose 6 phosphate dehydrogenase preserves mouse pancreatic beta cells function until late in life. *Free Radical Biology & Medicine*, *164*, 149–153. https://doi.org/10.1016/j.freeradbiomed.2020.12.439.

Ding, Y., Yamada, S., Wang, K. Y., Shimajiri, S., Guo, X., Tanimoto, A., et al. (2010). Overexpression of peroxiredoxin 4 protects against high-dose streptozotocin-induced diabetes by suppressing oxidative stress and cytokines in transgenic mice. *Antioxidants & Redox Signaling*, *13*(10), 1477–1490. https://doi.org/10.1089/ars.2010.3137.

Dominguez, C., Ruiz, E., Gussinye, M., & Carrascosa, A. (1998). Oxidative stress at onset and in early stages of type 1 diabetes in children and adolescents. *Diabetes Care*, *21*(10), 1736–1742 (Retrieved from https://www.ncbi.nlm.nih.gov/pubmed/9773740).

Ghiselli, A., Laurenti, O., De Mattia, G., Maiani, G., & Ferro-Luzzi, A. (1992). Salicylate hydroxylation as an early marker of in vivo oxidative stress in diabetic patients. *Free Radical Biology & Medicine*, *13*(6), 621–626. (Retrieved from https://www.ncbi.nlm.nih.gov/pubmed/1459481).

Gopaul, N. K., Anggard, E. E., Mallet, A. I., Betteridge, D. J., Wolff, S. P., & Nourooz-Zadeh, J. (1995). Plasma 8-epi-PGF2 alpha levels are elevated in individuals with non-insulin dependent diabetes mellitus. *FEBS Letters*, *368*(2), 225–229. (Retrieved from https://www.ncbi.nlm.nih.gov/pubmed/7628610).

Grankvist, K., Marklund, S. L., & Taljedal, I. B. (1981). CuZn-superoxide dismutase, Mn-superoxide dismutase, catalase and glutathione peroxidase in pancreatic islets and other tissues in the mouse. *The Biochemical Journal*, *199*(2), 393–398. (Retrieved from https://www.ncbi.nlm.nih.gov/pubmed/7041886).

Gromer, S., Johansson, L., Bauer, H., Arscott, L. D., Rauch, S., Ballou, D. P., et al. (2003). Active sites of thioredoxin reductases: Why selenoproteins? *Proceedings of the National Academy of Sciences of the United States of America*, *100*(22), 12618–12623. https://doi.org/10.1073/pnas.2134510100.

Hanschmann, E. M., Petry, S. F., Eitner, S., Maresch, C. C., Lingwal, N., Lillig, C. H., et al. (2020). Paracrine regulation and improvement of beta-cell function by thioredoxin. *Redox Biology*, *34*, 101570. https://doi.org/10.1016/j.redox.2020.101570.

Henquin, J. C. (2011). The dual control of insulin secretion by glucose involves triggering and amplifying pathways in beta-cells. *Diabetes Research and Clinical Practice*, *93*(Suppl 1), S27–S31. https://doi.org/10.1016/S0168-8227(11)70010-9.

Hotta, M., Tashiro, F., Ikegami, H., Niwa, H., Ogihara, T., Yodoi, J., et al. (1998). Pancreatic beta cell-specific expression of thioredoxin, an antioxidative and anti-apoptotic protein, prevents autoimmune and streptozotocin-induced diabetes. *The Journal of Experimental Medicine*, *188*(8), 1445–1451. (Retrieved from https://www.ncbi.nlm.nih.gov/pubmed/9782121).

Hwang, J., Suh, H. W., Jeon, Y. H., Hwang, E., Nguyen, L. T., Yeom, J., et al. (2014). The structural basis for the negative regulation of thioredoxin by thioredoxin-interacting protein. *Nature Communications*, *5*, 2958. https://doi.org/10.1038/ncomms3958.

Ihara, Y., Toyokuni, S., Uchida, K., Odaka, H., Tanaka, T., Ikeda, H., et al. (1999). Hyperglycemia causes oxidative stress in pancreatic beta-cells of GK rats, a model of type 2 diabetes. *Diabetes*, *48*(4), 927–932. (Retrieved from https://www.ncbi.nlm.nih.gov/pubmed/10102716).

Jin, M. H., Shen, G. N., Jin, Y. H., Sun, H. N., Zhen, X., Zhang, Y. Q., et al. (2020). Peroxiredoxin I deficiency increases pancreatic betacell apoptosis after streptozotocin stimulation via the AKT/GSK3beta signaling pathway. *Molecular Medicine Reports*, *22*(3), 1831–1838. https://doi.org/10.3892/mmr.2020.11279.

Jitrapakdee, S., Wutthisathapornchai, A., Wallace, J. C., & MacDonald, M. J. (2010). Regulation of insulin secretion: Role of mitochondrial signalling. *Diabetologia*, *53*(6), 1019–1032. https://doi.org/10.1007/s00125-010-1685-0.

Kim, K., Kim, I. H., Lee, K. Y., Rhee, S. G., & Stadtman, E. R. (1988). The isolation and purification of a specific "protector" protein which inhibits enzyme inactivation by a thiol/Fe(III)/O2 mixed-function oxidation system. *The Journal of Biological Chemistry*, *263*(10), 4704–4711. (Retrieved from https://www.ncbi.nlm.nih.gov/pubmed/2895105).

Knoops, B., Clippe, A., Bogard, C., Arsalane, K., Wattiez, R., Hermans, C., et al. (1999). Cloning and characterization of AOEB166, a novel mammalian antioxidant enzyme of the peroxiredoxin family. *The Journal of Biological Chemistry*, *274*(43), 30451–30458. https://doi.org/10.1074/jbc.274.43.30451.

Kowluru, A. (2020). Oxidative stress in cytokine-induced dysfunction of the pancreatic beta cell: Known knowns and known unknowns. *Metabolites*, *10*(12). https://doi.org/10.3390/metabo10120480.

Lassegue, B., & Griendling, K. K. (2010). NADPH oxidases: Functions and pathologies in the vasculature. *Arteriosclerosis, Thrombosis, and Vascular Biology*, *30*(4), 653–661. https://doi.org/10.1161/ATVBAHA.108.181610.

Laurent, T. C., Moore, E. C., & Reichard, P. (1964). Enzymatic synthesis of deoxyribonucleotides. Iv. Isolation and characterization of thioredoxin, the hydrogen donor from *Escherichia coli* B. *The Journal of Biological Chemistry*, *239*, 3436–3444. (Retrieved from https://www.ncbi.nlm.nih.gov/pubmed/14245400).

Lee, S. R., Bar-Noy, S., Kwon, J., Levine, R. L., Stadtman, T. C., & Rhee, S. G. (2000). Mammalian thioredoxin reductase: Oxidation of the C-terminal cysteine/selenocysteine active site forms a thioselenide, and replacement of selenium with sulfur markedly reduces catalytic activity. *Proceedings of the National Academy of Sciences of the United States of America*, *97*(6), 2521–2526. https://doi.org/10.1073/pnas.050579797.

Leloup, C., Tourrel-Cuzin, C., Magnan, C., Karaca, M., Castel, J., Carneiro, L., et al. (2009). Mitochondrial reactive oxygen species are obligatory signals for glucose-induced insulin secretion. *Diabetes*, *58*(3), 673–681. https://doi.org/10.2337/db07-1056.

Lenzen, S. (2008). Oxidative stress: The vulnerable beta-cell. *Biochemical Society Transactions, 36*(Pt 3), 343–347. https://doi.org/10.1042/BST0360343.

Lenzen, S. (2017). Chemistry and biology of reactive species with special reference to the antioxidative defence status in pancreatic beta-cells. *Biochimica et Biophysica Acta, 1861*(8), 1929–1942. https://doi.org/10.1016/j.bbagen.2017.05.013.

Lenzen, S., Drinkgern, J., & Tiedge, M. (1996). Low antioxidant enzyme gene expression in pancreatic islets compared with various other mouse tissues. *Free Radical Biology & Medicine, 20*(3), 463–466. (Retrieved from https://www.ncbi.nlm.nih.gov/pubmed/8720919).

Lin, X., Xu, Y., Pan, X., Xu, J., Ding, Y., Sun, X., et al. (2020). Global, regional, and national burden and trend of diabetes in 195 countries and territories: An analysis from 1990 to 2025. *Scientific Reports, 10*(1), 14790. https://doi.org/10.1038/s41598-020-71908-9.

MacDonald, M. J., Fahien, L. A., Brown, L. J., Hasan, N. M., Buss, J. D., & Kendrick, M. A. (2005). Perspective: Emerging evidence for signaling roles of mitochondrial anaplerotic products in insulin secretion. *American Journal of Physiology. Endocrinology and Metabolism, 288*(1), E1–15. https://doi.org/10.1152/ajpendo.00218.2004.

Manta, B., Hugo, M., Ortiz, C., Ferrer-Sueta, G., Trujillo, M., & Denicola, A. (2009). The peroxidase and peroxynitrite reductase activity of human erythrocyte peroxiredoxin 2. *Archives of Biochemistry and Biophysics, 484*(2), 146–154. https://doi.org/10.1016/j.abb.2008.11.017.

Marinho, H. S., Real, C., Cyrne, L., Soares, H., & Antunes, F. (2014). Hydrogen peroxide sensing, signaling and regulation of transcription factors. *Redox Biology, 2*, 535–562. https://doi.org/10.1016/j.redox.2014.02.006.

Maritim, A. C., Sanders, R. A., & Watkins, J. B., 3rd. (2003). Diabetes, oxidative stress, and antioxidants: A review. *Journal of Biochemical and Molecular Toxicology, 17*(1), 24–38. https://doi.org/10.1002/jbt.10058.

Matschinsky, F. M. (1996). Banting lecture 1995. A lesson in metabolic regulation inspired by the glucokinase glucose sensor paradigm. *Diabetes, 45*(2), 223–241. Retrieved from https://www.ncbi.nlm.nih.gov/pubmed/8549869.

Mehmeti, I., Lortz, S., Elsner, M., & Lenzen, S. (2014). Peroxiredoxin 4 improves insulin biosynthesis and glucose-induced insulin secretion in insulin-secreting INS-1E cells. *The Journal of Biological Chemistry, 289*(39), 26904–26913. https://doi.org/10.1074/jbc.M114.568329.

Minn, A. H., Hafele, C., & Shalev, A. (2005). Thioredoxin-interacting protein is stimulated by glucose through a carbohydrate response element and induces beta-cell apoptosis. *Endocrinology, 146*(5), 2397–2405. https://doi.org/10.1210/en.2004-1378.

Murrell, G. A., Francis, M. J., & Bromley, L. (1990). Modulation of fibroblast proliferation by oxygen free radicals. *The Biochemical Journal, 265*(3), 659–665. https://doi.org/10.1042/bj2650659.

National Diabetes Statistics Report. (2014). *Estimates of diabetes and its burden in the United States.* (Retrieved from Atlanta, GA).

Neal, A., Rountree, A., Kernan, K., Van Yserloo, B., Zhang, H., Reed, B. J., et al. (2016). Real-time imaging of intracellular hydrogen peroxide in pancreatic islets. *The Biochemical Journal, 473*(23), 4443–4456. https://doi.org/10.1042/BCJ20160481.

Nourooz-Zadeh, J., Tajaddini-Sarmadi, J., McCarthy, S., Betteridge, D. J., & Wolff, S. P. (1995). Elevated levels of authentic plasma hydroperoxides in NIDDM. *Diabetes, 44*(9), 1054–1058. (Retrieved from https://www.ncbi.nlm.nih.gov/pubmed/7657028).

Olahova, M., & Veal, E. A. (2015). A peroxiredoxin, PRDX-2, is required for insulin secretion and insulin/IIS-dependent regulation of stress resistance and longevity. *Aging Cell, 14*(4), 558–568. https://doi.org/10.1111/acel.12321.

Pacifici, F., Arriga, R., Sorice, G. P., Capuani, B., Scioli, M. G., Pastore, D., et al. (2014). Peroxiredoxin 6, a novel player in the pathogenesis of diabetes. *Diabetes*, *63*(10), 3210–3220. https://doi.org/10.2337/db14-0144.

Pandolfi, P. P., Sonati, F., Rivi, R., Mason, P., Grosveld, F., & Luzzatto, L. (1995). Targeted disruption of the housekeeping gene encoding glucose 6-phosphate dehydrogenase (G6PD): G6PD is dispensable for pentose synthesis but essential for defense against oxidative stress. *The EMBO Journal*, *14*(21), 5209–5215. (Retrieved from https://www.ncbi.nlm.nih.gov/pubmed/7489710).

Patwari, P., Higgins, L. J., Chutkow, W. A., Yoshioka, J., & Lee, R. T. (2006). The interaction of thioredoxin with Txnip. Evidence for formation of a mixed disulfide by disulfide exchange. *The Journal of Biological Chemistry*, *281*(31), 21884–21891. https://doi.org/10.1074/jbc.M600427200.

Paula, F. M., Ferreira, S. M., Boschero, A. C., & Souza, K. L. (2013). Modulation of the peroxiredoxin system by cytokines in insulin-producing RINm5F cells: Down-regulation of PRDX6 increases susceptibility of beta cells to oxidative stress. *Molecular and Cellular Endocrinology*, *374*(1–2), 56–64. https://doi.org/10.1016/j.mce.2013.04.009.

Peskin, A. V., Low, F. M., Paton, L. N., Maghzal, G. J., Hampton, M. B., & Winterbourn, C. C. (2007). The high reactivity of peroxiredoxin 2 with H(2)O(2) is not reflected in its reaction with other oxidants and thiol reagents. *The Journal of Biological Chemistry*, *282*(16), 11885–11892. https://doi.org/10.1074/jbc.M700339200.

Pi, J., Bai, Y., Zhang, Q., Wong, V., Floering, L. M., Daniel, K., et al. (2007). Reactive oxygen species as a signal in glucose-stimulated insulin secretion. *Diabetes*, *56*(7), 1783–1791. https://doi.org/10.2337/db06-1601.

Poole, L. B. (2015). The basics of thiols and cysteines in redox biology and chemistry. *Free Radical Biology & Medicine*, *80*, 148–157. https://doi.org/10.1016/j.freeradbiomed.2014.11.013.

Portillo-Ledesma, S., Randall, L. M., Parsonage, D., Dalla Rizza, J., Karplus, P. A., Poole, L. B., et al. (2018). Differential kinetics of two-cysteine peroxiredoxin disulfide formation reveal a novel model for peroxide sensing. *Biochemistry*, *57*(24), 3416–3424. https://doi.org/10.1021/acs.biochem.8b00188.

Prentki, M., Matschinsky, F. M., & Madiraju, S. R. (2013). Metabolic signaling in fuel-induced insulin secretion. *Cell Metabolism*, *18*(2), 162–185. https://doi.org/10.1016/j.cmet.2013.05.018.

Rehman, A., Nourooz-Zadeh, J., Moller, W., Tritschler, H., Pereira, P., & Halliwell, B. (1999). Increased oxidative damage to all DNA bases in patients with type II diabetes mellitus. *FEBS Letters*, *448*(1), 120–122. (Retrieved from https://www.ncbi.nlm.nih.gov/pubmed/10217422).

Rhee, S. G., & Woo, H. A. (2011). Multiple functions of peroxiredoxins: Peroxidases, sensors and regulators of the intracellular messenger H(2)O(2), and protein chaperones. *Antioxidants & Redox Signaling*, *15*(3), 781–794. https://doi.org/10.1089/ars.2010.3393.

Rhee, S. G., Woo, H. A., Kil, I. S., & Bae, S. H. (2012). Peroxiredoxin functions as a peroxidase and a regulator and sensor of local peroxides. *The Journal of Biological Chemistry*, *287*(7), 4403–4410. https://doi.org/10.1074/jbc.R111.283432.

Robertson, R. P., Harmon, J., Tran, P. O., & Poitout, V. (2004). Beta-cell glucose toxicity, lipotoxicity, and chronic oxidative stress in type 2 diabetes. *Diabetes*, *53*(Suppl 1), S119–S124. (Retrieved from https://www.ncbi.nlm.nih.gov/pubmed/14749276).

Sakuraba, H., Mizukami, H., Yagihashi, N., Wada, R., Hanyu, C., & Yagihashi, S. (2002). Reduced beta-cell mass and expression of oxidative stress-related DNA damage in the islet of Japanese type II diabetic patients. *Diabetologia*, *45*(1), 85–96. https://doi.org/10.1007/s001250200009.

Schuit, F., De Vos, A., Farfari, S., Moens, K., Pipeleers, D., Brun, T., et al. (1997). Metabolic fate of glucose in purified islet cells. Glucose-regulated anaplerosis in beta cells. *The Journal of Biological Chemistry*, *272*(30), 18572–18579. Retrieved from https://www.ncbi.nlm.nih.gov/pubmed/9228023.

Shalev, A. (2014). Minireview: Thioredoxin-interacting protein: Regulation and function in the pancreatic beta-cell. *Molecular Endocrinology*, *28*(8), 1211–1220. https://doi.org/10.1210/me.2014-1095.

Shalev, A., Pise-Masison, C. A., Radonovich, M., Hoffmann, S. C., Hirshberg, B., Brady, J. N., et al. (2002). Oligonucleotide microarray analysis of intact human pancreatic islets: Identification of glucose-responsive genes and a highly regulated TGFbeta signaling pathway. *Endocrinology*, *143*(9), 3695–3698. https://doi.org/10.1210/en.2002-220564.

Shin, C. S., Moon, B. S., Park, K. S., Kim, S. Y., Park, S. J., Chung, M. H., et al. (2001). Serum 8-hydroxy-guanine levels are increased in diabetic patients. *Diabetes Care*, *24*(4), 733–737 (Retrieved from https://www.ncbi.nlm.nih.gov/pubmed/11315839).

Sies, H. (2017). Hydrogen peroxide as a central redox signaling molecule in physiological oxidative stress: Oxidative eustress. *Redox Biology*, *11*, 613–619. https://doi.org/10.1016/j.redox.2016.12.035.

Sies, H., & Chance, B. (1970). The steady state level of catalase compound I in isolated hemoglobin-free perfused rat liver. *FEBS Letters*, *11*(3), 172–176. https://doi.org/10.1016/0014-5793(70)80521-x.

Sies, H., & Jones, D. P. (2020). Reactive oxygen species (ROS) as pleiotropic physiological signalling agents. *Nature Reviews. Molecular Cell Biology*, *21*(7), 363–383. https://doi.org/10.1038/s41580-020-0230-3.

Sobotta, M. C., Liou, W., Stocker, S., Talwar, D., Oehler, M., Ruppert, T., et al. (2015). Peroxiredoxin-2 and STAT3 form a redox relay for H2O2 signaling. *Nature Chemical Biology*, *11*(1), 64–70. https://doi.org/10.1038/nchembio.1695.

Spindel, O. N., World, C., & Berk, B. C. (2012). Thioredoxin interacting protein: Redox dependent and independent regulatory mechanisms. *Antioxidants & Redox Signaling*, *16*(6), 587–596. https://doi.org/10.1089/ars.2011.4137.

Stadtman, E. R., & Levine, R. L. (2000). Protein oxidation. *Annals of the New York Academy of Sciences*, *899*, 191–208. https://doi.org/10.1111/j.1749-6632.2000.tb06187.x.

Stancill, J. S., Broniowska, K. A., Oleson, B. J., Naatz, A., & Corbett, J. A. (2019). Pancreatic beta-cells detoxify H2O2 through the peroxiredoxin/thioredoxin antioxidant system. *The Journal of Biological Chemistry*, *294*(13), 4843–4853. https://doi.org/10.1074/jbc.RA118.006219.

Stancill, J. S., & Corbett, J. A. (2021). The role of thioredoxin/peroxiredoxin in the beta-cell defense against oxidative damage. *Frontiers in Endocrinology*, *12*, 718235. https://doi.org/10.3389/fendo.2021.718235.

Stancill, J. S., Hansen, P. A., Mathison, A. J., Schmidt, E. E., & Corbett, J. A. (2022). Deletion of Thioredoxin Reductase Disrupts Redox Homeostasis and Impairs beta-Cell Function. *Function (Oxf)*, *3*(4), zqac034. https://doi.org/10.1093/function/zqac034.

Stancill, J. S., Happ, J. T., Broniowska, K. A., Hogg, N., & Corbett, J. A. (2020). Peroxiredoxin 1 plays a primary role in protecting pancreatic beta-cells from hydrogen peroxide and peroxynitrite. *American Journal of Physiology. Regulatory, Integrative and Comparative Physiology*, *318*(5), R1004–R1013. https://doi.org/10.1152/ajpregu.00011.2020.

Stancill, J. S., Kasmani, M. Y., Khatun, A., Cui, W., & Corbett, J. A. (2021). Single-cell RNA sequencing of mouse islets exposed to proinflammatory cytokines. *Life Science Alliance*, *4*(6). https://doi.org/10.26508/lsa.202000949.

Stocker, S., Van Laer, K., Mijuskovic, A., & Dick, T. P. (2018). The conundrum of hydrogen peroxide signaling and the emerging role of peroxiredoxins as redox relay hubs. *Antioxidants & Redox Signaling*, *28*(7), 558–573. https://doi.org/10.1089/ars.2017.7162.

Stone, J. R., & Yang, S. (2006). Hydrogen peroxide: A signaling messenger. *Antioxidants & Redox Signaling*, *8*(3–4), 243–270. https://doi.org/10.1089/ars.2006.8.243.

Subasinghe, W., Syed, I., & Kowluru, A. (2011). Phagocyte-like NADPH oxidase promotes cytokine-induced mitochondrial dysfunction in pancreatic beta-cells: Evidence for regulation by Rac1. *American Journal of Physiology. Regulatory, Integrative and Comparative Physiology*, *300*(1), R12–R20. https://doi.org/10.1152/ajpregu.00421.2010.

Tanaka, Y., Gleason, C. E., Tran, P. O., Harmon, J. S., & Robertson, R. P. (1999). Prevention of glucose toxicity in HIT-T15 cells and Zucker diabetic fatty rats by antioxidants. *Proceedings of the National Academy of Sciences of the United States of America*, *96*(19), 10857–10862. (Retrieved from https://www.ncbi.nlm.nih.gov/pubmed/10485916).

Tavender, T. J., Sheppard, A. M., & Bulleid, N. J. (2008). Peroxiredoxin IV is an endoplasmic reticulum-localized enzyme forming oligomeric complexes in human cells. *The Biochemical Journal*, *411*(1), 191–199. https://doi.org/10.1042/BJ20071428.

Tiedge, M., Lortz, S., Munday, R., & Lenzen, S. (1998). Complementary action of antioxidant enzymes in the protection of bioengineered insulin-producing RINm5F cells against the toxicity of reactive oxygen species. *Diabetes*, *47*(10), 1578–1585. (Retrieved from https://www.ncbi.nlm.nih.gov/pubmed/9753295).

Wolf, G., Aumann, N., Michalska, M., Bast, A., Sonnemann, J., Beck, J. F., et al. (2010). Peroxiredoxin III protects pancreatic ss cells from apoptosis. *The Journal of Endocrinology*, *207*(2), 163–175. https://doi.org/10.1677/JOE-09-0455.

Yasuda, M., Ohzeki, Y., Shimizu, S., Naito, S., Ohtsuru, A., Yamamoto, T., et al. (1999). Stimulation of in vitro angiogenesis by hydrogen peroxide and the relation with ETS-1 in endothelial cells. *Life Sciences*, *64*(4), 249–258. https://doi.org/10.1016/s0024-3205(98)00560-8.

Zhang, Z., Liew, C. W., Handy, D. E., Zhang, Y., Leopold, J. A., Hu, J., et al. (2010). High glucose inhibits glucose-6-phosphate dehydrogenase, leading to increased oxidative stress and beta-cell apoptosis. *The FASEB Journal*, *24*(5), 1497–1505. https://doi.org/10.1096/fj.09-136572.

Zhao, R. Z., Jiang, S., Zhang, L., & Yu, Z. B. (2019). Mitochondrial electron transport chain, ROS generation and uncoupling (review). *International Journal of Molecular Medicine*, *44*(1), 3–15. https://doi.org/10.3892/ijmm.2019.4188.

Zhao, F., & Wang, Q. (2012). The protective effect of peroxiredoxin II on oxidative stress induced apoptosis in pancreatic beta-cells. *Cell & Bioscience*, *2*(1), 22. https://doi.org/10.1186/2045-3701-2-22.

CHAPTER THREE

Molecular docking approaches and its significance in assessing the antioxidant properties in different compounds

Neha Srivastava[a], Prekshi Garg[b], Anurag Singh[c], and Prachi Srivastava[b,*]

[a]Excelra Knowledge Solution Pvt Ltd, NSL Arena, Uppal, Hyderabad, India
[b]Amity Institute of Biotechnology, Amity University, Lucknow, Uttar Pradesh, India
[c]Department of Biochemistry, University of Lucknow, Lucknow, Uttar Pradesh, India
*Corresponding author: e-mail address: psrivastava@amity.edu

Contents

Abstract

In the last few years, the significance of antioxidant compounds and their properties has attracted great interest from the scientific community. The role of an antioxidant in managing & regulating oxidative stress and also in the protection of the human body from severe adverse effects due to excess release of free radicles or reactive oxygen species (ROS) is remarkable. From aiding protection & combating severe illnesses such as cancer, neurodegeneration, aging, and diabetes to being a vital part of the treatment of SARs-CoV-19 is of great importance. Therefore, the study of anti-oxidants is of great importance in human sustenance. Additionally, molecular docking techniques and their various mathematical features help in understanding the molecular interactions of anti-oxidants based on their lowest binding energy. The evaluation of the binding score between two constituent molecules will provide insight as to the binding process and also suggest possible novel therapeutic targets for the treatment of diseases. In this chapter, we will discuss the significance of molecular docking techniques in the study of antioxidant compounds.

Vitamins and Hormones, Volume 121
ISSN 0083-6729
https://doi.org/10.1016/bs.vh.2022.09.005

1. Introduction

In the human body, several natural biological processes such as food digestion, breathing, drug metabolism, and conversion of fat into energy, all release harmful chemicals termed free radicles. The excess production of these free radicals set-off negative chain reactions which in turn choke the activity of numerous important enzymes, prevent various essential cellular processes inc. cell division and oxidative metabolism, and damage various cellular components such as cell membranes and genetic material (Kurutas, 2015). These free radicles are generally blocked by our body's anti-oxidant system. The anti-oxidant can hunt free radicles from the human body and work as a protective shield against diseases like cancer, cardiovascular, neurodegeneration, and diabetes. As free radicles are essential for life, there are, therefore, a variety of enzymatic mechanisms to reduce the damage free radicals cause, as well as to guard against their over-production (Finkel & Holbrook, 2000). Anti-oxidant mechanisms have a significant role in the protection of our body from various external and internal stresses. The proper balance between oxidant and anti-oxidant plays a key role in protecting against the negative effects of reactive oxygen species (ROS) (Bhattacharyya, Chattopadhyay, Mitra, & Crowe, 2014). The antioxidant system is divided into two categories: endogenous antioxidants & exogenous antioxidants. The endogenous system refers to enzymes including superoxide dismutase (SOD), catalase (CAT), glutathione peroxidase, or non-enzymatic compounds, such as bilirubin and albumin. The exogenous system refers to getting antioxidants via food & nutritional supplements, or pharmaceutical sources that trigger the failure of the endogenous antioxidant mechanism (Santos-Sánchez, Salas-Coronado, Villanueva-Cañongo, & Hernández-Carlos, 2019). The phenolic compounds like carotenoids and vitamins C and minerals such as selenium and zinc are a few important exogenous anti-oxidant compounds. Thus, the antioxidant system is vital for a healthy body. In recent years, various research findings suggest the importance of antioxidant compounds and their defense mechanism (Menchaca, Juárez-Portilla, & Zepeda, 2020). Therefore, the proper intake of these antioxidant compounds as a supplement will protect our bodies from the harmful effects of free radicals. Several in-vivo & in-vitro studies are ongoing to understand the mechanism of these natural & synthetic antioxidant compounds and their activities. Molecular docking techniques are a widely used approach to studying anti-oxidant compounds using various algorithms. Molecular docking techniques are quick & accurate and have high value in predicting the drug-target intermolecular interaction based on the lowest binding energy & molecular interaction. Various studies highlighted the significance of molecular docking studies in identifying novel therapeutic targets for antioxidant compounds as well as in the prediction of their antioxidant properties. Therefore, this chapter highlights molecular docking techniques & their implementation in the study of anti-oxidants.

2. Molecular docking techniques

The methodologies of molecular docking or molecular interactions intend to predict the best binding orientation of a ligand to a receptor. It proposes several suitable conformations, i.e., poses, of the ligand within the active or docking site of a receptor molecule. Thus, it becomes crucial that the three-dimensional structure of macromolecules (receptors) be present, and this could be either an experimentally validated structure or a computational predicted and validated model (Salmaso, 2018).

To select the best matching conformations or poses, a scoring function is required. The scoring function is used to distinctively identify putative correct binding matches and binders from an array of non-binders and other misfit poses. Essentially three scoring functions are known, viz.,

(a) **Force-field-based scoring function**—This function approximates the potential energy of the system with bonded and nonbonded components. Examples of this function are GoldScore and AutoDock
(b) **Empirical Scoring Function**—This function is the sum of different empirical energies viz., van der Waals, electrostatic, hydrogen bond, desolvation, entropy, etc. Examples are GlideScore, ChemScore, and PLANTSCHEMPLP.
(c) **Knowledge-based scoring functions**—These scoring functions work on the assumption that the interaction between ligand and protein receptor statistically more explored are correlated with certain favorable interactions. Examples of this function include DrugScore and GOLD/ASP.

Other than the aforementioned functions there are a huge number of new scoring functions being developed daily based on machine learning techniques, interaction fingerprints and quantum mechanical scores (Brooijmans & Kuntz, 2003; Eldridge, Murray, Auton, Paolini, & Mee, 1997; Friesner et al., 2006; Gohlke, Hendlich, & Klebe, 2000; Huang & Zou, 2010; Korb, Stützle, & Exner, 2009; Morris et al., 1998; Velec, Gohlke, & Klebe, 2005; Verdonk, Cole, Hartshorn, Murray, & Taylor, 2003).

3. Docking strategies based on the flexibility and rigidity of interacting components

Docking methodologies have been divided into three main categories depending upon the degree of flexibility of the interacting components, these docking types include rigid docking, semi-flexible docking, and flexible docking.

(a) Rigid Docking

In this type of docking protocol, both ligand and proteins are considered rigid entities, and only three translational and three rotational degrees of freedom are considered during the sampling process. This type of docking mechanism can easily be understood by imagining the scenario of the "lock & key" binding model. Rigid docking protocols are generally used for protein-protein docking (Mooij & Verdonk, 2005).

(b) Semi-flexible Docking

In this type of docking, rather than both components being rigid, one is flexible (the ligand) and the other protein component is rigid. Thus, in this case, the conformational degrees of freedom on the ligand molecule are taken into consideration while sampling, in addition to the six translational and rational ones. The assumption in place behind this type of docking protocol is that the rigid protein molecule will be able to recognize the ligand to be docked, but this assumption, though true for most cases, is yet not verified for all scenarios (Yuriev, Holien, & Ramsland, 2015). There are various algorithms in place for performing semi-flexible docking, which includes.

3.1 Systemic search techniques

In this technique, a set of discrete values is associated with each degree of freedom, and the values of each coordinate are explored in a combinatorial manner (Taylor, Jewsbury, & Essex, 2002). These methods are further divided into

I. **Exhaustive search**—In this, all the rotatable bonds of the ligands are examined systematically. Several constraints and termination criteria are established to avoid any combinatorial error. Docking software Glide (Huang & Zou, 2010) has an exhaustive search stage in it.

II. Fragmentation—In this method, the ligand is fragmented and rigid docking of the ligand fragments are done on the receptor-binding site followed by subsequent linking of the fragments. FlexX (Halgren et al., 2004) and Hammerhead (Rarey, Kramer, Lengauer, & Klebe, 1996) utilize fragmentation approaches.

III. Conformational Ensemble—An algorithm of rigid-body docking can be enriched with a little flexibility if an ensemble of the previous confirmation of ligand is docked in the target molecule. Examples: EUDOC & MS-DOCK (Pang, Perola, Xu, & Prendergast, 2001; Welch, Ruppert, & Jain, 1996)

3.2 Stochastic methods

Stochastic algorithms happen to change at random rather than systematically the values of the degrees of freedom. Now the advantage of such a random approach is that it helps in identifying an optimal pose fast but on the other hand, the lack of a systematic search of the fully available conformational space might lead to a true solution being missed. The increment in the number of iterations could significantly reduce the drawback of this methodology.

I. Monte Carlo (MC) Methodology—This approach is based on the Metropolis Monte Carlo Algorithm, which incorporates the criteria of acceptance in the docking search evolution. This is incorporated in AutoDock, ICM, QXP, MCDOCK, and AutoDock Vina (Abagyan, Totrov, & Kuznetsov, 1994; McMartin & Bohacek, 1997; Sauton, Lagorce, Villoutreix, & Miteva, 2008; Trott & Olson, 2010), etc.

II. Tabu Search methods—This algorithm aims to prevent exploring already sampled zones of the conformational space. There are some random modifications in the degree of freedom of the ligand components at each iteration. An example includes PSI-DOCK (Liu & Wang, 1999).

(c) Flexible Docking

Is based on the concept that **proteins** are not passively rigid throughout the binding process, and as such considers both ligand and protein as flexible counterparts. Over the years, some have rested on induced fit binding, while some on conformational selection. There is a great number of degrees of freedom induced by flexible docking, which makes the potential energy surface a function of numerous coordinates (Table 1).

Table 1 Summarized docking tools/software's based on type of scoring function and type of docking and algorithm they use.

S. No.	Name of software	Scoring function type	Reference
1.	GoldScore	Force-field based	Mooij and Verdonk (2005)
2.	AutoDock		Yuriev et al. (2015)
3.	GlideScore	Empirical	Sauton et al. (2008)
4.	ChemScore		Gohlke et al. (2000)
6.	DrugScore	Knowledge-based	Taylor et al. (2002)
7.	GOLD		Mooij and Verdonk (2005)

Docking softwares based on docking type and algorithm used

S. No.	Name of software	Docking type	Algorithm Used	Reference
1.	Glide	Semi-flexible	Exhaustive Search	Sauton et al. (2008)
2.	FlexX		Fragmentation	Trott and Olson (2010)
3.	EUDOC		Conformational Ensemble	McMartin and Bohacek (1997)
4.	MS-DOCK			Liu and Wang (1999)
5.	AutoDock		Monte Carlo methodology	Yuriev et al. (2015)
6.	ICM			Lobo, Patil, Phatak, and Chandra (2010)
7.	AutoDock Vina			Pei et al. (2006)
8.	QXP			Farouk, Mohsen, Ali, Shaaban, and Albaridi (2021)
9.	MCDOCK			Bassam, Abdur, Yasir, Saima, and Abdul (2021)
10.	PSI-DOCK		Tabu Search methods	Gabr, Bakr, Mostafa, El-Fishawy and El-Alfy (2019)

4. Docking studies to evaluate antioxidant activity of compounds

Antioxidants are molecules that are capable of neutralizing free radicals by donating their electrons (Pei et al., 2006). Free radicals, left unchecked, lead to oxidative stress that causes chronic diseases like cardiovascular diseases, cancer, and diabetes. The key roles played by antioxidant molecules include reducing oxidative stress and regulating redox homeostasis (Lobo et al., 2010). Thus, the antioxidant activity of compounds is very important for maintaining the overall health of an individual. There have been various docking studies done on molecules to analyze their antioxidant properties (Fig. 1).

One of the recent docking studies conducted by Bassam et al. (2021) revealed the antioxidant activity of seven isolated dinaphthodiospyrols obtained from *Diospyros lotus* Linn roots. Another study conducted by Gabr et al. (2019) aimed at the identification of antioxidant activity of seven phenolic compounds that were extracted from a hybrid plant known as Erythrina × neillii. In his study, he concluded that the phytoconstituents obtained from the hybrid were powerful and significant antioxidants. In another study conducted by Al-Salahi et al. (2019) docking studies were used to identify interactions between the binding models and the antioxidant activity of 15 derivatives of thioxobenzo[*g*]quinazoline. Another recent study conducted by Farouk et al. (2021) investigated the molecular docking of essential oils obtained from *Mentha spicata* L., *Mentha longifolia* L., and *Origanum majorana* L., cultivated in Madinah, Saudi Arabia. The docking study of the main volatile oils obtained from these species was done with NADPH oxidase and significant antioxidant ability was observed in these

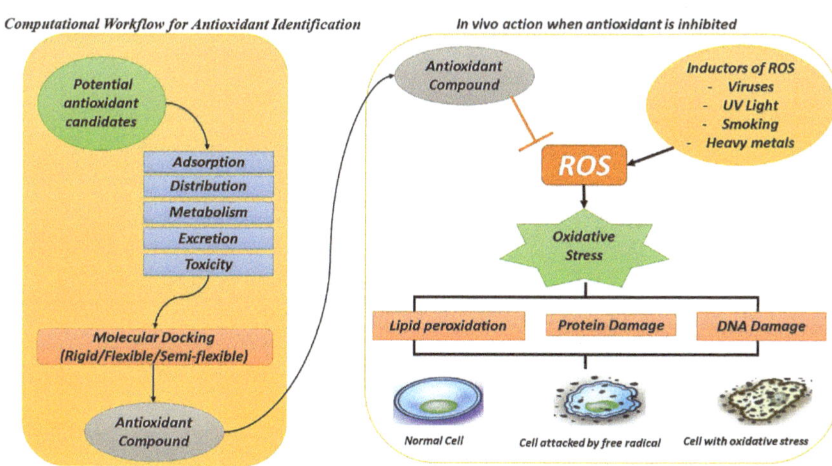

Fig. 1 The schematic representation of the workflow to identify potent antioxidant molecule from a pool of candidate molecules using molecular docking approach (left) and what is the relationship between Reactive Oxygen Species (ROS) and oxidative stress and how it leads to cell apoptosis/necrosis when antioxidant molecules are inhibited or not present (right).

essential oils. In another study conducted by Nayab et al. (2020), both in silico and in vitro investigations into the antioxidant activity of a-amino phosphonates were performed. During the in silico study, the researcher performed docking studies of the compounds with aromatase enzyme and obtained a docking score greater than −8 kJ/mol for each target compound. In another study conducted by Kalirajan, Rafick, Sankar, & Jubie (2012), docking studies were done on some novel isoxazole-substituted 9-anilinoacridine derivatives. These compounds were docked against nucleoside dsDNA through AutoDock Vina 4.0. Another study conducted by Shehzadi et al. (2018) revealed the activity of 5-[(4-chlorophenoxy) methyl]-1,3,4-oxadiazole-2-thiol as a radical scavenger and endogenous defense system inducer by conducting molecular docking studies. In one of the studies conducted by Rajendran, Nithya, Brindha Devi, and Kanakam (2017), hydroxy (diphenyl) acetic acid and its derivatives were studied for their antioxidant activity using molecular docking studies. The compounds were docked with nitric oxide synthase enzyme to reveal the amino acid interactions of the ligand with the active site pocket of the target. Another molecular docking study conducted on the essential oils of the aerial part of Lippia origanoides was done by da Silva et al. (2017). According to this study, the best antioxidant activity was displayed by thymol and (E)-nerolidol. A very interesting study was conducted by Yapati, Devineni, Chirumamilla, et al. (2016) wherein the docking studies were performed on metal complexes of 1-(benzo[d]thiazol-2-yl)thiourea to evaluate its antioxidant activity. Another study conducted by Kumar et al. (2016) also evaluated the antioxidant and antimicrobial activities of ethyl 2-(4-chlorobenzylidene)-3-oxobutanoate prepared in the laboratory through a condensation reaction. A study conducted by Rehman et al. (2021) claimed that the zinc metal carboxylates synthesized by them in the laboratory have a significant ability to manage Alzheimer's disease on anticholinesterase as well as antioxidant targets.

A study conducted on Diabetic retinopathy by Srivastava et al., (2014) suggested the potential role of D-pinitol, an herbal extract of *Glycine max* based on their binding pattern and molecular interaction. In addition, an Occuloinformatics study performed by Srivastava and Tiwari (2016) explained the usefulness of Ginkgolide, D-pinitol, Gugglesterones, Berberine and Curcumin based on their binding affinity and scoring function in ocular disorders.

In a recent study on Covid-19, the authors (Srivastava, Garg, Srivastava, & Seth, 2021) suggested the role of natural inhibitor, targeting the ACE2 receptor based on their lowest binding score, confirmation and H—H bond interaction. In addition, several antioxidant molecules are also reported to play a promising role in the treatment of various neurological disorders, helping to promote brain health. The network-based findings in Srivastava, Mishra, and Srivastava (2019) reveals the potential of beta amyrin, ajmaline, serpentine, urosolic, huperzine A phytochemicals against Attention deficit hyperactivity disorder (ADHD) & Autism spectrum disorder (ASD) neurodevelopmental disorders. In this study a protein-protein interaction (PPI) network-based model was built using gene candidates for ADHD & ASD to predict potential common drug targets for both. Further molecular docking studies were performed to identify potential drug leads. Based on binding affinity & scoring function, a potential inhibitor was obtained. Another Study also signifies the role of preferred orientation of receptor to ligand lead to form more stable complexes (Verma, Chauhan, Pankaj, Srivastva, & Srivastava, 2020). A review study by Srivastava and Tiwari (2021) highlights the Prophylactic role of anti-oxidants in neurological disorder using standard protocol of virtual screening and molecular docking. A study by Tiwari and Singh (2022) briefly describes the promising role and importance of molecular docking and virtual screening in drug discovery and possible therapeutic intervention. This review covers some of the more relevant molecular docking studies conducted on antioxidants. Currently, antioxidants are one of the most intensively researched group of compounds. Importantly, the application of molecular docking techniques has helped to make this expanding and important body of work more concrete, reliable, and accurate.

5. Future scope of the work

The fundamental metabolic processes in all living beings are known to be oxidative. However, due to the potential for over-production of free radicles associated with oxidation, these processes can lead to cellular damage. These damages can effect biomolecules like lipids, proteins, and DNA, or can even impact tissues, further reducing cellular protection. Several human diseases such as neurodegeneration, cardiovascular disease, ocular disease, diabetes and cancer are linked to oxidative stress. Additionally, the stress arisen by UV rays, Sunlight and radio radiation resulting in hyperpigmentation and sensitivity is also major health concern. Therefore, there is a pressing need for the use of interdisciplinary research to identify novel anti-oxidant molecules. Molecular docking opens new doors toward this approach. Molecular docking helps in understanding the intermolecular interactions that form between protein-protein or protein and small molecules. It deciphers the underlying molecular function of docking complexes based on their scoring function, binding affinity and confirmation. Molecular docking plays a vital role in exploring and evaluating the anti-xidant properties of small molecules to their receptor. This exciting new investigative approach will drive antioxidant research toward novel anti-oxidant lead design, targeting drug discovery and therapeutic interventions. In the post-pandemic era, Molecular docking approaches will play a vital role in quick, fast and accurate prediction of the anti-oxidant properties of small molecules leading to drug discovery. Therefore, through investigating different binding mode and affinity between anti-oxidants, molecular docking becomes a crucial precursor to experimental research.

6. Conclusion

It is now a well-established fact that the presence of antioxidant compounds in the human body in an adequate amount is needed for cells to be in a healthy state and fight off any oxidative stress. Antioxidants are known to prevent or delay some types of the cell damage caused by free radicals. They neutralize the free radicals generated in our body by donating their electrons. If the free radicals are not neutralized, they can lead to oxidative stress that can cause chronic diseases like cardiovascular diseases, cancer, and diabetes. Thus, it is advised by nutritionists and dieticians to make fruits and vegetables rich in antioxidants a part of your diet. The antioxidant activity of compounds present in fruits and vegetables maintains your overall health. Computational strategies have been known to ease the whole process of lead identification and validation and it is a very potent strategy concerning antioxidant identification and characterization. With the computational technique of molecular docking, we can easily identify antioxidants specific to our receptor protein of interest and also establish whether that particular antioxidant is inhibiting or activating the protein. Molecular docking is a very flexible technique, and can be performed in many different scenarios viz. if both the ligand and protein are rigid, we can go for rigid docking or if the ligand is flexible, but the receptor is rigid, we can move for the semi-flexible type, or if both the receptor and the ligand are flexible, we can easily switch to flexible docking. There are several software packages available that can help and assist us in our pursuit of identifying antioxidants using our technique of choice. The major challenge in molecular docking studies is receptor flexibility, especially the movement of the backbone and the secondary elements that are involved in ligand binding and catalytic activity. Molecular docking studies have made it possible to study the antioxidant activity of different compounds that can be extracted from various plant sources. The antioxidant activity of these compounds can further be exploited for various medical purposes. Thus, docking proves to be an amazing tool that aids in antioxidant-related studies.

References

Abagyan, R., Totrov, M., & Kuznetsov, D. (1994). ICM? A new method for protein modeling and design: Applications to docking and structure prediction from the distorted native conformation. *Journal of Computational Chemistry*, *15*, 488–506. https://doi.org/10.1002/jcc.540150503.

Al-Salahi, R., Taie, H. A. A., Bakheit, A. H., Marzouk, M., Almehizia, A. A., Herqash, R., et al. (2019). Antioxidant activities and molecular docking of 2-thioxobenzo[g] quinazoline derivatives. *Pharmacological Reports*, *71*(4), 695–700. ISSN 1734-1140. https://doi.org/10.1016/j.pharep.2019.04.003.

Bassam, O. A., Abdur, R., Yasir, A., Saima, N., & Abdul, W. (2021). Antibacterial, antifungal, antioxidant, and docking studies of potential Dinaphthodiospyrols from Diospyros lotus Linn roots. *ACS Omega*, *6*(8), 5878–5885. https://doi.org/10.1021/acsomega.0c0629.

Bhattacharyya, A., Chattopadhyay, R., Mitra, S., & Crowe, S. E. (2014). Oxidative stress: An essential factor in the pathogenesis of gastrointestinal mucosal diseases. *Physiological Reviews*, *94*, 329–354. https://doi.org/10.1152/physrev.00040.2012.

Brooijmans, N., & Kuntz, I. D. (2003). Molecular recognition and docking algorithms. *Annual Review of Biophysics and Biomolecular Structure*, *32*, 335–373. https://doi.org/10.1146/annurev.biophys.32.110601.142532.

da Silva, A. P., Silva, N. F., Andrade, E. H. A., Gratieri, T., Setzer, W. N., Maia, J. G. S., et al. (2017). Tyrosinase inhibitory activity, molecular docking studies and antioxidant potential of chemotypes of Lippia origanoides (Verbenaceae) essential oils. *PLoS One*, *12*(5), e0175598. https://doi.org/10.1371/journal.pone.0175598.

Eldridge, M. D., Murray, C. W., Auton, T. R., Paolini, G. V., & Mee, R. P. (1997). Empirical scoring functions: I. The development of a fast empirical scoring function to estimate the binding affinity of ligands in receptor complexes. *Journal of Computer-Aided Molecular Design*, *11*, 425–445. https://doi.org/10.1023/A:1007996124545.

Farouk, A., Mohsen, M., Ali, H., Shaaban, H., & Albaridi, N. (2021). Antioxidant activity and molecular docking study of volatile constituents from different aromatic Lamiaceous plants cultivated in Madinah Monawara, Saudi Arabia. *Molecules*, *26*, 4145. https://doi.org/10.3390/molecules26144145.

Finkel, T., & Holbrook, N. J. (2000). Oxidants, oxidative stress and the biology of ageing. *Nature*, *408*, 239–247. https://doi.org/10.1038/35041687.

Friesner, R. A., Murphy, R. B., Repasky, M. P., Frye, L. L., Greenwood, J. R., Halgren, T. A., et al. (2006). Extra precision glide: Docking and scoring incorporating a model of hydrophobic enclosure for protein-ligand complexes. *Journal of Medicinal Chemistry*, *49*, 6177–6196. https://doi.org/10.1021/jm051256o.

Gabr, S. K., Bakr, R. O., Mostafa, E. S., El-Fishawy, A. M., & El-Alfy, T. S. (2019). Antioxidant activity and molecular docking study of Erythrina × neillii polyphenolics. *South African Journal of Botany*, *121*, 470–477. ISSN 0254-6299. https://doi.org/10.1016/j.sajb.2018.12.011.

Gohlke, H., Hendlich, M., & Klebe, G. (2000). Knowledge-based scoring function to predict protein-ligand interactions. *Journal of Molecular Biology, 295*, 337–356. https://doi.org/10.1006/jmbi.1999.3371.

Halgren, T. A., Murphy, R. B., Friesner, R. A., Beard, H. S., Frye, L. L., Pollard, W. T., et al. (2004). Glide: A new approach for rapid, accurate docking and scoring. 2. Enrichment factors in database screening. *Journal of Medicinal Chemistry, 47*, 1750–1759. https://doi.org/10.1021/jm030644s.

Huang, S.-Y., & Zou, X. (2010). Advances and challenges in protein-ligand docking. *International Journal of Molecular Sciences, 11*, 3016–3034. https://doi.org/10.3390/ijms11083016.

Kalirajan, R., Rafick, M. H. M., Sankar, S., & Jubie, S. (2012). Docking studies, synthesis, characterization and evaluation of their antioxidant and cytotoxic activities of some novel isoxazole-substituted 9-anilinoacridine derivatives. *The Scientific World Journal, 2012*, 165258. 6 p. https://doi.org/10.1100/2012/165258.

Korb, O., Stützle, T., & Exner, T. E. (2009). Empirical scoring functions for advanced protein-ligand docking with PLANTS. *Journal of Chemical Information and Modeling, 49*, 84–96. https://doi.org/10.1021/ci800298z.

Kumar, D. A., Naveen, S., Vivek, H. K., Prabhuswamy, M., Lokanath, N. K., & Kumar, K. A. (2016). Synthesis, crystal and molecular structure of ethyl 2-(4-chlorobenzylidene)-3-oxobutanoate: Studies on antioxidant, antimicrobial activities and molecular docking. *Chemical Data Collections, 5–6*, 36–45. ISSN 2405-8300. https://doi.org/10.1016/j.cdc.2016.10.002.

Kurutas, E. B. (2015). The importance of antioxidants which play the role in cellular response against oxidative/nitrosative stress: Current state. *Nutrition Journal, 15*, 71.

Liu, M., & Wang, S. (1999). MCDOCK: A Monte Carlo simulation approach to the molecular docking problem. *Journal of Computer-Aided Molecular Design, 13*, 435–451. https://doi.org/10.1023/A:1008005918983.

Lobo, V., Patil, A., Phatak, A., & Chandra, N. (2010). Free radicals, antioxidants, and functional foods: Impact on human health. *Pharmacognosy Reviews, 4*(8), 118–126. https://doi.org/10.4103/0973-7847.70902.

McMartin, C., & Bohacek, R. S. (1997). QXP: Powerful, rapid computer algorithms for structure-based drug design. *Journal of Computer-Aided Molecular Design, 11*, 333–344. https://doi.org/10.1023/A:1007907728892.

Menchaca, T. M., Juárez-Portilla, C., & Zepeda, R. C. (2020). Past, present, and future of molecular docking. In V. Gaitonde, P. Karmakar, & A. Trivedi (Eds.), *Drug discovery and development—New advances* IntechOpen. https://doi.org/10.5772/intechopen.90921.

Mooij, W. T., & Verdonk, M. L. (2005). General and targeted statistical potentials for protein-ligand interactions. *Proteins, 61*, 272–287. https://doi.org/10.1002/prot.20588.

Morris, G. M., Goodsell, D. S., Halliday, R. S., Huey, R., Hart, W. E., Belew, R. K., et al. (1998). Automated docking using a Lamarckian genetic algorithm and an empirical binding free energy function. *Journal of Computational Chemistry, 19*, 1639–1662. https://doi.org/10.1002/(SICI)1096-987X(19981115)19:143.0.CO;2-B.

Nayab, R. S., Maddila, S., Krishna, M. P., Salam, J. J. T., Thaslim, B. S., Chintha, V., et al. (2020 Apr). In silico molecular docking and in vitro antioxidant activity studies of novel α-aminophosphonates bearing 6-amino-1,3-dimethyl uracil. *Journal of Receptor and Signal Transduction Research, 40*(2), 166–172. https://doi.org/10.1080/10799893.2020.1722166. Epub 2020 Feb 5. PMID: 32019395.

Pang, Y.-P., Perola, E., Xu, K., & Prendergast, F. G. (2001). EUDOC: A computer program for identification of drug interaction sites in macromolecules and drug leads from chemical databases. *Journal of Computational Chemistry, 22*, 1750–1771. https://doi.org/10.1002/jcc.1129.

Pei, J., Wang, Q., Liu, Z., Li, Q., Yang, K., & Lai, L. (2006). PSI-DOCK: Towards highly efficient and accurate flexible ligand docking. *Proteins, 62*, 934–946. https://doi.org/10.1002/prot.20790.

Rajendran, S., Nithya, G., Brindha Devi, P, & Kanakam, C. C. (2017). Docking antioxidant activity on hydroxy (diphenyl) aceticacid and its derivatives. *Asian Journal of Pharmaceutical and Clinical Research, 10*(7), 263–265. https://doi.org/10.22159/ajpcr.2017.v10i7.18299.

Rarey, M., Kramer, B., Lengauer, T., & Klebe, G. (1996). A fast flexible docking method using an incremental construction algorithm. *Journal of Molecular Biology, 261*, 470–489. https://doi.org/10.1006/jmbi.1996.0477.

Rehman, Z., Zubair, M., Ali, S., Shahid, K., Waseem, W., Naureen, H., et al. (2021). Zinc metal carboxylates as potential anti-Alzheimer's candidate: *In vitro* anticholinesterase, antioxidant and molecular docking studies. *Journal of Biomolecular Structure and Dynamics, 39*(3), 1044–1054. https://doi.org/10.1080/07391102.2020.1724569.

Salmaso, V. (2018). *Exploring protein flexibility during docking to investigate ligand-target recognition.* Ph.D. thesis Padova: University of Padova.

Santos-Sánchez, N. F., Salas-Coronado, R., Villanueva-Cañongo, C., & Hernández-Carlos, B. (2019). Antioxidant compounds and their antioxidant mechanism. In E. Shalaby (Ed.), *Antioxidants* (pp. 1–28). IntechOpen.

Sauton, N., Lagorce, D., Villoutreix, B. O., & Miteva, M. A. (2008). MS-DOCK: Accurate multiple conformation generator and rigid docking protocol for multi-step virtual ligand screening. *BMC Bioinformatics, 9*, 184. https://doi.org/10.1186/1471-2105-9-184.

Shehzadi, N., Hussain, K., Khan, M. T., Bukhari, N. I., Islam, M., Salman, M., et al. (2018). Radical scavenging and endogenous defence system inducing activities of 5-[(4-Chlorophenoxy)methyl]-1,3,4-oxadiazole-2-thiol: A novel antioxidant. *Indian Journal of Pharmaceutical Sciences, 80*(6), 1125–1135.

Srivastava, N., Garg, P., Srivastava, P., & Seth, P. K. (2021 Apr 23). A molecular dynamics simulation study of the ACE2 receptor with screened natural inhibitors to identify novel drug candidate against COVID-19. *PeerJ, 9*, e11171. https://doi.org/10.7717/peerj.11171. PMID: 33981493; PMCID: PMC8074842.

Srivastava, N., Mishra, B. N., & Srivastava, P. (2019). In-silico identification of drug lead molecule against pesticide exposed-neurodevelopmental disorders through network-based computational model approach. *Current Bioinformatics, 14*, 460–467.

Srivastava, P, & Tiwari, A. (2016). A new insight of herbal promises against ocular disorders: An occuloinformatics approach. *Current Topics in Medicinal Chemistry, 16*(6), 634–654. https://doi.org/10.2174/1568026615666150819105716.

Srivastava, P., & Tiwari, A. (2021). *The potential role of phytochemical in establishing prophylactic measurements against neurological diseases.* Elsevier. https://doi.org/10.1002/9781119682059.ch16.

Srivastava, P., Tiwari, A., Trivedi, A. C., Thakur, V., Pant, A. B., & Saxena, S. (2014). Virtual screening of natural and synthetic ligands against diabetic retinopathy by molecular interaction with Angiopoietin-2. *The Asia-Pacific Journal of Ophthalmology, 3*(4), 257–259. https://doi.org/10.1097/APO.0000000000000071. PMID: 26107766.

Taylor, R. D., Jewsbury, P. J., & Essex, J. W. (2002). A review of proteinsmall molecule docking methods. *Journal of Computer-Aided Molecular Design, 16*, 151–166. https://doi.org/10.1023/A:1020155510718.

Tiwari, A., & Singh, S. (2022). Computational approaches in drug designing. In *Bioinformatics* (pp. 207–217). Academic Press. https://doi.org/10.1016/B978-0-323-89775-4.00010-9.

Trott, O., & Olson, A. J. (2010). AutoDock Vina: Improving the speed and accuracy of docking with a new scoring function, efficient optimization, and multithreading. *Journal of Computational Chemistry, 31*, 455–461. https://doi.org/10.1002/jcc.21334.

Velec, H. F. G., Gohlke, H., & Klebe, G. (2005). DrugScore(CSD)-knowledgebased scoring function derived from small molecule crystal data with superior recognition rate of near-native ligand poses and better affinity prediction. *Journal of Medicinal Chemistry*, *48*, 6296–6303. https://doi.org/10.1021/jm050436v.

Verdonk, M. L., Cole, J. C., Hartshorn, M. J., Murray, C. W., & Taylor, R. D. (2003). Improved protein-ligand docking using GOLD. *Proteins*, *52*, 609–623. https://doi.org/10.1002/prot.10465.

Verma, A., Chauhan, S., Pankaj, V., Srivastva, N., & Srivastava, P. (2020). Network biology approaches to identify the drug lead molecule for neurodevelopmental disorders in human. *The Open Bioinformatics Journal*, *13*, 15–24. https://doi.org/10.2174187503620201301001 5.

Welch, W., Ruppert, J., & Jain, A. N. (1996). Hammerhead: Fast, fully automated docking of flexible ligands to protein binding sites. *Chemistry & Biology*, *3*, 449–462. https://doi.org/10.1016/S1074-5521(96)90093-9.

Yapati, H., Devineni, S. R., Chirumamilla, S., et al. (2016). Synthesis, characterization and studies on antioxidant and molecular docking of metal complexes of 1-(benzo[d]thiazol-2-yl)thiourea. *Journal of Chemical Sciences*, *128*, 43–51. https://doi.org/10.1007/s12039-015-0999-3.

Yuriev, E., Holien, J., & Ramsland, P. A. (2015). Improvements, trends, and new ideas in molecular docking: 2012-2013 in review. *Journal of Molecular Recognition*, *28*, 581–604. https://doi.org/10.1002/jmr.2471.

CHAPTER FOUR

Scavengome of an antioxidant

Attila Hunyadi[a,b,*], Orinhamhe G. Agbadua[a], Gergely Takács[c,d], and Gyorgy T. Balogh[c,e]

[a]Institute of Pharmacognosy, Interdisciplinary Excellence Centre, University of Szeged, Szeged, Hungary
[b]Interdisciplinary Centre for Natural Products, University of Szeged, Szeged, Hungary
[c]Department of Chemical and Environmental Process Engineering, Budapest University of Technology and Economics, Budapest, Hungary
[d]Mcule.com Ltd., Budapest, Hungary
[e]Department of Pharmacodynamics and Biopharmacy, University of Szeged, Szeged, Hungary
*Corresponding author: e-mail address: hunyadi.attila@szte.hu

Contents

Abstract

The term "scavengome" refers to the chemical space of all the metabolites that may be formed from an antioxidant upon scavenging reactive oxygen or nitrogen species (ROS/RNS). This chemical space covers a wide variety of free radical metabolites with drug discovery potential. It is very rich in structures representing an increased chemical complexity as compared to the parent antioxidant: a wide range of unusual heterocyclic structures, new C—C bonds, etc. may be formed. Further, in a biological environment, this increased chemical complexity is directly translated from the localized conditions of oxidative stress that determines the amounts and types of ROS/RNS present. Biomimetic oxidative chemistry provides an excellent tool to model chemical reactions between antioxidants and ROS/RNS. In this chapter, we provide an overview on the known metabolites obtained by biomimetic oxidation of a few selected natural antioxidants, i.e., a stilbene (resveratrol), a pair of hydroxycinnamates (caffeic acid and methyl caffeate), and a flavonol (quercetin), and discuss the drug discovery perspectives of the related chemical space.

Vitamins and Hormones, Volume 121
ISSN 0083-6729
https://doi.org/10.1016/bs.vh.2022.09.003

1. Antioxidants and their mechanism of action: Introduction of the scavengome concept

Plant polyphenols are among the best-studied dietary antioxidants, and their research continues to escalate. In PubMed, the search term "antioxidant AND polyphenol" yields over 30,000 hits in total, and over 3300 only for 2021. This everlasting popularity is undoubtedly due to the plethora of beneficial bioactivities attributed to such compounds in connection with a wide range of chronic diseases including diabetes, malignant tumors, and degenerative central nervous system (CNS) pathologies. This is no wonder, considering the well-known key role of oxidative stress in the pathomechanism and progression of such diseases.

Oxidative stress and its role in health and disease has been thoroughly reviewed over the years, and the field has recently gone through a paradigm shift toward recognizing the physiological importance of oxidative processes and re-defining oxidative stress as a loss of control and balance. The biologically most important reactive oxygen species (ROS) are hydrogen peroxide (H_2O_2) (Gough & Cotter, 2011), singlet oxygen ($^1O_2$) (Bauer, 2016), hydroxyl, alkoxyl and peroxyl radicals ($^{\bullet}OH$, $RO^{\bullet}$, and $ROO^{\bullet}$, respectively) (Dickinson & Chang, 2011), hypochlorous acid (HOCl) (Bauer, 2018), and superoxide anion radical ($O_2^{\bullet-}$) (Hayyan, Hashim, & AlNashef, 2016), and major reactive nitrogen species (RNS) are nitric oxide (NO), nitrogen dioxide (NO_2), and peroxynitrite ($ONOO^-$) (Nimse & Pal, 2015). The reactivity of these species varies; primary ROS/RNS, like H_2O_2, $O_2^{\bullet-}$, and NO are much less reactive than toxic secondary species, particularly $^{\bullet}OH$, $ONOO^-$, and HOCl, whose increasing amounts highly contribute to oxidative stress and related damage of macromolecules, cells, and tissues (Weidinger & Kozlov, 2015). The formation of ROS/RNS, and particularly the secondary species is determined by transition metal catalysis; in biological environment this is primarily connected to enzymatic processes, i.e., mitochondrial electron transport chain (ETC) complexes I and II, lipoxygenases, cyclooxygenases, xanthine oxidase, and cytochrome P monoxygenases (Dickinson & Chang, 2011; Weidinger & Kozlov, 2015).

There is now wide a consensus that small molecule antioxidants decrease oxidative stress mainly through modulating the intrinsic enzymatic defense that maintains the redox balance in the body (Forman, Davies, & Ursini, 2014; Ruskovska, Maksimova, & Milenkovic, 2020; Sies, 1993). However, while direct free radical scavenging is generally thought to have little in vivo relevance in decreasing the levels of toxic ROS/RNS, the interaction between such reactive species and oxidizable dietary antioxidants should also not be ignored. Naturally, such an interaction has a major impact on both reaction partners. The reactive intermediate that is formed from the antioxidant may then be reduced back in a redox cycle, couple with an appropriate small-molecule or macromolecule reaction partner, or get stabilized by intramolecular rearrangements. Because reactive antioxidant intermediates have a much more complex structure then ROS/RNS, their binding to macromolecules is also more specific, therefore such a "damage" may not necessarily cause toxic effects but can also be translated to pharmacological activity. A good example to this is the covalent binding of quinones to the thiol groups of cysteine residues, which then leads to complex signal transduction processes (e.g., through the activation of Keap1/Nrf2 signaling) and may result in beneficial effects through an adaptive antioxidant response (Kato & Suga, 2018; Kerimi & Williamson, 2018). Further, the many possible intra- or intermolecular rearrangements stabilizing reactive antioxidant intermediates unravel a rich chemical space of small-molecule metabolites that may then exert a non-covalent action on any druggable targets in the microenvironment where they have been formed. Even though some antioxidants, e.g., curcumin, may undergo oxidative fragmentation (Shen & Ji, 2009), free radical scavenging by an antioxidant more frequently yields more complex chemical structures; this is due to the many possibilities for radical coupling reactions forming unique new rings or ring systems, various heterocycles, etc. In our recent review, we outlined the potential drug discovery perspectives of this novel and orthogonal chemical space and named it as the "scavengome" (Hunyadi, 2019).

Diversity-oriented synthesis is a major drug discovery strategy. A wide variety of success-oriented chemical approaches have been pursued over the years to turn plain chemical diversity into a so-called biological performance diversity, i.e., to create chemical libraries that are not only diverse but also have a high pharmacological hit rate (Pavlinov, Gerlach, & Aldrich, 2019). In this context it is worth stressing that (i) any given antioxidant will manifest only a segment of its scavengome in a biological microenvironment under oxidative stress, and (ii) the formed metabolites contain chemical information directly translated from the local conditions of oxidative stress, i.e., the types and amounts of ROS/RNS present. Considering the co-evolution of the biochemical machinery (i.e., potential drug targets) of animals and humans with their food rich in plant secondary metabolites, polyphenols, etc. (i.e., potential drugs), it seems reasonable to postulate that scavengome of a dietary antioxidant is an organic part of the signaling network related to oxidative stress. This notion suggests that a diversity-oriented chemical approach to explore the scavengome is a promising strategy toward biological performance diversity.

In the followings, we aim to provide a summary of what is known about the scavengome of a few selected abundant natural antioxidants, i.e., a stilbene (resveratrol), a pair of hydroxycinnamates (caffeic acid and methyl caffeate), and a flavonol (quercetin). We approach this subject by overviewing the chemical complexity of their metabolites prepared through their oxidative chemical transformations, in perspective of the biomimetic value of the oxidants. We also provide some characteristic examples on the bioactivity of several metabolites and evaluate their positioning in the drug-like chemical space.

2. Biomimetic oxidative chemistry: Exploring the scavengome

For modeling biological processes with an in vitro chemical approach, it is especially important to characterize and classify the models based on the translational distance from the studied human physiological process. In the case of oxidative stress or antioxidant studies, model developments typically focus on mitochondrial, macrophage and neutrophil-related, and microsomal free radical formation processes (Bartesaghi & Radi, 2018). In this context, the classical approach basically distinguishes three main classes of biomimetic models (López-Alarcón & Denicola, 2013) that evaluate (i) the scavenging activity toward stable free radicals, e.g., 2,2-Diphenyl-1-picrylhydrazyl (DPPH$^{\bullet}$) and 2,2′-Azinobis-(3-ethylbenzothiazole-6-sulphonate) radical cation (ABTS$^{\bullet+}$), (ii) the reduction of metal ions, e.g., Ferric Reducing Antioxidant Power (FRAP) and Cupric ion-Reducing Antioxidant Capacity (CUPRAC) to evaluate the sample's capacity to reduce ferric or cupric ions in aqueous media, and (iii) the oxidation of low-density lipoprotein (LDL). In relation with the scavengome concept, a comprehensive work is not limited to classical and directly biorelevant oxidative stress studies on the transformations of selected model compounds. In our novel approach to introduce potentially diverse oxidative chemical model systems in exploring a broad chemical space of oxidized antioxidant metabolites, we use a hierarchical classification system in our related studies, and this classification will also serve as an organizing principle to this chapter. This is based on biocompatibility or translational goodness in terms of bioequivalence, i.e., oxidative transformation of an antioxidant by any chemical system can be classified as a biorelevant (A), biomimetic (B), or biomimetic-related chemical (C) oxidation. Accordingly, we consider biorelevant (A) any in vitro chemical models that directly provide oxidative agents/free radicals present in the body, such as $^{\bullet}OH$: metalloporphyrin/H_2O_2 (Mazoir, Benharref, Vaca, Reina, & González-Coloma, 2020), Fe^{2+}/H_2O_2 (Miller, Chang, Wegeberg, McKenzie, & Waite, 2021), H_2O_2: Cu^{2+}/ascorbic acid (Shen et al., 2021), and $ONOO^-$ (Ferrer-Sueta et al., 2018), and chemical systems for which there is significant experimental evidence of their suitability for biological oxidative stress (DPPH, AAPH, AIBN AMVN) (Marano et al., 2021; Takatsuka, Goto, Kobayashi, Otsuka, & Shimada, 2022) are also included in this group. We classify as biomimetic (B) the chemical systems in which the oxidative reaction medium contains an aqueous component in addition to a cosolvent (in some cases dissolved O_2 as a prooxidant) providing a medium with better suitability to the physiological system. The last class, i.e., chemical related to biomimetic (C), differs from class B only in the anhydrous organic solvent medium, which may still provide relevant models of oxidative processes within biological membranes. In some cases, we also classify oxidative conditions using biologically less relevant pH values into class C. In the model systems class B and C, three subgroups can be distinguished: (a) oxygen atom donors (H_2O_2, *t*-BuOOH, O_2 in alkaline media) (Costa, Vega-Aguilar, & Rodrigues, 2021; Wenz et al., 2019), (b) electron donors (e.g., $NaIO_4$, $K_3Fe(CN)_6$, Ru^{3+}, Cu^{2+}, Fe^{2+}/Fe^{3+}, Au^{3+}) (Galletti, Martelli, Prandini, Colucci, & Giacomini, 2018; Pradhan et al., 2020; Wu et al., 2021; Yang et al., 2016), and (c) agents that mimic the nitrating capacity of peroxynitrite (e.g., $NaNO_2$) (Ozyurt & Otles, 2020; Vossen & De Smet, 2015). Fig. 1

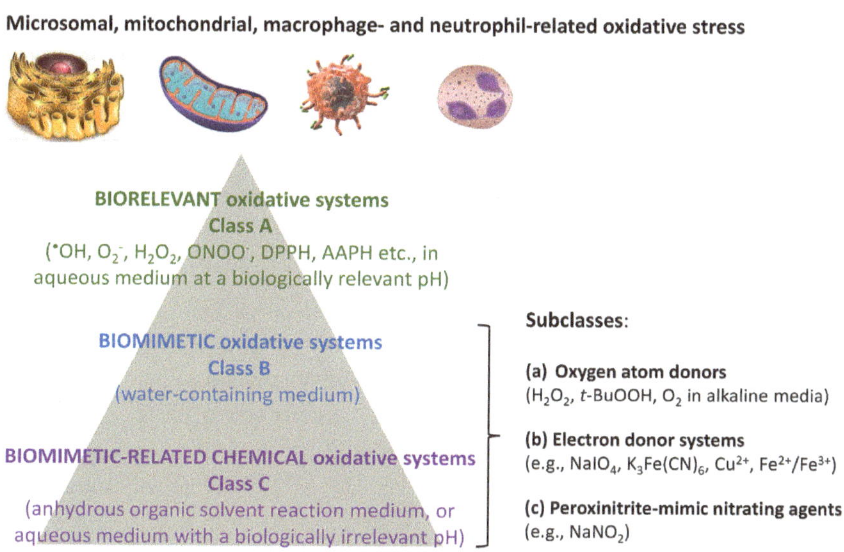

Fig. 1 Hierarchical classification of oxidative chemical systems based on their bioequivalence value to model ROS/RNS scavenging reactions.

shows our hierarchical classification of oxidative chemical model systems according to their applicability to explore chemical metabolite space within the scavengome concept.

3. Oxidative transformations of selected antioxidants

3.1 Resveratrol (I)

Resveratrol (*trans*-3,5,4′-trihydroxystilbene; compound **I**) is a natural polyphenol and phytoalexin, and probably the most popular dietary antioxidant due to its well-known presence in red wine. Other than that, it is also present in many common foods including, e.g., a variety of berries, tomato skin, peanuts, pistachios and cocoa (Dybkowska, Sadowska, Świderski, Rakowska, & Wysocka, 2018). Resveratrol was reported to have a myriad of beneficial bioactivities in health and disease (Baur & Sinclair, 2006), and its antioxidant properties are at the hallmark of these. Resveratrol can efficiently scavenge various types of ROS/RNS, resulting in radical intermediates stabilized by the delocalization of electrons between the two aromatic rings and the unsaturated methylene bridge joining them (Karlsson, Emgard, Brundin, & Burkitt, 2000). This gives ample opportunities to C—C coupling at various regions of the resveratrol molecule. Interestingly, in a biological environment there is a high chance that the most likely reaction partner to such couplings would be another resveratrol molecule; this is because of the strong self-association of resveratrol fixed by strong π-π stacking in aqueous solution (Bonechi, Martini, Magnani, & Rossi, 2008; Velu, Di Meo, Trouillas, Sancho-Garcia, & Weber, 2013). This may be the reason why biorelevant/biomimetic oxidation of resveratrol very frequently results in various dimers. The various oxidative reaction conditions applied are summarized in Table 1, and chemical structures of the metabolites obtained are presented in Fig. 2.

The biomimetic oxidation of stilbenes may lead to various regioselective couplings depending on the oxidant, solvent, and substitution (Velu et al., 2008). Compound **I/1** was a major dimer formed by several biorelevant and biomimetic oxidants, and it was also a common oxidized product of resveratrol when it was reacted with various transition metals in less biomimetic experimental setups. An impressive, nearly quantitative (97%) yield of **I/1** was achieved when resveratrol was oxidized by AgOAc or Ag_2O (Sako et al., 2004). Similarly high (>90%) yields of **I/1** were obtained with $FeCl_3$ in acetone or MnO_2 in CH_2Cl_2 at 25 °C, while ε-viniferin (**I/16**)

Table 1 Oxidative transformations of resveratrol (**I**).

Reagents	Experimental conditions	Metabolites obtained	References
Biorelevant			
$NaNO_2$	0.1 M phosphate buffer, pH 3.0, 37 °C	**I/1–11**	Panzella et al. (2006)
Peroxynitrite	0.1 M, pH 7.4, phosphate-buffered ethanol solution	Unspecified dimers, nitro- and dinitro-derivatives	Holthoff et al. (2010)
DPPH	MeOH	**I/1**, **I/12**	Wang, Jin, and Ho (1999)
Biomimetic			
$K_3Fe(CN)_6$	pH 5.5 phosphate buffered CH_3CN	**I/1**	Sako, Hosokawa, Ito, and Iinuma (2004)
$FeCl_3$	EtOH(aq)	**I/1**, **I/12**, **I/16**	Shingai, Fujimoto, Nakamura, and Masuda (2011)
Formic acid	Reflux	**I/13–15**	Li, Huang, Lin, and Zhou (2003)
Ruthenium chloride	MeOH(aq), 35 °C	**I/16**, **I/17**	Yadav et al. (2019)
Biomimetic-related chemical			
AgOAc, Ag_2O, Ag_2CO_3, $AgNO_3$, $Mn(OAc)_3$ CuOAc, $Cu(OAc)_2$	Dry MeOH, 50 °C	**I/1**	Sako et al. (2004)
$Tl(NO_3)_3$	MeOH, −50 °C	**I/16**	Takaya et al. (2005)
$Tl(NO_3)_3$	MeOH, −30 °C	**I/16**	
$K_3[Fe(CN)_6]/K_2CO_3$	MeOH, 25 °C	**I/1**, **I/16**, **I/18**	
$Ce(SO_4)_2$	MeOH, −50 °C	**I/1**, **I/16**, **I/18**	
$FeCl_3$	Acetone, 25 °C	**I/1**, **I/16**, **I/18**	
MnO_2	CH_2Cl_2, 25 °C	**I/1**, **I/18**	
$K_3[Fe(CN)_6]$, sodium acetate	Aqueous acetone under reflux	**I/1**, **I/18–22**	Xie et al. (2015)
AgOAc or $FeCl_3$	EtOH	**I/16**, **I/23–25**	El Khawand et al. (2020)

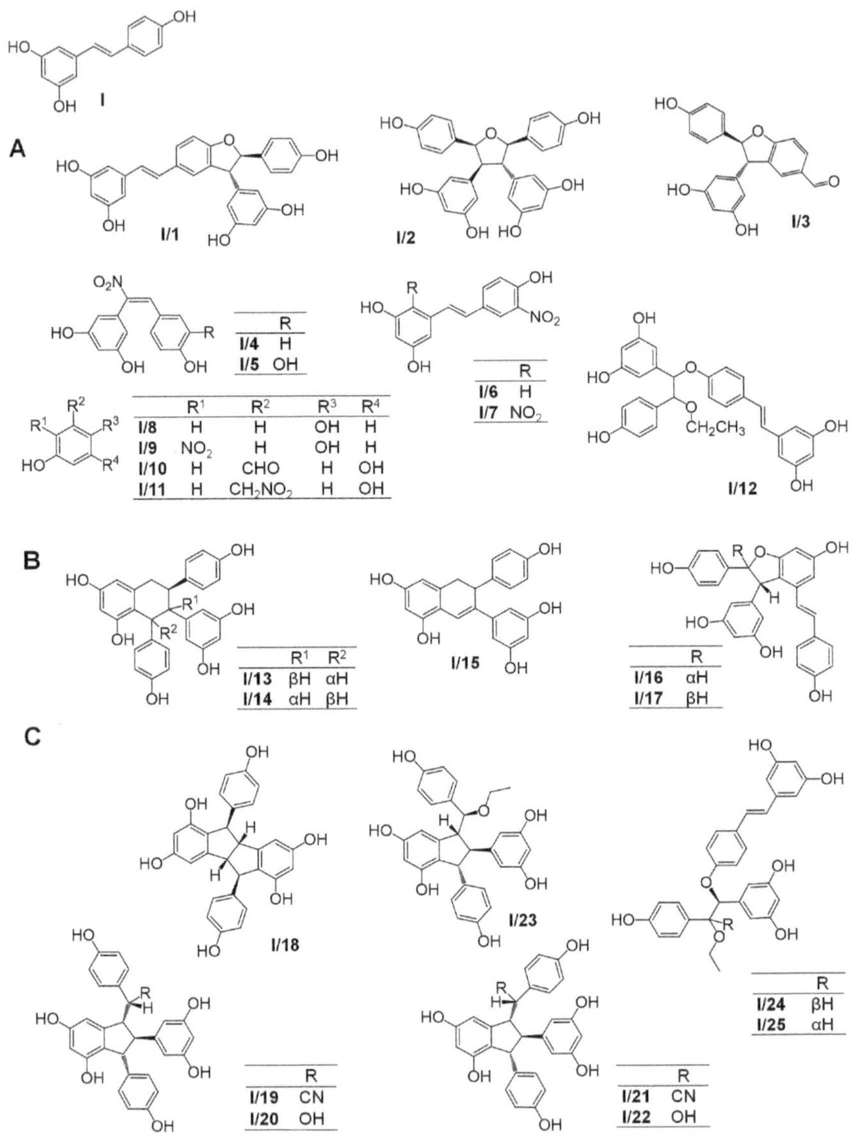

Fig. 2 Oxidized resveratrol derivatives formed in reactions with biorelevant (A), biomimetic (B), or biomimetic-related chemical oxidants (C).

became the major product with no detectable formation of **I/1** when resveratrol was oxidized by thallium(III) nitrate at −50 °C (Takaya et al., 2005).

In a broader context of the scavengome, it is of interest that stressing grapevine leaves (var. Chasselas) by *Plasmopara viticola* (downy mildew) infection or UV-C irradiation also leads to the oxidative dimerization of resveratrol to form a mixture of products including the *E*/*Z* isomers of δ- and ε-viniferins (**I/1** and **I/16–17**) (Pezet et al., 2003). Both *P. viticola* infection and UV-C irradiation induces oxidative stress in grapevine leaves (Carvalho, Vidigal, & Amâncio, 2015). While resveratrol dimers may certainly have been biosynthesized in the used experimental setup through various stress-activated enzymatic processes, the involvement of direct free radical scavenging may also not be excluded.

Oxidizing resveratrol by AgOAc in ethanol resulted in oxidative coupling to form **I/16** and three additional dimers, **I/23–I/25** (El Khawand et al., 2020). It may be worth stressing that, while the presence of ethanol as a solvent makes these conditions of less use to model pathophysiological ROS/RNS scavenging, such oxidative processes can naturally occur in wine that also contains transition metals, mainly iron and copper (Płotka-Wasylka, Frankowski, Simeonov, Polkowska, & Namieśnik, 2018).

Concerning the bioactivity of oxidized resveratrol metabolites, many of these compounds are widely recognized as phytoalexins, i.e., antibiotic compounds produced by plants in response to environmental stress stimuli (Jeandet, 2015). Other than that, a wide range of bioactivities of resveratrol dimers have been reported, including antiinflammatory, antioxidant, neuroprotective, and vascular protective activities. Related literature on ε-viniferin (**I/16**) has been most recently reviewed; this compound appears to exhibit particularly strong bioactivities against inflammatory and oxidative stress both in vitro and in vivo (Beaumont, Courtois, Atgié, Richard, & Krisa, 2022).

Unlike resveratrol and its open-ring dimer (**I/12**), δ- and ε-viniferins **I/1** and **I/16** have 5-LOX inhibitory activity (Shingai et al., 2011). Additional anti-inflammatory activities of some oxidized resveratrol metabolites were also reported in models where resveratrol was found less active or inactive. Compounds **I/1**, **I/23** and **I/24** reduced the NO production in LPS-induced RAW 264.7 macrophages with around twice the potency of resveratrol (IC_{50} values <10 μM) (El Khawand et al., 2020). At 10 μM concentrations, compounds **I/13** and **I/14** inhibited leukotriene B_4 (LTB_4) biosynthesis, inhibited the adhesion of HL-60 cells to capillary endothelial cells, and acted as LTD_4 antagonists (Li et al., 2003).

Both δ- and ε-viniferins (**I/1** and **I/16**) exerted significant protective functions on vascular endothelial cells (Wu, Nakamoto, Hisatome, Yoshida, & Miyazaki, 2020), and compound **I/1** was reported as a potent antioxidant against hemoglobin (Hb)-induced oxidative stress in erythrocytes, and to protect Hb from methemoglobin formation (Ficarra et al., 2016).

3.2 Caffeic acid (II) and methyl caffeate (III)

Caffeic acid is widely distributed in plants, and it is the most abundant hydroxycinnamic acid in many fruits and vegetables including, e.g., black chokeberries, lingonberries, dates, plums, olives, and potatoes, beverages, e.g., coffee, wines, and beers, seeds, like sunflower seeds, and a wide range of herbs and spices, like common sage, thyme, oregano, etc. It may occur in free form or glycosides, and it is valued for a wide range of beneficial bioactivities (Birková, 2020). While it is a very small and simple molecule, it is considered a highly valuable building block in the design of new bioactive compounds (Touaibia, Jean-François, & Doiron, 2011). Both caffeic acid (**II**) and its naturally occurring methyl ester (**III**) have been described as chain-breaking, radical scavenging antioxidants due to their hydrogen atom or electron donating capacity and to the ability to delocalize/stabilize the resulting phenoxyl radical (Teixeira, Gaspar, Garrido, Garrido, & Borges, 2013). Caffeic acid is susceptible to autoxidation and it was reported to react with transition metals, glutathione and other thiols (cysteine, thioglycolic acid or thiocresol), and ascorbic acid (Cilliers & Singleton, 1990; Heleno, Martins, Queiroz, & Ferreira, 2015). The behavior of compounds **II** and **III** under oxidative conditions have been studied in various experimental setups; these are summarized in Table 2, and chemical structures of the metabolites obtained are presented in Fig. 3.

Unsurprisingly, the *o*-quinone derivative **II/1** was frequently observed as the primary intermediate during the oxidative coupling of caffeic acid (Antolovich et al., 2004; Tazaki et al., 2003), and thus it was reported as a common product in several oxidative transformations. As an interesting biomimetic element of the reaction setup, **II** was subjected to oxidation in the presence of L-cysteine (Bassil et al., 2005). The thiol group of cysteine residues plays a central role in redox signaling and stress response (Fra, Yoboue, & Sitia, 2017; Paulsen & Carroll, 2013). Conjugation of caffeic acid with L-cysteine took place to yield **II/13** and this somewhat (by ca. 7%) enhanced the antiradical efficiency compared to that of caffeic acid (Bassil et al., 2005). More importantly, however, this also demonstrates the potential of reactive caffeic acid intermediates, particularly **II/1**, to modulate proteins' function through covalent coupling to accessible cysteine residues. Bubbling oxygen gas through the alkaline aqueous solution of caffeic acid led to the formation of compounds **II/7–10** (Cilliers & Singleton, 1991). Several related 1,4-benzodioxane-type neolignans were obtained when **II** was transformed by a non-specified crude mixture of pear enzymes,

Table 2 Oxidative transformations of caffeic acid **(II)** and methyl caffeate **(III)**.

Reagents	Experimental conditions	Metabolites obtained	References
Biorelevant			
Iron-porphyrin Fe(TDCPPS)Cl, H_2O_2	MeOH(aq)	**II/1–4**	Šmejkalová and Piccolo (2006) and Šmejkalová, Piccolo, and Spiteller (2006)
Peroxynitrite	50 mM phosphate buffer, pH 7	**II/1**	Rice-Evans and Miller (1996)
Peroxynitrite	Continuous-flow, $ONOO^-$ prepared in situ, NaCl and glycine buffer present	**III/6**	Fási et al. (2020)
Fenton oxidation	Aq. solution of Fe^{2+}/EDTA (1:1) and 3% H_2O_2	**II/5**, **II/6**	Antolovich et al. (2004)
2,2′-Azobis(2,4-dimethylvaleronitrile) (AMVN)	CH_3CN, reaction in the presence and absence of ethyl linoleate, 37 °C	**III/1–5**	Masuda et al. (2014)
AAPH	CH_3CN–water (2.5:1, v/v), 60 °C	**III/6**, **III/7**	Fási et al. (2020)
Biomimetic			
Oxygen	Bubbled through aqueous solution containing KOH	**II/2**, **II/3**, **II7–10**	Cilliers and Singleton (1991)
$NaIO_4$	Aqueous solution ranging from pH 2 to 7	**II/11**, **II/12**	Fulcrand, Cheminat, Brouillard, and Cheynier (1994)
$NaIO_4$	Citric acid solution (pH 3.5), reaction in the absence and presence of L-cysteine, NaOH to adjust the pH	**II/1**, **II/13**	Bassil, Makris, and Kefalas (2005)
$NaIO_4$	CH_3CN–water (5:1, v/v)	**II/5**, **II/6**	Antolovich et al. (2004)
$NaIO_4$	Water	**II/14**, **III/8**	Tazaki, Kawabata, and Fujita (2003)
$FeCl_3$	Caffeic acid in EtOH with the oxidant dissolved in water	**II/15**	Masuda et al. (2014)

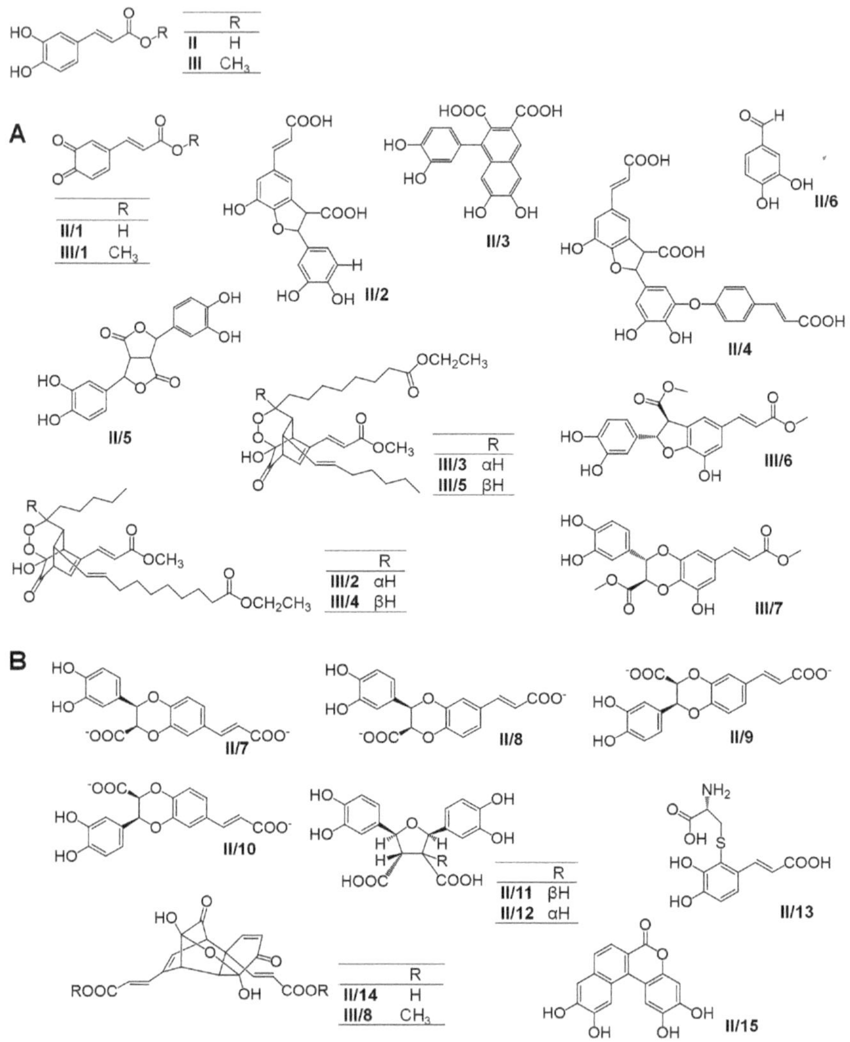

Fig. 3 Oxidized caffeic acid or methyl caffeate derivatives formed in reactions with biorelevant (A) or biomimetic (B) oxidants.

and this study revealed the enhanced COX-2 inhibitory activity of **II/7** ($IC_{50}=7.5\,\mu M$) and **II/8** ($IC_{50}=25.6\,\mu M$) as compared to the inactive parent compound **II** (Bae & Kim, 2012). Among the oxidized metabolites of caffeic acid derivatives, perhaps the most significant bioactivity changes were observed for compound **III/6**. This dihydrobenzofuran lignan was formed when methyl caffeate (**III**) was oxidized in biorelevant (Fási et al., 2020) or biomimetic (Pieters et al., 1999) experimental setups. Compound **III/6** was identified as a highly promising antitumor agent in various in vitro and in vivo experimental models. Its 2*R*,3*R* enantiomer (but not the inactive 2*S*,3*S*) acted as an anti-tubulin agent as potent ($IC_{50} = 6.0\,\mu M$) as combretastatin A-4 (Pieters et al., 1999), and exerted strong antiangiogenic activity apparently without interfering with fibroblast growth factor-2 (FGF-2) or vascular endothelial growth factor (VEGF) (Apers et al., 2002). Further, the racemate of **III/6** was found to suppress 4T1 tumor metastasis in mice at a dose of 100 μg/kg through increasing the interleukin-25 (IL-25) secretion of tumor-associated fibroblasts (Yin et al., 2016). In our previous work, we investigated the possible formation of **III/6** in HeLa, SiHa, MCF-7 and MDA-MB-231 cancer cells treated with **III** and incubated with or without *t*-BuOOH-induced oxidative stress. While no direct evidence was found for an in-situ formation of **III/6**, the cytotoxic activity of **III** was altered by oxidative stress in a way that coincided with the cell line specificity of compound **III/6** (Fási et al., 2020).

3.3 Quercetin (IV)

Quercetin is one of the most common dietary flavonols due to its widespread occurrence in many common fruits and vegetables, including apples, blueberries, cherries, onions, lettuce, Chinese cabbage, etc., as well as in popular beverages like black tea and red wine (Dabeek & Marra, 2019). The free radical scavenging properties of quercetin have been thoroughly studied; it is well known to possess ROS/RNS scavenging properties due to (i) the *o*-dihydroxy (catechol) structure in its B ring; (ii) the 2,3-double bond conjugated with the 4-keto function; and (iii) the additional presence of both 3- and 5-OH groups (Buchner, Krumbein, Rohn, & Kroh, 2006; Makris & Rossiter, 2002a, 2002b; Rice-Evans & Miller, 1996). The oxidation of quercetin has been studied in many experimental models including enzymatic, electrochemical, chemical, and microbial systems, and a wide spectrum of products were identified, arising from dimerization, skeleton alteration and/or decomposition (Zhou & Sadik, 2008). Herein, we discuss the chemical models; experimental setups are summarized in Table 3, and structures of the metabolites are shown in Fig. 4.

When studying the effect of L-ascorbic acid (Bobolaki et al., 2018) and L-cysteine (Photiades et al., 2020) on the Cu^{2+}-induced oxidation of quercetin (**IV**) under several experimental conditions (pH and temperature), three oxidized products (**IV/5**, **IV/6** and **IV/1**) were tentatively identified

Table 3 Oxidative transformations of quercetin (**IV**).

Reagents	Experimental conditions	Metabolites obtained	References
Biorelevant			
Cu^{2+}	Phosphate-citrate buffer solution (pH 7.5), reaction in the presence of L-ascorbic acid	**IV/1**	Bobolaki, Photiades, Grigorakis, and Makris (2018)
Cu^{2+}	Phosphate-citrate buffer solution (pH 7.5), reaction in the presence and absence of L-cysteine	**IV/1**, **IV/5–8**	Photiades, Grigorakis, and Makris (2020)
DPPH	MeOH	**IV/2**	Furusawa et al. (2003)
Biomimetic			
2,2′-azobis-isobutyronitrile (AIBN)	CH_3CN, 60 °C	**IV/1–4**	Krishnamachari, Levine, and Paré (2002)
$NaNO_2$	KCl-HCl solution (pH 1–3)	**IV/1**	Takahama, Yamamoto, Hirota, and Oniki (2003)
$K_3[Fe(CN)_6]$	Aqueous phosphate buffer (0.05 M, pH 7.4)	**IV/2**, **IV/9**	Chervyakovsky et al. (2008)
t-BuOOH	*t*-BuOH–water (2:1, *v*/v), reaction in NaOH, neutralized with sodium phosphate buffer (pH 7.4)	**IV/2**	Mallepu, Potlapally, and Gollapalli (2018)
$K_3[Fe(CN)_6]$, $NaHCO_3$	70% CH_3CN	**IV/6**, other unspecified products	Makris and Rossiter (2002a, 2002b)
$CuSO_4/H_2O_2$ $NaIO_4$	70% aq. CH_3CN	**IV/6**, other unspecified products	Makris and Rossiter (2002b)

Table 3 Oxidative transformations of quercetin (**IV**).—cont'd

Reagents	Experimental conditions	Metabolites obtained	References
Fe^{3+}, Fe^{2+} and Cu^{2+}	70:30 (v/v) mixtures of aqueous acetic acid and different organic modifiers; (CH_3CN, CH_3OH and EtOH)	**IV/1**	Jungbluth, Rühling, and Ternes (2000)
$CuSO_4$	CH_3CN–water (8:2, v/v), pH adjusted from 4.5 to 2 at the end of the reaction	**IV/1**	Gülşen, Makris, and Kefalas (2007) and Gülşen, Turan, Makris, and Kefalas (2007)
$K_3[Fe(CN)_6]$, Na_2CO_3	CH_3CN	**IV1**, **IV/2**	Gülşen, Makris, and Kefalas (2007) and Gülşen, Turan, et al. (2007)
Air	NaOH (pH 12)	**IV/1**, **IV/5**, **IV/6**, **IV/11–15**	Fuentes, Atala, Pastene, Carrasco-Pozo, and Speisky (2017)
$[AuCl_4]^-$	0.1 M $HClO_4$ aqueous solution (1:1, v:v)	**IV/11**, **IV/19**, **IV/21**	Bondžić et al. (2013)
Biomimetic-related chemical			
$CuC1_2$	MeOH and EtOH	**IV/16–21**	Utaka and Takeda (1985)
Ru + Ag + NaOH $PdCl_2$ K_2OsCl_6 Solid Pt and Au in aqua regia.	MeOH	**IV/1**	Balcerzak, Kopacz, Kosiorek, Swiecicka, and Kus (2004)
$PbCl_2$	MeOH	**IV/1**	Cornard, Dangleterre, and Lapouge (2005)

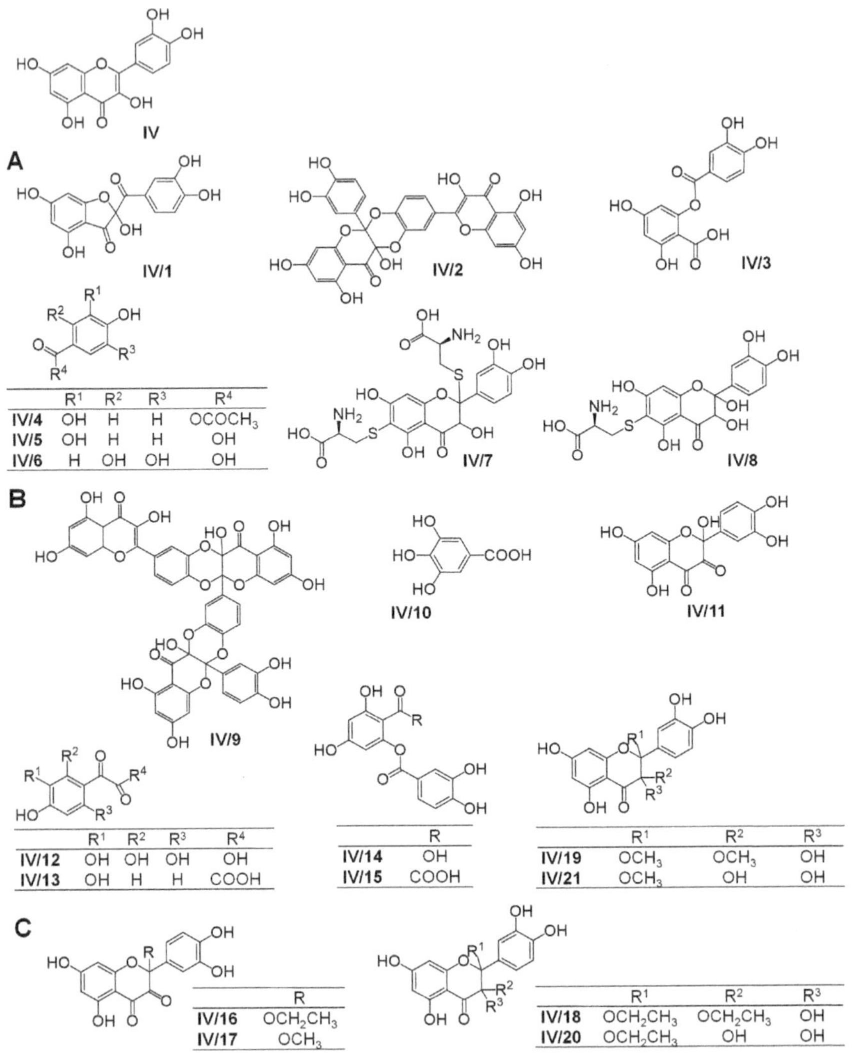

Fig. 4 Oxidized quercetin derivatives formed in reactions with biorelevant (A), biomimetic (B) or biomimetic-related chemical oxidants (C).

by LC-DAD-MS. The product profile was pH-dependent, and when the oxidation was performed in presence of L-cysteine (Photiades et al., 2020), two tentatively assigned quercetin/L-cysteine adducts (**IV/7** and IV/**8**) were also observed. It was concluded that both ascorbyl and cysteinyl radicals may take an active part in the oxidation of quercetin. In view of the scavengome concept, this is an interesting example for the chemical environment significantly affecting the product profile connected to ROS/RNS scavenging by a small-molecule antioxidant. The dimer **IV/2** was observed as a main oxidation product in a photooxidation experiment, in which quercetin was oxidized by in situ formed *tert*-butoxyl radicals (*t*-BuO$^•$) in a solvent mixture of *t*-BuOH water (2:1 *v*/v) containing *t*-BuOOH (Mallepu et al., 2018). Since the excess *t*-BuOH supposedly scavenged the $^•$OH radicals also formed from the photolysis, results of this experiment suggest that such a dimerization of quercetin may also take place when it reacts with alkoxyl radicals. This dimer **IV/2** is also naturally present in the brownish scale of onion, and it was reported to exhibit stronger anti-platelet activity than quercetin (**IV**), and to rigidify liposomal and surface membrane bilayers (Furusawa et al., 2003). The membrane activity of **IV/2** may also be the reason for its increased anti-proliferative activity as compared to that of **IV**, based on its enhanced capacity to altering essential physicochemical properties of tumor cells needed for their proliferation, metastasis and malignancy (Furusawa et al., 2006). Interestingly, several oxidized products of quercetin act as more potent antioxidants than quercetin itself (Vásquez-Espinal et al., 2019). In particular, compound **IV/1** was shown to exhibit improved antioxidant characteristics compared to **IV** (Gülşen, Makris, & Kefalas, 2007; Gülşen, Turan, et al., 2007). Further, it is of particular interest that a whole mixture of autoxidized quercetin metabolites was found to exert a ca. 20-fold stronger cytoprotective activity than quercetin on Hs68 and Caco2 cells exposed to indomethacin, and a metabolite, identified by ESI-MS/MS as **IV/1**, was 200-times stronger cytoprotective agent than its parent compound (Fuentes et al., 2017). These results strongly support the notion that the oxidized metabolites of quercetin are of potential value for drug discovery.

4. Drug discovery value of the chemical metabolite space of I–IV

To place the selected four antioxidants and their oxidized derivatives in the field of medicinal chemistry, we studied them by a popular method in cheminformatics, the t-distributed stochastic neighbor embedding (t-SNE) (Van der Maaten & Hinton, 2008) for dimensional reduction and visualization. The t-SNE defines a probabilistic distribution of the data points in the high dimensional space, in our case in the space of extended-connectivity fingerprints (ECFP4) (Rogers & Hahn, 2010), then defines a similar probability distribution in the low dimensional space. In the last step it minimizes

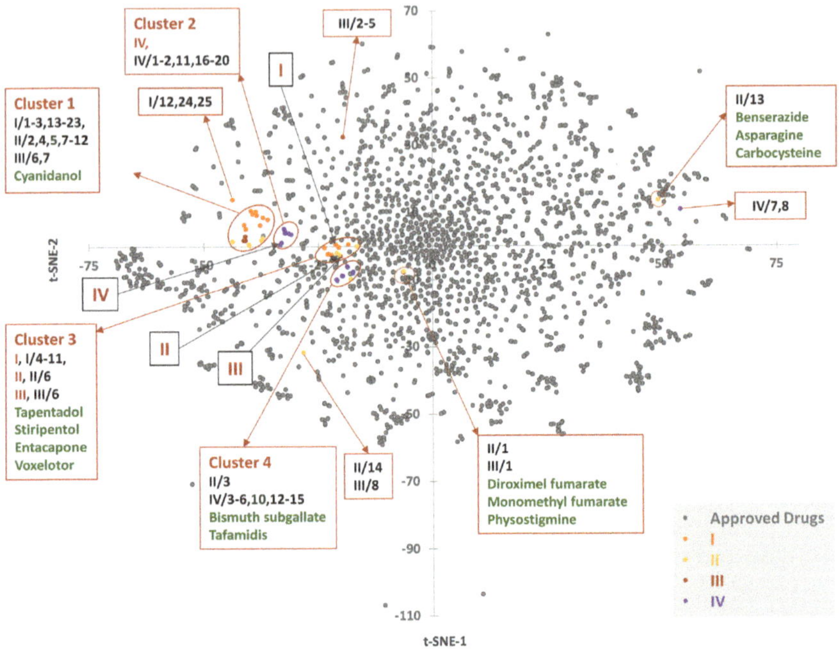

Fig. 5 t-SNE plot of compounds **I–IV**, their oxidized derivatives, and approved drugs with a molecular mass below 700 Da. A total of 2378 approved drugs are plotted that were retrieved from the DrugBank database. Each sphere represents a compound, and color codes are assigned to approved drugs (gray), and compound groups **I**, **II**, **III**, and **IV** (orange, yellow, red, and purple, respectively).

the Kullback-Leibler divergence between the two distributions and provides a 2D projection of the high-dimensional space. In this projection, we can state that objects similar in the high dimensional space fall close to each other in the low dimensional space with high probability. Fig. 5 shows the t-SNE plot of the distribution of compounds **I–IV** and their metabolites over the chemical space of approved drugs (Wishart et al., 2017) that have a molar mass below 700 Da, a total of 2378 compounds. Groups I to IV contain parent compounds and derivatives of resveratrol (**I**), caffeic acid (**II**), methylcaffeate (**III**), and quercetin (**IV**). Parent compounds are also highlighted separately in Fig. 5 for an easier identification.

The investigated antioxidant compounds and their derivatives could be classified into four clusters and six additional small groups or singletons based on their structural similarity on the t-SNE map. Cluster groups that are located far from the origin of the t-SNE plot and in a space underrepresented by approved drugs can be considered unique, and thus to represent a new medicinal chemistry space. A good example for this is Cluster 1, which included only one known drug, Cyanidanol ((+)–catechin). This part of the t-SNE space is overall underrepresented. Another structurally valuable group is Cluster 2, which is close to Cluster 1 and contains only quercetin (**IV**) and its derivatives without any approved drugs. Clusters 3 and 4 are already much closer to the origin of the t-SNE plot, so that several structurally similar drugs could be included in them. It should be noted that Cluster 3 contains three of the parent compounds (**I–III**) and the active substances Tapentadol (opioid analgesic) (Tzschentke et al., 2006), Stiripentol (anticonvulsant) (Quilichini, Chiron, Ben-Ari, & Gozlan, 2006), Entacapone (reversible inhibitor of COMT) (Chong & Mersfelder, 2000), and Voxelotor (hemoglobin oxygen-affinity modulator) (Hutchaleelaha et al., 2019); this predicts a variety of expectable pharmacological effects for compounds placed within this cluster. In terms of structural exclusivity, it is important to highlight the caffeic acid derivatives **II/14** and **III/8**, which are also located in the under-represented part of the t-SNE space. Their chemical exclusivity is presumably explained by the presence of a rigid heterocycle formed by oxidative cyclization and the relatively high number of oxygen atoms. Similarly, methyl-caffeate derivatives (**III/2–5**) and quercetin derivatives (**IV/7–8**) were identified as small groups providing new chemical entities. However, the cysteine-coupled derivative of caffeic acid (**II/13**) was identified as a chemically represented independent singleton in a cluster with Benserazide (DOPA decarboxylase inhibitor) (Shen, Kannari, Yamato, Arai, & Matsunaga, 2003), Asparagine, and Carbocysteine (mucolytic) (Zheng et al., 2008). Furthermore, the quinone derivatives of **II** and **III** (**II/1**, **III/1**) were identified as small groups, and their cluster group includes Diroximel fumarate (immunosuppressant) (Kourakis et al., 2020), Monomethyl fumarate (alters the NFE2L2 transcription factor) (Kourakis et al., 2020), and Physostigmine (reversible cholinesterase inhibitor) (Arens, Shah, Al-Abri, Olson, & Kearney, 2018). Overall, the t-SNE map shows that although three of the four antioxidant parent compounds are in a common cluster (Cluster 1: **I–III**), their derivatives are scattered on the t-SNE map and distributed in a diverse way in the drug-like chemical space. Based on their appearance in several underrepresented areas of the t-SNE map and on the identified drugs in their cluster groups, a variety of pharmacological effects are expected from derivatives of compounds **I–IV** formed by biomimetic oxidation.

5. Summary

The scavengome concept focuses on the drug discovery potential of metabolites that may be formed when small-molecule antioxidants scavenge reactive oxygen or nitrogen species. Naturally, any attempt to systematically explore such metabolites' chemical space requires an extensive use of biomimetic oxidative chemistry from a diversity-oriented perspective. It is also necessary to mention that hardly anything is currently known about the true biological relevance of ROS/RNS-oxidized metabolites of antioxidants, which may be formed at very low concentrations in vivo. A given biological microenvironment's impact on the reactions' chemoselectivity is also mostly unknown, it is therefore hard to decipher which metabolite patterns may be expressed in vivo and at what specificity concerning cellular compartments and cell types, tissues, and organs. Accordingly, even an in-depth study of the scavengome may have significant challenges and limitations in providing new and meaningful explanations to the complex bioactivity of the parent antioxidant.

The examples discussed above demonstrate that antioxidants' transformation by a variety of biorelevant, biomimetic, or biomimetic-related chemical oxidants yields various product patterns and therefore significantly expands chemical space. At the same time, the high diversity of such product patterns is accompanied by a considerable overlap between products that are achievable through different chemical approaches, i.e., several less biomimetic oxidative conditions lead to the formation of products that are common with those formed in fully biorelevant models. This confers a plasticity to possible yield optimization initiatives for the selective preparation of any potential lead compounds identified from the scavengome.

The very high increase in the chemical complexity of oxidized metabolites as compared with that of their parent antioxidants (Figs. 2–4), and their spreading in the drug-like chemical space (Fig. 5) strongly suggest a great drug discovery value for such compounds. Accordingly, it may be postulated that a diversity-oriented exploration of the scavengome of antioxidants is a valid strategy to achieve an increase in biological performance diversity, and that using the scavengome concept as a guiding principle may serve as a novel strategy for antioxidant-inspired drug discovery.

Acknowledgments

The authors acknowledge support from the National Research, Development, and Innovation Office (NKFIH), Hungary (K-134704), and project no. TKP2021-EGA-32, implemented with the support provided by the Ministry of Innovation and Technology of Hungary from the NKFIH, financed under the TKP2021-EGA funding scheme, and project no. 1010984 implemented with the support provided by the NKFIH, financed under the KDP-2020 funding scheme.

References

Antolovich, M., Bedgood, D. R., Jr., Bishop, A. G., Jardine, D., Prenzler, P. D., & Robards, K. (2004). LC-MS investigation of oxidation products of phenolic antioxidants. *Journal of Agricultural and Food Chemistry*, *52*(4), 962–971. https://doi.org/10.1021/jf0349883.

Apers, S., Paper, D., Bürgermeister, J., Baronikova, S., Van Dyck, S., Lemière, G., et al. (2002). Antiangiogenic activity of synthetic dihydrobenzofuran lignans. *Journal of Natural Products*, *65*(5), 718–720. https://doi.org/10.1021/np0103968.

Arens, A. M., Shah, K., Al-Abri, S., Olson, K. R., & Kearney, T. (2018). Safety and effectiveness of physostigmine: A 10-year retrospective review. *Clinical Toxicology*, *56*(2), 101–107. https://doi.org/10.1080/15563650.2017.1342828.

Bae, J. S., & Kim, T. H. (2012). Enzymatic transformation of caffeic acid with enhanced cyclooxygenase-2 inhibitory activity. *Bioorganic & Medicinal Chemistry Letters*, *22*(2), 793–796. https://doi.org/10.1016/j.bmcl.2011.12.072.

Balcerzak, M., Kopacz, M., Kosiorek, A., Swiecicka, E., & Kus, S. (2004). Spectrophotometric studies of the interaction of noble metals with quercetin and quercetin-5′-sulfonic acid. *Analytical Sciences*, *20*(9), 1333–1337. https://doi.org/10.2116/analsci.20.1333.

Bartesaghi, S., & Radi, R. (2018). Fundamentals on the biochemistry of peroxynitrite and protein tyrosine nitration. *Redox Biology*, *14*, 618–625. https://doi.org/10.1016/j.redox.2017.09.009.

Bassil, D., Makris, D. P., & Kefalas, P. (2005). Oxidation of caffeic acid in the presence of l-cysteine: Isolation of 2-S-cysteinylcaffeic acid and evaluation of its antioxidant properties. *Food Research International*, *38*(4), 395–402. https://doi.org/10.1016/j.foodres.2004.10.009.

Bauer, G. (2016). The antitumor effect of singlet oxygen. *Anticancer Research*, *36*(11), 5649–5663. https://doi.org/10.21873/anticanres.11148.

Bauer, G. (2018). HOCl and the control of oncogenesis. *Journal of Inorganic Biochemisstry*, *179*, 10–23. https://doi.org/10.1016/j.jinorgbio.2017.11.005.

Baur, J. A., & Sinclair, D. A. (2006). Therapeutic potential of resveratrol: The in vivo evidence. *Nature Reviews. Drug Discovery*, *5*(6), 493–506. https://doi.org/10.1038/nrd2060.

Beaumont, P., Courtois, A., Atgié, C., Richard, T., & Krisa, S. (2022). In the shadow of resveratrol: Biological activities of epsilon-viniferin. *Journal of Physiology and Biochemistry*. https://doi.org/10.1007/s13105-022-00880-x.

Birková, A. (2020). Caffeic acid: A brief overview of its presence, metabolism, and bioactivity. *Bioactive Compounds in Health and Disease*, *3*, 74. https://doi.org/10.31989/bchd.v3i4.692.

Bobolaki, N., Photiades, A., Grigorakis, S., & Makris, D. P. (2018). Empirical kinetic modelling of the effect of l-ascorbic acid on the cu(II)-induced oxidation of quercetin. *ChemEngineering*, *2*(4), 46. https://doi.org/10.3390/chemengineering2040046.

Bondžić, A. M., Lazarević-Pašti, T. D., Bondžić, B. P., Čolović, M. B., Jadranin, M. B., & Vasić, V. M. (2013). Investigation of reaction between quercetin and Au(iii) in acidic media: Mechanism and identification of reaction products. *New Journal of Chemistry*, *37*(4), 901–908. https://doi.org/10.1039/C2NJ40742F.

Bonechi, C., Martini, S., Magnani, A., & Rossi, C. (2008). Stacking interaction study of trans-resveratrol (trans-3,5,4′-trihydroxystilbene) in solution by nuclear magnetic resonance and Fourier transform infrared spectroscopy. *Magnetic Resonance in Chemistry*, *46*(7), 625–629. https://doi.org/10.1002/mrc.2217.

Buchner, N., Krumbein, A., Rohn, S., & Kroh, L. W. (2006). Effect of thermal processing on the flavonols rutin and quercetin. *Rapid Communications in Mass Spectrometry*, *20*(21), 3229–3235. https://doi.org/10.1002/rcm.2720.

Carvalho, L. C., Vidigal, P., & Amâncio, S. (2015). Oxidative stress homeostasis in grapevine (Vitis vinifera L.). *Frontiers in Environmental Science*, *3*. https://doi.org/10.3389/fenvs.2015.00020.

Chervyakovsky, E. M., Bolibrukh, D. A., Kurovskii, D. L., Gilep, A. A., Vlasova, T. M., Kurchenko, V. P., et al. (2008). Oligomeric oxidation products of the flavonoid quercetin. *Chemistry of Natural Compounds*, *44*, 427–431. https://doi.org/10.1007/s10600-008-9092-1.

Chong, B. S., & Mersfelder, T. L. (2000). Entacapone. *Annals of Pharmacotherapy*, *34*(9), 1056–1065. https://doi.org/10.1345/aph.19328.

Cilliers, J. J. L., & Singleton, V. L. (1990). Caffeic acid autoxidation and the effects of thiols. *Journal of Agricultural and Food Chemistry*, *38*(9), 1789–1796. https://doi.org/10.1021/jf00099a002.

Cilliers, J. J. L., & Singleton, V. L. (1991). Characterization of the products of nonenzymic autoxidative phenolic reactions in a caffeic acid model system. *Journal of Agricultural and Food Chemistry*, *39*(7), 1298–1303. https://doi.org/10.1021/jf00007a021.

Cornard, J. P., Dangleterre, L., & Lapouge, C. (2005). Computational and spectroscopic characterization of the molecular and electronic structure of the Pb(II)−quercetin complex. *The Journal of Physical Chemistry A*, *109*(44), 10044–10051. https://doi.org/10.1021/jp053506i.

Costa, C. A. E., Vega-Aguilar, C. A., & Rodrigues, A. E. (2021). Added-value chemicals from lignin oxidation. *Molecules*, *26*(15), 4602. https://doi.org/10.3390/molecules26154602.

Dabeek, W. M., & Marra, M. V. (2019). Dietary quercetin and Kaempferol: Bioavailability and potential cardiovascular-related bioactivity in humans. *Nutrients*, *11*(10), 2288. https://doi.org/10.3390/nu11102288.

Dickinson, B. C., & Chang, C. J. (2011). Chemistry and biology of reactive oxygen species in signaling or stress responses. *Nature Chemical Biology*, 7(8), 504–511. https://doi.org/10.1038/nchembio.607.

Dybkowska, E., Sadowska, A., Świderski, F., Rakowska, R., & Wysocka, K. (2018). The occurrence of resveratrol in foodstuffs and its potential for supporting cancer prevention and treatment. A review. *Rocz Panstw Zakl Hig (Annals of the National Institute of Hygiene)*, *69*(1), 5–14 (PMID: 29517181).

El Khawand, T., Valls Fonayet, J., Da Costa, G., Hornedo-Ortega, R., Jourdes, M., Franc, C., et al. (2020). Resveratrol transformation in red wine after heat treatment. *Food Research International*, *132*, 109068. https://doi.org/10.1016/j.foodres.2020.109068.

Fási, L., Latif, A. D., Zupkó, I., Lévai, S., Dékány, M., Béni, Z., et al. (2020). AAPH or peroxynitrite-induced biorelevant oxidation of methyl caffeate yields a potent antitumor metabolite. *Biomolecules*, *10*(11). https://doi.org/10.3390/biom10111537.

Ferrer-Sueta, G., Campolo, N., Trujillo, M., Bartesaghi, S., Carballal, S., Romero, N., et al. (2018). Biochemistry of peroxynitrite and protein tyrosine nitration. *Chemical Reviews*, *118*(3), 1338–1408. https://doi.org/10.1021/acs.chemrev.7b00568.

Ficarra, S., Tellone, E., Pirolli, D., Russo, A., Barreca, D., Galtieri, A., et al. (2016). Insights into the properties of the two enantiomers of trans-δ-viniferin, a resveratrol derivative: Antioxidant activity, biochemical and molecular modeling studies of its interactions with hemoglobin. *Molecular BioSystems*, *12*(4), 1276–1286. https://doi.org/10.1039/c5mb00897b.

Forman, H. J., Davies, K. J., & Ursini, F. (2014). How do nutritional antioxidants really work: Nucleophilic tone and Para-hormesis versus free radical scavenging in vivo.

Free Radical Biology & Medicine, *66*, 24–35. https://doi.org/10.1016/j.freeradbiomed.2013.05.045.

Fra, A., Yoboue, E. D., & Sitia, R. (2017). Cysteines as redox molecular switches and targets of disease. *Frontiers in Molecular Neuroscience*, *10*. https://doi.org/10.3389/fnmol.2017.00167.

Fuentes, J., Atala, E., Pastene, E., Carrasco-Pozo, C., & Speisky, H. (2017). Quercetin oxidation paradoxically enhances its antioxidant and Cytoprotective properties. *Journal of Agricultural and Food Chemistry*, *65*(50), 11002–11010. https://doi.org/10.1021/acs.jafc.7b05214.

Fulcrand, H., Cheminat, A., Brouillard, R., & Cheynier, V. (1994). Characterization of compounds obtained by chemical oxidation of caffeic acid in acidic conditions. *Phytochemistry*, *35*(2), 499–505. https://doi.org/10.1016/S0031-9422(00)94790-3.

Furusawa, M., Tsuchiya, H., Nagayama, M., Tanaka, T., Nakaya, K.-I., & Iinuma, M. (2003). Anti-platelet and membrane-rigidifying flavonoids in brownish scale of onion. *Journal of Health Science*, *49*(6), 475–480. https://doi.org/10.1248/jhs.49.475.

Furusawa, M., Tsuchiya, H., Nagayama, M., Tanaka, T., Oyama, M., Ito, T., et al. (2006). Cell growth inhibition by membrane-active components in brownish scale of onion. *Journal of Health Science*, *52*(5), 578–584. https://doi.org/10.1248/jhs.52.578.

Galletti, P., Martelli, G., Prandini, G., Colucci, C., & Giacomini, D. (2018). Sodium periodate/TEMPO as a selective and efficient system for amine oxidation. *RSC Advances*, *8*(18), 9723–9730. https://doi.org/10.1039/C8RA01365A.

Gough, D. R., & Cotter, T. G. (2011). Hydrogen peroxide: A Jekyll and Hyde signalling molecule. *Cell Death & Disease*, *2*, e213. https://doi.org/10.1038/cddis.2011.96.

Gülşen, A., Makris, D. P., & Kefalas, P. (2007). Biomimetic oxidation of quercetin: Isolation of a naturally occurring quercetin heterodimer and evaluation of its in vitro antioxidant properties. *Food Research International*, *40*(1), 7–14. https://doi.org/10.1016/j.foodres.2006.07.009.

Gülşen, A., Turan, B., Makris, D. P., & Kefalas, P. (2007). Copper(II)-mediated biomimetic oxidation of quercetin: Generation of a naturally occurring oxidation product and evaluation of its in vitro antioxidant properties. *European Food Research and Technology*, *225*, 435–441. https://doi.org/10.1007/s00217-006-0437-3.

Hayyan, M., Hashim, M. A., & AlNashef, I. M. (2016). Superoxide ion: Generation and chemical implications. *Chemical Reviews*, *116*(5), 3029–3085. https://doi.org/10.1021/acs.chemrev.5b00407.

Heleno, S. A., Martins, A., Queiroz, M. J., & Ferreira, I. C. (2015). Bioactivity of phenolic acids: Metabolites versus parent compounds: A review. *Food Chemistry*, *173*, 501–513. https://doi.org/10.1016/j.foodchem.2014.10.057.

Holthoff, J. H., Woodling, K. A., Doerge, D. R., Burns, S. T., Hinson, J. A., & Mayeux, P. R. (2010). Resveratrol, a dietary polyphenolic phytoalexin, is a functional scavenger of peroxynitrite. *Biochemical Pharmacology*, *80*(8), 1260–1265. https://doi.org/10.1016/j.bcp.2010.06.027.

Hunyadi, A. (2019). The mechanism(s) of action of antioxidants: From scavenging reactive oxygen/nitrogen species to redox signaling and the generation of bioactive secondary metabolites. *Medicinal Research Reviews*, *39*(6), 2505–2533. https://doi.org/10.1002/med.21592.

Hutchaleelaha, A., Patel, M., Washington, C., Siu, V., Allen, E., Oksenberg, D., et al. (2019). Pharmacokinetics and pharmacodynamics of voxelotor (GBT440) in healthy adults and patients with sickle cell disease. *British Journal of Clinical Pharmacology*, *85*(6), 1290–1302. https://doi.org/10.1111/bcp.13896.

Jeandet, P. (2015). Phytoalexins: Current progress and future prospects. *Molecules*, *20*(2), 2770–2774. https://doi.org/10.3390/molecules20022770.

Jungbluth, G., Rühling, I., & Ternes, W. (2000). Oxidation of flavonols with Cu(II), Fe(II) and Fe(III) in aqueous media. *Journal of the Chemical Society, Perkin Transactions*, *2*(9), 1946–1952. https://doi.org/10.1039/B002323J.

Karlsson, J., Emgard, M., Brundin, P., & Burkitt, M. J. (2000). Trans-resveratrol protects embryonic mesencephalic cells from tert-butyl hydroperoxide: electron paramagnetic resonance spin trapping evidence for a radical scavenging mechanism. *Journal of Neurochemistry*, *75*(1), 141–150. https://doi.org/10.1046/j.1471-4159.2000.0750141.x.

Kato, Y., & Suga, N. (2018). Covalent adduction of endogenous and food-derived quinones to a protein: Its biological significance. *Journal of Clinical Biochemistry and Nutrition*, *62*(3), 213–220. https://doi.org/10.3164/jcbn.18-26.

Kerimi, A., & Williamson, G. (2018). Differential impact of flavonoids on redox modulation, bioenergetics, and cell signaling in Normal and tumor cells: A comprehensive review. *Antioxidants & Redox Signaling*, *29*(16), 1633–1659. https://doi.org/10.1089/ars.2017.7086.

Kourakis, S., Timpani, C. A., de Haan, J. B., Gueven, N., Fischer, D., & Rybalka, E. (2020). Dimethyl fumarate and its esters: A drug with broad clinical utility? *Pharmaceuticals*, *13*(10), 306. https://doi.org/10.3390/ph13100306.

Krishnamachari, V., Levine, L. H., & Paré, P. W. (2002). Flavonoid oxidation by the radical generator AIBN: A unified mechanism for quercetin radical scavenging. *Journal of Agricultural and Food Chemistry*, *50*(15), 4357–4363. https://doi.org/10.1021/jf020045e.

Li, X.-M., Huang, K.-S., Lin, M., & Zhou, L.-X. (2003). Studies on formic acid-catalyzed dimerization of isorhapontigenin and of resveratrol to tetralins. *Tetrahedron*, *59*(24), 4405–4413. https://doi.org/10.1016/S0040-4020(03)00623-9.

López-Alarcón, C., & Denicola, A. (2013). Evaluating the antioxidant capacity of natural products: A review on chemical and cellular-based assays. *Analytica Chimica Acta*, *763*, 1–10. https://doi.org/10.1016/j.aca.2012.11.051.

Makris, D. P., & Rossiter, J. T. (2002a). An investigation on structural aspects influencing product formation in enzymic and chemical oxidation of quercetin and related flavonols. *Food Chemistry*, *77*(2), 177–185. https://doi.org/10.1016/S0308-8146(01)00333-8.

Makris, D. P., & Rossiter, J. T. (2002b). Hydroxyl free radical-mediated oxidative degradation of quercetin and morin: A preliminary investigation. *Journal of Food Composition and Analysis*, *15*(1), 103–113. https://doi.org/10.1006/jfca.2001.1030.

Mallepu, R., Potlapally, L., & Gollapalli, V. L. (2018). Photo-oxidation of some flavonoids with photochemically generated t-BuO• radicals in a t-BuOH water system using a kinetic approach. *Journal of the Chinese Chemical Society*, *65*(10), 1266–1273. https://doi.org/10.1002/jccs.201700342.

Marano, S., Minnelli, C., Ripani, L., Marcaccio, M., Laudadio, E., Mobbili, G., et al. (2021). Insights into the antioxidant mechanism of newly synthesized Benzoxazinic Nitrones: In vitro and in silico studies with DPPH model radical. *Antioxidants*, *10*(8), 1224. https://doi.org/10.3390/antiox10081224.

Masuda, T., Shingai, Y., Takahashi, C., Inai, M., Miura, Y., Honda, S., et al. (2014). Identification of a potent xanthine oxidase inhibitor from oxidation of caffeic acid. *Free Radical Biology & Medicine*, *69*, 300–307. https://doi.org/10.1016/j.freeradbiomed.2014.01.016.

Mazoir, N., Benharref, A., Vaca, L., Reina, M., & González-Coloma, A. (2020). Optimization of insecticidal triterpene derivatives by biomimetic oxidations with hydrogen peroxide and Iodosobenzene catalyzed by MnIII and FeIII porphyrin complexes. *Chemistry & Biodiversity*, *17*(9), e2000287. https://doi.org/10.1002/cbdv.202000287.

Miller, C. J., Chang, Y., Wegeberg, C., McKenzie, C. J., & Waite, T. D. (2021). Kinetic analysis of H2O2 activation by an Iron(III) complex in water reveals a Nonhomolytic generation pathway to an Iron(IV)oxo complex. *ACS Catalysis*, *11*(2), 787–799. https://doi.org/10.1021/acscatal.0c02877.

Nimse, S. B., & Pal, D. (2015). Free radicals, natural antioxidants, and their reaction mechanisms. *RSC Advances*, *5*(35), 27986–28006. https://doi.org/10.1039/C4RA13315C.

Ozyurt, V. H., & Otles, S. (2020). Investigation of the effect of sodium nitrite on protein oxidation markers in food protein suspensions. *Journal of Food Biochemistry*, *44*(3), e13152. https://doi.org/10.1111/jfbc.13152.

Panzella, L., De Lucia, M., Amalfitano, C., Pezzella, A., Evidente, A., Napolitano, A., et al. (2006). Acid-promoted reaction of the stilbene antioxidant resveratrol with nitrite ions: Mild phenolic oxidation at the 4'-hydroxystiryl sector triggering nitration, dimerization, and aldehyde-forming routes. *The Journal of Organic Chemistry*, *71*(11), 4246–4254. https://doi.org/10.1021/jo060482i.

Paulsen, C. E., & Carroll, K. S. (2013). Cysteine-mediated redox signaling: Chemistry, biology, and tools for discovery. *Chemical Reviews*, *113*(7), 4633–4679. https://doi.org/10.1021/cr300163e.

Pavlinov, I., Gerlach, E. M., & Aldrich, L. N. (2019). Next generation diversity-oriented synthesis: A paradigm shift from chemical diversity to biological diversity. *Organic & Biomolecular Chemistry*, *17*(7), 1608–1623. https://doi.org/10.1039/C8OB02327A.

Pezet, R., Perret, C., Jean-Denis, J. B., Tabacchi, R., Gindro, K., & Viret, O. (2003). δ-Viniferin, a resveratrol Dehydrodimer: One of the major stilbenes synthesized by stressed grapevine leaves. *Journal of Agricultural and Food Chemistry*, *51*(18), 5488–5492. https://doi.org/10.1021/jf030227o.

Photiades, A., Grigorakis, S., & Makris, D. P. (2020). Kinetics and modeling of L-cysteine effect on the Cu(II)-induced oxidation of quercetin. *Chemical Engineering Communications*, *207*(2), 139–152. https://doi.org/10.1080/00986445.2019.1574767.

Pieters, L., Van Dyck, S., Gao, M., Bai, R., Hamel, E., Vlietinck, A., et al. (1999). Synthesis and biological evaluation of dihydrobenzofuran lignans and related compounds as potential antitumor agents that inhibit tubulin polymerization. *Journal of Medicinal Chemistry*, *42*(26), 5475–5481. https://doi.org/10.1021/jm990251m.

Płotka-Wasylka, J., Frankowski, M., Simeonov, V., Polkowska, Ż., & Namieśnik, J. (2018). Determination of metals content in wine samples by inductively coupled plasma-mass spectrometry. *Molecules*, *23*(11), 2886. https://doi.org/10.3390/molecules23112886.

Pradhan, B., Engelhard, C., Van Mulken, S., Miao, X., Canters, G. W., & Orrit, M. (2020). Single electron transfer events and dynamical heterogeneity in the small protein azurin from Pseudomonas aeruginosa. *Chemical Science*, *11*(3), 763–771. https://doi.org/10.1039/C9SC05405G.

Quilichini, P. P., Chiron, C., Ben-Ari, Y., & Gozlan, H. (2006). Stiripentol, a putative antiepileptic drug, enhances the duration of opening of GABAA-receptor channels. *Epilepsia*, *47*(4), 704–716. https://doi.org/10.1111/j.1528-1167.2006.00497.x.

Rice-Evans, C. A., & Miller, N. J. (1996). Antioxidant activities of flavonoids as bioactive components of food. *Biochemical Society Transactions*, *24*(3), 790–795. https://doi.org/10.1042/bst0240790.

Rogers, D., & Hahn, M. (2010). Extended-connectivity fingerprints. *Journal of Chemical Information and Modeling*, *50*(5), 742–754. https://doi.org/10.1021/ci100050t.

Ruskovska, T., Maksimova, V., & Milenkovic, D. (2020). Polyphenols in human nutrition: From the in vitro antioxidant capacity to the beneficial effects on cardiometabolic health and related inter-individual variability—An overview and perspective. *British Journal of Nutrition*, *123*(3), 241–254. https://doi.org/10.1017/S0007114519002733.

Sako, M., Hosokawa, H., Ito, T., & Iinuma, M. (2004). Regioselective oxidative coupling of 4-hydroxystilbenes: Synthesis of resveratrol and epsilon-viniferin (E)-dehydrodimers. *Journal of Organic Chemistry*, *69*(7), 2598–2600. https://doi.org/10.1021/jo035791c.

Shen, J., Griffiths, P. T., Campbell, S. J., Utinger, B., Kalberer, M., & Paulson, S. E. (2021). Ascorbate oxidation by iron, copper and reactive oxygen species: Review, model development, and derivation of key rate constants. *Scientific Reports*, *11*(1), 7417. https://doi.org/10.1038/s41598-021-86477-8.

Shen, L., & Ji, H. F. (2009). Insights into the inhibition of xanthine oxidase by curcumin. *Bioorganic & Medicinal Chemistry Letters*, *19*(21), 5990–5993. https://doi.org/10.1016/j.bmcl.2009.09.076.

Shen, H., Kannari, K., Yamato, H., Arai, A., & Matsunaga, M. (2003). Effects of benserazide on L-DOPA-derived extracellular dopamine levels and aromatic L-amino acid decarboxylase activity in the striatum of 6-hydroxydopamine-lesioned rats. *Tohoku Journal of Experimental Medicine*, *199*(3), 149–159. https://doi.org/10.1620/tjem.199.149.

Shingai, Y., Fujimoto, A., Nakamura, M., & Masuda, T. (2011). Structure and function of the oxidation products of polyphenols and identification of potent lipoxygenase inhibitors from Fe-catalyzed oxidation of resveratrol. *Journal of Agricultural and Food Chemistry*, *59*(15), 8180–8186. https://doi.org/10.1021/jf202561p.

Sies, H. (1993). Strategies of antioxidant defense. *European Journal of Biochemistry*, *215*(2), 213–219. https://doi.org/10.1111/j.1432-1033.1993.tb18025.x.

Šmejkalová, D., & Piccolo, A. (2006). Rates of oxidative coupling of humic phenolic monomers catalyzed by a biomimetic iron-porphyrin. *Environmental Science & Technology*, *40*(5), 1644–1649. https://doi.org/10.1021/es051760i.

Šmejkalová, D., Piccolo, A., & Spiteller, M. (2006). Oligomerization of humic phenolic monomers by oxidative coupling under biomimetic catalysis. *Environmental Science & Technology*, *40*(22), 6955–6962. https://doi.org/10.1021/es061127c.

Takahama, U., Yamamoto, A., Hirota, S., & Oniki, T. (2003). Quercetin-dependent reduction of salivary nitrite to nitric oxide under acidic conditions and interaction between quercetin and ascorbic acid during the reduction. *Journal of Agricultural and Food Chemistry*, *51*(20), 6014–6020. https://doi.org/10.1021/jf021253+.

Takatsuka, M., Goto, S., Kobayashi, K., Otsuka, Y., & Shimada, Y. (2022). Evaluation of pure antioxidative capacity of antioxidants: ESR spectroscopy of stable radicals by DPPH and ABTS assays with singular value decomposition. *Food Bioscience*, *48*, 101714. https://doi.org/10.1016/j.fbio.2022.101714.

Takaya, Y., Terashima, K., Ito, J., He, Y.-H., Tateoka, M., Yamaguchi, N., et al. (2005). Biomimic transformation of resveratrol. *Tetrahedron*, *61*(43), 10285–10290. https://doi.org/10.1016/j.tet.2005.08.023.

Tazaki, H., Kawabata, J., & Fujita, T. (2003). Novel oxidative dimer from caffeic acid. *Bioscience, Biotechnology, and Biochemistry*, *67*(5), 1185–1187. https://doi.org/10.1271/bbb.67.1185.

Teixeira, J., Gaspar, A., Garrido, E. M., Garrido, J., & Borges, F. (2013). Hydroxycinnamic acid antioxidants: An electrochemical overview. *BioMed Research International*, *2013*, 251754. https://doi.org/10.1155/2013/251754.

Touaibia, M., Jean-François, J., & Doiron, J. (2011). Caffeic acid, a versatile pharmacophore: An overview. *Mini-Reviwes in Medicinal Chemistry*, *11*(8), 695–713. https://doi.org/10.2174/138955711796268750.

Tzschentke, T. M., De Vry, J., Terlinden, R., Hennies, H. H., Lange, C., Strassburger, W., et al. (2006). Tapentadol hydrochloride: Analgesic mu-opioid receptor agonist noradrenaline reuptake inhibitor. *Drugs of the Future*, *31*(12), 1053–1061. https://doi.org/10.1358/dof.2006.031.12.1047744.

Utaka, M., & Takeda, A. (1985). Copper(II)-catalysed oxidation of quercetin and 3-hydroxyflavone. *Journal of the Chemical Society, Chemical Communications*, *24*, 1824–1826. https://doi.org/10.1039/C39850001824.

Van der Maaten, L., & Hinton, G. (2008). Visualizing data using t-SNE. *Journal of Machine Learning Research*, *9*(11), 2579–2605. https://www.jmlr.org/papers/v9/vandermaaten08a.html.

Vásquez-Espinal, A., Yañez, O., Osorio, E., Areche, C., García-Beltrán, O., Ruiz, L. M., et al. (2019). Theoretical study of the antioxidant activity of quercetin oxidation products. *Frontiers in Chemistry*, 7. https://doi.org/10.3389/fchem.2019.00818.

Velu, S. S., Buniyamin, I., Ching, L. K., Feroz, F., Noorbatcha, I., Gee, L. C., et al. (2008). Regio- and stereoselective biomimetic synthesis of oligostilbenoid dimers from resveratrol analogues: Influence of the solvent, oxidant, and substitution. *Chemistry – A European Journal*, *14*(36), 11376–11384. https://doi.org/10.1002/chem.200801575.

Velu, S. S., Di Meo, F., Trouillas, P., Sancho-Garcia, J. C., & Weber, J. F. (2013). Regio- and stereocontrolled synthesis of oligostilbenoids: Theoretical highlights at the supramolecular level. *Journal of Natural Products*, *76*(4), 538–546. https://doi.org/10.1021/np300705p.

Vossen, E., & De Smet, S. (2015). Protein oxidation and protein nitration influenced by sodium nitrite in two different meat model systems. *Journal of Agricultural and Food Chemistry*, *63*(9), 2550–2556. https://doi.org/10.1021/jf505775u.

Wang, M., Jin, Y., & Ho, C. T. (1999). Evaluation of resveratrol derivatives as potential antioxidants and identification of a reaction product of resveratrol and 2,2-diphenyl-1-picryhydrazyl radical. *Journal of Agricultural and Food Chemistry*, *47*(10), 3974–3977. https://doi.org/10.1021/jf990382w.

Weidinger, A., & Kozlov, A. V. (2015). Biological activities of reactive oxygen and nitrogen species: Oxidative stress versus signal transduction. *Biomolecules*, *5*(2), 472–484. https://doi.org/10.3390/biom5020472.

Wenz, C., Faust, D., Linz, B., Turmann, C., Nikolova, T., & Dietrich, C. (2019). Cell–cell contacts protect against t-BuOOH-induced cellular damage and ferroptosis in vitro. *Archives of Toxicology*, *93*(5), 1265–1279. https://doi.org/10.1007/s00204-019-02413-w.

Wishart, D. S., Feunang, Y. D., Guo, A. C., Lo, E. J., Marcu, A., Grant, J. R., et al. (2017). DrugBank 5.0: A major update to the DrugBank database for 2018. *Nucleic Acids Research*, *46*(D1), D1074–D1082. https://doi.org/10.1093/nar/gkx1037.

Wu, C. W., Nakamoto, Y., Hisatome, T., Yoshida, S., & Miyazaki, H. (2020). Resveratrol and its dimers ε-viniferin and δ-viniferin in red wine protect vascular endothelial cells by a similar mechanism with different potency and efficacy. *The Kaohsiung Journal of Medical Sciences*, *36*(7), 535–542. https://doi.org/10.1002/kjm2.12199.

Wu, S., Zhou, D., Geng, F., Dong, J., Su, L., Zhou, Y., et al. (2021). Metal-free oxidative condensation of Catechols, aldehydes and NH4OAc towards Benzoxazoles. *Advanced Synthesis & Catalysis*, *363*(14), 3607–3614. https://doi.org/10.1002/adsc.202100249.

Xie, J. S., Wen, J., Wang, X. F., Zhang, J. Q., Zhang, J. F., Kang, Y. L., et al. (2015). Potassium hexacyanoferrate (III)-catalyzed dimerization of hydroxystilbene: Biomimetic synthesis of Indane stilbene dimers. *Molecules*, *20*(12), 22662–22673. https://doi.org/10.3390/molecules 201219872.

Yadav, M. K., Mailar, K., Nagarajappa Masagalli, J., Chae, S. W., Song, J. J., & Choi, W. J. (2019). Ruthenium chloride-induced oxidative cyclization of trans-resveratrol to (±)-ε-viniferin and antimicrobial and antibiofilm activity against *Streptococcus pneumoniae*. *Frontiers in Pharmacology*, *10*, 890. https://doi.org/10.3389/fphar.2019.00890.

Yang, Y., Bai, B., Jin, M., Xu, Z., Zhang, J., Li, W., et al. (2016). Fluorescent imaging of Au3 + in living cells with two new high selective Au3 + probes. *Biosensors and Bioelectronics*, *86*, 939–943. https://doi.org/10.1016/j.bios.2016.07.112.

Yin, S.-Y., Jian, F.-Y., Chen, Y.-H., Chien, S.-C., Hsieh, M.-C., Hsiao, P.-W., et al. (2016). Induction of IL-25 secretion from tumour-associated fibroblasts suppresses mammary tumour metastasis. *Nature Communications*, 7(1), 11311. https://doi.org/10.1038/ncomms11311.

Zheng, J.-P., Kang, J., Huang, S.-G., Chen, P., Yao, W.-Z., Yang, L., et al. (2008). Effect of carbocisteine on acute exacerbation of chronic obstructive pulmonary disease (PEACE study): A randomized placebo-controlled study. *Lancet*, *371*(9629), 2013–2018. https://doi.org/10.1016/s0140-6736(08)60869-7.

Zhou, A., & Sadik, O. A. (2008). Comparative analysis of quercetin oxidation by electrochemical, enzymatic, autoxidation, and free radical generation techniques: A mechanistic study. *Journal of Agricultural and Food Chemistry*, *56*(24), 12081–12091. https://doi.org/10.1021/jf802413v.

CHAPTER FIVE

The antioxidant glutathione

Diana A. Averill-Bates*
Département des sciences biologiques (Center of Excellence in Orphan Diseases Research - Courtois Foundation (CERMO-FC), Research Group in Environmental Toxicology (TOXEN)), Université du Québec à Montréal, Montréal, QC, Canada
*Corresponding author: e-mail address: averill.diana@uqam.ca

Contents

Abstract

Reduced glutathione (GSH) is an essential non-enzymatic antioxidant in mammalian cells. GSH can act directly as an antioxidant to protect cells against free radicals and pro-oxidants, and as a cofactor for antioxidant and detoxification enzymes such as glutathione peroxidases, glutathione S-transferases, and glyoxalases. Glutathione peroxidases detoxify peroxides by a reaction that is coupled to GSH oxidation to glutathione disulfide (GSSG). GSSG is converted back to GSH by glutathione reductase and cofactor NADPH. GSH can regenerate vitamin E following detoxification reactions of vitamin E with lipid peroxyl radicals ($LOO^{\bullet}$). GSH is a cofactor for GST during detoxification of electrophilic substances and xenobiotics. Dicarbonyl stress induced by methylglyoxal and glyoxal is alleviated by glyoxalase enzymes and GSH. GSH regulates redox signaling through reversible oxidation of critical protein cysteine residues by

Vitamins and Hormones, Volume 121
ISSN 0083-6729
https://doi.org/10.1016/bs.vh.2022.09.002

S-glutathionylation. GSH is involved in other cellular processes such as protein folding, protecting protein thiols from oxidation and crosslinking, degradation of proteins with disulfide bonds, cell cycle regulation and proliferation, ascorbate metabolism, apoptosis and ferroptosis.

Abbreviations

AP-1	activator protein-1
AGE	advanced glycation endproduct
ADH	aldehyde dehydrogenase
AKR	aldoketo reductase
RO$^{\bullet}$	alkoxyl radical
ARE	antioxidant response element
ASK1	apoptosis signaling kinase 1
CHAC	cation transport regulator-like protein
JNK	c-jun N-terminal kinase
CYP	cytochrome P450
DDT	dichloro-diphenyl-trichloroethane
EpRE	electrophile response element
ERK	extracellular signal-related kinase
FoxO	forkhead Box class O
GRE	glucocorticoid-responsive elements
G6PD	glucose 6-phosphate dehydrogenase
γ-GCT	gamma-glutamyl cyclotransferase
Grx	glutaredoxin
GSH	glutathione, reduced
GSSG	glutathione disulfide
GPx	glutathione peroxidase
GR	glutathione reductase
GS	glutathione synthetase
GCL	glutamate cysteine ligase
γ-GCS	γ-glutamylcysteine synthetase
GGT	γ-glutamyltranspeptidase
GAPDH	glyceraldehyde 3-phosphate dehydrogenase
H_2O_2	hydrogen peroxide
OH$^{\bullet}$	hydroxyl radical
4-HNE	4-hydroxynonenal
HOCl	hypochlorous acid
HIF1α	hypoxia-inducible factor 1α
LOOH	lipid hydroperoxide
LOH	lipid alcohol
MnSOD	manganese SOD
MRE	metal response element
MG	methylglyoxal
MAPK	mitogen-activated protein kinase
MEKK1	mitogen-activated protein kinase/extracellular signal regulated kinase kinase kinase 1
MRP1	multidrug resistance-associated protein 1
NOX	NADPH oxidase

$NO^•$	nitric oxide
GSNO	nitrosothiol
NF-κB	nuclear factor kappa-light-chain-enhancer of activated B cells
Nrf2	nuclear factor erythroid 2–related factor 2
PPP	pentose phosphate pathway
$ROO^•$	peroxyl radical
$ONOO^-$	peroxynitrite
PI3K	phosphoinositide 3-kinase
PL-PUFA	phospholipid polyunsaturated fatty acid
PL-PUFA-LOOH	phospholipid polyunsaturated fatty acid lipid hydroperoxide
Protein-SH	protein cysteine thiols
PKB	protein kinase B
Protein-SOH	protein sulfenic acid
ROS	reactive oxygen species
RAGE	receptor for advanced glycation end products
SERCA	Sarco/endoplasmic reticulum calcium ATPase
SSAO	semicarbazide amino oxidase
$^1O_2$	singlet molecular oxygen
$O_2^{•-}$	superoxide anion
SOD	superoxide dismutase
UCP3	uncoupling protein-3
UPR	unfolded protein response
XRE	xenobiotic-responsive

1. Introduction

Reactive oxygen species (ROS) is an umbrella term for a group of highly reactive molecules derived from oxygen (Jones & Sies, 2015; Sies & Jones, 2020). ROS include free radicals with an unpaired electron in their outer orbital, and non-radical species. Oxygen-derived free radicals include the superoxide anion ($O_2^{•-}$), the hydroxyl radical ($OH^•$), and the peroxyl ($ROO^•$) and alkoxyl radicals ($RO^•$) (R = lipid or protein). Non-radical ROS molecules include hydrogen peroxide (H_2O_2), hypochlorous acid (HOCl), singlet molecular oxygen ($^1O_2$), and organic hydroperoxides (ROOH) (Halliwell & Gutteridge, 2015; Sies & Jones, 2020; Redza-Dutordoir & Averill-Bates, 2016, 2021). ROS are highly reactive and indiscriminate in their reactions with different biological targets such as proteins, lipids, membranes, and nucleic acids. However, the chemical reactivity of each individual ROS is very different and spans up to 11 orders of magnitude for second-order rate constants (Sies & Jones, 2020).

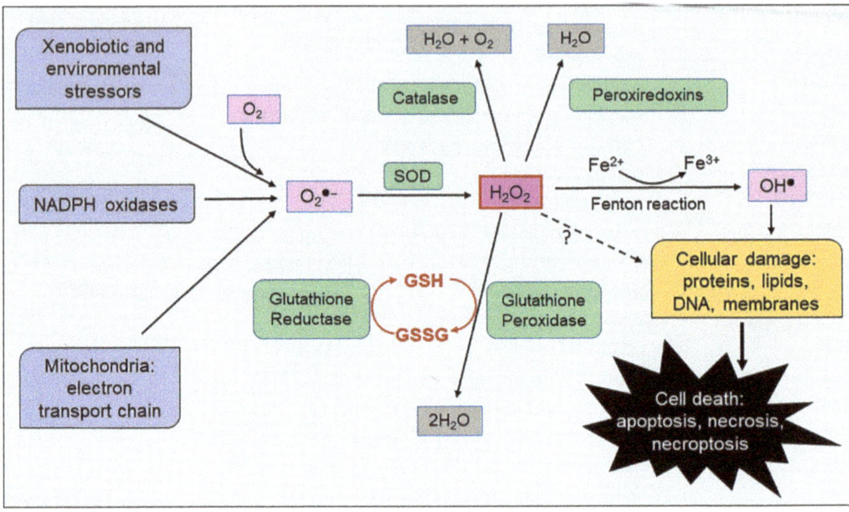

Fig. 1 Overview of cellular ROS production and detoxification by antioxidants. ROS are produced by intracellular and extracellular sources. Superoxide ($O_2^{\bullet-}$) undergoes dismutation by superoxide dismutase (SOD) to produce H_2O_2, which generates the hydroxyl radical ($OH^{\bullet}$) through metal-mediated Fenton or Haber Weiss chemistry, with metal ions such as Fe^{2+} and Cu^{1+}. $OH^{\bullet}$ causes damage to proteins, lipids, nucleic acids, and biological membranes, which can culminate in cell death. H_2O_2 is detoxified by several antioxidants including reduced glutathione (GSH) and the glutathione redox cycle, catalase, and the peroxiredoxins (Prxs).

ROS are produced in livings organisms through endogenous and exogenous sources (Fig. 1). They are produced through normal cellular metabolism in mitochondria, the cytosol, plasma membrane, endoplasmic reticulum, nucleus, and other organelles (Mailloux, 2020a; Sies & Jones, 2020). Human cells contain about 41 enzymes that produce H_2O_2 and/or $O_2^{\bullet-}$ (Sies & Jones, 2020). ROS, including H_2O_2 and $O_2^{\bullet-}$, are also produced by external factors, known as the "exposome" (Sies & Jones, 2020). External factors include pharmaceutical drugs, nutrients, inflammatory cytokines, environmental stressors such as pesticides, aldehydes, heavy metals, chlorinated by-products, and chemicals found in smoke, and ultraviolet and ionizing radiation (Redza-Dutordoir & Averill-Bates, 2021).

Two main endogenous sources of superoxide are from leaks in the mitochondrial electron transport chain at Complexes I, II and III, and by NADPH oxidases (NOX) at the plasma membrane (Bedard & Krause, 2007; Dickinson & Chang, 2011; Mailloux, 2020a). In mitochondria, superoxide is converted to hydrogen peroxide (H_2O_2), spontaneously and by the action of the superoxide dismutase (SOD) isoenzyme manganese SOD (MnSOD) (Halliwell & Gutteridge, 2015) (Fig. 1). Superoxide that is produced in the cytosol is converted to H_2O_2 by the copper-zinc isoenzyme CuZnSOD. Intracellular H_2O_2 is produced mainly in mitochondria, the cytosol and at the plasma membrane (Heppner, Janssen-Heininger, & van der Vliet, 2017; Mailloux, 2020a; Redza-Dutordoir & Averill-Bates, 2021). In addition to the SOD isoenzymes, H_2O_2 is produced by oxidase enzymes such as glucose, urate, xanthine, amine, L-amino acid, D-amino acid, and lysyl oxidases (Halliwell & Gutteridge, 2015; Mailloux, 2020a; Sies & Jones, 2020). Once produced in mitochondria, it is likely that H_2O_2 can cross cell membranes into the cytosolic compartment, by diffusion and through aquaporin channels (Bienert & Chaumont, 2014).

1.1 Hydrogen peroxide

The role of ROS in normal processes at physiological levels is known as "oxidative eustress" (Sies & Jones, 2020). The intracellular H_2O_2 concentration is tightly controlled at approximately 1–100 nM (Sies & Jones, 2020). At physiological levels, H_2O_2 is essential for many cellular processes. These include proliferation, cell cycle progression, migration and adhesion, adaptive stress responses, activation of transcription factors, redox regulation by oxidation and reduction, immune function, energy sensing and bioenergetics, wound healing, differentiation, embryonic development, and angiogenesis (Circu & Aw, 2010; Sies & Jones, 2020; Mailloux, 2020a; Redza-Dutordoir & Averill-Bates, 2016; Sies, 2017; Winterbourn, 2018). H_2O_2 is now considered to be a second messenger and an important cellular signaling molecule (Mailloux, 2020a). It can regulate cellular metabolism in response to changes in the local cellular environment, such as nutritional status, lifestyle, and irradiation (Mailloux, 2020a; Sies & Jones, 2020).

At physiological levels, ROS can activate multiple signaling pathways (Bae, Oh, Rhee, & Yoo, 2011; Circu & Aw, 2010; Sies & Jones, 2020; Priya Dharshini, Vishnupriya, Sakthivel, & Rasmi, 2020; Redza-Dutordoir & Averill-Bates, 2021; Sies, 2017). These include the three mitogen-activated protein kinases (MAPK): c-jun N-terminal (JNK), p38, and extracellular signal-related (ERK) kinases (Houle et al., 2003; Huot, Houle, Marceau, & Landry, 1997; Torres & Forman, 2003), apoptosis signaling kinase 1 (ASK1) (Sinha, Das, Pal, & Sil, 2013), protein kinase B (AKT), and phosphoinositide 3-kinase (PI3K) (Scherz-Shouval & Elazar, 2011). Moreover, ROS activate transcription factors such as nuclear factor kappa-light-chain-enhancer of activated B cells (NF-κB) (Song et al., 2017), nuclear factor-erythroid factor 2-related factor 2 (Nrf2) (Glory & Averill-Bates, 2016), activator protein 1 (AP-1), hypoxia-inducible factor 1α (HIF-1α), Forkhead Box class O (FoxO) (Priya Dharshini et al., 2020), and the p53 tumor suppressor (Pallepati & Averill-Bates, 2010).

However, at higher doses above about 100 nM, H_2O_2 is toxic and causes non-specific damage to biological molecules such as lipids, proteins, membranes, and nucleic acids (Halliwell & Gutteridge, 2015; Redza-Dutordoir & Averill-Bates, 2021; Sies & Jones, 2020) (Fig. 1). These reactions can give rise to inactivated and aggregated proteins, DNA mutations, and damaged membranes. This molecular damage can be reversible or irreversible and is known as "oxidative distress" (Sies & Jones, 2020). The definition of oxidative stress (distress) is "a serious imbalance between the generation of ROS and antioxidant defenses in favor of ROS, causing excessive oxidative damage to biomolecules" (Halliwell, 2011). H_2O_2 can inactivate certain enzymes by oxidizing their active site cysteine residues (Mailloux, 2020a). These include caspase enzymes involved in apoptosis, the glycolytic enzyme glyceraldehyde 3-phosphate dehydrogenase (GAPDH) and peroxiredoxins (Prx) (Mailloux, 2020a). H_2O_2 causes cell death by apoptosis, and at higher doses, by necrosis (Pallepati & Averill-Bates, 2010; Redza-Dutordoir & Averill-Bates, 2021).

1.2 Antioxidants

Aerobic organisms are equipped with a variety of antioxidants that are essential for their survival. Antioxidants are defined as "any substance that delays, prevents or removes oxidative damage to a target molecule" (Halliwell & Gutteridge, 2015). They protect different molecular targets such as proteins, lipids, nucleic acids, and membranes from oxidative damage. They are present at varying concentrations in distinct tissues, cell subtypes, and subcellular compartments. Different antioxidants target different ROS, and they have distinct reaction rates, and different mechanisms of action.

Antioxidants are divided into those with enzymatic and non-enzymatic activity. The SOD isoenzymes catalytically convert superoxide anion to H_2O_2 (Fridovich, 1975) (Fig. 1). Antioxidant enzymes such as catalase, Prxs and glutathione/glutathione peroxidases (GPx) convert H_2O_2 to non-toxic products (Cox, Winterbourn, & Hampton, 2010) (Fig. 1). The enzymes with the highest rates of reaction with H_2O_2 are GPxs and Prxs, followed by catalase (Deponte, 2013; Winterbourn, 2018).

Non-enzymatic antioxidants that occur *in vivo* include molecules that sequester metal ions such as copper and iron, heme oxygenase (Galaris, Barbouti, & Pantopoulos, 2019), lipoic acid, uric acid, coenzyme Q and bilirubin (Cox et al., 2010; Halliwell, 2011). Antioxidants from dietary sources such as fruits and vegetables include vitamins A, C (ascorbate), D and E, carotenoids, selenium, niacin, riboflavin, thiamine, folic acid, flavonoids, and polyphenols (Halliwell & Gutteridge, 2015). Glutathione is an extremely important non-enzymatic antioxidant that is the main theme of this chapter. Emphasis will be on the role of glutathione as an antioxidant in mammalian cells.

1.3 Oxidative distress and the redox equilibrium

The production of ROS in living systems exists in equilibrium with the multiple antioxidant defenses (Redza-Dutordoir & Averill-Bates, 2016). The redox equilibrium ensures that physiological, non-toxic levels of ROS are present in cells. If the production of ROS exceeds the capacity of the antioxidant defenses, the redox balance is perturbed, giving rise to oxidative distress (Sies & Jones, 2020). A decrease in levels of antioxidants due to mutations or insufficient dietary sources can lead to oxidative distress (Redza-Dutordoir & Averill-Bates, 2016). On the other hand, oxidative distress can arise from excessive ROS production due to increased exposure to xenobiotics and oxygen, and improper activation of phagocytic cells during inflammation (Halliwell & Gutteridge, 2015). The consequences of oxidative distress are oxidative damage to biological molecules such as proteins, lipids, carbohydrates, nucleic acids, and organelles such as mitochondria. Serious oxidative injury ultimately results in cell death by apoptosis and/or necrosis (Circu & Aw, 2010; Redza-Dutordoir & Averill-Bates, 2016).

2. Glutathione

2.1 Structure and synthesis

The reduced form of glutathione is GSH, whereas glutathione disulfide (GSSG) is the oxidized form. Unfortunately, GSSG is often referred to as oxidized GSH, but the correct name is glutathione disulfide. GSH (γ-L-glutamyl-L-cysteinylglycine) is a tripeptide composed of cysteine, glycine, and glutamic acid (Lu, 2013) (Fig. 2). GSSG is formed by a disulfide bridge between the thiol groups of the cysteine residues of two GSH molecules (Fig. 2). The predominant form is GSH, which represents at least 98% of total cellular glutathione (Akerboom, Bilizer, & Sies, 1982). The high ratio

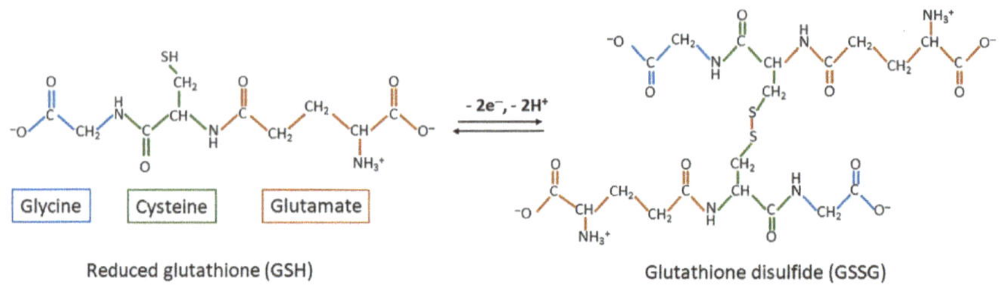

Fig. 2 Structures of reduced glutathione (GSH) and glutathione disulfide (GSSG). The tripeptide glutathione is composed of glycine, cysteine, and glutamate. GSSG is composed of two molecules of GSH connected by a disulfide bond through their cysteine residues.

of GSH:GSSG establishes the intracellular redox potential (Forman, Zhang, & Rinna, 2009). In eukaryotic cells, GSH is localized mainly in the cytosol (80–85%) and is also present in mitochondria (10–15%), the nucleus and the endoplasmic reticulum (Halliwell & Gutteridge, 2015; Lu, 2013).

GSH is synthesized mainly in the liver. It is synthesized in the cytosol of animal cells from its three constituent amino acids (Lu, 2013). GSH is transported from the cytosol into mitochondria and is present in the intermembrane space and matrix (Halliwell & Gutteridge, 2015). Key factors for GSH biosynthesis are availability of the precursor amino acid cysteine, and activity of the rate limiting enzyme glutamate cysteine ligase (GCL; EC 6.3.2.2), also known as γ-glutamylcysteine synthetase (γ-GCS). The second enzyme involved in GSH biosynthesis is glutathione synthetase (GS; EC 6.3.2.3). These two enzymatic reactions each require ATP.

The first reaction in GSH biosynthesis is the production of γ-glutamylcysteine from glutamine and cysteine, catalyzed by GCL (Fig. 3, reaction 1). This reaction is feedback inhibited by GSH (Halliwell & Gutteridge, 2015). GCL is composed of two subunits, a catalytic heavy subunit (GCLC), and a regulatory light subunit (GCLM) (Seelig, Simondsen, & Meister, 1984). GCLM does not possess enzymatic activity. However, GCLM plays a regulatory role by decreasing the K_m of GCL for glutamate, while increasing the K_i for GSH (Huang, Anderson, & Meister, 1993).

GCLC possesses enzymatic activity and is regulated by many factors (Lu, 2013). Among these factors, heat shock and oxidative distress cause increased levels of GSH, GCLC activity and GCLC transcription (Pallepati & Averill-Bates, 2010; Tchouagué, Grondin, Glory, & Averill-Bates, 2019). The consensus elements for NF-κB, Sp-1, activator protein-1 (AP-1), AP-2, the metal response element (MRE), and the antioxidant response

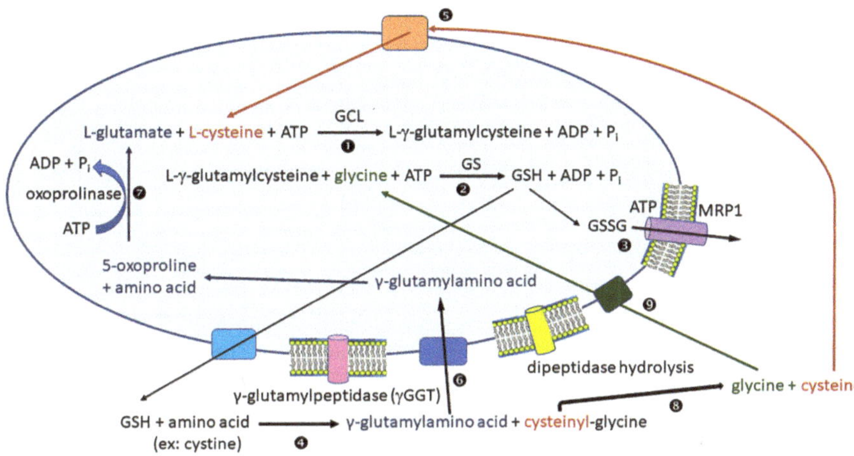

Fig. 3 Glutathione synthesis and the γ-glutamyl cycle. The GSH tripeptide is synthesized by two enzymes, GCL and GS. The γ-glutamyl cycle ensures that GSH provides a continuous supply of cysteine. GSH undergoes hydrolysis extracellularly by γ-GGT at the plasma membrane, producing γ-glutamylamino acid and cysteinyl-glycine. Cysteine, glycine, and γ-glutamylamino acid enter cells through distinct protein transporters. γ-Glutamylamino acid is converted to 5-oxoproline, which produces L-glutamate by the action of ATP-dependent oxoprolinase.

element (ARE)/electrophile response element (EpRE) are present in the promoter for human GCLC. The ARE is an important element for many inducers of GCLC. These include the transcription factor Nrf2 (Glory & Averill-Bates, 2016).

The second reaction of GSH biosynthesis, catalyzed by GS, is the formation of GSH from γ-glutamylcysteine and glycine (Fig. 3, reaction 2). GS contains two identical subunits and there is no feedback inhibition by GSH.

Cysteine is obtained mainly from the diet, protein breakdown, the surrounding fluids and from methionine in the liver through *trans*-sulfuration (Halliwell & Gutteridge, 2015). It is unstable extracellularly and readily oxidizes to cystine. Cystine enters cells through the x_c^- system (Fig. 3, reaction 5), where it is reduced to cysteine (Lu, 2013).

Continuous turnover of GSH occurs in cells. Its half-life is estimated to be 3 h in rat liver, 4 days in human red blood cells, and minutes in the plasma (Halliwell & Gutteridge, 2015). In several cell types, the turnover of GSH involves the ATP-dependent export of GSSG through the multidrug resistance-associated protein-1 (MRP1) transporter (Fig. 3, reaction 3). GSH is secreted continuously from the liver into the plasma and bile. GSH in the plasma is the source of substrates for GSH synthesis in other tissues (Halliwell & Gutteridge, 2015).

2.2 Degradation of glutathione

The γ-glutamyl bond confers stability to GSH, which makes it resistant to degradation by intracellular peptidases (Bachhawat & Kaur, 2017). GSH is hydrolyzed extracellularly by the enzyme γ-glutamyltranspeptidase (γGGT) on the plasma membrane of certain cells (Meister & Anderson, 1983) (Fig. 3, reaction 4). The γ-glutamyl cycle allows GSH to be a continuous source of cysteine, which is taken up by cells (Meister, 1988) (Fig. 3, reaction 5). The cycle involves the export of GSH from cells, and γGGT then transfers the γ-glutamyl group of GSH to an amino acid (e.g., cystine, methionine, glutamine) (Lu, 2013) (Fig. 3, reaction 4). This forms the reaction products γ-glutamylamino acid and cysteinyl-glycine. The γ-glutamylamino acid is transported into cells and converted to 5-oxoproline, releasing the amino acid (Fig. 3, reaction 6). The 5-oxoproline is converted to L-glutamate by oxoprolinase (Halliwell & Gutteridge, 2015) (Fig. 3, reaction 7). The cysteinyl-glycine is hydrolyzed by dipeptidases on the plasma membrane to produce cysteine and glycine (Fig. 3, reaction 8). These amino acids are transported back into cells and contribute to synthesis of GSH (Fig. 3, reactions 5 and 9).

New enzymes have been shown to degrade glutathione in the cytosol of mammalian cells (Desideri, Ciccarone, & Ciriolo, 2019). The cation transport regulator-like protein (CHAC) family function as γ-glutamyl cyclotransferases (γ-GCT) that specifically degrade glutathione to 5-oxoproline and Cys-Gly. The action of CHAC enzymes is completed by 2-oxoprolinase to produce glutamate, and by cysteinylglycine peptidases that generate cysteine and glycine.

3. The many roles of glutathione as an antioxidant

Reduced glutathione (GSH) is an essential non-enzymatic antioxidant that is present at millimolar concentrations (1–10 mM) in all mammalian cells (Halliwell & Gutteridge, 2015). It is involved in many essential cellular processes such as antioxidant defenses, as a cofactor for several enzymes, detoxification of xenobiotics, protecting protein thiols from oxidation and crosslinking, protein folding, degradation of proteins with disulfide bonds, cell cycle regulation, proliferation, differentiation, cell communication, DNA repair, ascorbate metabolism, redox signaling, apoptosis, energy production, radioprotection, and the immune response (Fig. 4). GSH is important for the storage of cysteine through the γ-glutamyl cycle

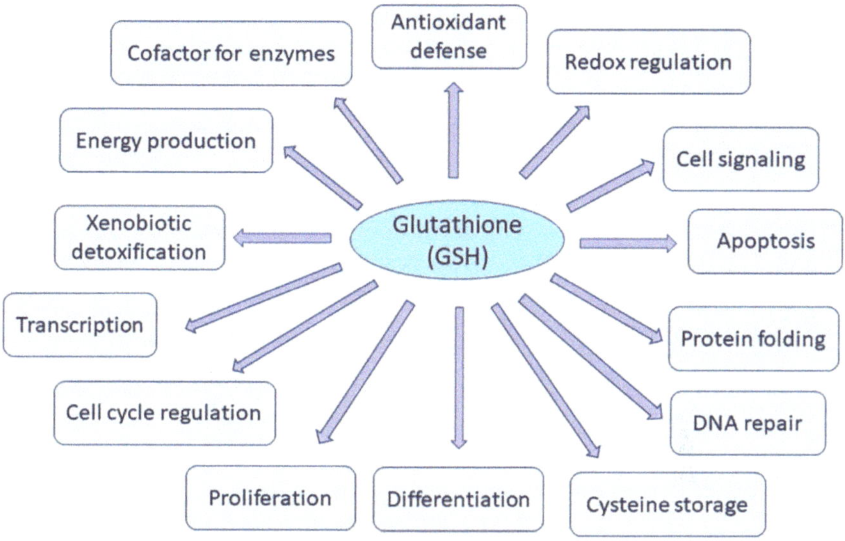

Fig. 4 Cellular functions of glutathione. GSH plays an important role in a variety of vital cellular functions, in addition to its action as an antioxidant.

(Meister, 1988). It plays an important antioxidant role in mitochondria by protecting against pro-oxidants that are generated physiologically through the electron transport chain (Mailloux, 2020a), as well as from exogenous sources.

3.1 Direct antioxidant action

As an antioxidant, GSH can react directly with a variety of free radicals and pro-oxidants through non-enzymatic reactions. GSH has been shown to react *in vitro* with $OH^{\bullet}$, peroxynitrite ($ONOO^-$), $NO_2^{\bullet}$, HOCl, $RO^{\bullet}$, $RO_2^{\bullet}$, $CO_3^{\bullet-}$ and singlet oxygen ($^1O_2$) (Deponte, 2017; Halliwell & Gutteridge, 2015). The reaction of GSH and other thiols with carbon-centered radicals, transition metals, and oxygen radicals ($OH^{\bullet}$, $RO^{\bullet}$, $RO_2^{\bullet}$, $CO_3^{\bullet-}$) leads to formation of the thiyl radical $NO^{\bullet}$ (Halliwell & Gutteridge, 2015). The detection of these reactions *in vivo* is not feasible, but they are likely to occur because GSH is present at mM concentrations inside cells. However, even at millimolar concentrations, non-enzymatic reactions of GSH with superoxide and H_2O_2 would be too slow to compete with the high-rate constants of enzymatic reactions for their removal (Deponte, 2017). GSH reacts with $ONOO^-$ to produce nitrosothiol (GSNO), which is known to generate nitric oxide ($NO^{\bullet}$). GSH also reacts with protein cysteine thiols by S-glutathionylation (Mailloux, 2020b) (Section 4), with methylglyoxal (Rabbani, Xue, & Thornalley, 2016) (Section 3.5) and with cytotoxic aldehydes such as acrolein (Agostinelli, Przybytkowski, & Averill-Bates, 1996) and 4-hydroxynonenal (Halliwell & Gutteridge, 2015).

In addition to acting as a direct antioxidant, GSH is a cofactor for several antioxidant and detoxification enzymes such as GPxs, glutathione S-transferases (GST), and glyoxalases (Sections 3.3–3.5).

3.2 Regeneration of vitamins E and C

Vitamin E is the main cellular lipid-soluble small molecule antioxidant. It can protect cell membranes against lipid radicals such as the peroxyl radical, $LOO^{\bullet}$ (Buettner, 1993). During this reaction, $LOO^{\bullet}$ is converted to the lipid hydroperoxide LOOH, while vitamin E accepts the unpaired electron and is converted to its free radical form (Fig. 5). Vitamin E can cooperate with vitamin C, a water-soluble small molecule antioxidant, to protect cell membranes against lipid peroxidation. Vitamin C can react with the vitamin E radical to regenerate vitamin E, which allows vitamin E to maintain its function as a free radical chain-breaking antioxidant (Fig. 5). During this reaction, vitamin C accepts the unpaired electron from vitamin E and is converted to the vitamin C free radical.

GSH plays an important role in the regeneration of vitamins E and C (Buettner, 1993). GSH can also react with the vitamin E radical to regenerate vitamin E and restore its antioxidant function (Fig. 5). Similarly,

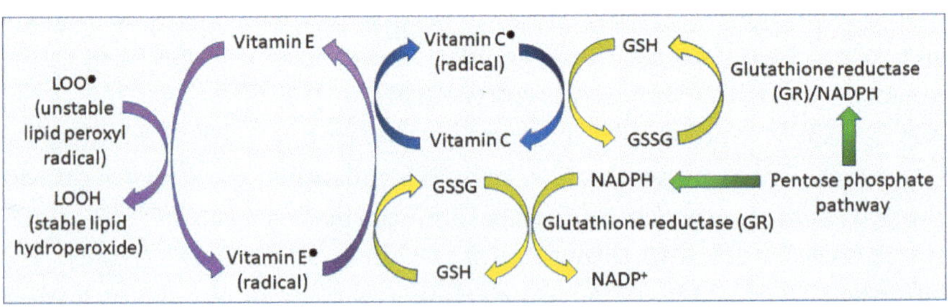

Fig. 5 Recycling of vitamin E and vitamin C by GSH. Vitamin E reacts with lipid peroxyl radicals that are converted to lipid hydroperoxides. This reaction converts vitamin E to the vitamin E radical. GSH or vitamin C can react with the vitamin E radical to regenerate vitamin E. GSH and vitamin C undergo oxidation to GSSG and the vitamin C radical, respectively. GSH can regenerate vitamin C and undergoes oxidation to GSSG. GSH is regenerated from GSSG by the action of GR and NADPH produced by the pentose phosphate pathway.

GSH can react with the vitamin C radical to regenerate vitamin C. During these reactions, GSH undergoes oxidation to GSSG, which is reduced back to GSH by the action of glutathione reductase (GR), with NADPH as cofactor (Fig. 5).

3.3 Glutathione peroxidases

An important antioxidant function of GSH is achieved by the elimination of hydrogen peroxide and other peroxides by the action of GPxs (Deponte, 2013; Halliwell & Gutteridge, 2015). GPxs are a ubiquitous family of enzymes that detoxify H_2O_2 and other peroxides by a reaction that is coupled to the oxidation of GSH to GSSG (Fig. 6). The conversion of GSSG back to GSH occurs by the action of glutathione reductase (GR) through a reaction that requires the electron donor NADPH + H^+ (Fig. 6). The two main cellular sources of NADPH are glucose metabolism through the pentose phosphate pathway (PPP), and the conversion of malate to pyruvate by the malic enzyme (Nelson & Cox, 2021; Patra & Hay, 2014) (Fig. 6). The key reaction step in the PPP that generates NADPH is catalyzed by the enzyme glucose 6-phosphate dehydrogenase (G6PD) (Fig. 6). The rate of the PPP is limited by the quantity of $NADP^+$.

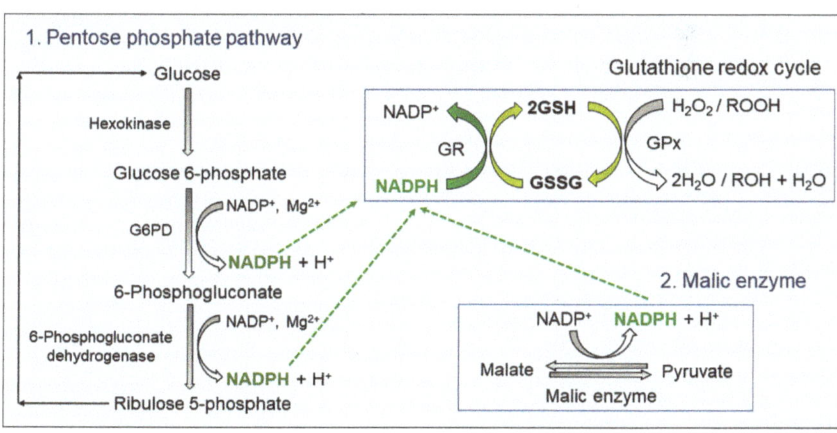

Fig. 6 Detoxification of hydrogen peroxide and other peroxides by the glutathione redox cycle. Peroxides are detoxified by their reaction with GSH, which is catalyzed by GPx. GSH undergoes oxidation to GSSG and is regenerated by catalytic activity of GR, with NADPH as cofactor. The two main cellular sources of NAPDH are (1) glucose metabolism through the pentose phosphate pathway, and (2) the conversion of malate to pyruvate catalyzed by the malic enzyme. R = protein or lipid. Glutathione peroxidase (GPx); Glutathione reductase (GR); Glucose-6-phosphate dehydrogenase (G6PD).

The reduction of GSSG by GR decreases the ratio of NADPH/$NADP^+$, which accelerates the rate of the PPP to replace NADPH (Halliwell & Gutteridge, 2015). By maintaining a reducing environment with a high ratio of NADPH to $NADP^+$, and a high ratio of GSH to GSSG, cells can prevent or undo oxidative damage to cellular targets.

The mammalian GPx family has 7 or 8 isoenzymes, of which 4 or 5 have selenium atoms at their active sites (Deponte, 2013). The classical isoenzyme is the cytosolic GPx1. GPx1, GPx2 and GPx3 can all detoxify H_2O_2 and lipid hydroperoxides (LOOH) through reactions that require GSH (Fig. 6). GPx4 (phospholipid hydroperoxide glutathione peroxidase 4) also requires GSH to detoxify cholesterol hydroperoxides, and it inhibits membrane lipid peroxidation due to its hydrophobic nature. Intestinal GPx2 is in the cytosol and GPx3 is found in plasma. GPx4 is found in the cytosol, nucleus, and mitochondria. GPx4 exists in both a free and membrane-bound form, which is coherent with its preference for lipid hydroperoxides (Deponte, 2013).

GPx4 is also involved in ferroptosis, which is an iron-dependent form of regulated cell death (Hirschhorn & Stockwell, 2019; Stockwell et al., 2017) (Fig. 7). Ferroptosis is caused by the accumulation of lipid-based ROS. Cell death by ferroptosis occurs when GSH-dependent lipid peroxide repair systems are overwhelmed and/or non-functional. GPx4 is a major enzymatic inhibitor of ferroptosis. Cystine, the oxidized form of cysteine, is present in the extracellular space. System x_c^- is a cystine importer that is coupled to glutamate export (Fig. 7). Once it enters the cell, cystine is converted to cysteine that is used to synthesize GSH. GSH is a cofactor for GPx4, for the elimination of ROS and Fe^{2+}-produced lipid peroxides generated in phospholipid polyunsaturated fatty acids (PL-PUFA) (Fig. 7). Potentially toxic lipid hydroperoxides (LOOH) are converted to non-toxic lipid alcohols (LOH) by GPx4.

The small molecule erastin inhibits the cellular import of cystine (Stockwell et al., 2017) (Fig. 7). This leads to GSH depletion and the inactivation of GPx4. This results in overwhelming levels of lipid peroxidation and cell death by ferroptosis.

Given that cysteine availability can limit GSH synthesis, some cells use the *trans*-sulfuration pathway to biosynthesize cysteine from methionine. This bypasses the requirement for cystine import through the x_c^- system.

NADPH and selenium are factors that affect cellular sensitivity to ferroptosis (Stockwell et al., 2017). NADPH is required as a cellular reductant to eliminate lipid hydroperoxides, while selenium is essential for GPx4

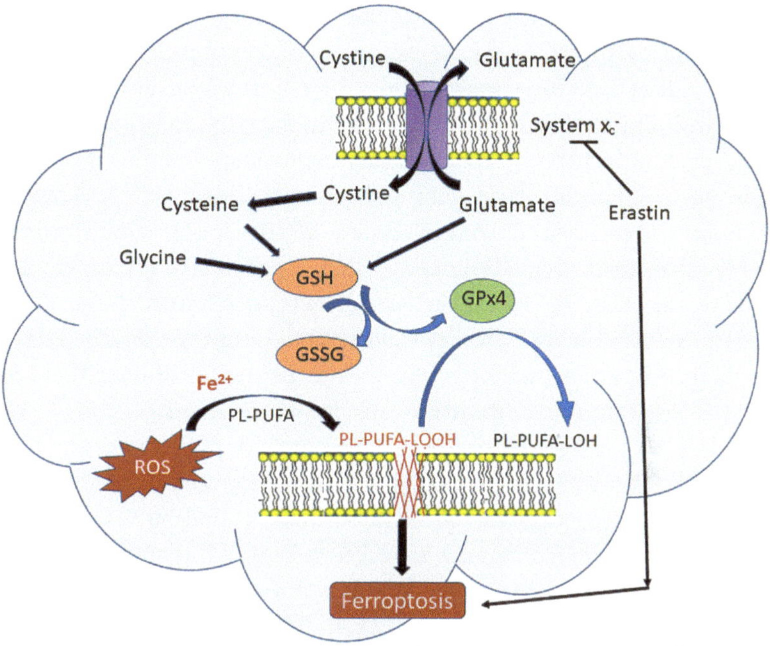

Fig. 7 Inhibition of lipid peroxidation and ferroptosis by GSH and GPx4. Increased levels of ROS and metals such as ferrous iron (Fe^{2+}) cause oxidative damage to membrane phospholipid polyunsaturated fatty acids (PL-PUFA), a process known as lipid peroxidation. This causes holes in the cell membrane and cell death by ferroptosis. The action of GSH and GPx4 converts lipid hydroperoxides (PL-PUFA-LOOH) to PL-PUFA-LOH and protects cells against ferroptosis.

synthesis. Nrf2, which controls levels of GSH and NADPH, also promotes resistance to ferroptosis (Sun et al., 2016).

3.4 Glutathione S-transferases

GSH is a ubiquitous nucleophile that acts as a cofactor for GSTs to convert a variety of electrophilic substances such as carcinogens, mutagens, and teratogens, under physiological conditions (Deponte, 2013). The products of these GSH-mediated reactions are often less toxic and more soluble, which facilitates their subsequent metabolism and/or excretion.

The GST (EC 2.5.1.18) superfamily consists of isoenzymes that catalyze the conjugation of GSH to a wide variety of non-polar, electrophilic molecules. These molecules can be of endogenous or exogenous origin (Kalinina, Chernov, & Novichkova, 2014). The latter, known as xenobiotics, consist of drugs, environmental pollutants and pesticides (Redza-Dutordoir & Averill-Bates, 2016, 2021). Some examples of xenobiotics that react with GSTs are aldehydes (Agostinelli et al., 1996), quinone compounds (e.g., Adriamycin), chloroform, BCNU, organic nitrates, aflatoxin, dichloro-diphenyl-trichloroethane (DDT), bromobenzene, chemicals found in cigarette smoke, naphthalene, and paracetamol (Halliwell & Gutteridge, 2015; Redza-Dutordoir & Averill-Bates, 2016, 2021). GSTs participate in cellular defenses by eliminating these potentially toxic molecules (Scheme 1, reaction 1). However, under certain circumstances, the GSH conjugates can themselves be toxic (Halliwell & Gutteridge, 2015). The exposure to high doses of xenobiotics can lead to depletion of liver GSH levels, therefore decreasing the antioxidant capacity of the liver.

Certain GSTs eliminates organic hydroperoxides by converting them to alcohols, GSSG and water through conjugation with GSH (Hiratsuka, Yamane, Yamazaki, Ozawa, & Watabe, 1997; Prabhu, Reddy, Jones, Liken, & Reddy, 2004) (Scheme 1, reaction 2). GST reduces hydroperoxides of polyunsaturated higher fatty acids, phospholipids, and cholesterol (Hiratsuka et al., 1997; Prabhu et al., 2004). This represents GPx-like activity, formerly known as non-selenium GPx activity (Halliwell & Gutteridge, 2015). H_2O_2 is not catalyzed by this reaction.

In mammals, GST is present in most organs, but the highest content is found in the liver. The GSH conjugates are often excreted from the liver into bile through ATP-dependent efflux pumps, such as MRP1 (Deponte, 2017; Halliwell & Gutteridge, 2015). The GST isoenzymes are found in the cytosol, mitochondria and microsomes. The cytosolic isoenzymes account for about 90% of cellular GST activity (Kalinina et al., 2014). The cytosolic GSTs are divided into seven classes (α, μ, π, θ, ζ, ω, and σ) and are comprised of 17 isoenzymes (Wu & Dong, 2012; Hayes, Flanagan, & Jowsey, 2005). The homology within the same class of GST varies from 40% to 90%, whereas there is <25% homology between the different classes (Kalinina et al., 2014). The primary structure of the conserved N-terminal domain

$$\mathrm{RX + GSH \xrightarrow{GST} RSG + HX} \qquad \text{Reaction 1}$$

$$\mathrm{ROOH + 2GSH \xrightarrow{GST} ROH + GSSG + H_2O} \qquad \text{Reaction 2}$$

Scheme 1 GST-catalyzed conjugation of xenobiotics (RX) and lipid hydroperoxides (ROOH) with GSH. Reaction 1: RX–xenobiotic; Reaction 2: R = lipid or protein.

of GST contains the catalytic residues tyrosine, cysteine, or serine (Hayes et al., 2005). This is the GSH binding site (G site).

GSTs are divided into two subgroups. Y-GSTs use a tyrosine residue in their active site for the activation of GSH (α, μ, π, and σ classes). S/C-GSTs bind to GSH through a serine (ϕ, τ, θ, and ζ classes) or a cysteine (β and ω classes) residue (Kalinina et al., 2014). The C-terminal domain is the co-substrate binding site (H site), and its structure is variable between different classes of GSTs. These different structures allow for the wide variety of GST substrates and diversity of functions. The cytosolic and mitochondrial GSTs are homodimers or heterodimers. For heterodimers, the subunits are from the same class of GSTs.

Levels of GSTs can be increased by exposure to xenobiotics. These compounds can act by increasing the transcription of GST genes through the ARE, xenobiotic-response element (XRE), or the glucocorticoid-response (GRE) element of the promoter region (Hayes & Pulford, 1995; Higgins & Hayes, 2011; Kalinina et al., 2014).

The tyrosine residue of Y-GSTs can be inactivated by exposure to $ONOO^-$, which causes nitration of protein tyrosine residues. Nitration prevents binding of GSH to the critical tyrosine residue of GST (Halliwell & Gutteridge, 2015).

Among the substrates of GSTs are products of lipid peroxidation, such as 4-hydroxynonenal (4-HNE) and acrolein (Halliwell & Gutteridge, 2015; Tanel & Averill-Bates, 2007). These reactive aldehydes cause covalent modification of cellular proteins and DNA by alkylation (Kehrer & Biswal, 2000). The GSTA4-4-catalyzed conjugation of these toxic aldehydes with GSH prevents these modifications (Kalinina et al., 2014). During oxidative distress, nucleotides undergo oxidation to produce propenals and hydroperoxides that are substrates of GSTP1-1. The highly reactive aldehyde acrolein, which is a major component of smoke from cigarettes, and house and forest fires, causes particularly potent depletion of cellular GSH levels (Agostinelli et al., 1996; Tanel & Averill-Bates, 2007).

GSTP isoenzymes can catalyze the glutathionylation of certain proteins, whereas the GSTO1-1 class can reverse glutathionylation (Halliwell & Gutteridge, 2015) (Section 4).

3.5 Dicarbonyl stress and glyoxalases

3.5.1 Dicarbonyl stress

Highly reactive dicarbonyl metabolites such as methylglyoxal (MG) (CH_3CH-CHO), glyoxal (CHO-CHO), and 3-deoxyglucosone contribute to dicarbonyl stress. Dicarbonyl stress is a dysfunctional state where MG, glyoxal, and 3-deoxyglucosone accumulate abnormally due to an increase in their formation and a decrease in activity of their detoxification systems (Rabbani et al., 2016).

MG and glyoxal are commonly found in foods. MG is produced during metabolism from the triose phosphate glycolytic intermediates, glyceraldehyde 3-phosphate and dihydroxyacetone phosphate (Rabbani et al., 2016) (Fig. 8). The formation of MG through glucose triose metabolism is enhanced under conditions of increased glucose metabolism during diabetes-associated hyperglycemia, decreased removal of glyceraldehyde 3-phosphate due to decreased activity of the pentose phosphate pathway, and increased dependence on anaerobic glycolysis and associated metabolic flux during hypoxia. Other sources of MG are the oxidation of acetone by the cytochrome P450 (CYP) enzyme CYP2E1 during ketone catabolism, lipid peroxidation, the degradation of proteins glycated by glucose, and the degradation of monosaccharides (Rabbani et al., 2016). MG is also produced by threonine catabolism from the oxidation of aminoacetone by semicarbazide amino oxidase (SSAO) (Lyles & Chalmers, 1992). Glyoxal is produced by the degradation of glucose.

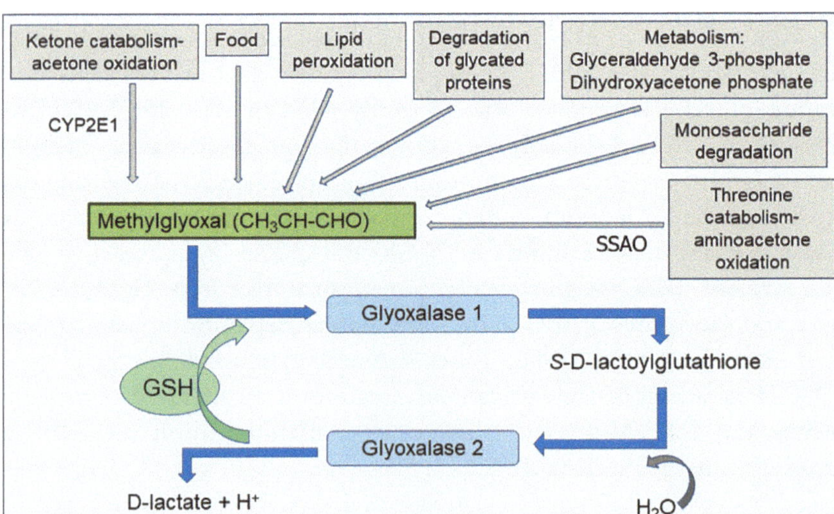

Fig. 8 Sources of methylglyoxal and metabolism by the glyoxalase enzymes. Methylglyoxal is produced in cells by a variety of intracellular and extracellular sources. It undergoes a reaction with GSH and is converted to *S*-D-lactoylglutathione and D-lactate in a two-step reaction system catalyzed by glyoxalases 1 and 2, respectively.

MG and glyoxal are metabolized by the glutathione-dependent glyoxalase system, which is a protective, cytoplasmic enzymatic system that maintains these dicarbonyls at tolerable levels in cells. Physiological levels of MG and glyoxal are 50–150 nM in human plasma, and 1–4 μM in mammalian cells (Rabbani et al., 2016).

At higher doses, dicarbonyl metabolites react rapidly with proteins, DNA, and phospholipids, leading to dicarbonyl stress, and ultimately cellular and tissue dysfunction (Halliwell & Gutteridge, 2015). These are non-oxidative reactions and therefore antioxidants are not therapeutic candidates to alleviate dicarbonyl stress (Rabbani, Xue, & Thornalley, 2021).

MG and glyoxal generate adducts of DNA. These include deoxyguanosine-derived imidazopurinones of DNA, GdG and MGdG-nucleotide advanced glycation endproducts (AGE)s (Rabbani et al., 2016). MGdG is the major nucleotide AGE found in biological systems. These nucleotide AGEs have been implicated in causing DNA strand breaks and mutagenesis (Thornalley et al., 2010).

MG and glyoxal bind covalently to arginine, cysteine, and lysine residues of proteins to generate AGEs (Halliwell & Gutteridge, 2015). Dicarbonyl glycation of arginine and cysteine residues is particularly problematic because these amino acids are important in the functional sites of many proteins (Halliwell & Gutteridge, 2015). MG is the most reactive of these dicarbonyl molecules and reacts mainly with arginine residues to produces dihydroxyimidazolidine and hydroimidazolone adducts. The most important AGE produced in physiological systems is the arginine-derived hydroimidazolone, MG-H1 (Rabbani et al., 2021). Minor AGEs are formed from lysine residues of proteins: Nε(1-carboxyethyl)lysine and Nε-carboxymethyl-lysine are formed from MG and glyoxal, respectively (Rabbani et al., 2016). MG-H1 often forms at arginine residues at the protein surface. Protein glycation of arginine residues leads to loss of the positive charges of the side chain guanidino group, an increase in molecular weight of the arginine residue, and a loss of electrostatic interactions (Rabbani et al., 2021). Arginine salt bridge interactions (e.g., Arg+/Glu- pairs), which play an important role in protein folding and stability, are lost by modification with MG. Interactions between arginine and tyrosine residues are important for intermolecular bonding at protein–protein interfaces, and these are lost by reaction with MG.

A wide variety of cellular proteins can be modified by MG. Proteins that were modified by glycation with MG appeared to exhibit structural distortion, instability and misfolding (Rabbani et al., 2021). The formation of

MG-H1 at arginine residues at the protein surface leads to an increase in hydrophobicity at the protein surface. Hydrophobic groups are usually located in the interior of globular proteins and away from the hydrophilic environment (Nelson & Cox, 2021). The location of hydrophobic groups at the protein surface is a signal for ubiquitination and targeting of proteins for degradation by the proteosome.

Recent studies have implicated the accumulation of glycated proteins with activation of the unfolded protein response (UPR) and inflammation (Rabbani et al., 2021). Protein glycation leads to protein misfolding and perturbs protein homeostasis. The UPR may be activated to remove MG-modified proteins. However, the mechanisms linking protein glycation to the UPR are not clear. The UPR senses misfolded/unfolded proteins and activates a process of protein refolding and repair, or proteolysis to eliminate damaged proteins (Redza-Dutordoir & Averill-Bates, 2016).

Protein glycation can result in a loss of protein function and a decrease in their half-life (Rabbani et al., 2021). Protein glycation at arginine residues causes cellular dysfunction, such as mitochondrial protein dysfunction, mitochondrial activation of apoptosis, increased ROS generation, expression of inflammatory proteins, and anoikis and cell detachment from the extracellular matrix.

AGEs and dicarbonyl stress have been implicated in aging, and diseases such as obesity, diabetes, cardiovascular disease, end stage renal failure, non-alcoholic fatty liver disease, and MG-protein modifications in the aging human lens (Rabbani et al., 2016, 2021). Dicarbonyl stress plays a role in cancer. Glyoxalase 1 is a tumor suppressor protein that is overexpressed in tumors and is also a mediator of multidrug resistance to chemotherapeutic agents. MG and protein glycation are increased during obesity, diabetes, and vascular complications of diabetes such as kidney disease, retinopathy, neuropathy, and cardiovascular disease. Therapeutic agents, including a combination of *trans*-resveratrol and hesperetin, which counter dicarbonyl stress by inducing glyoxalase I, are undergoing development and clinical trials (Rabbani et al., 2021). The strategies aim to (i) prevent MG formation, (ii) eliminate MG and (iii) increase expression of MG-metabolizing enzymes such as glyoxalase 1.

3.5.2 Glyoxalases

MG reacts non-enzymatically with GSH and this complex is converted into *S*-D-lactoylglutathione ($CH_3CHOHCOSG$) by glyoxalase 1 (EC 4.4.1.5) (Deponte, 2013) (Fig. 8). *S*-D-lactoylglutathione is converted into the 2-hydroxycarboxylic acid D-lactate by glyoxalase 2 (EC 3.1.2.6), which regenerates GSH. Glyoxal is also a substrate for glyoxalase 1. Minor metabolism of MG and glyoxal occurs by aldehyde dehydrogenases (ADH) and aldoketo reductases (AKR) (Rabbani et al., 2016). 3-Deoxyglucosone is metabolized by ADHs and AKRs, but not by glyoxalase 1.

Glyoxalases 1 and 2 are present in most organisms (Deponte, 2017; Rabbani et al., 2016). They belong to the superfamily of vicinal oxygen chelate enzymes, which contain a βαβββ-motif that is required for metal ion binding (Deponte, 2013). Glyoxalase 1 is a cytosolic metal-dependent isomerase. Human glyoxalase 1 isoenzymes are either small homodimers (more common) or a large monomer, containing Zn^{2+} (or Fe^{2+}) at their two active sites, and a GSH-binding site. Glyoxalase 2 is a thioesterase that contains a conserved zinc/metal-binding motif at the active site (Deponte, 2013). Glyoxalase 2 usually contains two Zn^{2+} ions or a Zn^{2+}/Fe^{2+} ion pair at the binding site, and a GSH-binding site. Mn^{2+} and Fe^{3+} can also form ion pairs at the active site.

The basal and inducible expression of glyoxalase 1, ADHs and AKRs are mainly regulated by the stress-responsive transcription factor Nrf2 through the ARE (Rabbani et al., 2016). Nrf2, therefore, plays a cytoprotective role by increasing the basal and inducible expression and activities of the glyoxalase 1, ADH, and AKR enzymes to decrease dicarbonyl stress. On the other hand, HIF1α is a negative regulator of glyoxalase 1 expression during hypoxia. Hypoxia appears to be an important driver of dicarbonyl stress because it increases MG formation by anaerobic glycolysis and decreases glyoxalase I expression through HIF1α (Rabbani et al., 2016). The receptor for advanced glycation end products (RAGE) is also a negative regulator of glyoxalase expression.

Glyoxalase 1 activity is dependent on cellular levels of GSH (Rabbani et al., 2016). Depletion of GSH leads to increased levels of MG and glyoxal, and enhanced dicarbonyl stress.

The mitochondrial pool of GSH is essential for preserving the matrix redox homeostasis, in terms of maintaining the redox state of mitochondrial protein thiols, protecting against ROS, and defending the integrity of mitochondrial DNA and membranes against oxidative damage (Armeni et al., 2014; Mailloux, 2020a). GSH is synthesized in the cytosol and must be transported actively into mitochondria. However, *S*-D-lactoylglutathione can serve as an alternative supply of mitochondrial glutathione (Armeni et al., 2014). *S*-D-lactoylglutathione can enter mitochondria, where it undergoes hydrolysis by the mitochondrial glyoxalase II enzyme to produce GSH and D-lactate. Mitochondrial glyoxalase II was located in the inter-membrane space and mitochondrial matrix.

4. S-glutathionylation and role of glutathione in redox regulation

4.1 S-glutathionylation and redox regulation

Glutathione has long been considered to act as a redox buffer, which equilibrates with thiol disulfides to control their oxidation state (Winterbourn, 2019). However, as knowledge has evolved, changes in cellular redox proceed more by kinetics than thermodynamics. Thus, while different redox systems interact, they are not in equilibrium with one another (Kemp, Go, & Jones, 2008).

The GSH:GSSG ratio is high under physiological conditions, usually at about 100:1. While most cellular glutathione is in the form of GSH, some glutathione exists as mixed disulfides with molecules containing thiol groups (Halliwell & Gutteridge, 2015). These molecules include cysteine, coenzyme A and protein thiols. During oxidative distress, the ability of cells to reduce GSSG to GSH can be compromised. This leads to either GSSG export out of cells (Fig. 3, reaction 3), or the reaction of GSSG with protein thiol groups to form mixed disulfides (Protein-SSG) (Mailloux, 2020b). Oxidative distress can, therefore, lead to cellular depletion of GSH.

Protein thiol groups are highly conserved in proteins and are often crucial for protein function (Poole, 2015; Ulrich & Jakob, 2019). Cysteine residues are usually found in critical structural and functional regions of proteins, where they are required for catalytic, metal binding and/or redox regulatory activity (Poole, 2015; Ulrich & Jakob, 2019). Cysteines are highly polarizable and can undergo a variety of nucleophilic reactions. The rate of their nucleophilic reactions increases substantially when the thiol group is deprotonated to form a thiolate anion (Fig. 9, reaction 3). Cells contain a high amount of protein thiol groups, which are considered as an antioxidant pool. There is a balance between their antioxidant function and the potentially damaging consequences of thiol oxidation by reactive species on protein function.

GSH is an important regulator of redox cell signaling. This can occur through the reversible oxidation of critical cysteine residues in proteins. The oxidation of GSH pools by H_2O_2 produces GSSG, which leads to S-glutathionylation of proteins (Mailloux, 2020b) (Fig. 9, reaction 1). GSSG can bind reversibly to the thiol groups of protein cysteines (Protein-SH) to form glutathionylated proteins (Protein-SSG) through a process called S-glutathionylation (Dalle-Donne, Rossi, Colombo, Giustarini, & Milzani, 2009) (Fig. 9, reaction 2).

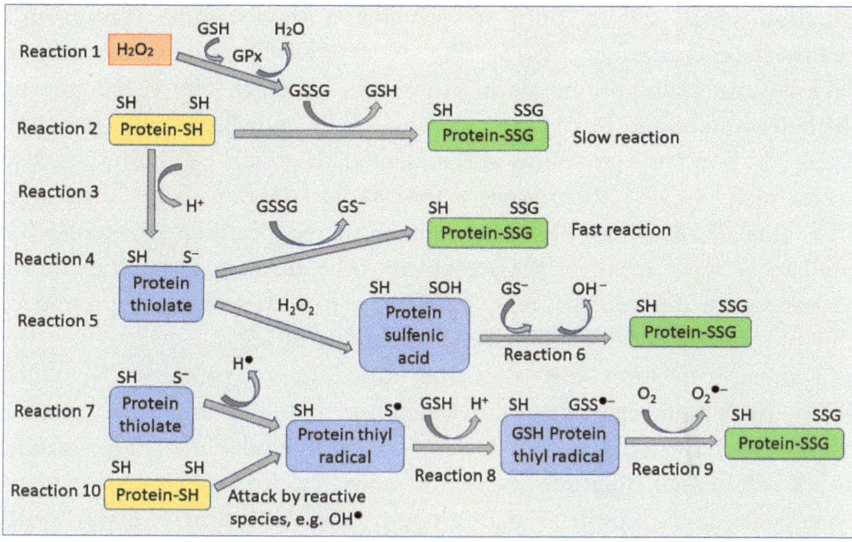

Fig. 9 S-glutathionylation of proteins. Reaction 1: GSSG is produced by detoxification of H_2O_2 by GSH and GPx. Reaction 2: A protein-SH reacts slowly with GSSG to produce a S-glutathionylated protein (Protein-SSG). Reaction 3: Under physiological conditions, protein cysteine thiols form thiolates (Protein-S^-), which react rapidly with GSSG to produce Protein-SSG (reaction 4). Reaction 5: Oxidation of protein cysteine thiolate to protein sulfenic acid (Protein-SOH), which reacts with GSH to produce Protein-SSG (reaction 6). Reaction 7: Protein cysteine thiolate is converted to a protein thiyl radical (Protein-$S^\bullet$), which reacts with GSH to produce a GSH Protein thiyl radical (Protein-$GSS^{\bullet-}$) (reaction 8). Reaction 9: Protein-$GSS^{\bullet-}$ reacts with oxygen to produce Protein-SSG and superoxide. Reaction 10: Attack of protein thiols by highly reactive species produces the protein thiyl radical (Protein-$S^\bullet$).

S-glutathionylation can proceed either enzymatically or non-enzymatically (Mailloux, Jin, & Willmore, 2014). Non-enzymatic S-glutathionylation reactions generally occur during oxidative distress and are unspecific. S-glutathionylation of proteins can proceed by three main reaction pathways (Fig. 9, reactions 4, 5 & 7). The first pathway involves a disulfide thiol exchange between GSSG and a protein thiolate ion (Protein-cysteine S^-), leading to the formation of a glutathionylated protein (Protein-SSG) and GSH (Fig. 9, reaction 4). The reaction of GSSG with un-ionized thiol groups (Protein-SH) (Fig. 9, reaction 2) is substantially slower than with thiolate ions (Protein-S^-) (Halliwell & Gutteridge, 2015) (Fig. 9, reaction 4). The second pathway involves the two-electron oxidation of a Protein-cysteine S^- to produce a sulfenic acid (Protein-SOH) (Fig. 9, reaction 5), which reacts with GSSG to form a glutathionylated protein (Protein-SSG). (Fig. 9, reaction 6). Pathway 3 involves the one-electron reduction of a protein thiol to produce a thiyl radical (Protein-S$^{\bullet}$) (Halliwell & Gutteridge, 2015) (Fig. 9, reaction 7), which reacts with GSH to form a glutathione anion protein thiyl radical (Fig. 9, reaction 8). The latter is stabilized by oxygen to produce a Protein-SSG (Fig. 9, reaction 9). In addition, protein thiols are subjected to attack by highly reactive species such as OH$^{\bullet}$ to produce a thiyl radical (Halliwell & Gutteridge, 2015) (Fig. 9, reaction 10). The protein mixed disulfide (Protein-SSG) could react with another thiolate group. The thiolate group can be on the same or a different protein, giving rise to either an intra-protein or inter-protein disulfide (Halliwell & Gutteridge, 2015).

S-glutathionylation can alter the functions of proteins (Mailloux, 2020b) by either activating or deactivating them. Examples of proteins that are deactivated by S-glutathionylation include adenylate cyclase, Na^+-K^+-ATPase, isocitrate dehydrogenase, calcium channels, enolase, glyceraldehyde-3-phosphate dehydrogenase, phosphofructokinase, pyruvate dehydrogenase, alpha-ketoglutarate dehydrogenase, mitogen-activated protein kinase/extracellular signal regulated kinase kinase kinase 1 (MEKK1) and mitochondrial uncoupling proteins (Cross & Templeton, 2004; Lillig & Berndt, 2013; Mailloux, 2020b). Further, the reversibility of S-glutathionylation modifications prevents the irreversible deactivation of critical metabolic enzymes by H_2O_2 (Mailloux, 2020b). These studies suggest that S-glutathionylation is a temporary inhibitor of important metabolic pathways, such as glycolysis, the citric acid cycle, oxidative phosphorylation, fatty acid combustion and amino acid metabolism (Mailloux, 2020b). This would allow diversion of energy toward pathways that produce NADPH and detoxify ROS. ROS inhibition and production of NADPH restores the redox buffering capacity of cellular GSH pools. These changes are sensed by glutaredoxins (Grx), and activate their deglutathionylase activities, leading to deglutathionylation of key proteins, and reactivation of their catabolic activities (Section 4.2).

Rare cases of activation by S-glutathionylation include sarco/endoplasmic reticulum calcium ATPase (SERCA), 12-lipoxygenase, carbonic anhydrase, and some calcium channels (Adachi et al., 2004; Halliwell & Gutteridge, 2015; Lillig & Berndt, 2013). S-glutathionylation can modulate the properties of cytoskeletal proteins such as actin and affect cell morphology. Other proteins that are regulated by S-glutathionylation include heat shock protein 70, caspase-3, protein tyrosine phosphatase 1B, Complex I of the inner mitochondrial membrane, and NF-κB signaling through I-κB kinase (Lillig & Berndt, 2013).

S-glutathionylation is an important post-translational modification of proteins. Protein cysteine oxidation involves post-translational modifications such as the formation of disulfides and sulfenic acids (Heppner et al., 2017). Many transcription factors and signaling molecules have critical cysteine residues that can be oxidized. This allows for the regulation of protein function by ROS, and regulation of cell signaling processes by GSH (Marinho, Real, Cyrne, Soares, & Antunes, 2014). S-glutathionylation plays an important role in redox signaling and can regulate transcription, energy metabolism, ion channel activity, cytoskeletal assembly, migration, proliferation, phagocytosis, calcium signaling and protein folding (Lillig & Berndt, 2013; Mailloux, 2020b). In mitochondria, S-glutathionylation acts as a redox sensor for the regulation of fission/fusion, bioenergetics, solute transport and proton leaks, and apoptosis (Mailloux, 2020b).

Protein mixed disulfides of glutathione can accumulate in the cytosol and mitochondria of cells subjected to oxidative distress (Mailloux et al., 2014). To avoid this, cells maintain low intracellular levels of GSSG under normal conditions by the action of GR (Fig. 6). S-glutathionylation of most protein cysteines generally occurs when GSSG levels are increased during oxidative distress, resulting in a decreased ratio of GSH:GSSG from 100:1 to about 1:1 (Gallogly & Mieyal, 2007). However, c-Jun is susceptible to S-glutathionylation by GSSG at a GSH:GSSG ratio of 13 (Klatt et al., 1999).

S-glutathionylation can protect sensitive thiol groups from further irreversible oxidation and could prevent loss of GSH under conditions of oxidative distress. Protein-SSG (Section 4.2) and protein sulfenic groups (Protein-SOH) can be regenerated by GSH (Fig. 9, reaction 6). However, protein sulfenic acid groups are unstable and can undergo further irreversible oxidation to sulfinic (Protein-SO_2H) and sulfonic acids (Protein-SO_3H), and the formation of protein crosslinks, which are difficult to repair. Sulfenic acids are now considered to be signaling molecules, whereas the formation of sulfinic and sulfonic acids is considered to occur under stress conditions (Heppner et al., 2017).

The action of H_2O_2 as a second messenger is thought to occur through thiol-based modification of target proteins (Sies & Jones, 2020). This could arise through the oxidation of protein cysteine thiols (Protein-SH) to form reactive protein sulfenic acids (Protein-SOH), which can either activate or deactivate protein function (Heppner et al., 2017). However, this reaction is controversial because H_2O_2 reacts slowly with most protein thiols (Winterbourn, 2018). Therefore, it is not entirely clear how thiols become oxidized during redox signal transmission. This reaction is likely to be outcompeted by much faster reactions of H_2O_2 with GSH, and by detoxification reactions of H_2O_2 by GPxs (Section 3.3) and Prxs (Cox et al., 2010; Forman et al., 2017; Heppner et al., 2017; Mailloux, 2020b). Furthermore, it appears unlikely that the formation of protein sulfenics would occur at physiological doses of H_2O_2. However, the rate of reaction of H_2O_2 with protein cysteine thiolate groups is several orders of magnitude faster than that with protein thiols (Section 4.1) (Sies & Jones, 2020). This reaction can alter protein function and is more likely to be involved in biological signaling.

The role of other enzymes in S-glutathionylation cannot be ruled out. It was reported that GSTs Pi and Mu were able to S-glutathionylate and activate AMP-activated protein kinase (AMPK) (Klaus et al., 2013). GST-Pi can modify several proteins by S-glutathionylation in the cytosol, and potentially other compartments such as the nucleus and mitochondria (Zhang, Ye, Singh, Townsend, & Tew, 2018).

4.2 Protein deglutathionylation

The reverse reaction of reversible S-glutathionylation is deglutathionylation, with GSH as substrate. The main enzyme responsible for deglutathionylation in human is Grx1. Grxs belong to the thioredoxin superfamily of proteins (Berndt, Christ, Rouhier, & Mühlenhoff, 2021). Grxs reduce oxidized thiols such as disulfides and glutathionylated thiols. Grx, also known as thiol-transferase, is a thiol-disulfide oxidoreductase that catalyzes deglutathionylation reactions (Gallogly & Mieyal, 2007) (Fig. 10, reaction 1). Oxidized Grxs are

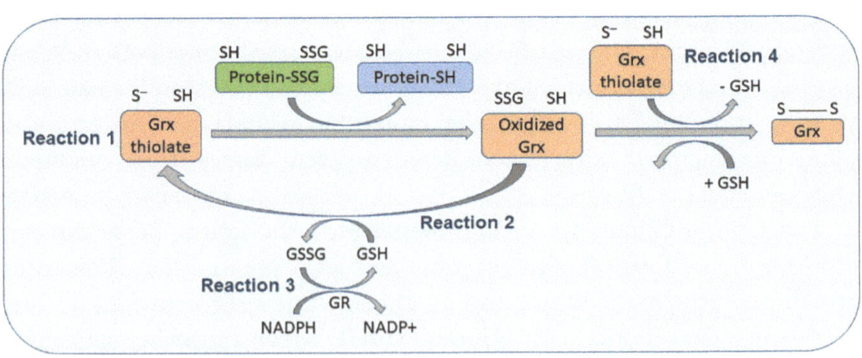

Fig. 10 Deglutathionylation of proteins by Grx. Reaction 1: Grx thiolate (Grx-S^-) reacts with Protein-SSG to produce Protein-SH and Grx-SSG. The Grx C-terminal can cross-react with Grx-SSG to produce GSH and a Grx intramolecular disulfide bridge (Grx-S—S). Reaction 2: Grx-SSG reacts with GSH to produce a Grx-thiolate and GSSG. GSH is regenerated from GSSG by GR and NADPH.

reduced by GSH, GR and NADPH (Fig. 10, reaction 2). Grxs catalyze thiol-disulfide interchange and can repair excess protein S-glutathionylation caused by oxidative distress.

Grxs play a role in many cellular functions such as protein folding, energy production, transcription, proliferation, detoxification, sulfate assimilation, cytoskeletal arrangement, protection against apoptosis, and embryonic development (Lillig & Berndt, 2013). They are found in the cytosol, nucleus, and mitochondria (Lillig & Berndt, 2013). Grx plays an important role in mitochondria because they cannot expel GSSG. The isoenzyme Grx2 is present in mitochondria and has been shown to deglutathionylate proteins such as Complex I and uncoupling protein-3 (UCP3) (Mailloux et al., 2014).

The active site motif of dithiol Grxs is Cys-x-x-Cys (Lillig & Berndt, 2013). The N-terminal active site cysteine of Grxs has a very low pK_a and is solvent accessible, whereas the C-terminal cysteine is buried and inaccessible. Due to the low pK_a, the N-terminal cysteine of Grxs is present as a thiolate group at physiological pH, which renders the catalytic reactions of Grxs more efficient (Yang & Wells, 1991). There are two steps in the enzymatic reaction (Mailloux et al., 2014). The first reaction is a thiol/disulfide exchange reaction, where the Grx N-terminal cysteine deglutathionylates Protein-SSG, producing Protein-SH and a Grx mixed disulfide (Grx-SSG) (Fig. 10, reaction 1). In the second reaction, Grx-SSG binds to GSH and the glutathionyl moiety is removed, producing GSSG and regenerating Grx (Fig. 10, reaction 2). GSSG is reduced back to GSH by the action of GR and NADPH through the glutathione redox cycle (Fig. 10, reaction 3; Fig. 6). The Grx C-terminal thiol can cross-react with Grx-SSG to produce GSH and an intramolecular disulfide bridge (Fig. 10, reaction 4). The disulfide bridge can be reversed by GSH, producing Grx-SSG.

Grx2 can also catalyze S-glutathionylation (Lillig & Berndt, 2013; Mailloux, 2020b). However, Grxs favor the reduction rather than the formation of mixed disulfides between protein thiols and GSH. However, when the GSH:GSSG ratio decreases, mixed disulfides are formed by Grx2-mediated S-glutathionylation.

The cycle of S-glutathionyation and deglutathionylation, catalyzed by GST-pi and Grx, respectively, is a reversible protein cysteine modification. This process is specific and responds to changes in the intracellular and extracellular environments (Mailloux, 2020b).

5. Conclusions and future directions

Despite tens of thousands of articles about glutathione and glutathione-dependent enzymes in the literature spanning numerous decades, many fundamental mechanistic questions remain unanswered about the role of glutathione metabolism at the cellular and organismal levels. Glutathione is an important antioxidant in aerobic organisms and has long been recognized for detoxifying H_2O_2 and other pro-oxidants. GSH can react directly as an antioxidant and acts as cofactor for several antioxidant and detoxification enzymes such as GPx, GST and glyoxalases. In addition, glutathione plays multiple roles in essential cellular processes such as protein folding, protection of protein thiols from oxidation and crosslinking, degradation of proteins with disulfide bonds, cell cycle regulation, cell proliferation, ascorbate metabolism, redox signaling, apoptosis, and energy production.

The reversible oxidation of protein cysteine residues through S-glutathionylation is an important regulatory mechanism in cellular redox signaling. Cysteine residues usually occur in critical structural and functional regions of proteins, where they are required for catalytic, metal binding and/or redox regulatory activity. Protein cysteine oxidation by ROS involves post-translational modifications such as the formation of protein disulfides (Protein-SSG) and protein sulfenic acids (Protein-SOH). These thiol-based modifications appear to protect protein cysteine residues from irreversible oxidation, and to regulate protein activity in response to pro-oxidants. However, the oxidation of protein thiols is controversial because ROS such as H_2O_2 react slowly with most protein thiols. The reaction kinetics are slow compared to more favorable reactions of H_2O_2 with abundant thiol-based reductants such as Prxs or GSH. The functional and/or structural consequences of these different oxidative events among the thousands of cysteine-containing cellular proteins are still not completely understood. Further understanding of the significance of protein thiol oxidation needs to consider the unique physical properties of different cell types, the distinctive properties of subcellular compartments and redox communication between them, the diffusion of ROS in different intracellular environments, reaction kinetics, the unique structural and physical properties of different thiol-containing proteins, and thiol oxidation in the context of multiprotein signaling complexes.

Acknowledgments

The author wishes to thank the Natural Sciences and Engineering Research Council of Canada (NSERC) (# 36725-16) (DAB) for financial support, and Dr. Mélanie Grondin for assistance in the preparation of several figures.

References

Adachi, T., Weisbrod, R. M., Pimentel, D. R., Ying, J., Sharov, V. S., Schoneich, C., et al. (2004). S-Glutathiolation by peroxynitrite activates SERCA during arterial relaxation by nitric oxide. *Nature Medicine, 10*(11), 1200–1207. https://doi.org/10.1038/nm1119.

Agostinelli, E., Przybytkowski, E., & Averill-Bates, D. A. (1996). Glucose, glutathione, and cellular response to spermine oxidation products. *Free Radical Biology & Medicine, 20*(5), 649–656. https://doi.org/10.1016/0891-5849(95)02149-3.

Akerboom, T. P. M., Bilizer, M., & Sies, H. (1982). The relationship of biliary GSSG efflux and intracellular GSSG content in perfused rat liver. *Journal of Biological Chemistry, 257*(8), 4248–4252.

Armeni, T., Cianfruglia, L., Piva, F., Urbanelli, L., Luisa Caniglia, M., Pugnaloni, A., et al. (2014). S-D-Lactoylglutathione can be an alternative supply of mitochondrial glutathione. *Free Radical Biology & Medicine, 67*, 451–459. https://doi.org/10.1016/j.freeradbiomed.2013.12.005.

Bachhawat, A. K., & Kaur, A. (2017). Glutathione degradation. *Antioxidants & Redox Signaling, 27*(15), 1200–1216. https://doi.org/10.1089/ars.2017.7136.

Bae, Y. S., Oh, H., Rhee, S. G., & Yoo, Y. D. (2011). Regulation of reactive oxygen species generation in cell signaling. *Molecules and Cells, 32*(6), 491–509. https://doi.org/10.1007/s10059-011-0276-3.

Bedard, K., & Krause, K. H. (2007). The NOX family of ROS-generating NADPH oxidases: Physiology and pathophysiology. *Physiological Reviews, 87*(1), 245–313. https://doi.org/10.1152/physrev.00044.2005.

Berndt, C., Christ, L., Rouhier, N., & Mühlenhoff, U. (2021). Glutaredoxins with iron-sulphur clusters in eukaryotes—Structure, function and impact on disease. *Biochimica et Biophysica Acta, Bioenergetics, 1862*(1), 148317. https://doi.org/10.1016/j.bbabio.2020.148317.

Bienert, G. P., & Chaumont, F. (2014). Aquaporin-facilitated transmembrane diffusion of hydrogen peroxide. *Biochimica et Biophysica Acta - General Subjects, 1840*(5), 1596–1604. https://doi.org/10.1016/j.bbagen.2013.09.017.

Buettner, G. R. (1993). The pecking order of free radicals and antioxidants: Lipid peroxidation, alpha-tocopherol, and ascorbate. *Archives of Biochemistry and Biophysics, 300*(2), 535–543. https://doi.org/10.1006/abbi.1993.1074.

Circu, M. L., & Aw, T. Y. (2010). Reactive oxygen species, cellular redox systems, and apoptosis. *Free Radical Biology & Medicine, 48*(6), 749–762. https://doi.org/10.1016/j.freeradbiomed.2009.12.022.

Cox, A. G., Winterbourn, C. C., & Hampton, M. B. (2010). Mitochondrial peroxiredoxin involvement in antioxidant defence and redox signalling. *Biochemical Journal, 425*(2), 313–325. https://doi.org/10.1042/BJ20091541.

Cross, J. V., & Templeton, D. J. (2004). Oxidative stress inhibits MEKK1 by site-specific glutathionylation in the ATP-binding domain. *Biochemical Journal, 381*(Pt 3), 675–683. https://doi.org/10.1042/BJ20040591.

Dalle-Donne, I., Rossi, R., Colombo, G., Giustarini, D., & Milzani, A. (2009). Protein S-glutathionylation: A regulatory device from bacteria to humans. *Trends in Biochemical Sciences, 34*(2), 85–96. https://doi.org/10.1016/j.tibs.2008.11.002.

Deponte, M. (2013). Glutathione catalysis and the reaction mechanisms of glutathione-dependent enzymes. *Biochimica et Biophysica Acta, General Subjects, 1830*(5), 3217–3266. https://doi.org/10.1016/j.bbagen.2012.09.018.

Deponte, M. (2017). The incomplete glutathione puzzle: Just guessing at numbers and figures? *Antioxidants & Redox Signaling, 27*(15), 1130–1161. https://doi.org/10.1089/ars.2017.7123.

Desideri, E., Ciccarone, F., & Ciriolo, M. R. (2019). Targeting glutathione metabolism: Partner in crime in anticancer therapy. *Nutrients, 11*(8), 1926. https://doi.org/10.3390/nu11081926.

Dickinson, B. C., & Chang, C. J. (2011). Chemistry and biology of reactive oxygen species in signaling or stress responses. *Nature Chemical Biology*, 7(8), 504–511. https://doi.org/10.1038/nchembio.607.

Forman, H. J., Davies, M. J., Krämer, A. C., Miotto, G., Zaccarin, M., Zhang, H., et al. (2017). Protein cysteine oxidation in redox signaling: Caveats on sulfenic acid detection and quantification. *Archives of Biochemistry and Biophysics, 617*, 26–37. https://doi.org/10.1016/j.abb.2016.09.013.

Forman, H. J., Zhang, H., & Rinna, A. (2009). Glutathione: Overview of its protective roles, measurement, and biosynthesis. *Molecular Aspects of Medicine, 30*(1–2), 1–12. https://doi.org/10.1016/j.abb.2016.09.013.

Fridovich, I. (1975). Superoxide dismutases. *Annual Review of Biochemistry, 44*, 147–159. https://doi.org/10.1146/annurev.bi.44.070175.001051.

Galaris, D., Barbouti, A., & Pantopoulos, K. (2019). Iron homeostasis and oxidative stress: An intimate relationship. *Biochimica et Biophysica Acta, Molecular Cell Research, 1866*(12), 118535. https://doi.org/10.1016/j.bbamcr.2019.118535.

Gallogly, M. M., & Mieyal, J. J. (2007). Mechanisms of reversible protein glutathionylation in redox signaling and oxidative stress. *Current Opinion in Pharmacology*, 7(4), 381–391. https://doi.org/10.1016/j.coph.2007.06.003.

Glory, A., & Averill-Bates, D. A. (2016). The antioxidant transcription factor Nrf2 contributes to the protective effect of mild thermotolerance (40°C) against heat shock-induced apoptosis. *Free Radical Biology & Medicine, 99*, 485–497. https://doi.org/10.1016/j.freeradbiomed.2016.08.032.

Halliwell, B. (2011). Free radicals and antioxidants—Quo vadis? *Trends in Pharmacological Sciences, 32*(3), 125–130. https://doi.org/10.1016/j.tips.2010.12.002.

Halliwell, B., & Gutteridge, J. M. C. (2015). Antioxidant defences synthetised *in vivo*. In B. Halliwell, & J. M. C. Gutteridge (Eds.), *Free radicals in biology and medicine* (5th ed., pp. 77–152). Oxford, UK: Oxford University Press. https://doi.org/10.1093/acprof:oso/9780198717478.001.0001.

Hayes, J. D., Flanagan, J. U., & Jowsey, I. R. (2005). Glutathione transferases. *Annual Review of Pharmacology and Toxicology, 45*, 51–88. https://doi.org/10.1146/annurev.pharmtox.45.120403.095857.

Hayes, J. D., & Pulford, D. J. (1995). The glutathione S-transferase supergene family: Regulation of GST and the contribution of the isoenzymes to cancer chemoprotection and drug resistance. *Critical Reviews in Biochemistry and Molecular Biology, 30*(6), 445–600. https://doi.org/10.3109/10409239509083491.

Heppner, D. E., Janssen-Heininger, Y. M. W., & van der Vliet, A. (2017). The role of sulfenic acids in cellular redox signaling: Reconciling chemical kinetics and molecular detection strategies. *Archives of Biochemistry and Biophysics, 616*, 40–46. https://doi.org/10.1016/j.abb.2017.01.008.

Higgins, L. G., & Hayes, J. D. (2011). Mechanisms of induction of cytosolic and microsomal glutathione transferase (GST) genes by xenobiotics and pro-inflammatory agents. *Drug Metabolism Reviews, 43*(2), 92–137. https://doi.org/10.3109/03602532.2011.567391.

Hiratsuka, A., Yamane, H., Yamazaki, S., Ozawa, N., & Watabe, T. (1997). Subunit Ya-specific glutathione peroxidase activity toward cholesterol 7-hydroperoxides of glutathione S-transferases in cytosols from rat liver and skin. *Journal of Biological Chemistry, 272*(8), 4763–4769. https://doi.org/10.1074/jbc.272.8.4763.

Hirschhorn, T., & Stockwell, B. R. (2019). The development of the concept of ferroptosis. *Free Radical Biology & Medicine, 133*, 130–143. https://doi.org/10.1016/j.freeradbiomed.2018.09.043.

Houle, F., Rousseau, S., Morrice, N., Luc, M., Mongrain, S., Turner, C. E., et al. (2003). Extracellular signal-regulated kinase mediates phosphorylation of tropomyosin-1 to promote cytoskeleton remodeling in response to oxidative stress: Impact on membrane blebbing. *Molecular Biology of the Cell, 14*(4), 1418–1432. https://doi.org/10.1091/mbc.e02-04-0235.

Huang, C., Anderson, M. E., & Meister, A. (1993). Amino acid sequence and function of the light subunit of rat kidney γ-glutamylcysteine synthetase. *Journal of Biological Chemistry, 268*(27), 20578–20583.

Huot, J., Houle, F., Marceau, F., & Landry, J. (1997). Oxidative stress-induced actin reorganization mediated by the p38 mitogen-activated protein kinase/heat shock protein 27 pathway in vascular endothelial cells. *Circulation Research, 80*(3), 383–392. https://doi.org/10.1161/01.res.80.3.383.

Jones, D. P., & Sies, H. (2015). The redox code. *Antioxidants & Redox Signaling, 23*(9), 734–746. https://doi.org/10.1089/ars.2015.6247.

Kalinina, E. V., Chernov, N. N., & Novichkova, M. D. (2014). Role of glutathione, glutathione transferase, and glutaredoxin in regulation of redox-dependent processes. *Biochemistry (Mosc), 79*(13), 1562–1583. https://doi.org/10.1134/S0006297914130082.

Kehrer, J. P., & Biswal, S. S. (2000). The molecular effects of acrolein. *Toxicological Sciences, 57*(1), 6–15. https://doi.org/10.1093/toxsci/57.1.6.

Kemp, M., Go, Y. M., & Jones, D. P. (2008). Nonequilibrium thermodynamics of thiol/disulfide redox systems: A perspective on redox systems biology. *Free Radical Biology & Medicine, 44*(6), 921–937. https://doi.org/10.1016/j.freeradbiomed.2007.11.008.

Klatt, P., Molina, E. P., De Lacoba, M. G., Padilla, C. A., Martinez-Galesteo, E., Barcena, J. A., et al. (1999). Redox regulation of c-Jun DNA binding by reversible S-glutathiolation. *FASEB Journal, 13*(12), 1481–1490. https://doi.org/10.1096/fasebj.13.12.1481.

Klaus, A., Zorman, S., Berthier, A., Polge, C., Ramirez, S., Michelland, S., et al. (2013). Glutathione S-transferases interact with AMP-activated protein kinase: Evidence for S-glutathionylation and activation in vitro. *PLoS One, 8*(5), e62497. https://doi.org/10.1371/journal.pone.0062497.

Lillig, C. H., & Berndt, C. (2013). Glutaredoxins in thiol/disulfide exchange. *Antioxidants & Redox Signaling, 18*(13), 1654–1665. https://doi.org/10.1089/ars.2012.5007.

Lu, S. C. (2013). Glutathione synthesis. *Biochimica et Biophysica Acta - General Subjects, 1830*(5), 3143–3153. https://doi.org/10.1016/j.bbagen.2012.09.008.

Lyles, G. A., & Chalmers, J. (1992). The metabolism of aminoacetone to methylglyoxal by semicarbazide-sensitive amine oxidase in human umbilical artery. *Biochemical Pharmacology, 43*(7), 1409–1414. https://doi.org/10.1016/0006-2952(92)90196-p.

Mailloux, R. J. (2020a). An update on mitochondrial reactive oxygen species production. *Antioxidants (Basel), 9*(6), 472. https://doi.org/10.3390/antiox9060472.

Mailloux, R. J. (2020b). Protein S-glutathionylation reactions as a global inhibitor of cell metabolism for the desensitization of hydrogen peroxide signals. *Redox Biology, 32*, 101472. https://doi.org/10.1016/j.redox.2020.101472.

Mailloux, R. J., Jin, X., & Willmore, W. G. (2014). Redox regulation of mitochondrial function with emphasis on cysteine oxidation reactions. *Redox Biology, 2*, 123–139. https://doi.org/10.1016/j.redox.2013.12.011.

Marinho, H. S., Real, C., Cyrne, L., Soares, H., & Antunes, F. (2014). Hydrogen peroxide sensing, signaling and regulation of transcription factors. *Redox Biology*, *2*, 535–562. https://doi.org/10.1016/j.redox.2014.02.006.

Meister, A. (1988). Glutathione. In I. M. Arias, W. B. Jakoby, H. Popper, D. Schachter, & D. A. Shafritz (Eds.), *The liver: Biology and pathobiology* (2nd ed., pp. 401–417). New York: Raven Press.

Meister, A., & Anderson, M. E. (1983). Glutathione. *Annual Review of Biochemistry*, *52*, 711–760. https://doi.org/10.1146/annurev.bi.52.070183.003431.

Nelson, D. J., & Cox, M. M. (2021). *Lehninger principles of biochemistry* (8th ed., p. 1096). New York: W.H. Freeman, Macmillan Learning.

Pallepati, P., & Averill-Bates, D. (2010). Mild thermotolerance induced at 40 degrees C increases antioxidants and protects HeLa cells against mitochondrial apoptosis induced by hydrogen peroxide: Role of p53. *Archives of Biochemistry and Biophysics*, *495*(2), 97–111. https://doi.org/10.1016/j.abb.2009.12.014.

Patra, K. C., & Hay, N. (2014). The pentose phosphate pathway and cancer. *Trends in Biochemical Sciences*, *39*(8), 347–354. https://doi.org/10.1016/j.tibs.2014.06.005.

Poole, L. B. (2015). The basics of thiols and cysteines in redox biology and chemistry. *Free Radical Biology & Medicine*, *80*, 148–157. https://doi.org/10.1016/j.freeradbiomed.2014.11.013.

Prabhu, K. S., Reddy, P. V., Jones, E. C., Liken, A. D., & Reddy, C. C. (2004). Characterization of a class alpha glutathione S-transferase with glutathione peroxidase activity in human liver microsomes. *Archives of Biochemistry and Biophysics*, *424*(1), 72–80. https://doi.org/10.1016/j.abb.2004.02.002.

Priya Dharshini, L. C., Vishnupriya, S., Sakthivel, K. M., & Rasmi, R. R. (2020). Oxidative stress responsive transcription factors in cellular signalling transduction mechanisms. *Cellular Signalling*, *72*, 109670. https://doi.org/10.1016/j.cellsig.2020.109670.

Rabbani, N., Xue, M., & Thornalley, P. J. (2016). Dicarbonyls and glyoxalase in disease mechanisms and clinical therapeutics. *Glycoconjugate Journal*, *33*(4), 513–525. https://doi.org/10.1007/s10719-016-9705-z.

Rabbani, N., Xue, M., & Thornalley, P. J. (2021). Dicarbonyl stress, protein glycation and the unfolded protein response. *Glycoconjugate Journal*, *38*(3), 331–340. https://doi.org/10.1007/s10719-021-09980-0.

Redza-Dutordoir, M., & Averill-Bates, D. A. (2016). Activation of apoptosis signalling pathways by reactive oxygen species. *Biochimica et Biophysica Acta, Molecular Cell Research*, *1863*(12), 2977–2992. https://doi.org/10.1016/j.bbamcr.2016.09.012.

Redza-Dutordoir, M., & Averill-Bates, D. A. (2021). Interactions between reactive oxygen species and autophagy: Special issue: Death mechanisms in cellular homeostasis. *Biochimica et Biophysica Acta, Molecular Cell Research*, *1868*(8), 119041. https://doi.org/10.1016/j.bbamcr.2021.119041.

Scherz-Shouval, R., & Elazar, Z. (2011). Regulation of autophagy by ROS: Physiology and Pathology. *Trends in Biochemical Sciences*, *36*(1), 30–38. https://doi.org/10.1016/j.tibs.2010.07.007.

Seelig, G. F., Simondsen, R. P., & Meister, A. (1984). Reversible dissociation of γ-glutamylcysteine synthetase into two subunits. *Journal of Biological Chemistry*, *259*(15), 9345–9347.

Sies, H. (2017). Hydrogen peroxide as a central redox signaling molecule in physiological oxidative stress: Oxidative eustress. *Redox Biology*, *11*, 613–619. https://doi.org/10.1016/j.redox.2016.12.035.

Sies, H., & Jones, D. P. (2020). Reactive oxygen species (ROS) as pleiotropic physiological signalling agents. *Nature Reviews in Molecular Cell Biology*, *21*(7), 363–383. https://doi.org/10.1038/s41580-020-0230-3.

Sinha, K., Das, J., Pal, P. B., & Sil, P. C. (2013). Oxidative stress: The mitochondria-dependent and mitochondria-independent pathways of apoptosis. *Archives of Toxicology*, *87*(7), 1157–1180. https://doi.org/10.1007/s00204-013-1034-4.

Song, C., Mitter, S. K., Qi, X., Beli, E., Rao, H. V., Ding, J., et al. (2017). Oxidative stress-mediated NFkappaB phosphorylation upregulates p62/SQSTM1 and promotes retinal pigmented epithelial cell survival through increased autophagy. *PLoS One*, *12*, e0171940. https://doi.org/10.1371/journal.pone.0171940.

Stockwell, B. R., Friedmann Angeli, J. P., Bayir, H., Bush, A. I., Conrad, M., Dixon, S. J., et al. (2017). Ferroptosis: A regulated cell death nexus linking metabolism, redox biology, and disease. *Cell*, *171*(2), 273–285. https://doi.org/10.1016/j.cell.2017.09.021.

Sun, X., Ou, Z., Chen, R., Niu, X., Chen, D., Kang, R., et al. (2016). Activation of the p62-Keap1-NRF2 pathway protects against ferroptosis in hepatocellular carcinoma cells. *Hepatology*, *63*(1), 173–184. https://doi.org/10.1002/hep.28251.

Tanel, A., & Averill-Bates, D. A. (2007). Inhibition of acrolein-induced apoptosis by the antioxidant N-acetylcysteine. *Journal of Pharmacology and Experimental Therapeutics*, *321*(1), 73–83. https://doi.org/10.1124/jpet.106.114678.

Tchouagué, M., Grondin, M., Glory, A., & Averill-Bates, D. (2019). Heat shock induces the cellular antioxidant defenses peroxiredoxin, glutathione and glucose 6-phosphate dehydrogenase through Nrf2. *Chemico-Biological Interactions*, *310*, 108717. https://doi.org/10.1016/j.cbi.2019.06.030.

Thornalley, P. J., Waris, S., Fleming, T., Santarius, T., Larkin, S. J., Winklhofer-Roob, B. M., et al. (2010). Imidazopurinones are markers of physiological genomic damage linked to DNA instability and glyoxalase 1-associated tumour multidrug resistance. *Nucleic Acids Research*, *38*(16), 5432–5442. https://doi.org/10.1093/nar/gkq306.

Torres, M., & Forman, H. J. (2003). Redox signaling and the MAP kinase pathways. *BioFactors*, *17*(1–4), 287–296. https://doi.org/10.1002/biof.5520170128.

Ulrich, K., & Jakob, U. (2019). The role of thiols in antioxidant systems. *Free Radical Biology & Medicine*, *140*, 14–27. https://doi.org/10.1016/j.freeradbiomed.2019.05.035.

Winterbourn, C. C. (2018). Biological production, detection, and fate of hydrogen peroxide. *Antioxidants & Redox Signaling*, *29*(6), 541–551. https://doi.org/10.1089/ars.2017.7425.

Winterbourn, C. C. (2019). Regulation of intracellular glutathione. *Redox Biology*, *22*, 101086. https://doi.org/10.1016/j.redox.2018.101086.

Wu, B., & Dong, D. (2012). Human cytosolic glutathione transferases: Structure, function, and drug discovery. *Trends in Pharmacological Sciences*, *33*(12), 656–668. https://doi.org/10.1016/j.tips.2012.09.007.

Yang, Y. F., & Wells, W. W. (1991). Identification and characterization of the functional amino acids at the active center of pig liver thioltransferase by site-directed mutagenesis. *Journal of Biological Chemistry*, *266*(19), 12759–12765.

Zhang, J., Ye, Z. W., Singh, S., Townsend, D. M., & Tew, K. D. (2018). An evolving understanding of the S-glutathionylation cycle in pathways of redox regulation. *Free Radical Biology & Medicine*, *120*, 204–216. https://doi.org/10.1016/j.freeradbiomed.2018.03.038.

CHAPTER SIX

Beneficial antioxidant effects of Coenzyme Q10 on reproduction

Maria Fernanda Hornos Carneiro[a,*] and Monica P. Colaiácovo[b,*]
[a]Department of Pharmacy, Faculty of Chemistry and Pharmacy, Pontificia Universidad Católica de Chile, Santiago, Chile
[b]Department of Genetics, Blavatnik Institute, Harvard Medical School, Boston, MA, United States
*Corresponding authors: e-mail address: maria.hornos@uc.cl; mcolaiacovo@genetics.med.harvard.edu

Contents

Abstract

This chapter focuses on preclinical and clinical studies conducted in recent years that contribute to increasing knowledge on the role of Coenzyme Q10 in female reproductive health. General aspects of CoQ10, such as its role as an antioxidant and in mitochondrial bioenergetics are considered. The age-dependent decline in human female reproductive potential is associated with cellular mitochondrial dysfunction and oxidative stress, and in some cases accompanied by a decrease in CoQ10 levels. Herein, we discuss experimental and clinical evidence on CoQ10 protective effects on reproductive health. We also address the potential of supplementation with this coenzyme to rescue reprotoxicity induced by exposure to environmental xenobiotics. This review not only contributes to our general understanding of the effects of aging on female reproduction but also provides new insights into strategies promoting reproductive health. The use of CoQ10 supplementation can improve reproductive performance through the scavenging of reactive oxygen species and free radicals. This strategy can constitute a low-risk and low-cost strategy to attenuate the impact on fertility related to aging and exposure to environmental chemicals.

Abbreviations

ART assisted reproductive technology
ATP adenosine triphosphate
BPA bisphenol A
CAT catalase
Cd cadmium

Vitamins and Hormones, Volume 121
ISSN 0083-6729
https://doi.org/10.1016/bs.vh.2022.10.004

CO	crossover
CoQ	Coenzyme Q
CoQ10	Coenzyme Q10
DDR	DNA damage response
DSBR	DNA double-strand break repair
EDCs	endocrine disrupting chemicals
ERs	estrogen receptors
FSHR	follicle-stimulating hormone receptors
GPx	glutathione peroxidase
GSH	glutathione
IVF	in vitro fertilization
IVM	in vitro maturation
LOAEL	lowest-observed-adverse-effect-level
NOAEL	no-observed-adverse-effect-level
OS	oxidative stress
PCNA	proliferating cell nuclear antigen
RONS	reactive oxygen and nitrogen species
SOD	superoxide dismutase
TRx	thioredoxin

1. Introduction

In comparison to anaerobic organisms that produce their energy from glucose, aerobic organisms have the advantage of producing much more energy during cellular respiration (Scott, 2005). However, this occurs at the expense of constantly generating reactive oxygen and nitrogen species (RONS) that contain unpaired electrons and are highly unstable. Depending on the level, the presence of RONS in biological systems can result either in homeostatic effects or in harmful interactions with biological molecules (Halliwell & Gutteridge, 2007; Indo et al., 2015). The physiological roles of low levels of RONS have been recognized for some time and include their microbicidal activity in neutrophils, and their role as short-lived second messengers in signal transduction (Dröge, 2002). However, either high levels of RONS, or when the antioxidant defense system is overloaded, can generate a state of intracellular imbalance known as oxidative stress (OS) resulting in injury to the genetic material (mitochondrial or nuclear), lipids from cellular membranes, and proteins (Moldogazieva, Lutsenko, & Terentiev, 2018; Srinivas, Tan, Vellayappan, & Jeyasekharan, 2019; Weidinger & Kozlov, 2015). Fortunately, antioxidant defense systems, consisting of either enzymatic or nonenzymatic components, can rapidly inactivate free radicals and other RONS (Halliwell & Gutteridge, 2007).

Enzymatic antioxidants include superoxide dismutase (SOD), glutathione peroxidase (GPx), catalase (CAT), and thioredoxin (TRx). Examples of endogenous nonenzymatic antioxidants include substances such as vitamins (A, C, E), flavonoids, carotenoids, polyphenols, glutathione (GSH), uric acid, melatonin, bilirubin, polyamines, allyl sulfides, curcumin, and the focus of this chapter: coenzyme Q10 (Mirończuk-Chodakowska, Witkowska, & Zujko, 2018; Nwose & Ewing, 2009).

Coenzyme Q10 (CoQ10), also known as ubidecaquinone or ubiquinone-10 (2,3-dimethoxy-5-methyl-6-multiprenyl-1,4-benzoquinone) is an endogenous antioxidant found in cells of almost all aerobic organisms (Overvad et al., 1999). In fact, the alternative name of Coenzyme Q (ubiquinone) refers to the ubiquitous presence of this family of proteins in living organisms and the existence of a benzoquinone group in their chemical structures (Acosta et al., 2016). Ubiquinone-10, or simply CoQ10, was first identified in 1940 (Greenberg & Frishman, 1988) and later isolated from beef heart mitochondria (Crane, Hatefi, Lester, & Widmer, 1957).

CoQ10 is synthesized in most human tissues via the following steps: (1) synthesis of the benzoquinone ring from 4-hydroxybenzoate derived from either tyrosine or phenylalanine amino acids, which confers the redox properties of the molecule; (2) synthesis of the polyisoprenoid "tail," responsible for its lipophilicity, from acetyl-coenzyme A, through the mevalonate pathway; and (3) the assembly of these two structures to finally form CoQ10 (Acosta et al., 2016; Kawamukai, 2016; Wang & Hekimi, 2019).

The benzoquinone structure exists in three redox states: the fully oxidized form known as CoQ10, the radical semiquinone intermediate (CoQ10H·), and the reduced form called ubiquinol ($CoQ10H_2$) (Gutierrez-Mariscal et al., 2020). The majority of CoQ10 *in vivo* is in the reduced form (Ernster & Forsmark-Andrée, 1993), which also makes this vitamin-like compound an important antioxidant for counteracting free radicals and RONS in addition to regenerating antioxidants such as GSH, α-tocopherol, and ascorbate (Alahmar & Naemi, 2022; Echtay, Winkler, & Klingenberg, 2000; Kagan & Quinn, 2000; Karamali & Gholizadeh, 2022; Kucharská et al., 2021; Lunetta & Roman, 2008; Nwose & Ewing, 2009).

There are several isoforms of Coenzyme Q (CoQ) and they can be distinguished based on the number of isoprenoid side-chains they present (Fig. 1). For instance, humans and most mammals have CoQ10 as its most prevalent Coenzyme Q (Karaayak & El Nehir, 2011; Ramasarma, 1985; Schou-Pedersen, Schemeth, & Lykkesfeldt, 2019), which refers to the existence of 10 isoprene repeats in the molecule of CoQ. Rats, mice,

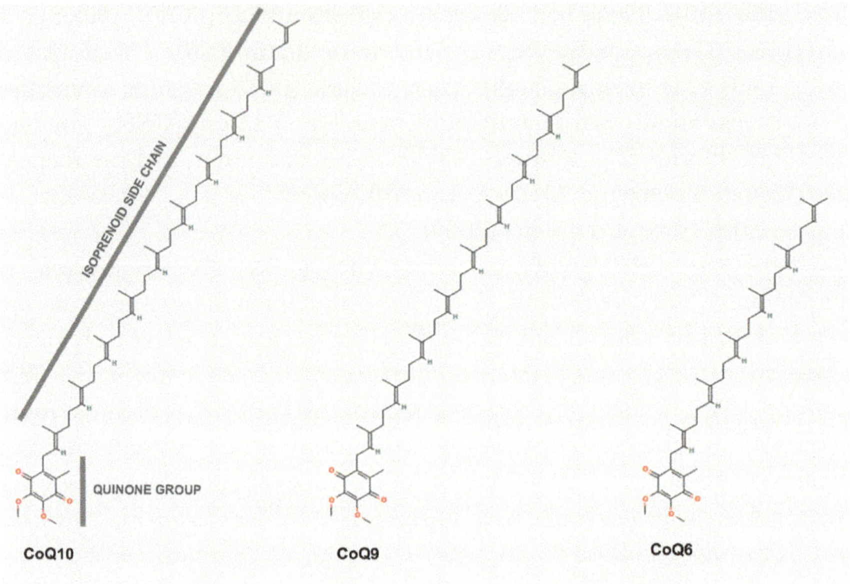

Fig. 1 Coenzymes Q (CoQ10, CoQ9, and CoQ6) chemical structures.

and guinea pigs, in turn, have Coenzyme Q9 (CoQ9) as the primary form (i.e., CoQ with 9 isoprene repeats), whereas Coenzyme Q6 (CoQ6), Coenzyme Q7 (CoQ7), and Coenzyme Q8 (CoQ8)—i.e., Quinone Coenzymes having 6, 7 and 8 isoprene repeats, respectively—are the predominant CoQ isoforms found in yeast and bacteria (Overvad et al., 1999).

Several proteins are needed to synthetize CoQ. For instance, the addition of isopentenyl diphosphates, derived from the mevalonate pathway, to the 4-hydroxybenzoic acid head to form the polyisoprenoid tail is catalyzed by the polyprenil diphosphate synthase. CoQ2 is needed to interpose the conjugation of the benzoquinone ring to the side chain, whereas CoQ3, CoQ5, CoQ6 and CoQ7 modify the benzoquinone structure by reactions of methylation, decarboxylation and hydroxylation (Kawamukai, 2016). There are also other proteins with different regulatory functions in CoQ biosynthesis. For instance, it has been proposed that ADCK1, ADCK2, ADCK5 and PTC7p may have the capacity to increase/reduce the CoQ biosynthetic rate by posttranslational modifications of some CoQ biosynthetic proteins, e.g., CoQ7, or by other unknown mechanisms (Kawamukai, 2016).

Mutations in any of the genes directly involved in CoQ10 synthesis can cause primary CoQ10 deficiency, a clinically and genetically heterogeneous disorder (Hargreaves, Heaton, & Mantle, 2020). Meanwhile, mutations in genes not directly related but also important to CoQ10 biosynthesis can result in secondary CoQ10 deficiency. Either primary or secondary CoQ10 deficiency can culminate in mitochondrial electron transport chain defects resulting in: impaired ATP production and antioxidant defense; defective pyrimidine, amino acid, and fatty acid metabolism; augmented inflammation; problems in lysosomal pH maintenance as well as in the regulation of the mitochondrial membrane potential (Crane, 2001; Hargreaves, 2014; Heaton, Heales, Rahman, Sexton, & Hargreaves, 2020; Hancock, Liu, Alvord, & Phang, 2015; Salvi & Gadda, 2013; Schmelzer et al., 2008). For instance, it has been shown that primary CoQ10 deficiency can arise from mutations in COQ1, involved in polyisoprenoid chain synthesis, and COQ2, that has a role in bond formation between the isoprenoid chain and the benzoquinone ring (Forsgren et al., 2004). Although the factors responsible for inducing a deficit in CoQ10 status in secondary CoQ10 deficiencies are currently not completely understood (Hargreaves et al., 2020), problems in the mevalonate pathway (for instance, in patients with the hereditary disorder phenylketonuria, due to high phenylalanine concentrations) or low blood levels of vitamin B6 (reported in patients with the rare genetic disorder called mucopolysaccharidosis I) are conditions associated with CoQ10 deficiency (Montero et al., 2019). Clinically, the resulting decrease in CoQ10 levels manifests as encephalomyopathy, nephropathy, cerebellar ataxia and atrophy, and isolated myopathy (Emmanuele et al., 2012).

CoQ10 is a lipid-soluble compound found in the membranes of various organelles such as the endoplasmic reticulum, peroxisomes, lysosomes, vesicles, and mitochondria (Heaton et al., 2020). Due to its primary function in energy production, the highest concentration of CoQ10 is found at the mitochondrias inner membrane (Overvad et al., 1999). CoQ10, together with cytochrome *c*, are itinerant electron-transporting molecules that connect five multisubunit enzymatic complexes (complex I–V) distributed in the inner mitochondrial membrane (Fig. 2). This shuttling between the complexes of the respiratory chain conducted by CoQ10 is facilitated by the lipophilicity of its polyisoprenoid chain, responsible for positioning CoQ10 within the mid-plane of the lipid bilayer of distinct types of cell membranes (Hargreaves et al., 2020). In its reduced form (ubiquinol), CoQ10 acts as an important electron-transporting coenzyme in the body by transferring electrons in the electron transport chain from complexes I and II to complex III in the mitochondrion (Fig. 2), establishing a proton gradient across the inner mitochondrial membranes that is coupled to

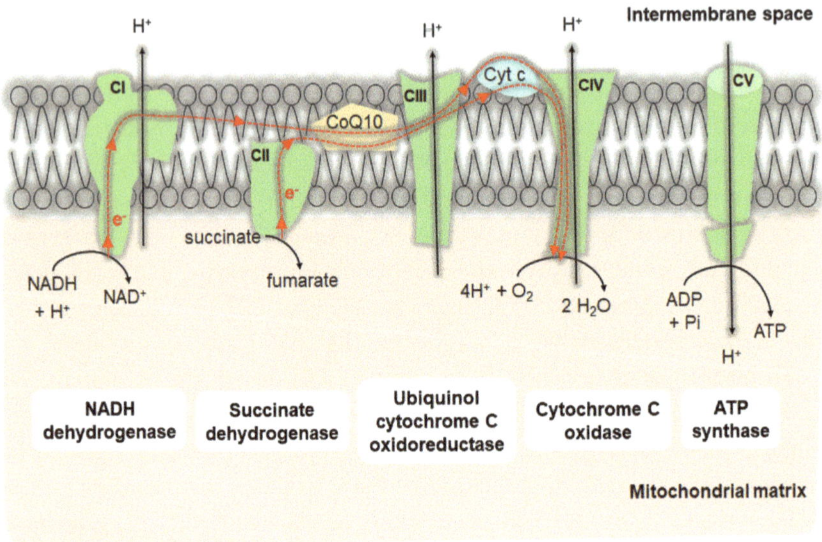

Fig. 2 Multisubunit oxidative phosphorylation complexes in mitochondria. Distributed in the inner mitochondrial membrane is an efficient arrangement of five multisubunit enzymatic complexes designated as NADH dehydrogenase (Complex I, CI), succinate dehydrogenase (Complex II, CII), ubiquinol-cytochrome c oxidoreductase (Complex III, CIII), cytochrome c oxidase (Complex IV, CIV), and ATP synthase (Complex V, CV). Shown is a representation of a classical electron transfer chain.

synthesis of adenosine triphosphate (ATP) by ATPase (complex V) (Larsen & Clarke, 2002).

In humans, CoQ10 is ubiquitous although its levels are higher in organs with high rates of metabolic activity such as the heart, kidneys, and liver (110, 70, and 60 μg/g tissue, respectively), and its lowest concentration is detected in the lungs (8 μg/g) (Ernster & Forsmark-Andrée, 1993). While the majority of CoQ10 in the human body is in the reduced form, there are exceptions such as in brain and lung tissues where it is mostly found as ubiquinone (Ernster & Forsmark-Andrée, 1993). Although CoQ10 is produced *in vivo*, it can also be obtained through the diet. Studies have shown that endogenous synthesis and a well-balanced diet provide appropriate CoQ10 levels in healthy people, while CoQ10 concentrations in tissues decline with age and some disease states (Ernster & Forsmark-Andrée, 1993). The main nutritional sources of CoQ10 include fatty fish (i.e., salmon, tuna and sardines), organ meats such as liver and beef heart, whole grains, and green vegetables such as spinach and broccoli (Pravst, Zmitek, & Zmitek, 2010; Román-Pizarro, Fernández-Romero, & Gómez-Hens, 2017). At present, there is uncertainty regarding the daily requirement for CoQ10, but it is estimated as 500 mg/day. Approximately 5 mg of CoQ10 is obtained daily from the diet while the remaining daily requirement is synthesized endogenously (Weber, Bysted, & Hłlmer, 1997).

A well-balanced diet very often supplies sufficient amounts of CoQ10, but supplementation may be beneficial in particular situations. CoQ10 is commercially available as a supplement in several pharmaceutical forms, including capsules (soft gel and hard shell), oral sprays, tablets as well as skin-care formulations (Knott et al., 2015; Lee, Huang, Chen, & Lin, 2012).

Medicinal use of CoQ10 is intended for treating serious mitochondrial disorders and other metabolic syndromes that are caused by mutations in coenzyme Q10 biosynthetic genes (Acosta et al., 2016; Neergheen et al., 2017; Xie, Jiang, & Guo, 2021). The recommended daily dose of CoQ10 supplementation ranges from 5 to 30 mg/kg (Marin & Haas, 2015; Pineda et al., 2010; Rötig et al., 2000), with 1200 and 3000 mg/day being the highest daily doses reported for long- and short-term supplementation, respectively (Ferrante et al., 2005; Huntington Study Group, 2001; Shults et al., 2002). With regards to safety, no serious side effects of chronic CoQ10 supplementation have been reported; however, some individuals have reported gastrointestinal adverse effects, such as nausea, diarrhea, changes in appetite, stomachaches, and abdominal discomfort, especially with doses $\geq$200 mg/day (Hendler & Rorvik, 2008). The no-observed- or lowest-observed-adverse-effect-level (NOAEL or LOAEL) resulting from CoQ10 consumption by humans has not been identified (Hathcock & Shao, 2006). Since the safety of CoQ10 supplementation in pregnant/lactating women or children has not been fully established, more data is needed to support supplementation in these cases.

2. CoQ10 supplementation promotes female reproductive health

2.1 Maternal age effect and oxidative stress

Current research on CoQ10 supplementation focuses on its role in the treatment of neurodegenerative, cardiovascular, and metabolic diseases including cancer as well as other conditions such as infertility. Infertility affects millions of individuals globally with current estimates indicating that approximately 15% of couples (about 48.5 million couples) experience reproductive issues (Agarwal, Mulgund, Hamada, & Chyatte, 2015). Fertility declines with increased age in both men and women, and in fact, one of the hallmarks of reproductive aging is a deterioration in energy metabolism in the reproductive tissues, especially the gonads. However, the effects of age are more premature and evident in women (US Department of Health and Human Services, 2022). Moreover, older women have fewer eggs available and are more susceptible to medical conditions that could result in infertility (Practice Committee of the American Society for Reproductive Medicine, 2017).

Diploid germ cells (2 N) carry the genomic blueprint that is transmitted from one generation to the next, and undergo the specialized cell division program of meiosis, during which the chromosome number is reduced by half resulting in the formation of haploid eggs and sperm (1 N) that upon fertilization will reconstitute proper ploidy. Errors in any key step of meiosis, such as homologous chromosome pairing, synapsis, DNA double-strand break repair (DSBR) and crossover (CO) formation, or in the activation of the DNA damage response (DDR) pathway, can result in errors in chromosome segregation leading to an increased rate of germ cell apoptosis and/or aneuploidy (cells carrying an incorrect number of chromosomes) (Láscarez-Lagunas, Martinez-Garcia, & Colaiácovo, 2022). Aneuploidy is associated with birth defects (e.g., Down Syndrome), clinically recognized miscarriages, stillbirths, tumorigenesis, and infertility (Mandrioli, Belpoggi, Silbergeld, & Perry, 2016; Webster & Schuh, 2017). One of the causes of female infertility is the oxidative damage known to escalate with age (Meldrum et al., 2016). Studies in flies have demonstrated that oxidative stimuli in *Drosophila* oocytes during meiotic prophase result in a significant increase in chromosome segregation errors due to premature loss of sister chromatid cohesion (Perkins, Das, Panzera, & Bickel, 2016; Perkins et al., 2019). In addition, it has been extensively shown that optimal energy production in mitochondria is essential for ideal oocyte maturation, fertilization, and embryonic development (Rahimi Darehbagh, Khalafi, Allahveisi, & Habiby, 2022). Therefore, the age-dependent decline in human female reproductive potential may be connected with increased oxidative damage and/or mitochondrial dysfunction, as these organelles are necessary for proper meiotic spindle assembly, chromosome segregation, maturation, fertilization, and embryo development (Sasaki et al., 2019; Wang, Wang, Zou, & Xu, 2009). Insufficient levels of antioxidants in the cell, such as CoQ10, could lead to reduced mitochondrial respiration, resulting in lower ATP synthesis, increased levels of RONS and augmented oxidative stress, resulting in mitochondrial damage and dysfunction (Rodríguez-Varela & Labarta, 2021). This scenario corresponds to a positive feedback loop, in which lower

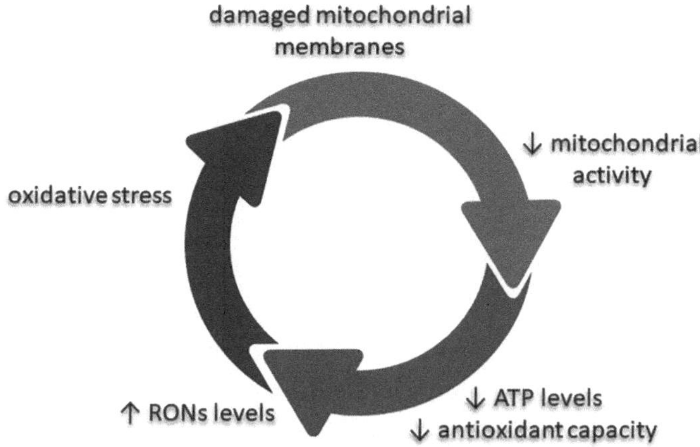

Fig. 3 Positive feedback loop between cellular mitochondrial dysfunction and oxidative stress. *Adapted from Rodríguez-Varela, C., & Labarta, E. (2021). Does coenzyme Q10 supplementation improve human oocyte quality? International Journal of Molecular Sciences 22(17):9541. doi: 10.3390/ijms22179541.*

mitochondrial activity generates redox imbalance, which as a consequence induces injury to the mitochondrial membrane impacting the activity of these organelles (Fig. 3).

Although both mitochondrial dysfunction and oxidative stress have been associated with the maternal age effect, oxidative stress can be caused by or be the cause of mitochondrial dysfunction (Rodríguez-Varela & Labarta, 2020), and insufficient CoQ10 levels may contribute to generating both of these. Therefore, given CoQ10's key roles in mitochondrial homeostasis as an important antioxidant, and that reproduction is an essential biological process for the perpetuation of the species, we will next address the effects of CoQ10 in promoting reproductive health, with a focus on evidence of CoQ10 supplementation promoting or restoring female fertility.

2.1.1 Experimental evidence

Oral preovulatory CoQ10 treatment (22 mg/kg) has shown encouraging results on the ovaries of aged mice, such as improvements in the ovarian reserve and oocyte quality along with the normalization of oocyte mitochondrial parameters (in relation to young controls) (Ben-Meir et al., 2015). These improvements encompass increases in the number and the mitochondrial activity of cumulus cells surrounding the oocyte, and have been observed *in vivo* even at lower levels of CoQ10 supplementation (84 μg/kg) (Ben-Meir et al., 2019). Supplementation with CoQ10 has also been shown to rescue oxidative stress damage in the oocytes of young mice (Boots, Boudoures, Zhang, Drury, & Moley, 2016), rats (Ozcan et al., 2016), and nematodes (Hornos Carneiro et al., 2020).

Experimental *ex vivo* studies show higher maturation rates in immature bovine oocytes/mouse oocytes treated with follicular fluid from infertile women with endometriosis in *in vitro* maturation (IVM) medium supplemented with CoQ10 (40 μM/50 μM, respectively) (Abdulhasan et al., 2017; Romero et al., 2020). Several studies have described potential mechanisms by which CoQ10 act as a beneficial supplement to counteract infertility using ovine (Heydarnejad et al., 2019), porcine (Yang et al., 2020), and mice oocytes (Xing et al., 2022). For instance, ovine oocytes cultured with 15–50 μM of CoQ10 showed an increase in blastocyst formation and hatching rates; improved distribution, relative mass, and membrane potential of mitochondria; decreased levels of reactive oxygen species and of oocytes with misaligned chromosomes after spindle assembly (Heydarnejad et al., 2019). These phenotypes in CoQ10-treated oocytes were accompanied by significantly reduced expression of CASPASE3 and BAX and augmented expression of GDF9, an oocyte-specific growth factor. Mice oocytes treated with 10 μM of CoQ10 showed attenuation in aging-induced abnormalities (including mitochondrial dysfunction, RONS accumulation, spindle abnormalities, and apoptosis in postovulatory aged oocytes) accompanied by downregulation of the mRNA and protein levels for SIRT4, a protein that is indispensable for DNA repair, control of inflammation and antioxidative defense. SIRT4 knockdown was shown to rescue aging-induced dysfunction of mitochondria, whereas SIRT4 overexpression decreased the efficiency of CoQ10 in rescuing mitochondria dysfunction (Heydarnejad et al., 2019). However, more investigation of the effects of CoQ10 supplementation on proteins involved in mitochondrial biogenesis and mitophagy is needed. CoQ10 (50 μM) has also been shown to promote nuclear maturation of pig oocytes and accelerate their development to the blastocyst stage, and to boost mitochondrial function, therefore increasing the levels of mitochondrial membrane potential, ATP production, and CoQ6 transcripts (Yang et al., 2020). CoQ10 treatment also reduced the levels of reactive oxygen species and apoptotic oocytes, corroborating previous findings on the antioxidant potential of this coenzyme. To date, only one study has reported absence of beneficial effects of CoQ10 on reproductive health (Maside et al., 2019). Overall, the role of CoQ10 as a stabilizing agent upon redox imbalances leads to improvement of female reproductive health, by resulting in lower rates of apoptosis and meiotic abnormalities, as well as better mitochondrial function and reproductive performance.

2.1.2 Clinical studies

Clinical *ex vivo* studies have evaluated the effect of CoQ10 supplementation on oocyte maturation rates and post meiotic aneuploidy rates during IVM of human oocytes. One study compared the outcomes from cultivating oocytes in IVM media for 24–48 h with or without 50 μmol/L CoQ10 from patients aged $\geq$ 38 years ($n = 45$) and $\leq$ 30 years ($n = 18$) undergoing *in vitro* fertilization (IVF) (Ma et al., 2020). This intervention resulted in significantly increased oocyte maturation rates and reduced oocyte and chromosome aneuploidy frequencies in women aged 38–46 years. An earlier study by the same researchers showed that supplementation with CoQ10 (50 μM) of oocytes from women between 35 and 46 years of age decreased the post-meiotic oocyte aneuploidy rate and increased the number of mitochondria (Ma, Hu, Ma, Cui, & Liu, 2018). These studies support CoQ10 as a promising strategy for rescuing defects generated from the age-related exposure to an oxidative follicular environment in humans.

Better embryo morphokinetic parameters and higher pregnancy rates have been documented in women under 41 years of age with higher CoQ10 levels in their follicular fluid, regardless of their age (Akarsu, Gode, Isik, Dikmen, & Tekindal, 2017). Therefore, deficits in cellular CoQ10 may contribute to infertility in women. Although some studies on women undergoing IVF treatment have failed to detect a beneficial effect of CoQ10 (Bentov, Hannam, Jurisicova, Esfandiari, & Casper, 2014; Caballero, Fiameni, Valcarcel, & Buzzi, 2016; Giannubilo et al., 2018), others report a correlation between elevated levels of CoQ10 and improved fertility (El Refaeey, Selem, & Badawy, 2014; Gat et al., 2016; Sen Sharma, 2017; Xu et al., 2018). More detailed information on the effects of CoQ10 taken orally, prior to any assisted reproductive technology (ART), or as a culture media adjuvant during IVF treatment can be found in (Rodríguez-Varela & Labarta, 2021).

2.1.3 Studies showing beneficial effects of CoQ10 on reprotoxicity

Endocrine disrupting chemicals (EDCs) are contaminants ubiquitously present in our environment and have been receiving intense attention in recent years due to their potential for a variety of toxicological effects. These are mostly man-made chemicals used as additives in the manufacture of several consumer products. Examples include perfluoroalkyl substances, phthalate esters, alkylphenols, pesticides and biocides, metals such as cadmium, lead, and mercury, dioxins, brominated and chlorinated substances among many others (La Merrill et al., 2020). Several EDCs exhibit nonmonotonic dose-response curves which challenges the establishment of safe exposure limits (Vandenberg et al., 2012). Therefore, legislation to ban EDCs and protect pregnant women and young children is urgently needed and may need to be revised and adjusted on a regular basis based on new developments and scientific findings (Kiess & Haeusler, 2021).

EDCs can act as hormone agonists or antagonists, alter hormone receptor expression, modify hormone synthesis, metabolism and transport, interfere with signaling mechanisms, induce epigenetic modifications in hormone-producing or hormone-responsive cells, and consequently disrupt hormonal homeostasis and developmental processes (La Merrill et al., 2020). In addition, environmental contaminants and xenobiotics we encounter on a daily basis are able to generate oxidative damage (Rocha, Asimakopoulos, Barbosa, & Kannan, 2017). Interactions of EDCs with hormone receptors such as estrogen receptors (ERs) can result in the formation of RONS leading to oxidative and nitrosative stress that ultimately can affect reproductive health (Hamid, Junaid, & Pei, 2021; Henderson & Colaiácovo, 2021). The generation of reactive oxygen species and reactive nitrogen species (ROS and RNS) can also occur through detoxification reactions and altered expression of antioxidant enzymes as documented for several EDCs. This may lead to an altered microenvironment surrounding the oocytes and, together with the aging process, constitute an explanation for fertility problems (Skakkebæk et al., 2022).

It is known that some EDCs can impair mitochondrial bioenergetics, biogenesis and dynamics, reducing the OXPHOS capacity, uncoupling respiration, and generating increased mitochondrial-derived reactive oxygen species in metabolic tissues such as liver (Azevedo et al., 2019). Although signs of mitochondrial dysfunction resulting from exposure to EDCs have been reported in reprotoxicity studies (Hornos Carneiro et al., 2020; Jia et al., 2019), there are limited mechanistic studies investigating mitochondrial function following EDC exposure in mammalian oocytes. For instance, exposure to melamine, a resin used for producing thermosetting plastics for kitchenware, results in abnormal mitochondrial distribution in the oocyte, affecting its maturation (Duan, Dai, Wang, Liu, & Sun, 2015). Exposure to the widely used plasticizer and endocrine disruptor bisphenol A (BPA) at 100 μM induces the abnormal distribution of mitochondria, which are observed clustered or dispersed in the oocyte instead of aggregated around the spindle (Pan et al., 2021). This change in mitochondrial localization pattern results in an insufficient supply of ATP (required for meiosis), which obstructs oocyte development. While more information is needed on how mitochondrial energy metabolism is altered in EDC-induced female reprotoxicity, CoQ10 could be a viable strategy to rescue this mitochondrial dysfunction due to its antioxidant potential and importance in the oxidative phosphorylation process.

We recently showed that CoQ10 supplementation is able to counteract the reproductive toxicity induced by BPA in the nematode *Caenorhabditis elegans*, in part by decreasing the elevated levels of DNA breaks resulting from BPA-induced oxidative damage in the germline (Hornos Carneiro et al., 2020). Age-synchronized worms were placed on DMSO/BPA-containing plates seeded with *Escherichia coli* strain OP50 for approximately 53 h (L1 through L4 stage). At the L4 stage, worms were exposed to DMSO, DMSO + CoQ10, BPA, or BPA + CoQ10 in liquid culture for 24 h and then washed and collected for evaluation. CoQ10 significantly reduced the elevated levels of BPA-induced germ cell apoptosis, checkpoint kinase 1 activation, and DNA double-strand breaks in the germline, as well as the rate of chromosome defects observed in the oocytes at diakinesis. Furthermore, CoQ10 treatment decreased mitochondrial dysfunction, oxidative stress, and mRNA levels of antioxidant enzymes in the germline of BPA-exposed worms. This study showed for the first time that CoQ10 supplementation is able to diminish the levels of aneuploid embryos and BPA-induced defects observed in early embryonic divisions.

Cadmium (Cd) is a potent toxicant that produces oxidative imbalance by generating RONS and therefore increasing lipid peroxidation affecting intracellular glutathione levels (Genchi, Sinicropi, Lauria, Carocci, & Catalano, 2020). It affects the ubiquitin/ATP-dependent proteolytic pathway. CoQ10 supplementation has also been proven to ameliorate cadmium (Cd)-induced deleterious effects on reproductive performance of Japanese quail (*Coturnix japonica*) (Rafieian-Naeini, Zhandi, Sadeghi, Yousefi, & Benson, 2021). In this study, birds were fed during 8 weeks with basal diet (control group), basal diet contaminated with Cd (10 mg/kg; Cd exposed group), or basal diet containing both Cd and CoQ10 (900 mg/kg diet; CdQ10). Co-treatment with CoQ10 decreased the mortality rate and improved the decreased egg production and hatching rate seen in Cd-treated quails compared to controls. This was accompanied by a reduction in infiltration of inflammatory cells in the reproductive tissue and RNA levels for HSP70. A subsequent study also reported beneficial effects of CoQ10 supplementation on egg production, egg mass, eggshell thickness and egg yolk color intensity in Japanese quail (Rafieian-Naeini et al., 2022).

The protective effect of CoQ10 has also been evaluated against the reprotoxicity of some medications. Given that chemotherapy produces significant toxic effects on folliculogenesis, leading to irreversible premature ovarian failure, Delkhosh et al. (2019) assessed the protective potential of CoQ10 on oocytes following chemotherapy with cyclophosphamide. NMRI mice were divided into four groups: control, CoQ10, cyclophosphamide, and cyclophosphamide + CoQ10. CoQ10 rescued the decreased mRNA expression of follicle-stimulating hormone receptors (FSHR) and proliferating cell nuclear antigen (PCNA), IVF rate, and embryo development induced by cyclophosphamide. CoQ10 treatment also decreased ROS formation and the number of atretic follicles in the ovary of CTX-treated mice (Delkhosh et al., 2019).

Ovarian stem cells derived from ovarian surface epithelium can be a target for treating infertility due to ovarian aging. Lee et al. (2021) examined the effect of CoQ10 on ovarian stem cells using a 4-vinylcyclohexene diepoxide (VCD)-induced murine model of ovarian failure. Female C57BL/6 mice were divided into four groups: control, CoQ10 (150 mg/kg/day), VCD (160 mg/kg/day) and VCD + CQ10. VCD was dissolved in saline and administered intraperitoneally for 5 days whereas CoQ10 was given orally for 14 days. Following treatment regimen, the number of surviving follicles (from primordial to antral follicles) as well as the number of zygotes retrieved and embryo development rate to blastocyst were significantly greater in the VCD + CoQ10 group than in the VCD group. CoQ10 supplementation (VCD + CoQ10 group) also augmented the levels of serum anti-müllerian hormone and ovarian mRNA levels of Bmp-15, Gdf-9, and c-Kit, genes related to ovarian aging, seen decreased in the VCD-treated group. In contrast, serum ROS level was significantly lower in the VCD + CoQ10 group than in the VCD group (Lee et al., 2021). This indicates that CoQ10 is able to improve reproductive performance while reducing ROS levels in a model of ovarian failure. Table 1 summarizes the findings on the protective role of CoQ10 on chemical-induced reprotoxicity observed *in vivo*.

3. Potential pharmacologic interactions of CoQ10 supplementation

Although the use of CoQ10 has been described as an effective and safe strategy to postpone oocyte aging, its use can generate adverse effects due to interactions with other substances and their biological targets. For instance, CoQ10 supplements were found to decrease the anticoagulant efficacy of warfarin in longitudinal studies (Engelsen, Nielsen, & Winther, 2002; Shalansky, Lynd, Richardson, Ingaszewski, & Kerr, 2007). Moreover,

Table 1 Overview of *in vivo* experimental reports on the beneficial effects of CoQ10 supplementation on reprotoxicity induced by different xenobiotics

Reprotoxic agent, dose and (main) route of exposure	Dose of CoQ10	Exposure regimen	Experimental model	Outcomes	Mechanism of CoQ10-induced protective effects	References
BPA (500 μM), oral	100 mg/mL	BPA from L1 to L4 (larval stages), co-exposure BPA-CoQ10 for 24 h	*Caenorhabditis elegans*	CoQ10 decreased BPA-induced elevated levels of germ cell apoptosis, phosphorylated checkpoint kinase 1, meiotic double-strand breaks, and chromosome defects in diakinesis oocytes. CoQ10 supplementation ameliorated BPA-induced oxidative stress, mitochondrial dysfunction, and altered gene expression of antioxidant enzymes in the germline, as well as reduced the levels of aneuploid embryos and defects observed in early embryonic divisions	CoQ10 rescues the reprotoxicity induced by BPA in part by neutralizing DNA damage resulting from oxidative stress	Hornos Carneiro et al. (2020)
Cadmium (10 mg/kg), subcutaneously	900 mg/kg, via diet	8 weeks of CoQ10 supplementation (starting at the 6th week of life), Cd injection at weeks 10 and 11	*Female Japanese quail (Coturnix japonica) aged 6 weeks*	CoQ10-supplemented quails produced more eggs with higher hatchling quality, had a reduction in infiltration of inflammatory cells and in RNA levels of HSP70 expressed in reproductive tissue	Although not reported, this study suggests that CoQ10 is able to counteract Cd-induced oxidative stress	Rafieian-Naeini et al. (2021)
Cadmium (10 mg/kg), subcutaneously	900 mg/kg, via diet	8 weeks of CoQ10 supplementation (starting at the 6th week of life), Cd injection at weeks 10 and 11	*Female Japanese quail (Coturnix japonica) aged 6 weeks*	CoQ10 supplementation partially mitigated the Cd-induced decrease in egg production, egg mass and eggshell thickness. Egg yolk color intensity was enhanced by CoQ10		Rafieian-Naeini et al. (2022)

Continued

Table 1 Overview of *in vivo* experimental reports on the beneficial effects of CoQ10 supplementation on reprotoxicity induced by different xenobiotics—cont'd

Reprotoxic agent, dose and (main) route of exposure	Dose of CoQ10	Exposure regimen	Experimental model	Outcomes	Mechanism of CoQ10-induced protective effects	References
Cyclophosphamide (20 mg/kg), intraperitonially	22 mg/kg, orally	Cyclophosphamide was given intraperitonially for 21 days whereas CoQ10 was supplemented every other day for 21 days	NMRI female mice aged 8–10 weeks	CoQ10 rescued the decreased mRNA expression of follicle-stimulating hormone receptors (FSHR) and proliferating cell nuclear antigen (PCNA), IVF rate, and embryo development induced by cyclophosphamide	CoQ10 acts as an antioxidant to decrease ROS formation and the number of atretic follicles in the ovary of cyclophosphamide-treated mice	Delkhosh et al. (2019)
4-Vinylcyclohexene diepoxide (VCD, 160 mg/kg), intraperitonially	150 mg/kg, orally	CoQ10 administered for 14 days (day 1 to 14), VCD administered for 5 days (day 1 to 5)	Female C57BL/6 mice aged 6 weeks	The numbers of surviving follicles, zygotes retrieved and embryo development rate to blastocyst were significantly greater in the VCD + CoQ10 group than in the VCD group. CoQ10 supplementation (VCD + CoQ10 group) also augmented the levels of serum anti-müllerian hormone and ovarian mRNA levels of Bmp-15, Gdf-9 and c-Kit. Serum ROS level was significantly lower in the VCD + CoQ10 group than in the VCD group	CoQ10 is able to improve reproductive performance while reducing ROS levels in a model of ovarian failure	Lee et al. (2021)

experimental analysis showed that the increased clearance values of warfarin in rats treated with CoQ10 were attributable to acceleration of certain metabolic pathways and renal excretion of the warfarin enantiomers (Zhou & Chan, 2001; Zhou, Zhou, & Chan, 2005). Therefore, the range of adverse effects of CoQ10 supplementation in patients using warfarin and other low volume of distribution drugs need to be evaluated in detail, mostly through randomized trials, and these patients should be advised to avoid using this supplement due to potential interactions.

Another recognized potential interaction of CoQ10 involves cholesterol-lowering medications or statins which can decrease circulating CoQ10 concentrations. This is explained by the statin-induced inhibition of 3-hydroxy-3-methylglutaryl (HMG)-CoA reductase which is an enzyme common to the biosynthetic pathways of both CoQ10 and cholesterol (Davidson et al., 1997; Folkers et al., 1990). However, there is no evidence that this could generate adverse effects in statin-treated patients. In fact, a recent report suggests that since low circulating levels of CoQ10 may exacerbate chronic heart failure, CoQ10 supplements may improve symptoms and slow progression of chronic heart diseases (Claxton et al., 2022). As statins are thought to block the production of coenzyme Q10, supplementation with this antioxidant is considered particularly advantageous for cardiac patients. Also, recent experimental data has shown that co-administration of CoQ10 with doxorubicin, a cardiotoxic antitumoral, results in protection against doxorubicin-induced cardiomyopathy (Shabaan, Mostafa, El-Desoky, & Arafat, 2021) and neurotoxicity and behavioral disturbances (Okudan, Belviranlı, & Sezer, 2022), through the modulation of proapoptotic and anti-apoptotic gene expression and acetylcholinesterase (AChE) activity, respectively. New independent and well-designed clinical trials employing CoQ10 are warranted to confirm the above findings.

4. Conclusion

Here, we summarized the general features of CoQ10, its role and dynamics in our body, and both experimental and clinical evidence of CoQ10 protective effects on reproductive health. Since aging-related decay in fertility in women is not only related to endogenous factors but also external agents, we also presented information on the potential of supplementation with this coenzyme to rescue the reprotoxicity induced by exposure to environmental xenobiotics. Studies show that both the age-dependent decline in human female reproductive potential and reprotoxicity are potentially a result of cellular mitochondrial dysfunction and oxidative stress. The mechanisms by which CoQ10 supplementation improves reproductive health involve the scavenging of reactive oxygen species and free radicals. This review not only contributes to our general understanding of the phenomenon of reproductive aging but also provides new insights into strategies for promoting reproductive health. Approximately 11% of women of reproductive age in the United States experience fertility problems (Chandra, Copen, & Stephen, 2013). Moreover, there is a trend for delayed childbearing in many Western nations (Breart, 1997; Ely & Hamilton, 2018). In this context, CoQ10 supplementation can represent a low-risk and low-cost strategy to attenuate the impact of age and exposure to environmental chemicals on fertility.

References

Abdulhasan, M. K., Li, Q., Dai, J., Abu-Soud, H. M., Puscheck, E. E., & Rappolee, D. A. (2017). CoQ10 increases mitochondrial mass and polarization, ATP and Oct4 potency levels, and bovine oocyte MII during IVM while decreasing AMPK activity and oocyte death. *Journal of Assisted Reproduction and Genetics*, *34*(12), 1595–1607. https://doi.org/10.1007/s10815-017-1027-y.

Acosta, M. J., Vazquez Fonseca, L., Desbats, M. A., Cerqua, C., Zordan, R., Trevisson, E., et al. (2016). Coenzyme Q biosynthesis in health and disease. *Biochimica et Biophysica Acta*, *1857*(8), 1079–1085. https://doi.org/10.1016/j.bbabio.2016.03.036.

Agarwal, A., Mulgund, A., Hamada, A., & Chyatte, M. R. (2015). A unique view on male infertility around the globe. *Reproductive Biology and Endocrinology*, *13*, 37. https://doi.org/10.1186/s12958-015-0032-1.

Akarsu, S., Gode, F., Isik, A. Z., Dikmen, Z. G., & Tekindal, M. A. (2017). The association between coenzyme Q10 concentrations in follicular fluid with embryo morphokinetics and pregnancy rate in assisted reproductive techniques. *Journal of Assisted Reproduction and Genetics*, *34*(5), 599–605. https://doi.org/10.1007/s10815-017-0882-x.

Alahmar, A. T., & Naemi, R. (2022). Predictors of pregnancy and time to pregnancy in infertile men with idiopathic oligoasthenospermia pre- and post-coenzyme Q10 therapy. *Andrologia*, *e14385*. https://doi.org/10.1111/and.14385.

Azevedo, L. F., Porto Dechandt, C. R., Rocha, C., Hornos Carneiro, M. F., Alberici, L. C., & Barbosa, F., Jr. (2019). Long-term exposure to bisphenol A or S promotes glucose intolerance and changes hepatic mitochondrial metabolism in male Wistar rats. *Food and Chemical Toxicology*, *132*, 110694. https://doi.org/10.1016/j.fct.2019.110694.

Ben-Meir, A., Burstein, E., Borrego-Alvarez, A., Chong, J., Wong, E., Yavorska, T., et al. (2015). Coenzyme Q10 restores oocyte mitochondrial function and fertility during reproductive aging. *Aging Cell*, *14*(5), 887–895. https://doi.org/10.1111/acel.12368.

Ben-Meir, A., Kim, K., McQuaid, R., Esfandiari, N., Bentov, Y., Casper, R. F., et al. (2019). Co-enzyme Q10 supplementation rescues cumulus cells dysfunction in a maternal aging model. *Antioxidants (Basel)*, *8*(3). https://doi.org/10.3390/antiox8030058.

Bentov, Y., Hannam, T., Jurisicova, A., Esfandiari, N., & Casper, R. F. (2014). Coenzyme Q10 supplementation and oocyte aneuploidy in women undergoing IVF-ICSI treatment. *Clinical Medicine Insights: Reproductive Health*, *8*, 31–36. https://doi.org/10.4137/cmrh.s14681.

Boots, C. E., Boudoures, A., Zhang, W., Drury, A., & Moley, K. H. (2016). Obesity-induced oocyte mitochondrial defects are partially prevented and rescued by supplementation with co-enzyme Q10 in a mouse model. *Human Reproduction, 31*(9), 2090–2097. https://doi.org/10.1093/humrep/dew181.

Breart, G. (1997). Delayed childbearing. *European Journal of Obstetrics, Gynecology, and Reproductive Biology, 75*(1), 71–73. https://doi.org/10.1016/s0301-2115(97)00190-5.

Caballero, T., Fiameni, F., Valcarcel, A., & Buzzi, J. (2016). Dietary supplementation with coenzyme Q10 in poor responder patients undergoing IVF-ICSI treatment. *Fertility and Sterility, 106*(3), e58. https://doi.org/10.1016/j.fertnstert.2016.07.177.

Chandra, A., Copen, C. E., & Stephen, E. H. (2013). Infertility and impaired fecundity in the United States, 1982-2010: Data from the National Survey of family growth. *National Health Statistics Reports, Report(67)*, 1–18.

Claxton, L., Simmonds, M., Beresford, L., Cubbon, R., Dayer, M., Gottlieb, S. S., et al. (2022). Coenzyme Q10 to manage chronic heart failure with a reduced ejection fraction: A systematic review and economic evaluation. *Health Technology Assessment, 26*(4), 1–128. https://doi.org/10.3310/KVOU6959.

Crane, F. L. (2001). Biochemical functions of coenzyme Q10. *Journal of the American College of Nutrition, 20*, 591–598. https://doi.org/10.1080/07315724.2001.10719063.

Crane, F. L., Hatefi, Y., Lester, R. L., & Widmer, C. (1957). Isolation of a quinone from beef heart mitochondria. *Biochimica et Biophysica Acta, 25*(1), 220–221. https://doi.org/10.1016/0006-3002(57)90457-2.

Davidson, M., McKenney, J., Stein, E., Schrott, H., Bakker-Arkema, R., Fayyad, R., et al. (1997). Comparison of one-year efficacy and safety of atorvastatin versus lovastatin in primary hypercholesterolemia. Atorvastatin Study Group I. *The American Journal of Cardiology, 79*(11), 1475–1481. https://doi.org/10.1016/s0002-9149(97)00174-4.

Delkhosh, A., Delashoub, M., Tehrani, A. A., Bahrami, A. M., Niazi, V., Shoorei, H., et al. (2019). Upregulation of FSHR and PCNA by administration of coenzyme Q10 on cyclophosphamide-induced premature ovarian failure in a mouse model. *Journal of Biochemical and Molecular Toxicology, 33*(11), e22398. https://doi.org/10.1002/jbt.22398.

Dröge, W. (2002). Free radicals in the physiological control of cell function. *Physiological Reviews, 82*(1), 47–95. https://doi.org/10.1152/physrev.00018.2001.

Duan, X., Dai, X. X., Wang, T., Liu, H. L., & Sun, S. C. (2015). Melamine negatively affects oocyte architecture, oocyte development and fertility in mice. *Human Reproduction, 30*, 1643–1652. https://doi.org/10.1093/humrep/dev091.

Echtay, K. S., Winkler, E., & Klingenberg, M. (2000). Coenzyme Q is an obligatory cofactor for uncoupling protein function. *Nature, 408*(6812), 609–613. https://doi.org/10.1038/35046114.

El Refaeey, A., Selem, A., & Badawy, A. (2014). Combined coenzyme Q10 and clomiphene citrate for ovulation induction in clomiphene-citrate-resistant polycystic ovary syndrome. *Reproductive Biomedicine Online, 29*(1), 119–124. https://doi.org/10.1016/j.rbmo.2014.03.011.

Ely, D. M., & Hamilton, B. E. (2018). Trends in fertility and mother's age at first birth among rural and metropolitan counties: United States, 2007-2017. *NCHS Data Brief, 323*, 1–8.

Emmanuele, V., López, L. C., Berardo, A., Naini, A., Tadesse, S., Wen, B., et al. (2012). Heterogeneity of coenzyme Q10 deficiency. *Archives of Neurology, 69*, 978–983. https://doi.org/10.1001/archneurol.2012.206.

Engelsen, J., Nielsen, J. D., & Winther, K. (2002). Effect of coenzyme Q10 and Ginkgo biloba on warfarin dosage in stable, long-term warfarin treated outpatients. A randomised, double blind, placebo-crossover trial. *Thrombosis and Haemostasis, 87*(6), 1075–1076.

Ernster, L., & Forsmark-Andrée, P. (1993). Ubiquinol: An endogenous antioxidant in aerobic organisms. *The Clinical Investigator*, *71*(8 Suppl), S60–S65. https://doi.org/10.1007/bf00226842.

Ferrante, K. L., Shefner, J., Zhang, H., Betensky, R., O'Brien, M., Yu, H., et al. (2005). Tolerance of high-dose (3,000 mg/day) coenzyme Q10 in ALS. *Neurology*, *65*(11), 1834–1836. https://doi.org/10.1212/01.wnl.0000187070.35365.d7.

Folkers, K., Langsjoen, P., Willis, R., Richardson, P., Xia, L. J., Ye, C. Q., et al. (1990). Lovastatin decreases coenzyme Q levels in humans. *Proceedings of the National Academy of Sciences of the United States of America*, *87*(22), 8931–8934. https://doi.org/10.1073/pnas.87.22.8931.

Forsgren, M., Attersand, A., Lake, S., Grünler, J., Swiezewska, E., Dallner, G., et al. (2004). Isolation and functional expression of human COQ2, a gene encoding a polyprenyl transferase involved in the synthesis of CoQ. *The Biochemical Journal*, *382*, 519–526. https://doi.org/10.1042/BJ20040261.

Gat, I., Blanco Mejia, S., Balakier, H., Librach, C. L., Claessens, A., & Ryan, E. A. (2016). The use of coenzyme Q10 and DHEA during IUI and IVF cycles in patients with decreased ovarian reserve. *Gynecological Endocrinology*, *32*(7), 534–537. https://doi.org/10.3109/09513590.2015.1137095.

Genchi, G., Sinicropi, M. S., Lauria, G., Carocci, A., & Catalano, A. (2020). The effects of cadmium toxicity. *International Journal of Environmental Research and Public Health*, *17*(11). https://doi.org/10.3390/ijerph17113782.

Giannubilo, S. R., Orlando, P., Silvestri, S., Cirilli, I., Marcheggiani, F., Ciavattini, A., et al. (2018). CoQ10 supplementation in patients undergoing IVF-ET: The relationship with follicular fluid content and oocyte maturity. *Antioxidants*, 7(10), 141. https://doi.org/10.3390/antiox7100141.

Greenberg, S. M., & Frishman, W. H. (1988). Coenzyme Q10: A new drug for myocardial ischemia? *The Medical Clinics of North America*, *72*(1), 243–258. https://doi.org/10.1016/s0025-7125(16)30792-1.

Gutierrez-Mariscal, F. M., Arenas-de Larriva, A. P., Limia-Perez, L., Romero-Cabrera, J. L., Yubero-Serrano, E. M., & López-Miranda, J. (2020). Coenzyme Q10 supplementation for the reduction of oxidative stress: Clinical implications in the treatment of chronic diseases. *International Journal of Molecular Sciences*, *21*(21), 7870. https://doi.org/10.3390/ijms21217870.

Halliwell, B., & Gutteridge, J. M. C. (2007). *Free radicals in biology and medicine*. New York: Oxford University Press.

Hamid, N., Junaid, M., & Pei, D.-S. (2021). Combined toxicity of endocrine-disrupting chemicals: A review. *Ecotoxicology and Environmental Safety*, *215*, 112136. https://doi.org/10.1016/j.ecoenv.2021.112136.

Hancock, C. N., Liu, W., Alvord, W. G., & Phang, J. M. (2015). Co-regulation of mitochondrial respiration by proline dehydrogenase/oxidase and succinate. *Amino Acids*, *48*, 859–872. https://doi.org/10.1007/s00726-015-2134-7.

Hargreaves, I., Heaton, R. A., & Mantle, D. (2020). Disorders of human coenzyme Q10 metabolism: An overview. *International Journal Molecular Sciences*, *21*(18), 6695. https://doi.org/10.3390/ijms21186695.

Hargreaves, I. P. (2014). Coenzyme Q10 as a therapy for mitochondrial disease. *The International Journal of Biochemistry & Cell Biology*, *49*, 105–111. https://doi.org/10.1016/j.biocel.2014.01.020.

Hathcock, J. N., & Shao, A. (2006). Risk assessment for coenzyme Q10 (Ubiquinone). *Regulatory Toxicology and Pharmacology*, *45*(3), 282–288. https://doi.org/10.1016/j.yrtph.2006.05.006.

Heaton, R. A., Heales, S., Rahman, K., Sexton, D. W., & Hargreaves, I. (2020). The effect of cellular coenzyme Q(10) deficiency on lysosomal acidification. *Journal of Clinical Medicine*, *9*(6). https://doi.org/10.3390/jcm9061923.

Henderson, A. L., & Colaiácovo, M. P. (2021). Exposure to phthalates: Germline dysfunction and aneuploidy. *Prenatal Diagnosis, 41*(5), 610–619. https://doi.org/10.1002/pd.5921.

Hendler, S. S., & Rorvik, D. R. (2008). *PDR for nutritional supplements* (2nd ed.). Montvale: Thomson Reuters.

Heydarnejad, A., Ostadhosseini, S., Varnosfaderani, S. R., Jafarpour, F., Moghimi, A., & Nasr-Esfahani, M. H. (2019). Supplementation of maturation medium with CoQ10 enhances developmental competence of ovine oocytes through improvement of mitochondrial function. *Molecular Reproduction and Development, 86*(7), 812–824. https://doi.org/10.1002/mrd.23159.

Hornos Carneiro, M. F., Shin, N., Karthikraj, R., Barbosa, F., Kannan, K., & Colaiacovo, M. P. (2020). Antioxidant CoQ10 restores fertility by rescuing bisphenol A-induced oxidative DNA damage in the *Caenorhabditis elegans* germline. *Genetics, 214*(2), 381–395. https://doi.org/10.1534/genetics.119.302939.

Huntington Study Group. (2001). A randomized, placebo-controlled trial of coenzyme Q10 and remacemide in Huntington's disease. *Neurology, 57*(3), 397–404. https://doi.org/10.1212/wnl.57.3.397.

Indo, H. P., Yen, H. C., Nakanishi, I., Matsumoto, K., Tamura, M., Nagano, Y., et al. (2015). A mitochondrial superoxide theory for oxidative stress diseases and aging. *Journal of Clinical Biochemistry and Nutrition, 56*(1), 1–7. https://doi.org/10.3164/jcbn.14-42.

Jia, Z., Wang, H., Feng, Z., Zhang, S., Wang, L., Zhang, J., et al. (2019). Fluorene-9-bisphenol exposure induces cytotoxicity in mouse oocytes and causes ovarian damage. *Ecotoxicology and Environmental Safety, 30*(180), 168–178. https://doi.org/10.1016/j.ecoenv.2019.05.019.

Kagan, V. E., & Quinn, P. J. (2000). *Coenzyme Q: Molecular mechanisms in health and disease* (1st ed.). Boca Raton, FL, USA: CRC Press.

Karaayak, P. E., & El Nehir, S. (2011). Changes in content of coenzyme Q10 in beef muscle, beef liver and beef heart with cooking and in vitro digestion. *Journal of Food Composition and Analysis, 24*(8), 1136–1140. https://doi.org/10.1016/j.jfca.2011.05.002.

Karamali, M., & Gholizadeh, M. (2022). The effects of coenzyme Q10 supplementation on metabolic profiles and parameters of mental health in women with polycystic ovary syndrome. *Gynecological Endocrinology, 38*(1), 45–49. https://doi.org/10.1080/09513590.2021.1991910.

Kawamukai, M. (2016). Biosynthesis of coenzyme Q in eukaryotes. *Bioscience, Biotechnology, and Biochemistry, 80*(1), 23–33. https://doi.org/10.1080/09168451.2015.1065172.

Kiess, W., & Haeusler, G. (2021). Endocrine-disrupting chemicals. *Best Practice & Research. Clinical Endocrinology & Metabolism, 35*(5), 101566. https://doi.org/10.1016/j.beem.2021.101566.

Knott, A., Achterberg, V., Smuda, C., Mielke, H., Sperling, G., Dunckelmann, K., et al. (2015). Topical treatment with coenzyme Q10-containing formulas improves skin's Q10 level and provides antioxidative effects. *BioFactors, 41*(6), 383–390. https://doi.org/10.1002/biof.1239.

Kucharská, J., Poništ, S., Vančová, O., Gvozdjáková, A., Uličná, O., Slovák, L., et al. (2021). Treatment with coenzyme Q10, omega-3-polyunsaturated fatty acids and their combination improved bioenergetics and levels of coenzyme Q9 and Q10 in skeletal muscle mitochondria in experimental model of arthritis. *Physiological Research, 70*(5), 723–733. https://doi.org/10.33549/physiolres.934664.

La Merrill, M. A., Vandenberg, L. N., Smith, M. T., Goodson, W., Browne, P., Patisaul, H. B., et al. (2020). Consensus on the key characteristics of endocrine-disrupting chemicals as a basis for hazard identification. *Nature Reviews Endocrinology, 16*(1), 45–57. https://doi.org/10.1038/s41574-019-0273-8.

Larsen, P. L., & Clarke, C. F. (2002). Extension of life-span in *Caenorhabditis elegans* by a diet lacking coenzyme Q. *Science, 295*(5552), 120–123. https://doi.org/10.1126/science.1064653.

Láscarez-Lagunas, L. I., Martinez-Garcia, M., & Colaiácovo, M. P. (2022). Loss, gain, and retention: Mechanisms driving late prophase I chromosome remodeling for accurate meiotic chromosome segregation. *Genes, 13*(3), 546. https://doi.org/10.3390/genes13030546.

Lee, B.-J., Huang, Y.-C., Chen, S.-J., & Lin, P.-T. (2012). Coenzyme Q10 supplementation reduces oxidative stress and increases antioxidant enzyme activity in patients with coronary artery disease. *Nutrition, 28*(3), 250–255. https://doi.org/10.1016/j.nut.2011.06.004.

Lee, H. J., Park, M. J., Joo, B. S., Joo, J. K., Kim, Y. H., Yang, S. W., et al. (2021). Effects of coenzyme Q10 on ovarian surface epithelium-derived ovarian stem cells and ovarian function in a 4-vinylcyclohexene diepoxide-induced murine model of ovarian failure. *Reproductive Biology and Endocrinology, 19*(1), 59. https://doi.org/10.1186/s12958-021-00736-x.

Lunetta, S., & Roman, M. (2008). Determination of coenzyme Q10 content in raw materials and dietary supplements by high-performance liquid chromatography-UV: Collaborative study. *Journal of AOAC International, 91*(4), 702–708.

Ma, L., Cai, L., Hu, M., Wang, J., Xie, J., Xing, Y., et al. (2020). Coenzyme Q10 supplementation of human oocyte in vitro maturation reduces postmeiotic aneuploidies. *Fertility and Sterility, 114*(2), 331–337. https://doi.org/10.1016/j.fertnstert.2020.04.002.

Ma, L., Hu, M., Ma, X., Cui, Y., & Liu, J. (2018). CoQ10 decreases aneuploidy rate and increases mitochondrial mass during in vitro maturation of human immature oocytes. *Fertility and Sterility, 110*(4), e312. https://doi.org/10.1016/j.fertnstert.2018.07.878.

Mandrioli, D., Belpoggi, F., Silbergeld, E. K., & Perry, M. J. (2016). Aneuploidy: A common and early evidence-based biomarker for carcinogens and reproductive toxicants. *Environmental Health, 15*(1), 97. https://doi.org/10.1186/s12940-016-0180-6.

Marin, S. E., & Haas, R. H. (2015). Coenzyme Q10 and the treatment of mitochondrial disease. In I. P. Hargreaves, & A. K. Hargreaves (Eds.), *Coenzyme Q10: From fact to fiction* (pp. 85–107). New York: Nova Science Publishers.

Maside, C., Martinez, C. A., Cambra, J. M., Lucas, X., Martinez, E. A., Gil, M. A., et al. (2019). Supplementation with exogenous coenzyme Q10 to media for in vitro maturation and embryo culture fails to promote the developmental competence of porcine embryos. *Reproductive Domestic Animals, 54*, 72–77. https://doi.org/10.1111/rda.13486.

Meldrum, D. R., Casper, R. F., Diez-Juan, A., Simon, C., Domar, A. D., & Frydman, R. (2016). Aging and the environment affect gamete and embryo potential: Can we intervene? *Fertility and Sterility, 105*(3), 548–559. https://doi.org/10.1016/j.fertnstert.2016.01.013.

Mirończuk-Chodakowska, I., Witkowska, A. M., & Zujko, M. E. (2018). Endogenous non-enzymatic antioxidants in the human body. *Advances in Medical Sciences, 63*(1), 68–78. https://doi.org/10.1016/j.advms.2017.05.005.

Moldogazieva, N. T., Lutsenko, S. V., & Terentiev, A. A. (2018). Reactive oxygen and nitrogen species-induced protein modifications: Implication in carcinogenesis and anticancer therapy. *Cancer Research, 78*(21), 6040–6047. https://doi.org/10.1158/0008-5472.can-18-0980.

Montero, R., Yubero, D., Salgado, M. C., González, M. J., Campistol, J., O'Callaghan, M. D. M., et al. (2019). Plasma coenzyme Q10 status is impaired in selected genetic conditions. *Scientific Reports, 9*, 1–8. https://doi.org/10.1038/s41598-018-37542-2.

Neergheen, V., Chalasani, A., Wainwright, L., Yubero, D., Montero, R., Artuch, R., et al. (2017). Coenzyme Q10 in the treatment of mitochondrial disease. *The Journal of Inborn Errors of Metabolism and Screening, 5*. https://doi.org/10.1177/2326409817707771.

Nwose, E. U., & Ewing, G. W. (2009). Computer diagnosis in cardiology: Oxidative stress hypothesis. *North American Journal of Medical Sciences, 1*(5), 220–225.

Okudan, N., Belviranlı, M., & Sezer, T. (2022). Potential protective effect of coenzyme Q10 on doxorubicin-induced neurotoxicity and behavioral disturbances in rats. *Neurochemical Research, 47*(5), 1280–1289. https://doi.org/10.1007/s11064-021-03522-8.

Overvad, K., Diamant, B., Holm, L., Holmer, G., Mortensen, S. A., & Stender, S. (1999). Coenzyme Q10 in health and disease. *European Journal of Clinical Nutrition, 53*(10), 764–770. https://doi.org/10.1038/sj.ejcn.1600880.

Ozcan, P., Ficicioglu, C., Kizilkale, O., Yesiladali, M., Tok, O. E., Ozkan, F., et al. (2016). Can coenzyme Q10 supplementation protect the ovarian reserve against oxidative damage? *Molecular & Cellular Proteomics, 33*(9), 1223–1230. https://doi.org/10.1007/s10815-016-0751-z.

Pan, M. H., Wu, Y. K., Liao, B. Y., Zhang, H., Li, C., Wang, J. L., et al. (2021). Bisphenol A exposure disrupts organelle distribution and functions during mouse oocyte maturation. *Frontiers in Cell and Development Biology, 23*(9), 661155. https://doi.org/10.3389/fcell.2021.661155.

Perkins, A. T., Das, T. M., Panzera, L. C., & Bickel, S. E. (2016). Oxidative stress in oocytes during midprophase induces premature loss of cohesion and chromosome segregation errors. *Proceedings of the National Academy of Sciences of the United States of America, 113*(44), E6823–e6830. https://doi.org/10.1073/pnas.1612047113.

Perkins, A. T., Greig, M. M., Sontakke, A. A., Peloquin, A. S., McPeek, M. A., & Bickel, S. E. (2019). Increased levels of superoxide dismutase suppress meiotic segregation errors in aging oocytes. *Chromosoma, 128*(3), 215–222. https://doi.org/10.1007/s00412-019-00702-y.

Pineda, M., Montero, R., Aracil, A., O'Callaghan, M. M., Mas, A., Espinos, C., et al. (2010). Coenzyme Q(10)-responsive ataxia: 2-year-treatment follow-up. *Movement Disorders, 25*(9), 1262–1268. https://doi.org/10.1002/mds.23129.

Practice Committee of the American Society for Reproductive Medicine, in collaboration with the Society for Reproductive Endocrinology and Infertility. (2017). Optimizing natural fertility: A committee opinion. *Fertility and Sterility, 107*(1), 52–58. https://doi.org/10.1016/j.fertnstert.2016.09.029.

Pravst, I., Zmitek, K., & Zmitek, J. (2010). Coenzyme Q10 contents in foods and fortification strategies. *Critical Reviews in Food Science and Nutrition, 50*, 269–280. https://doi.org/10.1080/10408390902773037.

Rafieian-Naeini, H. R., Zhandi, M., Sadeghi, M., Yousefi, A. R., & Benson, A. P. (2021). Effects of coenzyme Q10 on reproductive performance of laying Japanese quail (Coturnix japonica) under cadmium challenge. *Poultry Science, 100*(11), 101418. https://doi.org/10.1016/j.psj.2021.101418.

Rafieian-Naeini, H. R., Zhandi, M., Sadeghi, M., Yousefi, A. R., Marzban, H., & Benson, A. P. (2022). The effect of dietary coenzyme Q10 supplementation on egg quality and liver histopathology of layer quails under cadmium challenge. *Journal of Animal Physiology and Animal Nutrition*. https://doi.org/10.1111/jpn.13715.

Rahimi Darehbagh, R., Khalafi, B., Allahveisi, A., & Habiby, M. (2022). Effects of the mitochondrial genome on germ cell fertility: A review of the literature. *International Journal of Fertility & Sterility, 16*(2), 70–75. https://doi.org/10.22074/ijfs.2021.527076.1098.

Ramasarma, T. (1985). Natural occurrence and distribution of coenzyme Q. In G. Lenaz (Ed.), *Coenzyme Q. Biochemistry, biogenetics and clinical applications of Uniquinone* (pp. 67–81). New York: John Wiley and Sons.

Rocha, B. A., Asimakopoulos, A. G., Barbosa, F., Jr., & Kannan, K. (2017). Urinary concentrations of 25 phthalate metabolites in Brazilian children and their association with oxidative DNA damage. *Science of the Total Environment, 586*, 152–162. https://doi.org/10.1016/j.scitotenv.2017.01.193.

Rodríguez-Varela, C., & Labarta, E. (2020). Clinical application of antioxidants to improve human oocyte mitochondrial function: A review. *Antioxidants (Basel)*, *9*(12), 1197. https://doi.org/10.3390/antiox9121197.

Rodríguez-Varela, C., & Labarta, E. (2021). Does coenzyme Q10 supplementation improve human oocyte quality? *International Journal of Molecular Sciences*, *22*(17), 9541. https://doi.org/10.3390/ijms22179541.

Román-Pizarro, V., Fernández-Romero, J. M., & Gómez-Hens, A. (2017). Automatic determination of coenzyme Q10 in food using cresyl violet encapsulated into magnetoliposomes. *Food Chemistry*, *221*, 864–870. https://doi.org/10.1016/j.foodchem.2016.11.085.

Romero, S., Pella, R., Zorrilla, I., Berrío, P., Escudero, F., Pérez, Y., et al. (2020). Coenzyme Q10 improves the in vitro maturation of oocytes exposed to the intrafollicular environment of patients on fertility treatment. *JBRA Assisted Reproduction*, *24*(3), 283–288. https://doi.org/10.5935/1518-0557.20200003.

Rötig, A., Appelkvist, E. L., Geromel, V., Chretien, D., Kadhom, N., Edery, P., et al. (2000). Quinone-responsive multiple respiratory-chain dysfunction due to widespread coenzyme Q10 deficiency. *Lancet*, *356*(9227), 391–395. https://doi.org/10.1016/s0140-6736(00)02531-9.

Salvi, F., & Gadda, G. (2013). Human choline dehydrogenase: Medical promises and biochemical challenges. *Archives of Biochemistry and Biophysics 2013*, *537*(2), 243–252. https://doi.org/10.1016/j.abb.2013.07.018.

Sasaki, H., Hamatani, T., Kamijo, S., Iwai, M., Kobanawa, M., Ogawa, S., et al. (2019). Impact of oxidative stress on age-associated decline in oocyte developmental competence. *Frontiers in Endocrinology*, *10*. https://doi.org/10.3389/fendo.2019.00811.

Schmelzer, C., Lindner, I., Rimbach, G., Niklowitz, P., Menke, T., & Döring, F. (2008). Functions of coenzyme Q10in inflammation and gene expression. *BioFactors*, *32*, 179–183. https://doi.org/10.1002/biof.5520320121.

Schou-Pedersen, A. M. V., Schemeth, D., & Lykkesfeldt, J. (2019). Determination of reduced and oxidized coenzyme Q(10) in canine plasma and heart tissue by HPLC-ECD: Comparison with LC-MS/MS quantification. *Antioxidants (Basel)*, *8*(8), 253. https://doi.org/10.3390/antiox8080253.

Scott, C. (2005). Misconceptions about aerobic and anaerobic energy expenditure. *Journal of the International Society of Sports Nutrition*, *2*, 32–37. https://doi.org/10.1186/1550-2783-2-2-32.

Sen Sharma, D. (2017). Category C: Oral presentations: Fertility and reproductive medicine. *BJOG: An International Journal of Obstetrics & Gynaecology*, *124*(S1), 9–11. https://doi.org/10.1111/1471-0528.2_14571.

Shabaan, D. A., Mostafa, N., El-Desoky, M. M., & Arafat, E. A. (2021). Coenzyme Q10 protects against doxorubicin-induced cardiomyopathy via antioxidant and anti-apoptotic pathway. *Tissue Barriers*, *2019504*. https://doi.org/10.1080/21688370.2021.2019504.

Shalansky, S., Lynd, L., Richardson, K., Ingaszewski, A., & Kerr, C. (2007). Risk of warfarin-related bleeding events and supratherapeutic international normalized ratios associated with complementary and alternative medicine: A longitudinal analysis. *Pharmacotherapy*, *27*(9), 1237–1247. https://doi.org/10.1592/phco.27.9.1237.

Shults, C. W., Oakes, D., Kieburtz, K., Beal, M. F., Haas, R., Plumb, S., et al. (2002). Effects of coenzyme Q10 in early Parkinson disease: Evidence of slowing of the functional decline. *Archives of Neurology*, *59*(10), 1541–1550. https://doi.org/10.1001/archneur.59.10.1541.

Skakkebæk, N. E., Lindahl-Jacobsen, R., Levine, H., Andersson, A.-M., Jørgensen, N., Main, K. M., et al. (2022). Environmental factors in declining human fertility. *Nature Reviews Endocrinology*, *18*(3), 139–157. https://doi.org/10.1038/s41574-021-00598-8.

Srinivas, U. S., Tan, B. W. Q., Vellayappan, B. A., & Jeyasekharan, A. D. (2019). ROS and the DNA damage response in cancer. *Redox Biology*, *25*, 101084. https://doi.org/10.1016/j.redox.2018.101084.

US Department of Health and Human Services. (2022) Accessed on June, 28 of 2022, from https://www.nichd.nih.gov/health/topics/infertility/conditioninfo/common.

Vandenberg, L. N., Colborn, T., Hayes, T. B., Heindel, J. J., Jacobs, D. R., Jr., Lee, D. H., et al. (2012). Hormones and endocrine-disrupting chemicals: Low-dose effects and nonmonotonic dose responses. *Endocrine Reviews*, *33*(3), 378–455. https://doi.org/10.1210/er.2011-1050.

Wang, L. Y., Wang, D. H., Zou, X. Y., & Xu, C. M. (2009). Mitochondrial functions on oocytes and preimplantation embryos. *Journal of Zhejiang University. Science. B*, *10*(7), 483–492. https://doi.org/10.1631/jzus.B0820379.

Wang, Y., & Hekimi, S. (2019). The complexity of making ubiquinone. *Trends in Endocrinology & Metabolism*, *30*(12), 929–943. https://doi.org/10.1016/j.tem.2019.08.009.

Weber, C., Bysted, A., & Hłlmer, G. (1997). The coenzyme Q10 content of the average Danish diet. *International Journal for Vitamin and Nutrition Research*, *67*, 123–129. https://doi.org/10.1016/s0098-2997(97)00003-4.

Webster, A., & Schuh, M. (2017). Mechanisms of aneuploidy in human eggs. *Trends in Cell Biology*, *27*(1), 55–68. https://doi.org/10.1016/j.tcb.2016.09.002.

Weidinger, A., & Kozlov, A. V. (2015). Biological activities of reactive oxygen and nitrogen species: Oxidative stress versus signal transduction. *Biomolecules*, *5*(2), 472–484. https://doi.org/10.3390/biom5020472.

Xie, J., Jiang, J., & Guo, Q. (2021). Primary coenzyme Q10 deficiency-7 and pathogenic COQ4 variants: Clinical presentation, biochemical analyses, and treatment. *Frontiers in Genetics*, *12*, 776807. https://doi.org/10.3389/fgene.2021.776807.

Xing, X., Zhang, J., Zhang, J., Wang, Y., Wang, J., Kang, J., et al. (2022). Coenzyme Q10 supplement rescues postovulatory oocyte aging by regulating SIRT4 expression. *Current Molecular Pharmacology*, *15*(1), 190–203. https://doi.org/10.2174/1874467214666210420112819.

Xu, Y., Nisenblat, V., Lu, C., Li, R., Qiao, J., Zhen, X., et al. (2018). Pretreatment with coenzyme Q10 improves ovarian response and embryo quality in low-prognosis young women with decreased ovarian reserve: A randomized controlled trial. *Reproductive Biology and Endocrinology*, *16*(1), 29. https://doi.org/10.1186/s12958-018-0343-0.

Yang, C. X., Song, Z. Q., Pei, S., Yu, X. X., Miao, J. K., Liang, H., et al. (2020). Single cell RNA-seq reveals molecular pathways altered by 7, 12-dimethylbenz[a]anthracene treatment on pig oocytes. *Theriogenology*, *157*, 449–457. https://doi.org/10.1016/j.theriogenology.2020.08.020.

Zhou, S., & Chan, E. (2001). Effect of ubidecarenone on warfarin anticoagulation and pharmacokinetics of warfarin enantiomers in rats. *Drug Metabolism and Drug Interactions*, *18*(2), 99–122. https://doi.org/10.1515/dmdi.2001.18.2.99.

Zhou, Q., Zhou, S., & Chan, E. (2005). Effect of coenzyme Q10 on warfarin hydroxylation in rat and human liver microsomes. *Current Drug Metabolism*, *6*(2), 67–81. https://doi.org/10.2174/1389200053586091.

CHAPTER SEVEN

Antioxidants affect endoplasmic reticulum stress-related diseases

Tania Gómez-Sierra, Alexis Paulina Jiménez-Uribe, Ariadna Jazmín Ortega-Lozano, Karla Jaqueline Ramírez-Magaña, and José Pedraza-Chaverri*

Antioxidant Biochemistry Laboratory, Department of Biology, Faculty of Chemistry, National Autonomous University of Mexico (UNAM), Mexico City, Mexico

*Corresponding author: e-mail address: pedraza@unam.mx

Contents

Abstract

The endoplasmic reticulum (ER) is a complex multifunctional organelle that maintains cell homeostasis. Intrinsic and extrinsic factors alter ER functions, including the rate of protein folding that triggers the accumulation of misfolded proteins and alters homeostasis, thus generating stress in the ER, which activates the unfolded protein response (UPR) pathway to promote cell survival and restore their homeostasis; however, if the damage is not corrected, it could also trigger cell death. In addition, ER stress and oxidative stress are closely related because excessive production of reactive oxygen species (ROS), a well-known inducer of ER stress, promotes the accumulation of misfolded proteins; at the same time, the ER stress enhances ROS production, generating a pathological cycle. Furthermore, it has been described that the dysregulation of the UPR contributes to the progression of various diseases, so the use of compounds capable of regulating ER stress, such as antioxidants, has been used in several experimental models of diseases to alleviate the damage induced by the maladaptive signaling of

Vitamins and Hormones, Volume 121
ISSN 0083-6729
https://doi.org/10.1016/bs.vh.2022.10.006

the UPR, the mechanism of action of antioxidants generally is dose-dependent, and it is specific in each tissue and pathology, could decrease or enhance specific proteins of the UPR to have beneficial or detrimental effects.

1. The endoplasmic reticulum (ER)

The endoplasmic reticulum (ER) is a central organelle classified as rough and smooth based on the presence or absence of ribosomes, respectively. A more recent classification considered its morphology and spatial distribution within the cell, which includes the three main compartments: (1) the tubules, (2) the cisternae and, (3) the nuclear envelope; each one may or may not have associated ribosomes (Almanza et al., 2019). The functions of the ER include protein synthesis and transport, protein folding, calcium (Ca^{2+}) storage, lipid, and steroid synthesis, among others (Schwarz & Blower, 2016). Interestingly, this membranous organelle distributes throughout the cytoplasm making contact with other organelles, which gives insights into its great relevance in intercellular communication as is evidenced by the lipid synthesis in the smooth ER, which occurs near the site of its utilization as the plasma membrane, Golgi apparatus, mitochondria, and peroxisomes (Kannan, Lahiri, Liu, Choudhary, & Prinz, 2017; Quon et al., 2018) or Ca^{2+} content in the cytoplasm which is regulated by the ER and cause a great diversity of cellular processes, such as ion opening channels, cytoskeleton remodeling, activation of transcription factors, cell proliferation, and apoptosis, among others (Newton, Bootman, & Scott, 2016).

The ER is the main quality control of proteins within the cell. It has been estimated that one-third of the newly synthesized proteins target the ER to acquire their native state (Chevet, Cameron, Pelletier, Thomas, & Bergeron, 2001; Walter & Ron, 2011). The proteins are protected with different chaperones within the ER lumen to avoid their aggregation or nonspecific folding and assist folding, glycosylation, and protein quality control. Among the chaperones within the ER lumen are lectins belonging to the heat shock protein (Hsp) family, such as Hsp40, 70, 90, and 100. The master regulator of ER homeostasis is glucose-related protein 78 kDa (GRP78), also referred to as BiP, a member of the Hsp70 family; this chaperone binds rapidly to newly translocated proteins to the ER lumen to initially block the return to the cytosol and to guide it entirely into the ER lumen in an ATP-dependent manner, once completely inside

GRP78/BiP disassociates from proteins to begin the folding process (Rapoport, 2007; Wang, Groenendyk, & Michalak, 2015). The oxidizing environment in the ER favors the formation of disulfide bridges, which stabilize the tertiary structure and thus the assembly of the protein (Chevet et al., 2001; Walter & Ron, 2011). In addition, if the protein cannot reach its native conformation due to exposure of hydrophobic residues or incorrect formation of disulfide bonds, the ER activates a mechanism to preserve the quality control called ER-associated degradation (ERAD) pathway to eliminate misfolded proteins (Lederkremer & Glickman, 2005). ERAD is a quality control system in homeostatic conditions; however, if the accumulation of misfolded proteins exceeds the capacity of ERAD, a stress response in the ER activates the unfolded protein response (UPR).

1.1 ER stress and the unfolded protein response (UPR)

There are intrinsic and extrinsic factors such as genetic mutations, errors in transcription and translation, hypoxia, oxidative stress, hydric and thermal stress, and abnormalities in Ca^{2+} regulation caused by ischemia, hypoglycemia, heavy metals, chemotherapeutic agents or antibiotics, that alter ER functions including the folding rate of proteins, triggering the accumulation of misfolded proteins, which disrupts homeostasis to drive a stress response in the ER. The ER stress activates a signal pathway called the UPR, which is composed of three sensors: inositol-requiring enzyme 1 alpha (IREα), protein kinase RNA-like ER kinase (PERK), and the activating transcription factor 6 alpha (ATF6α), which together determine between cell survival that restores ER homeostasis and the consequent cellular homeostasis or cell death if the damage is not corrected. These proteins under homeostatic conditions are in an inactive state bound to GRP78/BiP; however, when the chaperone GRP78/BiP binds to misfolded proteins is released from IREα, PERK, ATF6α, and this activate the UPR (Almanza et al., 2019; Schönthal, 2012).

The IRE1α, the most conserved branch of the UPR, is a transmembrane protein with a N-terminal luminal domain that interacts with GRP78/BiP in homeostatic conditions, and a C-terminal that possess a kinase and a RNase domain localized in the cytosolic face (Ali et al., 2011). Under stress conditions, GRP78/BiP maintains IRE1α inactive; when misfolded proteins accumulate in the ER lumen, GRP78/BiP dissociates from IRE1α, which now interacts with Hsp47. The binding of IRE1α to Hsp47 appears to be a step that enhances the functions of IRE1α as a serine/threonine

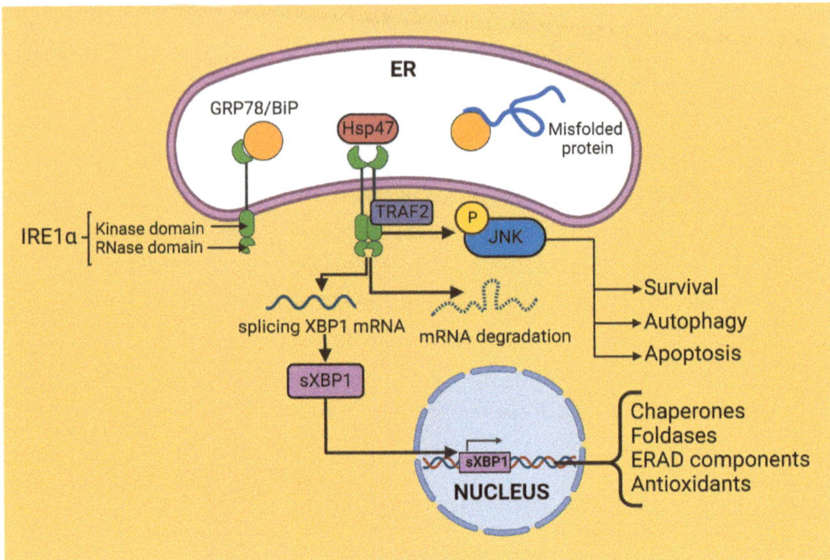

Fig. 1 IRE1α signaling pathway. Under ER stress, the misfolded proteins accumulate in the ER lumen, interact with GRP78/BiP and are released from IRE1α to dimerizes and autophosphorylate assisted by the chaperone Hsp47 and activate the IRE1α signaling pathway. The IRE1α kinase domain could recruit TRAF2 and phosphorylate JNK, which are involved in several pathways. The IRE1α RNase function splices the XBP1 mRNA and induces the expression of genes involved in restoring ER homeostasis and can also decrease the rate of protein transcription. *Figure created with Biorender.com.*

kinase and endonuclease (Fig. 1). For its complete activation, IRE1α requires to dimerize and autophosphorylates through its kinase domain to act as kinase and RNase, which could act directly on the mRNA of X-box-binding protein 1 (XBP1) to generate its active spliced form (sXBP1) or through the IRE1α-dependent regulated mRNA decay (RIDD) and decrease the rate of protein transcription (Maurel, Chevet, Tavernier, & Gerlo, 2014; Sepulveda et al., 2018). The sXBP1 is a transcription factor that promotes chaperones synthesis, foldases, and ERAD components. Interestingly, it has been reported that XBP1 also regulates the expression of antioxidant enzymes such as catalase (CAT) under oxidative stress conditions (Liu et al., 2009; Martin et al., 2014).

On the other hand, the IRE1α kinase function, besides the autophosphorylation, also involves the c-Jun amino-terminal kinases (JNK) phosphorylation through the interaction with the tumor necrosis factor receptor-associated factor 2 (TRAF2) (Urano et al., 2000). JNK pathway modulates decision fate of the cell, at early stages promotes cell survival and autophagy, whereas sustained signaling could promote the activation of the intrinsic pathway of apoptosis (Ogata et al., 2006; Raciti, Lotti, Valia, Pulcinelli, & Di Renzo, 2012; Wei, Pattingre, Sinha, Bassik, & Levine, 2008; Weston & Davis, 2007).

The second signaling pathway of the UPR is PERK, a transmembrane protein with a N-terminal luminal domain that interacts with GRP78/BiP in homeostatic conditions. Under stress conditions, GRP78/BiP bind to misfolded proteins, and PERK can be activated (Bertolotti, Zhang, Hendershot, Harding, & Ron, 2000); it also seems that PERK can interact directly with misfolded proteins through hydrophobic interactions enhancing its activation (Wang, Li, Tao, & Sha, 2018). Like IRE1α, PERK dimerizes and phosphorylates its C-terminal cytoplasmic domain, which has a kinase function (Cui, Li, Ron, & Sha, 2011). PERK phosphorylates two main targets, the eukaryotic translation initiator factor 2 (eIF2), specifically the subunit (eIF2α), and the nuclear factor erythroid 2-related factor 2 (Nrf2) (Fig. 2). Under stress conditions, PERK can phosphorylate eIF2α, resulting in a global decrease in protein translation rate. In addition,

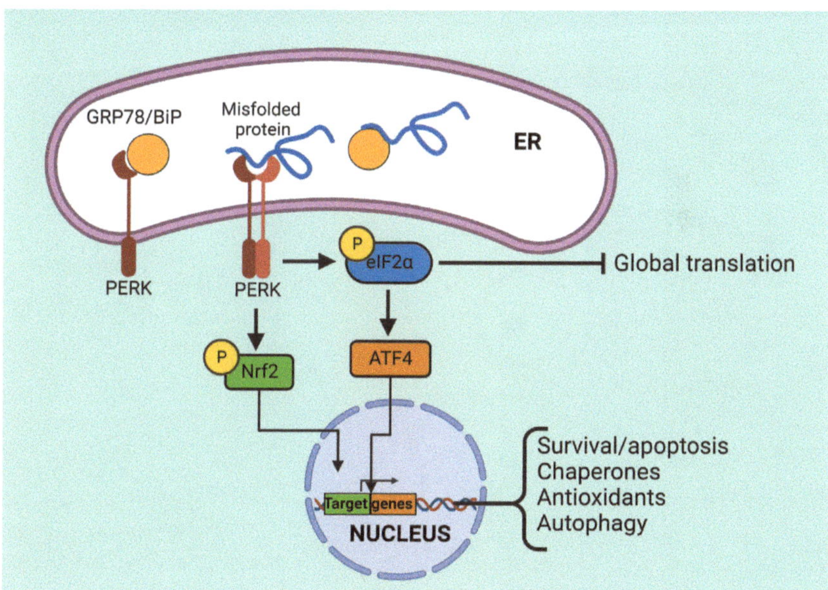

Fig. 2 PERK signaling pathway. Under ER stress, the misfolded protein accumulates in the ER lumen, interacts with GRP78/BiP and is releases from PERK. Misfolded proteins interact with PERK and dimerize, leading to its cytoplasmic kinase domain activation. PERK phosphorylates two main targets, eIF2α and Nrf2, to decrease the global protein translation and regulate antioxidant responses, respectively. *Figure created with Biorender.com.*

p-eIF2α increases the translation of the activating transcription factor 4 (ATF4), which interacts with other transcriptional factors (Pakos-Zebrucka et al., 2016). The ATF4 target genes include ERAD components, chaperones, antioxidant response, autophagy, amino acid metabolism, and ER expansion (Harding et al., 2003). The sustained activation of PERK/p-eIF2α/ATF4 induces the activation of CCAAT/enhancer-binding protein C/EBP homologous protein (CHOP). This transcription factor induces the expression of several pro-apoptotic genes and inhibits the expression of anti-apoptotic genes (Li, Guo, Tang, Jiang, & Chen, 2015). Moreover, the other target of PERK is Nrf2, a transcriptional factor that promotes the expression of antioxidant system components (Cullinan et al., 2003). In addition, one novel reported function of PERK is promoting mitochondrial hyperfusion to sustain its respiratory capacity (Lebeau et al., 2018).

Finally, the third branch of UPR is ATF6 α, under homeostatic conditions, ATF6α luminal domain is bound with GRP78/BiP; however, when misfolded proteins accumulate, ATF6α/GRP78/BiP complex dissociates, leading to GRP78/BiP interaction with misfolded proteins and ATF6α mobilization from ER to Golgi apparatus (Fig. 3) (Haze, Yoshida, Yanagi,

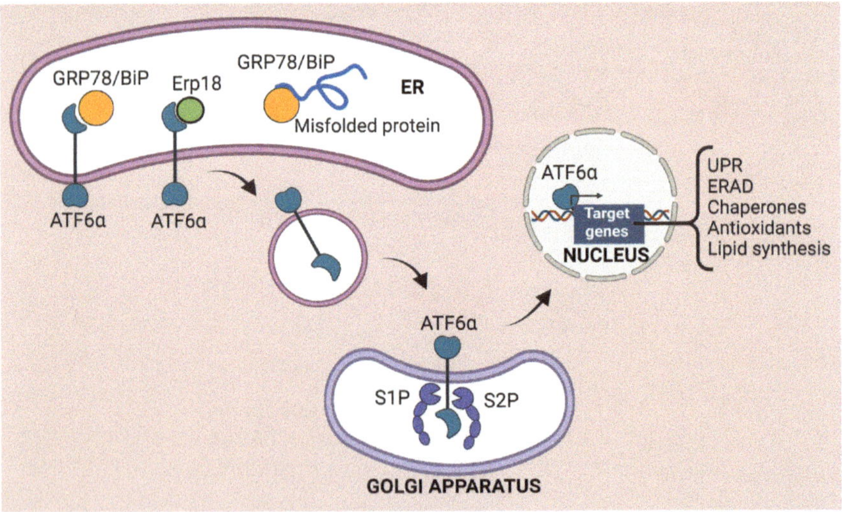

Fig. 3 ATF6α signaling pathway. Under ER stress, the misfolded protein accumulates in the ER lumen, interacts with GRP78/BiP, and is released from ATF6α, leading to its interaction with Erp18. ATF6α is then transported to the Golgi apparatus, where it is cleaved by proteases S1P and S2P to release its cytoplasmic domain, which acts as a transcription factor that induces expression genes involved in the restoration of ER homeostasis. *Figure created with Biorender.com.*

Yura, & Mori, 1999). This dissociation occurs actively (Shen, Snapp, Lippincott-Schwartz, & Prywes, 2005) and it is accompanied by the action of the ER protein 18 (ERp18), a member of the protein disulfide isomerase (PDI) family, that could interact with cysteine 467 of ATF6α, inducing a structural change, an essential step for the later ATF6α cleavage. The next step is the transport of ATF6α from the ER to Golgi apparatus, where ATF6α is cleavage by the action of site-1-protease (S1P) and site-2-protease (S2P), releasing the N-terminal domain, that functions as a transcriptional factor (Oka et al., 2019). Among the different target genes of ATF6α are ERAD components, ER biogenesis, and chaperone synthesis (Adachi et al., 2008; Bommiasamy et al., 2009; Maiuolo, Bulotta, Verderio, Benfante, & Borgese, 2011; Rutkowski & Kaufman, 2007).

2. ER stress and oxidative stress crosstalk

The ER has an oxidizing environment that favors the formation of disulfide bridges, which stabilize the tertiary structure and, therefore, the assembly of the protein (Chevet et al., 2001; Walter & Ron, 2011); this environment must be controlled to avoid alterations in the functions of the ER (Braakman & Hebert, 2013; Schwarz & Blower, 2016). Disulfide bond formation can occur naturally and spontaneously by the same polypeptide in a redox state-dependent manner; however, this mechanism is too slow (Malhotra & Kaufman, 2007). PDI catalyzes the reaction of disulfide bonds in the ER and can correct the disulfide bonds in misfolded proteins. During disulfide bond formation, the cysteine residue within the PDI active site accepts two electrons from substrates of the polypeptide chain. This transfer of electrons results in reducing the PDI active site and oxidation of the substrate (Malhotra & Kaufman, 2007; Zeeshan, Lee, Kim, & Chae, 2016). Moreover, PDI is oxidized by ER oxidoreductase 1 (ERO1); this is a flavin adenine dinucleotide-dependent reaction, where molecular oxygen is the last electron receptor to produce hydrogen peroxide (H_2O_2) and which is reduced by glutathione peroxidase (GPx) to form molecular oxygen and oxidized glutathione (GSSG) is also generated. This electron transfer has the potential to generate reactive oxygen species (ROS) and modify the ratio between glutathione (GSH) and GSSG, which contributes to oxidative stress and ER stress (Malhotra & Kaufman, 2007; Tu & Weissman, 2002; Zeeshan et al., 2016). Moreover, GSH maintains ER oxidoreductases in a reduced form to catalyze the reduction or isomerization of disulfide bonds of proteins, so any disturbance in GSH/GSSG ratio modifies protein folding and redox balance (Chakravarthi, Jessop, & Bulleid, 2006). Additionally, ROS can be produced by enzymes as lipoxygenases, nitric oxide synthase, xanthine oxidase, and nicotinamide adenine dinucleotide phosphate oxidase (NOX); enzymes that can be found in various cellular compartments; however, about 25% of cellular ROS are generated by protein folding in the ER (Forrester, Kikuchi, Hernandes, Xu, & Griendling, 2018; Zeeshan et al., 2016).

The presence of ROS due to protein folding in the ER causes Ca^{2+} release into the cytosol and increases Ca^{2+} within the mitochondria, which activates the dehydrogenases of the tricarboxylic acid cycle to promote adenosine triphosphate (ATP) production, open the mitochondrial permeability transition pore, release cytochrome *c*, and increase mitochondrial ROS production (Bhandary, Marahatta, Kim, & Chae, 2012; Cadenas & Davies, 2000; Rizzuto, De Stefani, Raffaello, & Mammucari, 2012). The depletion of Ca^{2+} levels in the ER leads to the rapid accumulation of misfolded proteins, which activates the UPR because ER chaperones require high levels of Ca^{2+} in the ER lumen to function correctly (Agellon & Michalak, 2017; Malhotra & Kaufman, 2007).

ROS overproduction can be detected by ER, which is very sensitive to changes in ROS levels through redox sensors, such as thiol groups of cysteines (Cao & Kaufman, 2013; Goligorsky, 2016; Mailloux, Jin, & Willmore, 2014; Ozgur, Uzilday, Iwata, Koizumi, & Turkan, 2018; Ursini, Maiorino, & Forman, 2016), which cause an accumulation of misfolded proteins and consequently increase ROS production, which leads to a vicious circle capable of activating the UPR (Cao & Kaufman, 2013; Liu et al., 2019; Malhotra & Kaufman, 2007; Ozgur et al., 2018). The relationship between ER stress and oxidative stress is evident, as shown in Fig. 4. In addition, the UPR is related to the antioxidant response, such as PERK, which directly phosphorylates Nrf2 to dissociates from its negative regulator Kelch-like ECH-associated protein 1 (Keap1) and elicits its nuclear translocation to induce the transcription of antioxidants enzyme genes and maintain redox homeostasis (Cullinan et al., 2003; Sarcinelli et al., 2020; Zhu et al., 2020). Moreover, this antioxidant response induced by PERK/eIF2α/ATF4 axis directly controls Nrf2 expression, enhancing its function and also increasing amino acid metabolism; thus, protein synthesis decreases and GSH production increases, which contributes to the antioxidant response (Sarcinelli et al., 2020; Zeeshan et al., 2016). Furthermore, it has been described that IRE1α can also contribute to antioxidant respond by activating Nrf2 (Sarcinelli et al., 2020).

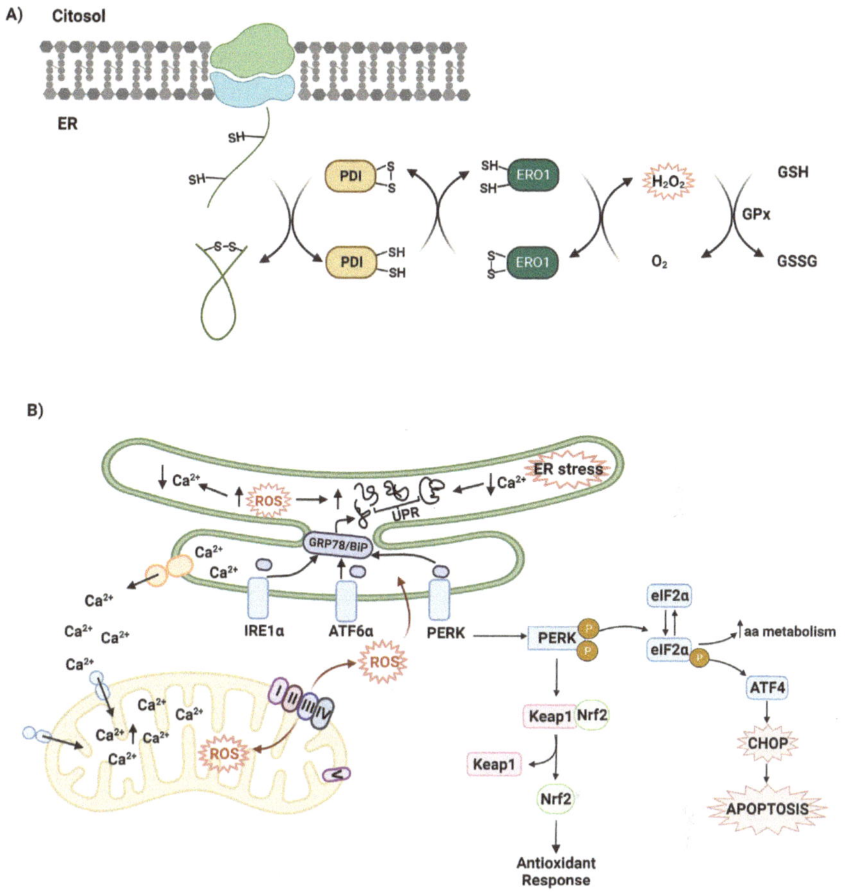

Fig. 4 ER stress and oxidative stress relationship. ROS production causes misfolded proteins that increases the activity of PDI/ERO1 and therefore increase the overproduction of ROS, which causes a decrease in the GSH/GSSG ratio and the levels of antioxidant enzymes. Therefore, ER stress could generate oxidative stress, contributing to ER stress. Increased ROS generation, inflammation, and cell death occur if both persist. *Figure created with Biorender.com.*

Finally, there is evidence that other ATF6α non-canonical target genes include those codifying for antioxidant enzymes such as CAT, peroxiredoxin 5, GPx 4, and 5, among others (Jin et al., 2017).

3. Antioxidants and regulation of the UPR

Antioxidants are substances that, at low concentrations, scavenge free radicals and inhibit the oxidation of carbohydrates, lipids, proteins, and

DNA (Neha, Haider, Pathak, & Yar, 2019; Rasheed & Fathima Abdul Azeez, 2019). Depending on the origin, they can be endogenous such as GSH, superoxide dismutase (SOD), bilirubin, L-carnitine, and melatonin, to name a few examples, or they can be exogenous as vitamins (A, C, D, and E), polyphenols (resveratrol, quercetin, curcumin) and also antioxidants could be synthetics such as butylated hydroxytoluene, butylated hidroxyanisole (BHA), propyl gallate (Neha et al., 2019; Sindhi et al., 2013). The antioxidant activity of these compounds can be directly scavenging free radicals as vitamin C that donate a hydrogen atom and form ascorbyl-free radical to neutralize ROS; antioxidants also can be indirectly inducing cytoprotective enzymes to inhibit ROS production, such as vitamin A that induces the activity of cytoprotective enzymes such as SOD and glutathione S-transferase (Olubukola Sinbad, Folorunsho, Olabisi, Abimbola Ayoola, & Johnson Temitope, 2019). In addition, some compounds that have directly and indirectly antioxidant activity and are known as bifunctional antioxidants, such as melatonin, which has an electron-rich aromatic indole ring that acts as an electron donor and can scavenge ROS and can also stimulate SOD, CAT, GPx, glutathione reductase, glucose-6-phosphate dehydrogenase and γ-glutamylcysteine synthetase activities (Chainy & Sahoo, 2020; Chovancova et al., 2017). The antioxidant activity of several compounds is widely described; however, in some compounds, studies are scarce as vitamin D, so it is suggested that the hydrophobic rings in vitamin D structure could scavenge ROS and prevent lipid peroxidation in the cell membrane (Wiseman, 1993), and also increases GSH, induces the expression of GPx, SOD and decreases the expression of NOX1 (Mokhtari, Hekmatdoost, & Nourian, 2016). Several studies have shown that intake or treatment with antioxidants reduces oxidative stress and can prevent or attenuate the progression of acute and chronic diseases such as cancer, diabetes, hypertension, neurodegenerative, kidney, and liver diseases (Neha et al., 2019; Rasheed & Fathima Abdul Azeez, 2019). Interestingly, in addition to the antioxidant activity of these compounds, some of them have others activities as antimicrobials, anti-inflammatory, and modulators of ER stress.

ER stress and oxidative stress are closely related; as mentioned above, there is a crosstalk and feedback loop of damage between them. Antioxidants could regulate redox balance and activate UPR to improve protein folding (Malhotra et al., 2008; Malhotra & Kaufman, 2007). However, the mechanism of regulation of UPR by these antioxidants is not completely elucidated and is also different in non-cancer cells and cancer cells; however, these compounds' results are of interest due to the therapeutic potential in various diseases. Several antioxidants such as apocynin, BHA, isoliquiritigenin, licochalcone A, mangiferin, α-mangostin, moryn hydrate, curcumin, epigallocatechin-3 gallate, piperine, melatonin, quercetin, resveratrol, sulforaphane, vitamin C, and vitamin D has been used to regulate ER stress in a wide range of diseases such as diabetes, endothelial dysfunction, kidney and liver diseases, Parkinson's disease and cancer (Farooqi et al., 2015; H. Liu et al., 2016; Malhotra et al., 2008; Reyes-Fermín et al., 2020).

4. Antioxidants and ER stress-related diseases

ER stress and UPR are activated under physiological and pathological conditions and exert an adaptive response to maintain proteostasis (Hetz, Zhang, & Kaufman, 2020; Lindholm, Korhonen, Eriksson, & Kõks, 2017). The UPR in each cell type has a specific role, and as a consequence, its deregulation contributes to the progression of various diseases (Bhattarai, Riaz, Kim, & Chae, 2021). Abnormal UPR signaling has been implicated in several diseases as Alzheimer's, Parkinson's, amyotrophic lateral sclerosis, atherosclerosis, rheumatoid arthritis, liver, renal, and lung diseases, diabetes, and cancer (Bhattarai et al., 2021; Hetz et al., 2020; Khanna, Agrawal, Chandra, & Dhawan, 2021; Lindholm et al., 2017). The study of the UPR in pathological conditions has revealed critical points of each disease. Therefore, ER modulators have been used to alleviate the damage induced by the maladaptive signaling of the UPR. This section discusses selected examples of antioxidants in ER stress-related diseases.

4.1 Cancer

Cancer is a disease that is the result of a set of genetic alterations, which result in the uncontrolled proliferation of cells that can evade growth suppressors, resistance to cell apoptosis, reprogramming of energy metabolism, evasion of the immune system, ability to induce angiogenesis, and metastasis (Hanahan & Weinberg, 2011). During the process of carcinogenesis, cancer cells are exposed to extrinsic (hypoxia, nutrient deprivation, and acidosis) and intrinsic (oncogenic activation and chromosome number alteration) perturbations that favor cell transformation and cancer progression. These perturbations alter ER proteostasis and trigger the UPR (Urra, Dufey, Avril, Chevet, & Hetz, 2016). In cancer cells, ER stress and the UPR is double-edged sword. Initially, the adaptive response of the UPR enhances tumor growth, angiogenesis, metabolic rewiring, promoting metastasis, and chemoresistance. However, the sustained ER stress and UPR induces apoptosis

(Hetz et al., 2020; Oakes, 2020; Yadav, Chae, Kim, & Chae, 2014); however, the sustained ER stress and UPR induces apoptosis (Chovancova et al., 2017; Fernández, Ordóñez, Reiter, González-Gallego, & Mauriz, 2015; Yadav et al., 2014). The increase of proliferation and invasion in tumors is associated with the overexpression of GRP78/BiP; the cytosolic portion of this protein can bind to pro-caspase 7 and BIK, inhibiting the release of cytochrome *c* from mitochondria, thus suppressing the apoptosis (Lee, 2007). In addition, overexpression of GRP78/BiP is associated with chemoresistance in cancer of the lung, bladder, stomach, breast, prostate, and gastric (Lee, 2007; Oakes, 2020; Urra et al., 2016; Walczak, Gradzik, Kabzinski, Przybylowska-Sygut, & Majsterek, 2019). As well as overexpression of PERK, IRE1α, ATF6α, and XBP1 are associated with progression, recurrence, worse prognosis, and chemoresistance in leukemia, lymphoma, myeloma, glioblastoma triple-negative breast cancer, stomach, colon, esophagus, lung, pancreas, liver, and prostate cancer (Oakes, 2020; Urra et al., 2016; Walczak et al., 2019).

On the other hand, cancer cells are characterized by increased ROS generation, mainly by the mitochondrial pathway. However, there is evidence that ER stress generates oxidative stress that favors cell proliferation and metabolic adaptations that contribute to drug resistance or promote an aggressive phenotype (Arfin et al., 2021; Chovancova et al., 2017). Therefore, ER stress is a potential target in the treatment of cancer. Antioxidants has been widely used as a protective factor against cancer (Chovancova et al., 2017). In general, they can modulate ROS production and have regulatory effects on ER stress and the response of UPR (Farooqi et al., 2015). A variety of antioxidants as baicalein, curcumin, nicotine, epigallocatechin 3-gallate, isoliquiritigenin, melatonin, resveratrol, quercetin, and vitamins C, D, and E has been used in experimental cancer models to promote apoptosis and prevent drug resistance (Choi et al., 2020; Chovancova et al., 2017; Farooqi et al., 2015; Liu et al., 2016; Montagnani Marelli et al., 2016; Wen, Eder, & Ringseis, 2020).

Melatonin has antiproliferative and cytotoxic effects in neuroblastoma, hepatocarcinoma, leukemia, colorectal cancer, lung cancer, breast cancer, and prostate cancer (Chovancova et al., 2017; Moreira et al., 2015). Melatonin increases the protein expression of GRP78/BiP, PERK/p-PERK, IRE1α, ATF6α, PDI, calnexin and, CHOP and decreases the Bcl-2/BAX ratio in cancer cells (Chok, Koh, Ng, Ng, & Chye, 2021; Fernández et al., 2015; Huang et al., 2021; Moreira et al., 2015).

Interestingly, it has been described that melatonin could also down-regulate the ATF6α signaling pathway and induce apoptosis in hepatocarcinoma cells (Bu et al., 2017), which shows that the effects of these compounds are different in each cell.

Furthermore, melatonin has been associated with decreased cytosolic Ca^{2+} in colorectal and ovarian carcinoma cells in a dose-dependent manner, through up-regulation of Ca^{2+} channels as the sodium-Ca^{2+} exchanger (NCX) and the inositol 1,4,5-triphosphate-receptor 1 (IP_3R1) which, increases apoptosis. Surprisingly, melatonin does not cause changes in cytosolic Ca^{2+} in non-cancer cells. It is important to note that the effects of melatonin can be variable, as they are dose-dependent. For example, in colorectal and ovarian carcinoma cells at 10 μM for 24 h acts as a weak scavenger because it decreases ~30% of ROS production, compared with non-cancer cells that decrease ~68% (Chovancova et al., 2017) and also acts as pro-oxidant to induces ROS production at 2.5 mM for 12 h until 48 h in colorectal carcinoma cells (Chok et al., 2021).

On the other hand, activating the PI3K/Akt signaling pathway is associated with chemoresistance in cancer cells. For example, doxorubicin and melatonin treatment decreases p-Akt and improves CHOP expression and, consequently, apoptosis in hepatocellular carcinoma cells, while tunicamycin-induced ER stress increases p-Akt and decreases CHOP expression and thus apoptosis, suggesting that ER stress is involved in tumor survival (Fan et al., 2013). Down-regulation of the three branches of the UPR decreases CHOP expression and favors tumor survival, but also autophagy promotes resistance to chemotherapy, which is activated by the PERK/p-eIF2α/ATF4 axis. In addition to increasing CHOP expression, melatonin also decreases the expression of LC3-II/LC3-I and beclin-1 and, therefore, autophagy that increases the sensitivity of hepatocellular carcinoma cells sorafenib, an anticancer drug (Zhou et al., 2019).

Vitamin D is an antioxidant used to regulate ER stress. In breast cancer, vitamin D attenuates PERK, ATF4, ATF6α, DDIT3, and sXBP1 mRNA levels (Wen et al., 2020) and also reduces the phosphorylation of IRE1α, JNK and, Bcl-2, these data suggesting that vitamin D has the potential to inhibit autophagy (Li et al., 2021), induces Ca^{2+}-mediated apoptosis and decrease the resistance to chemotherapy (Sergeev, 2012). Additionally, vitamin C inhibits the growth and metastasis of melanoma, gastric cancer, breast cancer, colorectal cancer, pancreatic cancer, and leukemia. Interestingly, vitamin C does not decrease ROS production in breast cancer but induces apoptosis by inducing caspase 12 and CHOP, activated by IRE1α and PERK/p-eIF2α/ATF4 axis (Choi et al., 2020). Like vitamin C, sulforaphane induces the increase of ROS in cancer; this increase of ROS is related to the induction of apoptosis of bladder cancer cells, as well as induction of activation of the GRP78/BiP and CHOP proteins of the ER stress pathway and the Nrf2 pathway (Jo, Kim, Kim, Park, & Choi, 2014; Xie, Chun, Rutz, & Blaheta, 2021).

Quercetin is a flavonoid that inhibits cell growth and induces apoptosis in cancer cells (Ma et al., 2018), and activates the three branches of the UPR in cervical and ovarian cancer cells (He et al., 2021; Yang et al., 2015). It has been observed that quercetin increases protein expression of GRP78/BiP, p-eIF2α/eIF2α, ATF4, IRE1α, p-JNK, XBP1s, ATF6α, and CHOP (Ma et al., 2018; Yang et al., 2015). As mentioned above, ER stress through the components of the UPR could activate autophagy and activate the JAK/STAT3 signaling pathway, which contributes to cell proliferation, angiogenesis, metastasis, and chemoresistance. The decrease in CHOP expression increases the expression of p-STAT3 and Bcl-2, and thus, cancer cells avoid apoptosis. In addition to activating the three branches of the UPR, quercetin suppresses phosphorylation of STAT3 and Bcl-2 protein expression in tumors. When quercetin was combined with cisplatin, drug sensitivity increased, and tumor weight decreased 89% (Yang et al., 2015). Finally, it is essential to note that these antioxidants, such as quercetin, can induce ROS production and increase Ca^{2+} levels, regulating ER stress and activating other signaling pathways, promoting apoptosis in cancer cells (Ma et al., 2018).

4.2 Metabolic syndrome

Metabolic syndrome is a public health concern that includes a variety of metabolic diseases such as obesity, diabetes, cardiovascular disease, and non-alcoholic fatty liver disease (NAFLD). These metabolic diseases are derived by excessive nutrient supply compared to energetic needs, particularly fatty acids and glucose, which exert cytotoxic effects on ER and induces ER stress (Ghemrawi, Battaglia-Hsu, & Arnold, 2018). The ER stress plays a key role in the pathogenesis of diabetes and its complications (Suganya, Dornadula, Chatterjee, & Mohanram, 2018) because it is related to insulin resistance, lipogenesis, gluconeogenesis, and inflammation (Ghemrawi et al., 2018; Mohan, Preetha Rani, Brown, Ayyappan, & Raghu, 2019), for which several antioxidants such as curcumin, proanthocyanidin, vitamin D, melatonin, resveratrol and quercetin, to name few examples, has been used to reduce ER stress in order to ameliorate complications associated to metabolic syndrome (Suganya et al., 2018).

The effect of vitamin D supplementation in metabolic syndrome is related to the reduction of ER stress; for example, in the hippocampus of diabetic rats decrease the expression of GRP78/BiP and PERK/eIF2α/ATF4/CHOP pathway in a dose dependent-manner that contributes to ameliorates pathological changes and then of the promotion to diabetic neuropathy (Guo et al., 2017). Also, vitamin D analogs decrease the expression of GRP78/BiP, p-eIF2α, p-JNK and sXBP1, and ROS production in endothelial cells, which has potential benefits for cardiovascular health in diabetes (Haas, Jafri, Wehmeier, Onstead-Haas, & Mooradian, 2016). Quercetin also improves endothelial function and decreases pancreatic β-cell dysfunction by decreasing oxidative stress and CHOP overexpression in the pancreatic islet of diabetic rats, which is associated with reduced levels of endothelin-1 and nitric oxide and also improves the antioxidant response (Suganya et al., 2018). Surprisingly, it has been shown that CHOP deletion in addition to preventing apoptosis mediated by the ER, also improves the capacity of β-cell to produce insulin (Song, Scheuner, Ron, Pennathur, & Kaufman, 2008), which is why it is a therapeutic target of interest to maintain the integrity of these cells.

The effects of compounds capable of regulating ER could affect various organs such as melatonin, which regulates ER stress in the liver, brain, and skeletal muscle in obese mice. In these animals, a decrease in IRE1α and p-eIF2α is observed. The ATF6α signaling pathway is notably altered, particularly in the brain, and is re-established in obese mice treated with melatonin and whit it a reduction in lipogenesis (Potes, De Luxán-Delgado, Rubio-González, Reiter, & Coto Montes, 2019). Furthermore, melatonin supplementation did not affect the expression of p-PERK in the kidney of diabetic Zücker fatty rats. However, it decreased the expression of GRP78/BiP, IRE1α/JNK and ATF6α, and apoptosis, suggesting that supplementation with melatonin could prevent kidney damage associated with obesity and diabetes (Aouichat et al., 2021). In the kidney of diabetic db/db mice is observed an increase in the expression of p-PERK, p-IRE1α, ATF6α, sXBP1, CHOP, and caspase-12 that is related to ER stress-mediated dysfunction in glomeruli and tubules (Cao et al., 2016).

In addition to the kidney, another organ damaged by metabolic syndrome is the liver. The alteration in the insulin receptor associated with obesity produces hepatic insulin resistance that causes a metabolic imbalance and produces the NAFLD, characterized by steatosis, oxidative stress, and inflammation. Interestingly, abnormalities in the three branches of the UPR are involved in the NAFLD (Cao & Kaufman, 2013; Lebeaupin et al., 2018; Liu & Li, 2018), so negative regulator of the UPR signaling such as resveratrol has been used to restore ER homeostasis and metabolic abnormalities in the liver (Chen et al., 2020). This antioxidant negatively regulates the three branches of the UPR, but particularly ATF6α, the master transcription factor in the regulation of triglycerides synthesis in hepatocytes, which also have a transcriptional effect of SIRT1, involved in the regulation of lipid and carbohydrate metabolism in the liver, and therefore, decrease the accumulation of lipid droplets in the liver (Zhou et al., 2018).

4.3 Neurodegenerative diseases

Ca^{2+} levels regulate synaptic function and cell viability in the cell. The ER is one of the primary storage sites for this ion, which participates in protein folding and as a second messenger. Hence, its homeostasis is essential (Bravo et al., 2013). ER chaperones require optimal Ca^{2+} concentrations to fold proteins so that depleting ER Ca^{2+} produces an accumulation of misfolded proteins in neurons leads to disruption of synapses function, apoptosis, and the development of some neurodegenerative diseases (Bravo et al., 2013; Ghemrawi & Khair, 2020; Lindholm, Wootz, & Korhonen, 2006). The accumulation of proteins such as a β-amyloid peptide, α-synuclein, and SOD1 are critical elements in Alzheimer's, Parkinson's, and amyotrophic lateral sclerosis diseases, respectively (Cabral-Miranda & Hetz, 2017; Colla, 2019; Lindholm et al., 2006; Sprenkle, Sims, Sánchez, & Meares, 2017), to mention some examples. In addition to inducing the formation of these aberrant proteins due to Ca^{2+} depletion, it also affects the interactions between the ER-mitochondria and the control of cell death (Lindholm et al., 2006), exposing the complexity of the associated damage mechanisms, which results in a vicious cycle that promotes diseases progression. Therefore, ER stress and IRE1α activation, and UPR induction favor neurotoxicity in these diseases (Oakes & Papa, 2015).

Additionally, ER stress in chronic neurodegenerative disorders often stimulates autophagy through the PERK/eIF2α/ATF4 signaling pathway. Furthermore, ATG5 y ATG7 have been shown to connect autophagy to ER stress through PERK signaling. Therefore, accumulation of aggregated proteins and impairment of UPR autophagy leads to accumulation of misfolded proteins and progression of neurodegenerative diseases (Ghemrawi & Khair, 2020).

One of the main therapeutic objectives in neurological diseases is to avoid the activation of the UPR. In dopaminergic neurons, the increase in chaperones (GRP78/BiP, GRP94) is related to the PERK/eIF2α/ATF4/CHOP axis and, consequently, cell death (Cabral-Miranda & Hetz, 2017; Colla, 2019). For example, crocin, a carotenoid compound from saffron, was shown to have an attenuating effect on ER stress, thus preventing ER Ca^{2+} release and inhibiting CHOP expression, decreasing apoptosis, and also significantly improving the learning and memory abilities in Alzheimer's disease rats (Hatziagapiou, Kakouri, Lambrou, Bethanis, & Tarantilis, 2019; Lin, Liu, & Yang, 2019). This negative regulation on ER stress and UPR exerts neuroprotective effects as evidenced in neuronal cells treated with vitamin D or resveratrol that decrease protein expression of GRP78/BiP, p-eIF2α, and CHOP, as well as mRNA levels of XBP1. Furthermore, these antioxidants' negative regulation of the UPR is related to the improvement of insulin signaling and the decrease of hyperphosphorylation, which contributes to neuronal damage. Interestingly, the combination of these antioxidants has a synergistic effect on the ER stress markers evaluated (Cheng et al., 2016). The inhibition of ER stress-mediated apoptosis in neuronal cells is associated with neuroprotective effects, it is important to note that not only the PERK/eIF2α/ATF4/CHOP axis plays an essential role in apoptosis; also the IRE1α/JNK and ATF6α signaling pathway can mediate cell survival. Antioxidants such as ginsenoside Rg1 downregulated IRE1α/JNK and reduced the expression of caspase 3, whereas taurine decreased the expression of p-IRE1α, CHOP, caspase 12, and ATF6α and thus ameliorating cognitive effects in animals (Gao & Xu, 2021).

4.4 Severe acute respiratory syndrome coronavirus-2 (SARS-CoV-2)

The SARS-CoV-2 virus is a positive single-stranded RNA genome with a lipid membrane and four constituent proteins: spike glycoprotein, nucleocapsid protein, membrane protein, and envelope protein. This virus has a high transmission and mortality rate and is characterized by a severe inflammatory response and oxidative stress (Aoe, 2020; Cross, Landis, Sehgal, & Payne, 2021; Forcados, Muhammad, Oladipo, Makama, & Meseko, 2021). The spike glycoprotein binds to angiotensin-converting enzyme 2 (ACE-2) proteins, expressed on the cell surface of the oral and nasal mucosa, nasopharynx, lung, stomach, small intestine, colon, skin, lymph nodes, thymus, bone marrow, spleen, liver, kidney, and brain. Successful binding of the virus to the host cell requires a conformational change of the spike glycoprotein that depends on the redox state of the host cell, which is determined by the disulfide-thiol balance; viral infection is associated with increased ROS production, leading to oxidation of the cysteine residue in proteins, increasing the affinity of spike glycoprotein for ACE-2 (Aoe, 2020; Boga & Coto-Montes, 2020; Forcados et al., 2021). Interestingly, elevated malondialdehyde levels and suppression of the Nrf2 gene have been observed in hospitalized patients of SARS-CoV-2, evidencing the role of oxidative stress in pathogenesis (Forcados et al., 2021). Furthermore, ER stress is also related to the pathogenesis of SARS-CoV-2, which uses the host cell's protein translation machinery to synthesize viral proteins that accumulate in the ER and cause the ER stress, evidenced by increased levels of GRP78/BiP and GRP94 mRNA and significant p-PERK leading the activation of the eIF2α/ATF4/CHOP axis in SARS-CoV-2-infected cells (Aoe, 2020; Banerjee, Czinn, Reiter, & Blanchard, 2020; Boga & Coto-Montes, 2020). Moreover, the IRE1α signaling pathway is also activated, and significant splicing of XBP1, apoptosis, and stimulation of NF-κB activity by TRAF2 is observed (Bartolini et al., 2021; Boga & Coto-Montes, 2020). Interestingly, ATF6α is not activated in SARS-CoV-2-infected cells (Aoe, 2020; Banerjee et al., 2020; Boga & Coto-Montes, 2020), suggesting this virus uses and regulate the host's UPR to maintain a favorable environment and maintain the infection.

Since the appearance of SARS-CoV-2, antiviral drugs, antibiotics, antimalarial drugs, antiparasitic and antiprotozoal drugs, serine protease inhibitors, and immunomodulators have been used as therapeutic management (Rangappa et al., 2020). While an effective vaccine is being developed and tested to control SARS-CoV-2, several treatments must be proposed and tested. Antioxidants such as vitamin C, vitamin D, melatonin, *N*-acetyl cysteine, GSH, curcumin, quercetin, and resveratrol have been considered adjuvant therapies for SARS-CoV-2 (Forcados et al., 2021). The protective action of these antioxidants against SARS-CoV-2 is unknown, although the efficacy and safety of these compounds have been widely demonstrated in various diseases. Interestingly, in SARS-CoV-2 patients, a significant decrease in serum and plasma vitamin C and D levels is observed (Forcados et al., 2021; Güven & Gültekin, 2021) and was associated with the severity of the disease (Güven & Gültekin, 2021). However, the studies indicate that vitamin C and D administration has no effects in SARS-CoV-2 patients (Güven & Gültekin, 2021; Rawat et al., 2021), but there are too many gaps in these studies so far cannot be controlled, such as the deficiency of these vitamins in SARS-CoV-2 patients as well as the doses used that probably do not allow to observe the beneficial effects of these vitamins. Another antioxidant widely used in various viral diseases is melatonin (Bahrampour Juybari, Pourhanifeh, Hosseinzadeh, Hemati, & Mehrzadi, 2020), which indirectly inhibits the coupling of ACE2 with SARS-CoV-2 during viral particle fusion through its inhibition of calmodulin (Bahrampour Juybari et al., 2020; Cardinali, Brown, & Pandi-Perumal, 2020; Cross et al., 2021). Surprisingly, it is described that patients who were on melatonin reduced the risk of being positive to SARS-CoV-2 (Jehi et al., 2020); the effects could be related to the antioxidant activity and modulator of ER stress, autophagy, and apoptosis (Forcados et al., 2021; Kleszczyński, Slominski, Steinbrink, & Reiter, 2020). However, the effects of these antioxidants and mechanisms in patients with SARS-COV2 are still being studied.

5. Concluding remarks and future directions

ER stress and the UPR signaling pathway maintain protein quality control, antioxidant response, metabolic imbalance, pathogen invasion, survival, and cell death. This signaling pathway is an objective of the study because it can be regulated to activate different responses. The disruption of the UPR could be a cause or consequence of the etiology or progression of various diseases. Abnormal UPR signaling has been linked to many diseases; it is essential to note that damage mechanisms involving ER stress and oxidative stress, mitochondrial dysfunction, DNA damage, and inflammation participate in the pathological cycle. The interruption of any of the components of this cycle will reduce the damage induced by each disease. Hence, antioxidants capable of regulating ER stress and oxidative stress are a promising therapeutic target because the effects described in experimental models of diseases are beneficial and have also revealed and clarified the unknown mechanism between ER stress and diseases, allowing more therapeutic strategies to be offered. The safety and efficacy of the antioxidants used to regulate ER stress have been tested; however, clinical trials are scarce and are necessary. In addition, it is essential to note that antioxidant treatments are preventives; some of them could delay the progression of the disease but are not curatives. In conclusion, the knowledge of the role of ER stress and each branch of the UPR in normal and pathological conditions and their interconnection will allow us to find and study compounds capable of regulating specific proteins of the UPR signaling pathway to be used as adjuvant treatment.

Acknowledgments

This work was supported by Consejo Nacional de Ciencia y Tecnología (CONACYT, grant number A1-S-7495), Programa de Apoyo a Proyectos de Investigación e Innovación Tecnológica (PAPIIT, grant numbers IN202219 and IN200922), and by Programa de Apoyo a la Investigación y al Posgrado (PAIP, grant number 5000-9105) of Faculty of Chemistry, UNAM.

References

Adachi, Y., Yamamoto, K., Okada, T., Yoshida, H., Harada, A., & Mori, K. (2008). ATF6 is a transcription factor specializing in the regulation of quality control proteins in the endoplasmic reticulum. *Cell Structure and Function*, *33*(1), 75–89. https://doi.org/10.1247/csf.07044.

Agellon, L. B., & Michalak, M. (2017). The endoplasmic reticulum and the cellular reticular network. In J. Krebs (Ed.), *Vol. 981. Membrane dynamics and calcium signaling* (pp. 61–76). Springer International Publishing. https://doi.org/10.1007/978-3-319-55858-5_4.

Ali, M. M. U., Bagratuni, T., Davenport, E. L., Nowak, P. R., Silva-Santisteban, M. C., Hardcastle, A., et al. (2011). Structure of the Ire1 autophosphorylation complex and implications for the unfolded protein response. *The EMBO Journal*, *30*(5), 894–905. https://doi.org/10.1038/emboj.2011.18.

Almanza, A., Carlesso, A., Chintha, C., Creedican, S., Doultsinos, D., Leuzzi, B., et al. (2019). Endoplasmic reticulum stress signalling—From basic mechanisms to clinical applications. *The FEBS Journal*, *286*(2), 241–278. https://doi.org/10.1111/febs.14608.

Aoe, T. (2020). Pathological aspects of COVID-19 as a conformational disease and the use of pharmacological chaperones as a potential therapeutic strategy. *Frontiers in Pharmacology*, *11*(1095), 1–11. https://doi.org/10.3389/fphar.2020.01095.

Aouichat, S., Navarro-Alarcon, M., Alarcón-Guijo, P., Salagre, D., Ncir, M., Zourgui, L., et al. (2021). Melatonin improves endoplasmic reticulum stress-mediated IRE1α pathway in Zücker diabetic fatty rat. *Pharmaceuticals*, *14*(3), 232. https://doi.org/10.3390/ph14030232.

Arfin, S., Jha, N. K., Jha, S. K., Kesari, K. K., Ruokolainen, J., Roychoudhury, S., et al. (2021). Oxidative stress in cancer cell metabolism. *Antioxidants (Basel, Switzerland)*, *10*(5), 642. https://doi.org/10.3390/antiox10050642.

Bahrampour Juybari, K., Pourhanifeh, M. H., Hosseinzadeh, A., Hemati, K., & Mehrzadi, S. (2020). Melatonin potentials against viral infections including COVID-19: Current evidence and new findings. *Virus Research*, *287*, 198108. https://doi.org/10.1016/j.virusres.2020.198108.

Banerjee, A., Czinn, S. J., Reiter, R. J., & Blanchard, T. G. (2020). Crosstalk between endoplasmic reticulum stress and anti-viral activities: A novel therapeutic target for COVID-19. *Life Sciences*, *255*, 117842. https://doi.org/10.1016/j.lfs.2020.117842.

Bartolini, D., Stabile, A. M., Vacca, C., Pistilli, A., Rende, M., Gioiello, A., et al. (2021). Endoplasmic reticulum stress and NF-kB activation in SARS-CoV-2 infected cells and their response to antiviral therapy. *IUBMB Life*, *74*(1), 93–100. https://doi.org/10.1002/iub.2537. iub.

Bertolotti, A., Zhang, Y., Hendershot, L. M., Harding, H. P., & Ron, D. (2000). Dynamic interaction of BiP and ER stress transducers in the unfolded-protein response. *Nature Cell Biology*, *2*(6), 326–332. https://doi.org/10.1038/35014014.

Bhandary, B., Marahatta, A., Kim, H.-R., & Chae, H.-J. (2012). An involvement of oxidative stress in endoplasmic reticulum stress and its associated diseases. *International Journal of Molecular Sciences*, *14*(1), 434–456. https://doi.org/10.3390/ijms14010434.

Bhattarai, K. R., Riaz, T. A., Kim, H.-R., & Chae, H.-J. (2021). The aftermath of the interplay between the endoplasmic reticulum stress response and redox signaling. *Experimental & Molecular Medicine, 53*(2), 151–167. https://doi.org/10.1038/s12276-021-00560-8.

Boga, J. A., & Coto-Montes, A. (2020). ER stress and autophagy induced by SARS-CoV-2: The targets for melatonin treatment. *Melatonin Research, 3*(3), 346–361. https://doi.org/10.32794/mr11250067.

Bommiasamy, H., Back, S. H., Fagone, P., Lee, K., Meshinchi, S., Vink, E., et al. (2009). ATF6alpha induces XBP1-independent expansion of the endoplasmic reticulum. *Journal of Cell Science, 122*(Pt 10), 1626–1636. https://doi.org/10.1242/jcs.045625.

Braakman, I., & Hebert, D. N. (2013). Protein folding in the endoplasmic reticulum. *Cold Spring Harbor Perspectives in Biology, 5*(5), a013201. https://doi.org/10.1101/cshperspect.a013201.

Bravo, R., Parra, V., Gatica, D., Rodriguez, A. E., Torrealba, N., Paredes, F., et al. (2013). Endoplasmic reticulum and the unfolded protein response. In *Vol. 301. International review of cell and molecular biology* (pp. 215–290). Elsevier. https://doi.org/10.1016/B978-0-12-407704-1.00005-1.

Bu, L.-J., Yu, H.-Q., Fan, L.-L., Li, X.-Q., Wang, F., Liu, J.-T., et al. (2017). Melatonin, a novel selective ATF-6 inhibitor, induces human hepatoma cell apoptosis through COX-2 downregulation. *World Journal of Gastroenterology, 23*(6), 986. https://doi.org/10.3748/wjg.v23.i6.986.

Cabral-Miranda, F., & Hetz, C. (2017). ER stress in neurodegenerative disease: From disease mechanisms to therapeutic interventions. *Endoplasmic Reticulum Stress in Diseases, 4*(1), 11–26. https://doi.org/10.1515/ersc-2017-0002.

Cadenas, E., & Davies, K. J. A. (2000). Mitochondrial free radical generation, oxidative stress, and aging 11This article is dedicated to the memory of our dear friend, colleague, and mentor Lars Ernster (1920–1998), in gratitude for all he gave to us. *Free Radical Biology and Medicine, 29*(3–4), 222–230. https://doi.org/10.1016/S0891-5849(00)00317-8.

Cao, S. S., & Kaufman, R. J. (2013). Targeting endoplasmic reticulum stress in metabolic disease. *Expert Opinion on Therapeutic Targets, 17*(4), 437–448. https://doi.org/10.1517/14728222.2013.756471.

Cao, A.-L., Wang, L., Chen, X., Wang, Y.-M., Guo, H.-J., Chu, S., et al. (2016). Ursodeoxycholic acid and 4-phenylbutyrate prevent endoplasmic reticulum stress-induced podocyte apoptosis in diabetic nephropathy. *Laboratory Investigation, 96*(6), 610–622. https://doi.org/10.1038/labinvest.2016.44.

Cardinali, D. P., Brown, G. M., & Pandi-Perumal, S. R. (2020). Can melatonin be a potential "silver bullet" in treating COVID-19 patients? *Diseases, 8*(4), 44. https://doi.org/10.3390/diseases8040044.

Chainy, G. B. N., & Sahoo, D. K. (2020). Hormones and oxidative stress: An overview. *Free Radical Research, 54*(1), 1–26. https://doi.org/10.1080/10715762.2019.1702656.

Chakravarthi, S., Jessop, C. E., & Bulleid, N. J. (2006). The role of glutathione in disulphide bond formation and endoplasmic-reticulum-generated oxidative stress. *EMBO Reports, 7*(3), 271–275. https://doi.org/10.1038/sj.embor.7400645.

Chen, Y., Zhang, H., Chen, Y., Zhang, Y., Shen, M., Jia, P., et al. (2020). Resveratrol alleviates endoplasmic reticulum stress–Associated hepatic steatosis and injury in mice challenged with tunicamycin. *Molecular Nutrition & Food Research, 64*(14), 2000105. https://doi.org/10.1002/mnfr.202000105.

Cheng, J., Xia, X., Rui, Y., Zhang, Z., Qin, L., Han, S., et al. (2016). The combination of 1α,25dihydroxyvitaminD3 with resveratrol improves neuronal degeneration by regulating endoplasmic reticulum stress, insulin signaling and inhibiting tau hyperphosphorylation in SH-SY5Y cells. *Food and Chemical Toxicology, 93*, 32–40. https://doi.org/10.1016/j.fct.2016.04.021.

Chevet, E., Cameron, P. H., Pelletier, M. F., Thomas, D. Y., & Bergeron, J. J. (2001). The endoplasmic reticulum: Integration of protein folding, quality control, signaling and degradation. *Current Opinion in Structural Biology, 11*(1), 120–124. https://doi.org/10.1016/s0959-440x(00)00168-8.

Choi, Y. K., Kang, J.-I., Han, S., Kim, Y. R., Jo, J., Kang, Y. W., et al. (2020). L-ascorbic acid inhibits breast Cancer growth by inducing IRE/JNK/CHOP-related endoplasmic reticulum stress-mediated p62/SQSTM1 accumulation in the nucleus. *Nutrients, 12*(5), 1351. https://doi.org/10.3390/nu12051351.

Chok, K. C., Koh, R. Y., Ng, M. G., Ng, P. Y., & Chye, S. M. (2021). Melatonin induces autophagy via reactive oxygen species-mediated endoplasmic reticulum stress pathway in colorectal Cancer cells. *Molecules, 26*(16), 5038. https://doi.org/10.3390/molecules26165038.

Chovancova, B., Hudecova, S., Lencesova, L., Babula, P., Rezuchova, I., Penesova, A., et al. (2017). Melatonin-induced changes in cytosolic calcium might be responsible for apoptosis induction in tumour cells. *Cellular Physiology and Biochemistry: International Journal of Experimental Cellular Physiology, Biochemistry, and Pharmacology, 44*(2), 763–777. https://doi.org/10.1159/000485290.

Colla, E. (2019). Linking the endoplasmic reticulum to Parkinson's disease and alpha-synucleinopathy. *Frontiers in Neuroscience, 13*, 560. https://doi.org/10.3389/fnins.2019.00560.

Cross, K. M., Landis, D. M., Sehgal, L., & Payne, J. D. (2021). Melatonin for the early treatment of COVID-19: A narrative review of current evidence and possible efficacy. *Endocrine Practice, 27*(8), 850–855. https://doi.org/10.1016/j.eprac.2021.06.001.

Cui, W., Li, J., Ron, D., & Sha, B. (2011). The structure of the PERK kinase domain suggests the mechanism for its activation. *Acta Crystallographica. Section D, Biological Crystallography, 67*(Pt 5), 423–428. https://doi.org/10.1107/S0907444911006445.

Cullinan, S. B., Zhang, D., Hannink, M., Arvisais, E., Kaufman, R. J., & Diehl, J. A. (2003). Nrf2 is a direct PERK substrate and effector of PERK-dependent cell survival. *Molecular and Cellular Biology, 23*(20), 7198–7209. https://doi.org/10.1128/MCB.23.20.7198-7209.2003.

Fan, L., Sun, G., Ma, T., Zhong, F., Lei, Y., Li, X., et al. (2013). Melatonin reverses tunicamycin-induced endoplasmic reticulum stress in human hepatocellular carcinoma cells and improves cytotoxic response to doxorubicin by increasing CHOP and decreasing Survivin. *Journal of Pineal Research, 55*(2), 184–194. https://doi.org/10.1111/jpi.12061.

Farooqi, A. A., Li, K.-T., Fayyaz, S., Chang, Y.-T., Ismail, M., Liaw, C.-C., et al. (2015). Anticancer drugs for the modulation of endoplasmic reticulum stress and oxidative stress. *Tumor Biology, 36*(8), 5743–5752. https://doi.org/10.1007/s13277-015-3797-0.

Fernández, A., Ordóñez, R., Reiter, R. J., González-Gallego, J., & Mauriz, J. L. (2015). Melatonin and endoplasmic reticulum stress: Relation to autophagy and apoptosis. *Journal of Pineal Research, 59*(3), 292–307. https://doi.org/10.1111/jpi.12264.

Forcados, G. E., Muhammad, A., Oladipo, O. O., Makama, S., & Meseko, C. A. (2021). Metabolic implications of oxidative stress and inflammatory process in SARS-CoV-2 pathogenesis: Therapeutic potential of natural antioxidants. *Frontiers in Cellular and Infection Microbiology, 11*, 654813. https://doi.org/10.3389/fcimb.2021.654813.

Forrester, S. J., Kikuchi, D. S., Hernandes, M. S., Xu, Q., & Griendling, K. K. (2018). Reactive oxygen species in metabolic and inflammatory signaling. *Circulation Research, 122*(6), 877–902. https://doi.org/10.1161/CIRCRESAHA.117.311401.

Gao, X., & Xu, Y. (2021). Therapeutic effects of natural compounds and small molecule inhibitors targeting endoplasmic reticulum stress in Alzheimer's disease. *Frontiers in Cell and Development Biology, 9*, 745011. https://doi.org/10.3389/fcell.2021.745011.

Ghemrawi, R., Battaglia-Hsu, S.-F., & Arnold, C. (2018). Endoplasmic reticulum stress in metabolic disorders. *Cell*, 7(6), 63. https://doi.org/10.3390/cells7060063.
Ghemrawi, R., & Khair, M. (2020). Endoplasmic reticulum stress and unfolded protein response in neurodegenerative diseases. *International Journal of Molecular Sciences*, *21*(17), 6127. https://doi.org/10.3390/ijms21176127.
Goligorsky, M. S. (2016). Oxidative stress and the kidney: Riding on the curve of hormesis. *Antioxidants & Redox Signaling*, *25*(3), 117–118. https://doi.org/10.1089/ars.2016.6794.
Guo, F., Yue, H., Wang, L., Ding, C., Wu, L., Wu, Y., et al. (2017). Vitamin D supplement ameliorates hippocampal metabolism in diabetic rats. *Biochemical and Biophysical Research Communications*, *490*(2), 239–246. https://doi.org/10.1016/j.bbrc.2017.06.028.
Güven, M., & Gültekin, H. (2021). The effect of high-dose parenteral vitamin D3 on COVID-19-related inhospital mortality in critical COVID-19 patients during intensive care unit admission: An observational cohort study. *European Journal of Clinical Nutrition*, *75*(9), 1383–1388. https://doi.org/10.1038/s41430-021-00984-5.
Haas, M. J., Jafri, M., Wehmeier, K. R., Onstead-Haas, L. M., & Mooradian, A. D. (2016). Inhibition of endoplasmic reticulum stress and oxidative stress by vitamin D in endothelial cells. *Free Radical Biology and Medicine*, *99*, 1–10. https://doi.org/10.1016/j.freeradbiomed.2016.07.020.
Hanahan, D., & Weinberg, R. A. (2011). Hallmarks of cancer: The next generation. *Cell*, *144*(5), 646–674. https://doi.org/10.1016/j.cell.2011.02.013.
Harding, H. P., Zhang, Y., Zeng, H., Novoa, I., Lu, P. D., Calfon, M., et al. (2003). An integrated stress response regulates amino acid metabolism and resistance to oxidative stress. *Molecular Cell*, *11*(3), 619–633. https://doi.org/10.1016/s1097-2765(03)00105-9.
Hatziagapiou, K., Kakouri, E., Lambrou, G. I., Bethanis, K., & Tarantilis, P. A. (2019). Antioxidant properties of crocus Sativus L. and its constituents and relevance to neurodegenerative diseases; focus on Alzheimer's and Parkinson's disease. *Current Neuropharmacology*, *17*(4), 377–402. https://doi.org/10.2174/1570159X16666180321095705.
Haze, K., Yoshida, H., Yanagi, H., Yura, T., & Mori, K. (1999). Mammalian transcription factor ATF6 is synthesized as a transmembrane protein and activated by proteolysis in response to endoplasmic reticulum stress. *Molecular Biology of the Cell*, *10*(11), 3787–3799. https://doi.org/10.1091/mbc.10.11.3787.
He, C., Lu, X., Li, J., Shen, K., Bai, Y., Li, Y., et al. (2021). The effect of quercetin on cervical cancer cells as determined by inducing tumor endoplasmic reticulum stress and apoptosis and its mechanism of action. *American Journal of Translational Research*, *13*(5), 5240–5247.
Hetz, C., Zhang, K., & Kaufman, R. J. (2020). Mechanisms, regulation and functions of the unfolded protein response. *Nature Reviews. Molecular Cell Biology*, *21*(8), 421–438. https://doi.org/10.1038/s41580-020-0250-z.
Huang, Y., Yuan, K., Tang, M., Yue, J., Bao, L., Wu, S., et al. (2021). Melatonin inhibiting the survival of human gastric cancer cells under ER stress involving autophagy and Ras-Raf-MAPK signalling. *Journal of Cellular and Molecular Medicine*, *25*(3), 1480–1492. https://doi.org/10.1111/jcmm.16237.
Jehi, L., Ji, X., Milinovich, A., Erzurum, S., Rubin, B. P., Gordon, S., et al. (2020). Individualizing risk prediction for positive coronavirus disease 2019 testing. *Chest*, *158*(4), 1364–1375. https://doi.org/10.1016/j.chest.2020.05.580.
Jin, J.-K., Blackwood, E. A., Azizi, K., Thuerauf, D. J., Fahem, A. G., Hofmann, C., et al. (2017). ATF6 decreases myocardial ischemia/reperfusion damage and links ER stress and oxidative stress signaling pathways in the heart. *Circulation Research*, *120*(5), 862–875. https://doi.org/10.1161/CIRCRESAHA.116.310266.
Jo, G. H., Kim, G.-Y., Kim, W.-J., Park, K. Y., & Choi, Y. H. (2014). Sulforaphane induces apoptosis in T24 human urinary bladder cancer cells through a reactive oxygen

species-mediated mitochondrial pathway: The involvement of endoplasmic reticulum stress and the Nrf2 signaling pathway. *International Journal of Oncology*, *45*(4), 1497–1506. https://doi.org/10.3892/ijo.2014.2536.

Kannan, M., Lahiri, S., Liu, L.-K., Choudhary, V., & Prinz, W. A. (2017). Phosphatidylserine synthesis at membrane contact sites promotes its transport out of the ER. *Journal of Lipid Research*, *58*(3), 553–562. https://doi.org/10.1194/jlr.M072959.

Khanna, M., Agrawal, N., Chandra, R., & Dhawan, G. (2021). Targeting unfolded protein response: A new horizon for disease control. *Expert Reviews in Molecular Medicine*, *23*, e1. https://doi.org/10.1017/erm.2021.2.

Kleszczyński, K., Slominski, A. T., Steinbrink, K., & Reiter, R. J. (2020). Clinical trials for use of melatonin to fight against COVID-19 are urgently needed. *Nutrients*, *12*(9), 2561. https://doi.org/10.3390/nu12092561.

Lebeau, J., Saunders, J. M., Moraes, V. W. R., Madhavan, A., Madrazo, N., Anthony, M. C., et al. (2018). The PERK arm of the unfolded protein response regulates mitochondrial morphology during acute endoplasmic reticulum stress. *Cell Reports*, *22*(11), 2827–2836. https://doi.org/10.1016/j.celrep.2018.02.055.

Lebeaupin, C., Vallée, D., Hazari, Y., Hetz, C., Chevet, E., & Bailly-Maitre, B. (2018). Endoplasmic reticulum stress signalling and the pathogenesis of non-alcoholic fatty liver disease. *Journal of Hepatology*, *69*(4), 927–947. https://doi.org/10.1016/j.jhep.2018.06.008.

Lederkremer, G. Z., & Glickman, M. H. (2005). A window of opportunity: Timing protein degradation by trimming of sugars and ubiquitins. *Trends in Biochemical Sciences*, *30*(6), 297–303. https://doi.org/10.1016/j.tibs.2005.04.010.

Lee, A. S. (2007). GRP78 induction in cancer: Therapeutic and prognostic implications. *Cancer Research*, *67*(8), 3496–3499. https://doi.org/10.1158/0008-5472.CAN-07-0325.

Li, Y., Cook, K. L., Yu, W., Jin, L., Bouker, K. B., Clarke, R., et al. (2021). Inhibition of antiestrogen-promoted pro-survival autophagy and tamoxifen resistance in breast Cancer through vitamin D receptor. *Nutrients*, *13*(5), 1715. https://doi.org/10.3390/nu13051715.

Li, Y., Guo, Y., Tang, J., Jiang, J., & Chen, Z. (2015). New insights into the roles of CHOP-induced apoptosis in ER stress. *Acta Biochimica et Biophysica Sinica*, *47*(2), 146–147. https://doi.org/10.1093/abbs/gmu128.

Lin, L., Liu, G., & Yang, L. (2019). Crocin improves cognitive behavior in rats with Alzheimer's disease by regulating endoplasmic reticulum stress and apoptosis. *BioMed Research International*, *2019*, 1–9. https://doi.org/10.1155/2019/9454913.

Lindholm, D., Korhonen, L., Eriksson, O., & Kõks, S. (2017). Recent insights into the role of unfolded protein response in ER stress in health and disease. *Frontiers in Cell and Development Biology*, *5*, 48. https://doi.org/10.3389/fcell.2017.00048.

Lindholm, D., Wootz, H., & Korhonen, L. (2006). ER stress and neurodegenerative diseases. *Cell Death and Differentiation*, *13*(3), 385–392. https://doi.org/10.1038/sj.cdd.4401778.

Liu, Y., Adachi, M., Zhao, S., Hareyama, M., Koong, A. C., Luo, D., et al. (2009). Preventing oxidative stress: A new role for XBP1. *Cell Death and Differentiation*, *16*(6), 847–857. https://doi.org/10.1038/cdd.2009.14.

Liu, L., & Li, X. (2018). Vitamin D3 alleviates high-fat diet–induced endoplasmic reticulum stress and insulin resistance in mice. *Diabetes*, *67*(Suppl. 1), 1986-P. https://doi.org/10.2337/db18-1986-P.

Liu, H., Yang, J., Li, L., Shi, W., Yuan, X., & Wu, L. (2016). The natural occurring compounds targeting endoplasmic reticulum stress. *Evidence-based Complementary and Alternative Medicine*, *2016*, 1–13. https://doi.org/10.1155/2016/7831282.

Liu, X., Zhang, R., Huang, L., Zheng, Z., Vlassara, H., Striker, G., et al. (2019). Excessive oxidative stress contributes to increased acute ER stress kidney injury in aged mice.

Oxidative Medicine and Cellular Longevity, 2019, 1–15. https://doi.org/10.1155/2019/2746521.

Ma, Y., Yao, C., Liu, H., Yu, F., Lin, J., Lu, K., et al. (2018). Quercetin induced apoptosis of human oral cancer SAS cells through mitochondria and endoplasmic reticulum mediated signaling pathways. *Oncology Letters, 15*(6), 9663–9672. https://doi.org/10.3892/ol.2018.8584.

Mailloux, R. J., Jin, X., & Willmore, W. G. (2014). Redox regulation of mitochondrial function with emphasis on cysteine oxidation reactions. *Redox Biology, 2*, 123–139. https://doi.org/10.1016/j.redox.2013.12.011.

Maiuolo, J., Bulotta, S., Verderio, C., Benfante, R., & Borgese, N. (2011). Selective activation of the transcription factor ATF6 mediates endoplasmic reticulum proliferation triggered by a membrane protein. *Proceedings of the National Academy of Sciences of the United States of America, 108*(19), 7832–7837. https://doi.org/10.1073/pnas.1101379108.

Malhotra, J. D., & Kaufman, R. J. (2007). Endoplasmic reticulum stress and oxidative stress: A vicious cycle or a double-edged sword? *Antioxidants & Redox Signaling, 9*(12), 2277–2294. https://doi.org/10.1089/ars.2007.1782.

Malhotra, J. D., Miao, H., Zhang, K., Wolfson, A., Pennathur, S., Pipe, S. W., et al. (2008). Antioxidants reduce endoplasmic reticulum stress and improve protein secretion. *Proceedings of the National Academy of Sciences of the United States of America, 105*(47), 18525–18530. https://doi.org/10.1073/pnas.0809677105.

Martin, D., Li, Y., Yang, J., Wang, G., Margariti, A., Jiang, Z., et al. (2014). Unspliced X-box-binding protein 1 (XBP1) protects endothelial cells from oxidative stress through interaction with histone deacetylase 3. *The Journal of Biological Chemistry, 289*(44), 30625–30634. https://doi.org/10.1074/jbc.M114.571984.

Maurel, M., Chevet, E., Tavernier, J., & Gerlo, S. (2014). Getting RIDD of RNA: IRE1 in cell fate regulation. *Trends in Biochemical Sciences, 39*(5), 245–254. https://doi.org/10.1016/j.tibs.2014.02.008.

Mohan, S., Preetha Rani, M. R., Brown, L., Ayyappan, P., & Raghu, K. G. (2019). Endoplasmic reticulum stress: A master regulator of metabolic syndrome. *European Journal of Pharmacology, 860*, 1–10. https://doi.org/10.1016/j.ejphar.2019.172553.172553.

Mokhtari, Z., Hekmatdoost, A., & Nourian, M. (2016). Antioxidant efficacy of vitamin D. *Journal of Parathyroid Disease, 5*(1), 11–16. https://jparathyroid.com/Article/JPD_20160924145139.

Montagnani Marelli, M., Marzagalli, M., Moretti, R. M., Beretta, G., Casati, L., Comitato, R., et al. (2016). Vitamin E δ-tocotrienol triggers endoplasmic reticulum stress-mediated apoptosis in human melanoma cells. *Scientific Reports, 6*(1), 30502. https://doi.org/10.1038/srep30502.

Moreira, A. J., Ordoñez, R., Cerski, C. T., Picada, J. N., García-Palomo, A., Marroni, N. P., et al. (2015). Melatonin activates endoplasmic reticulum stress and apoptosis in rats with diethylnitrosamine-induced hepatocarcinogenesis. *PLoS One, 10*(12), e0144517. https://doi.org/10.1371/journal.pone.0144517.

Neha, K., Haider, M. R., Pathak, A., & Yar, M. S. (2019). Medicinal prospects of antioxidants: A review. *European Journal of Medicinal Chemistry, 178*, 687–704. https://doi.org/10.1016/j.ejmech.2019.06.010.

Newton, A. C., Bootman, M. D., & Scott, J. D. (2016). Second messengers. *Cold Spring Harbor Perspectives in Biology, 8*(8), a005926. https://doi.org/10.1101/cshperspect.a005926.

Oakes, S. A. (2020). Endoplasmic reticulum stress signaling in cancer cells. *The American Journal of Pathology, 190*(5), 934–946. https://doi.org/10.1016/j.ajpath.2020.01.010.

Oakes, S. A., & Papa, F. R. (2015). The role of endoplasmic reticulum stress in human pathology. *Annual Review of Pathology: Mechanisms of Disease*, *10*(1), 173–194. https://doi.org/10.1146/annurev-pathol-012513-104649.

Ogata, M., Hino, S., Saito, A., Morikawa, K., Kondo, S., Kanemoto, S., et al. (2006). Autophagy is activated for cell survival after endoplasmic reticulum stress. *Molecular and Cellular Biology*, *26*(24), 9220–9231. https://doi.org/10.1128/MCB.01453-06.

Oka, O. B., van Lith, M., Rudolf, J., Tungkum, W., Pringle, M. A., & Bulleid, N. J. (2019). ERp18 regulates activation of ATF6α during unfolded protein response. *The EMBO Journal*, *38*(15), e100990. https://doi.org/10.15252/embj.2018100990.

Olubukola Sinbad, O., Folorunsho, A. A., Olabisi, O. L., Abimbola Ayoola, O., & Johnson Temitope, E. (2019). Vitamins as Antioxidants. *Journal of Food Science and Nutrition Research*, *2*(3), 214–235. https://doi.org/10.26502/jfsnr.2642-11000021.

Ozgur, R., Uzilday, B., Iwata, Y., Koizumi, N., & Turkan, I. (2018). Interplay between the unfolded protein response and reactive oxygen species: A dynamic duo. *Journal of Experimental Botany*, *69*(14), 3333–3345. https://doi.org/10.1093/jxb/ery040.

Pakos-Zebrucka, K., Koryga, I., Mnich, K., Ljujic, M., Samali, A., & Gorman, A. M. (2016). The integrated stress response. *EMBO Reports*, *17*(10), 1374–1395. https://doi.org/10.15252/embr.201642195.

Potes, Y., De Luxán-Delgado, B., Rubio-González, A., Reiter, R. J., & Coto Montes, A. M. (2019). Dose-dependent beneficial effect of melatonin on obesity; interaction of melatonin and leptin. *Melatonin Research*, *2*(1), 1–8. https://doi.org/10.32794/mr11250008.

Quon, E., Sere, Y. Y., Chauhan, N., Johansen, J., Sullivan, D. P., Dittman, J. S., et al. (2018). Endoplasmic reticulum-plasma membrane contact sites integrate sterol and phospholipid regulation. *PLoS Biology*, *16*(5), e2003864. https://doi.org/10.1371/journal.pbio.2003864.

Raciti, M., Lotti, L. V., Valia, S., Pulcinelli, F. M., & Di Renzo, L. (2012). JNK2 is activated during ER stress and promotes cell survival. *Cell Death & Disease*, *3*, e429. https://doi.org/10.1038/cddis.2012.167.

Rangappa, P., Dixit, S. B., Jog, S. A., Khatib, K. I., Bandopadhyay, S., Shrivastav, O., et al. (2020). Current approaches to COVID-19: Therapy and prevention. *Indian Journal of Critical Care Medicine*, *24*(9), 838–846. https://doi.org/10.5005/jp-journals-10071-23470.

Rapoport, T. A. (2007). Protein translocation across the eukaryotic endoplasmic reticulum and bacterial plasma membranes. *Nature*, *450*(7170), 663–669. https://doi.org/10.1038/nature06384.

Rasheed, A., & Fathima Abdul Azeez, R. (2019). A review on natural antioxidants. In C. Mordeniz (Ed.), *Traditional and complementary medicine*. https://doi.org/10.5772/intechopen.82636. Intech Open.

Rawat, D., Roy, A., Maitra, S., Gulati, A., Khanna, P., & Baidya, D. K. (2021). Vitamin C and COVID-19 treatment: A systematic review and meta-analysis of randomized controlled trials. *Diabetes and Metabolic Syndrome: Clinical Research and Reviews*, *15*(6), 102324. https://doi.org/10.1016/j.dsx.2021.102324.

Reyes-Fermín, L. M., Aparicio-Trejo, O. E., Avila-Rojas, S. H., Gómez-Sierra, T., Martínez-Klimova, E., & Pedraza-Chaverri, J. (2020). Natural antioxidants' effects on endoplasmic reticulum stress-related diseases. *Food and Chemical Toxicology*, *138*, 111229. https://doi.org/10.1016/j.fct.2020.111229.

Rizzuto, R., De Stefani, D., Raffaello, A., & Mammucari, C. (2012). Mitochondria as sensors and regulators of calcium signalling. *Nature Reviews Molecular Cell Biology*, *13*(9), 566–578. https://doi.org/10.1038/nrm3412.

Rutkowski, D. T., & Kaufman, R. J. (2007). That which does not kill me makes me stronger: Adapting to chronic ER stress. *Trends in Biochemical Sciences*, *32*(10), 469–476. https://doi.org/10.1016/j.tibs.2007.09.003.

Sarcinelli, C., Dragic, H., Piecyk, M., Barbet, V., Duret, C., Barthelaix, A., et al. (2020). ATF4-dependent NRF2 transcriptional regulation promotes antioxidant protection during endoplasmic reticulum stress. *Cancers*, *12*(3), 569. https://doi.org/10.3390/cancers12030569.

Schönthal, A. H. (2012). Endoplasmic reticulum stress: Its role in disease and novel prospects for therapy. *Scientifica*, *2012*, 1–26. https://doi.org/10.6064/2012/857516.

Schwarz, D. S., & Blower, M. D. (2016). The endoplasmic reticulum: Structure, function and response to cellular signaling. *Cellular and Molecular Life Sciences*, *73*(1), 79–94. https://doi.org/10.1007/s00018-015-2052-6.

Sepulveda, D., Rojas-Rivera, D., Rodríguez, D. A., Groenendyk, J., Köhler, A., Lebeaupin, C., et al. (2018). Interactome screening identifies the ER luminal chaperone Hsp47 as a regulator of the unfolded protein response transducer IRE1α. *Molecular Cell*, *69*(2), 238–252.e7. https://doi.org/10.1016/j.molcel.2017.12.028.

Sergeev, I. N. (2012). Vitamin D and cellular Ca2+ signaling in breast cancer. *Anticancer Research*, *32*(1), 299–302.

Shen, J., Snapp, E. L., Lippincott-Schwartz, J., & Prywes, R. (2005). Stable binding of ATF6 to BiP in the endoplasmic reticulum stress response. *Molecular and Cellular Biology*, *25*(3), 921–932. https://doi.org/10.1128/MCB.25.3.921-932.2005.

Sindhi, V., Gupta, V., Sharma, K., Bhatnagar, S., Kumari, R., & Dhaka, N. (2013). Potential applications of antioxidants—A review. *Journal of Pharmacy Research*, 7(9), 828–835. https://doi.org/10.1016/j.jopr.2013.10.001.

Song, B., Scheuner, D., Ron, D., Pennathur, S., & Kaufman, R. J. (2008). Chop deletion reduces oxidative stress, improves β cell function, and promotes cell survival in multiple mouse models of diabetes. *Journal of Clinical Investigation*, *118*(10), 3378–3389. https://doi.org/10.1172/JCI34587.

Sprenkle, N. T., Sims, S. G., Sánchez, C. L., & Meares, G. P. (2017). Endoplasmic reticulum stress and inflammation in the central nervous system. *Molecular Neurodegeneration*, *12*(1), 42. https://doi.org/10.1186/s13024-017-0183-y.

Suganya, N., Dornadula, S., Chatterjee, S., & Mohanram, R. K. (2018). Quercetin improves endothelial function in diabetic rats through inhibition of endoplasmic reticulum stress-mediated oxidative stress. *European Journal of Pharmacology*, *819*, 80–88. https://doi.org/10.1016/j.ejphar.2017.11.034.

Tu, B. P., & Weissman, J. S. (2002). The FAD- and O2-dependent reaction cycle of Ero 1-mediated oxidative protein folding in the endoplasmic reticulum. *Molecular Cell*, *10*(5), 983–994. https://doi.org/10.1016/S1097-2765(02)00696-2.

Urano, F., Wang, X., Bertolotti, A., Zhang, Y., Chung, P., Harding, H. P., et al. (2000). Coupling of stress in the ER to activation of JNK protein kinases by transmembrane protein kinase IRE1. *Science (New York, N.Y.)*, *287*(5453), 664–666. https://doi.org/10.1126/science.287.5453.664.

Urra, H., Dufey, E., Avril, T., Chevet, E., & Hetz, C. (2016). Endoplasmic reticulum stress and the hallmarks of cancer. *Trends in Cancer*, *2*(5), 252–262. https://doi.org/10.1016/j.trecan.2016.03.007.

Ursini, F., Maiorino, M., & Forman, H. J. (2016). Redox homeostasis: The Golden mean of healthy living. *Redox Biology*, *8*, 205–215. https://doi.org/10.1016/j.redox.2016.01.010.

Walczak, A., Gradzik, K., Kabzinski, J., Przybylowska-Sygut, K., & Majsterek, I. (2019). The role of the ER-induced UPR pathway and the efficacy of its inhibitors and inducers in the inhibition of tumor progression. *Oxidative Medicine and Cellular Longevity*, *2019*, 1–15. https://doi.org/10.1155/2019/5729710.

Walter, P., & Ron, D. (2011). The unfolded protein response: From stress pathway to homeostatic regulation. *Science*, *334*(6059), 1081–1086. https://doi.org/10.1126/science.1209038.

Wang, Q., Groenendyk, J., & Michalak, M. (2015). Glycoprotein quality control and endoplasmic reticulum stress. *Molecules*, *20*(8), 13689–13704. https://doi.org/10.3390/molecules200813689.

Wang, P., Li, J., Tao, J., & Sha, B. (2018). The luminal domain of the ER stress sensor protein PERK binds misfolded proteins and thereby triggers PERK oligomerization. *The Journal of Biological Chemistry*, *293*(11), 4110–4121. https://doi.org/10.1074/jbc.RA117.001294.

Wei, Y., Pattingre, S., Sinha, S., Bassik, M., & Levine, B. (2008). JNK1-mediated phosphorylation of Bcl-2 regulates starvation-induced autophagy. *Molecular Cell*, *30*(6), 678–688. https://doi.org/10.1016/j.molcel.2008.06.001.

Wen, G., Eder, K., & Ringseis, R. (2020). 1,25-hydroxyvitamin D3 decreases endoplasmic reticulum stress-induced inflammatory response in mammary epithelial cells. *PLoS One*, *15*(2), e0228945. https://doi.org/10.1371/journal.pone.0228945.

Weston, C. R., & Davis, R. J. (2007). The JNK signal transduction pathway. *Current Opinion in Cell Biology*, *19*(2), 142–149. https://doi.org/10.1016/j.ceb.2007.02.001.

Wiseman, H. (1993). Vitamin D is a membrane antioxidant. Ability to inhibit iron-dependent lipid peroxidation in liposomes compared to cholesterol, ergosterol and tamoxifen and relevance to anticancer action. *FEBS Letters*, *326*(1–3), 285–288. https://doi.org/10.1016/0014-5793(93)81809-e.

Xie, H., Chun, F. K.-H., Rutz, J., & Blaheta, R. A. (2021). Sulforaphane impact on reactive oxygen species (ROS) in bladder carcinoma. *International Journal of Molecular Sciences*, *22*(11), 5938. https://doi.org/10.3390/ijms22115938.

Yadav, R. K., Chae, S.-W., Kim, H.-R., & Chae, H. J. (2014). Endoplasmic reticulum stress and Cancer. *Journal of Cancer Prevention*, *19*(2), 75–88. https://doi.org/10.15430/JCP.2014.19.2.75.

Yang, Z., Liu, Y., Liao, J., Gong, C., Sun, C., Zhou, X., et al. (2015). Quercetin induces endoplasmic reticulum stress to enhance cDDP cytotoxicity in ovarian cancer: Involvement of STAT3 signaling. *The FEBS Journal*, *282*(6), 1111–1125. https://doi.org/10.1111/febs.13206.

Zeeshan, H., Lee, G., Kim, H.-R., & Chae, H.-J. (2016). Endoplasmic reticulum stress and associated ROS. *International Journal of Molecular Sciences*, *17*(3), 327. https://doi.org/10.3390/ijms17030327.

Zhou, B., Lu, Q., Liu, J., Fan, L., Wang, Y., Wei, W., et al. (2019). Melatonin increases the sensitivity of hepatocellular carcinoma to Sorafenib through the PERK-ATF4-Beclin1 pathway. *International Journal of Biological Sciences*, *15*(9), 1905–1920. https://doi.org/10.7150/ijbs.32550.

Zhou, R., Yi, L., Ye, X., Zeng, X., Liu, K., Qin, Y., et al. (2018). Resveratrol ameliorates lipid droplet accumulation in liver through a SIRT1/ ATF6-dependent mechanism. *Cellular Physiology and Biochemistry*, *51*(5), 2397–2420. https://doi.org/10.1159/000495898.

Zhu, Y. P., Zheng, Z., Hu, S., Ru, X., Fan, Z., Qiu, L., Zhang, Y., et al. (2020). Unification of opposites between two antioxidant transcription factors Nrf1 and Nrf2 in mediating distinct cellular responses to the endoplasmic Reticulum Stressor Tunicamycin. *Antioxidants*, *9*(4), 1–29. https://doi.org/10.3390/antiox9010004.

CHAPTER EIGHT

Hormone-linked redox status and its modulation by antioxidants

Dipak Kumar Sahoo[a,*] and Gagan B.N. Chainy[b]
[a]Department of Veterinary Clinical Sciences, College of Veterinary Medicine, Iowa States University, Ames, IA, United States
[b]Department of Biotechnology, Utkal University, Bhubaneswar, Odisha, India
*Corresponding author: e-mail address: dipaksahoo11@gmail.com

Contents

Abstract

Hormones have been considered as key factors involved in the maintenance of the redox status of the body. We are making considerable progress in understanding interactions between the endocrine system, redox status, and oxidative stress with the dynamics of life, which encompasses fertilization, development, growth, aging, and various pathophysiological states. One of the reasons for changes in redox states of vertebrates leading to oxidative stress scenario is the disruption of the endocrine system. Comprehending the dynamics of hormonal status to redox state and oxidative stress in living systems is challenging. It is more difficult to come to a unifying

Vitamins and Hormones, Volume 121
ISSN 0083-6729
https://doi.org/10.1016/bs.vh.2022.10.007

conclusion when some hormones exhibit oxidant properties while others have antioxidant features. There is a very limited approach to correlate alteration in titers of hormones with redox status and oxidative stress with growth, development, aging, and pathophysiological stress. The situation is further complicated when considering various tissues and sexes in vertebrates. This chapter discusses the beneficial impacts of hormones with antioxidative properties, such as melatonin, glucagon, insulin, estrogens, and progesterone, which protect cells from oxidative damage and reduce pathophysiological effects. Additionally, we discuss the protective effects of antioxidants like vitamins A, E, and C, curcumin, tempol, N-acetyl cysteine, α-lipoic acid, date palm pollen extract, resveratrol, and flavonoids on oxidative stress triggered by hormones such as aldosterone, glucocorticoids, thyroid hormones, and catecholamines. Inflammation, pathophysiology, and the aging process can all be controlled by understanding how antioxidants and hormones operate together to maintain cellular redox status. Identifying the hormonal changes and the action of antioxidants may help in developing new therapeutic strategies for hormonal imbalance-related disorders.

1. Introduction

In vertebrates, the endocrine system regulates general metabolism, development, growth, and differentiation, as well as the maintenance of physiological function and homeostasis, by producing and secreting directly into the bloodstream a discrete class of substances known collectively as hormones (Vargatu, 2016). In mammals, controlling antioxidant enzyme activity can be affected by factors such as age, hormonal state, and organ specificity (Harris, 1992). Because hormones control metabolic activities, metabolic activities require oxygen, and incomplete reduction of oxygen is the source of reactive oxygen species (ROS), it stands to reason that any change in the hormonal milieu of aerobic cells may significantly impact ROS production in living systems and thereby, may influence redox status. Redox status ascribes to the balance between prooxidants and antioxidants. Shifting the balance favoring prooxidants causes oxidative stress (OS) in living organisms (Sies, 1991). This imbalance could result from excessive ROS or failure of the antioxidant defense system (AODS). At the same time, OS refers to high amounts of ROS that cause damage to DNA, protein, or lipids, while redox biology describes low levels of ROS that activate signaling pathways to trigger biological processes. This demonstrates the hormesis present in the ROS response.

Hormesis, which refers to the paradoxical positive effects of low doses of stressors, is more accurately defined as the biphasic dose-effect or time-effect association of any substance (Li, Yang, & Sun, 2019). The immune system functions best when there are minimal levels of ROS. The inhibition of the immune system that results from lowering ROS levels prevents the activation of appropriate immunological responses. Tumorigenesis, autoimmunity, and the age-related decline in tissue regeneration are all under the control of signaling pathways that are regulated by redox signaling rather than OS. As a result of this paradigm shift, it is more challenging to interfere with redox biology by giving antioxidants, as this would impact redox biology during both normal and pathological reactions (Ristow et al., 2009). OS is implicated in several diseases, including hypertension, dyslipidemia, diabetes, neurological disorders, respiratory and inflammatory diseases, and cancer (Ilango et al., 2022; Sahoo et al., 2022). Interactions between hormones and oxidative metabolism and the potential repercussions of OS on the aging of endocrine organs show the connection of many pathways linking OS to aging. These mechanisms entail cell senescence, mitochondrial dysfunction, microRNA dysregulation, inflammation, and OS (Leisegang, Roychoudhury, Slama, & Finelli, 2021).

The genomic effects of hormones are mediated by their respective receptors, ligand-regulated nuclear receptor superfamily transcription factors, which might be either positive or negative. Positive regulation occurs when the receptor interacts directly with certain DNA sequences known as hormone response elements. For example, unliganded thyroid hormone receptors recruit specific corepressor proteins, which form a massive corepressor complex containing histone deacetylase and repress nearby genes by condensing chromatin. Liganded receptors undergo conformational changes, which result in the release of corepressors and the recruitment of coactivator proteins (p160, Glucocorticoid receptor-interacting protein, steroid receptor coactivator-1) (Szwarc, Kommagani, Lessey, & Lydon, 2014). Negative gene expression regulation is less well known. It is proposed that negative regulation occurs due to direct receptor binding to the promoter or the squelching of coregulators from other transcriptional factors. The molecular mechanisms that regulate OS and antioxidant status in different tissues and under different pathophysiological states are not fully understood and continue to be a topic of concern, even though it is known that a variety of hormones are responsible for this regulation. These findings are summarized here to better understand the dynamic interplay between hormones and AODS, which significantly impacts the internal homeostasis of organisms in both physiological and pathological states. This article also presents an overview of the mechanisms through which OS contributes to pathophysiology.

In contrast to the antioxidant characteristics of hormones such as melatonin, insulin, estrogen, and progesterone, hormones like thyroid hormone, corticosteroids, and catecholamines can increase the generation of free radicals and OS due to an imbalance in redox status. The function/impact of testosterone in OS/redox status varies according to tissue, species, gender, and physiological state. In this article, we make an effort to bring information available in the literature on OS, redox signaling/status, hormonal imbalance, antioxidant defense systems, and the factors that restrict the efficiency of antioxidants under one umbrella so that it will be possible to improve therapeutic intervention by understanding the link between how oxidants function, the limits of antioxidant therapy, and its potential benefits.

2. Hormones, redox status, and antioxidant defense system

Several hormones have the structural predisposition to operate as antioxidants or influence the numerous enzymatic and non-enzymatic antioxidant defense components, thereby modulating redox status. Cells are safeguarded against the damaging effects of OS via the evolutionarily conserved Keap1-Nrf2 pathway (Fig. 1) (Chew, Zhang, He, & Yu, 2021). Kelch-like ECH-associated protein (Keap1) is a BTB-Kelch protein that serves as a substrate adaptor for a Cul3-dependent E3 ubiquitin ligase complex (Cullinan, Gordan, Jin, Harper, & Diehl, 2004; Zhang et al., 2005). Keap1 was initially identified as a major inhibitor of the transcription factor nuclear factor erythroid 2-related factor 2 (Nrf2) and is crucial for its degradation under physiological circumstances (Kensler, Wakabayashi, & Biswal, 2009; Nguyen, Yang, & Pickett, 2004). Keap1-based Cul3 complex targets Nrf2 for ubiquitin-dependent degradation under basal circumstances (Cullinan et al., 2004). Several reactive cysteine residues in Keap1 are covalently changed in the presence of oxidative or electrophilic stresses, resulting in a conformational shift that prevents Nrf2 ubiquitination. Consequently, Nrf2 is stabilized and translocated to the nucleus, forming heterodimers with small Maf proteins (sMaf). The Nrf2/sMaf heterodimer binds specifically to a *cis*-acting enhancer called antioxidant response element (ARE) located on the promoters of genes encoding proteins involved in the cellular antioxidant response and activates the transcription of an array of antioxidant genes (Kensler et al., 2009; Li et al., 2008) (Fig. 1). The ARE-dependent gene activation in response to OS is mediated by the Keap1–Nrf2 complex, which has been proposed as a cellular redox

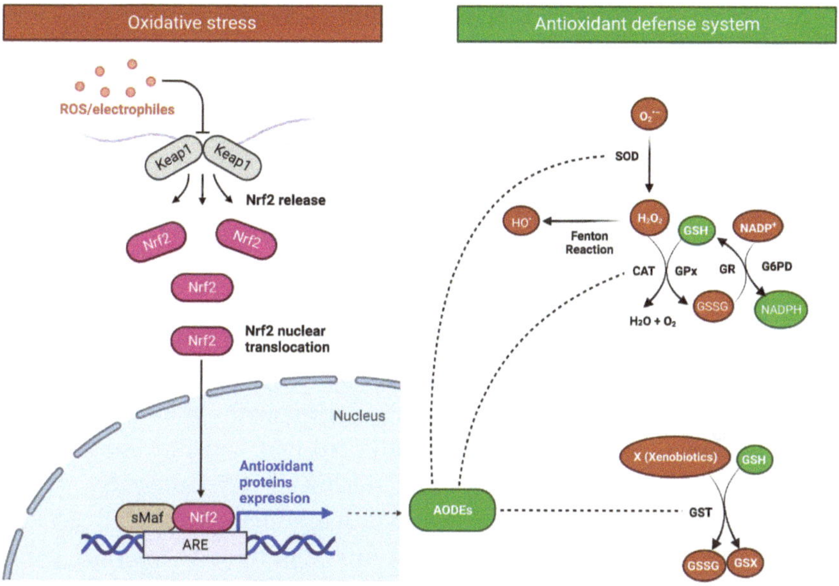

Fig. 1 An overview of the Keap1-Nrf2 signaling pathway and antioxidant defense system. SOD, superoxide dismutase; CAT, catalase; G6PD, glucose-6-phosphate dehydrogenase; GSH, reduced glutathione; GPx, GPx; GR, glutathione reductase; GST, glutathione S-transferase; GSSG, oxidized glutathione; $HO^{\bullet}$, hydroxyl radical; H_2O_2, hydrogen peroxide; $O_2^{\bullet -}$, superoxide anion radical; GSX, conjugate. The figure was created with BioRender.com.

sensor (Sun, Zhang, Chan, & Zhang, 2007). Many phytochemical protectants can likely elicit direct or indirect changes through modest adjustments in the redox state of cells due to the high thiol content of the Keap1 protein. A permanent upregulation of phase II enzymes such as glutathione-based conjugation enzymes, in contrast to the condition in the unchallenged cell, may emerge from maintaining this slight modification in the redox state, which may induce hormetic-type effects.

Antioxidants are able to neutralize ROS, reduce oxidizing species directly or scavenge oxygen and avoid Fenton reactions because of their ability to chelate transition metal ions. Frequently, antioxidant molecules combine two or more of these characteristics (Chen, Hsu, Tsai, Chen, & Drummen, 2013). Antioxidant enzymes (AOEs) (Fig. 1) neutralize free radicals by transforming them into less hazardous compounds (Chainy & Sahoo, 2020; Halliwell & Gutteridge, 2015). The enzymes superoxide dismutase (SOD), catalase (CAT), glutathione reductase (GR), and glutathione peroxidase (GPx) are among the most well-known examples of AOEs (Fig. 1)

(Chainy & Sahoo, 2020; Sahoo et al., 2022). While dismutation of superoxide anion radicals ($O_2^{\bullet -}$) to hydrogen peroxide (H_2O_2) and molecular oxygen (O_2) is catalyzed by SOD, H_2O_2 is converted to water and oxygen by CAT, an iron-containing hemoprotein (Chainy & Sahoo, 2020; Halliwell & Gutteridge, 2015; Sahoo, Jena, & Chainy, 2019; Sahoo, Roy, Bhanja, & Chainy, 2005). GPx is an enzyme that contains a selenium ion as a cofactor (Halliwell & Gutteridge, 2015; Sahoo & Roy, 2012), and in order for it to carry out the reaction that it catalyzes, it requires reduced glutathione (GSH), which is supplied by GR (Chainy & Sahoo, 2020; Sahoo et al., 2022). Scavenging ROS is significantly aided by the presence of small molecular non-enzymatic antioxidants such glutathione, vitamin E (α-tocopherol), retinol, and vitamin C (ascorbic acid), while ROS can be scavenged directly by GSH, NADPH, thioredoxin, vitamin E, vitamin C, and trace metals, such as selenium (Halliwell & Gutteridge, 2015). Tocopherol (vitamin E) and other free-radical terminators disrupt free-radical chain reactions by donating hydrogen atoms from hydroxyl groups, generating more stable species that do not stimulate oxidation, whereas oxygen scavengers, for example, ascorbic acid, employ their antioxidant mechanism by transferring hydrogen atoms and removing oxygen (Dziezak, 1986).

3. Methods to measure oxidative stress markers and their limitations

LPx is a major contributor to cell injury because it oxidizes cell membrane lipids through a complex radical chain reaction during OS (Halliwell & Gutteridge, 2015). Lipid peroxyl, hydroperoxides, alkoxyl radicals, and aldehydic fatty acid derivatives such as malondialdehyde and 4-hydroxynonenal are all reactive byproducts of LPx that can harm or alter proteins, DNA, and other macromolecules in addition to damaging cellular membranes (Marnett, 2002). No LPx assay was regarded as "perfect," despite the requirement of detecting LPx to study the potential impacts of OS. The most widespread method for measuring OS is the thiobarbituric acid-reacting substances (TBARS) assay, which analyzes malondialdehyde (MDA), a byproduct of LPx (Gutteridge & Halliwell, 1990). However, the TBARS assay is not specific for MDA, and MDA is not a specific product of LPx (Halliwell, 2000); for instance, base propenals, which are structural analogs of MDA, are formed during oxidative cleavage of DNA (Plastaras, Dedon, & Marnett, 2002). Measuring exhaled volatile alkanes, for example, ethane and pentane, is an additional method for detecting LPx *in vivo*.

Since these hydrocarbon gases are relatively insignificant end-products of peroxidation and their quantities are affected by the rate of breakdown of peroxides (Halliwell, 2000), their utility as a marker of endogenous LPx has been called into question. As ROS are responsible for many oxidative alterations in the DNA bases, urinary 8-oxo-7,8-dihydro-2′-deoxyguanosine (8-oxodG) is a frequently employed biomarker of OS (Loft et al., 2012). However, variability across chromatographic and ELISA methods hinders the interpretation of data, and this variability may increase if urine composition varies between individuals, causing assay interference. Unfortunately, the absolute amounts of 8-oxodG cannot be determined by ELISA due to its higher variability compared to chromatographic assays. Compared to ELISAs, which exhibited far more inter-laboratory variability and often overestimated levels, chromatographic tests demonstrated excellent consistency between urine samples taken from different patients (Barregard et al., 2013).

The oxidation of arachidonic acid catalyzed by free radicals produces F_2-isoprostanes (F_2-IPs). The measurement of F_2-IPs is widely used as a sensitive index of OS *in vivo* since it is a dependable marker of LPx (Montuschi, Barnes, & Roberts, 2004; Roberts & Morrow, 2000). F_2-isoprostanes have several benefits over currently used quantitative markers of OS (Roberts & Morrow, 2000) as they are formed *in vivo*, specific products of peroxidation, chemically stable, and unaffected by dietary lipid content (Roberts & Morrow, 2000). Unfortunately, the formation process of F_2-isoprostanes is inhibited at elevated oxygen tensions, making this marker of LPx insensitive in situations like hyperoxia-induced lung damage. However, this barrier has been removed with the finding of isofurans, which are substituted tetrahydrofuran ring products of LPx whose synthesis is preferred as oxygen tension increases (Fessel, Porter, Moore, Sheller, & Roberts, 2002).

In biological samples, the ratio of glutathione disulfide (GSSG) to reduced glutathione (GSH) is a commonly used indicator of OS. Earlier techniques for measuring GSSG utilized the chemical masking of GSH in test samples, followed by the enzymatic cycling of GSSG to GSH via GR and NADPH. In the cycling experiment, GSSG concentration was determined by oxidation of NADPH (Güntherberg & Rost, 1966) or production of 5-thio-2-nitrobenzoic acid (TNB), the reduced form of dithionitrobenzene (DTNB) (Sahoo, Roy, Bhanja, & Chainy, 2008; Sahoo, Roy, Chattopadhyay, & Chainy, 2007; Tietze, 1969). N-ethylmaleimide (NEM), a fast-acting masking agent, was employed (Güntherberg & Rost, 1966). However, it is well-documented that NEM inhibits GR, yet efforts to eliminate excess NEM have proven mostly infeasible. The challenge was avoided by switching the GSH masking agent from NEM to 2-vinyl pyridine (2VP) (Griffith, 1980). Since then, this technique has established the gold standard for GSSG analysis in various biological samples. The inexpensive cost and ease of use of this procedure are just two of its many benefits over newer analytical techniques. However, the approach may have issues due to the poor reaction time of 2VP with GSH. Complete GSH masking at room temperature takes around 1 h. GSH in the sample would likely become oxidized during this time due to simple exposure to ambient conditions (McGill & Jaeschke, 2015), leading to reports of erroneously high GSSG/GSH levels. It was recently found that NEM when combined with a simple solid phase extraction method, provides more precise results for determining GSSG (McGill & Jaeschke, 2015).

4. Hormones triggering oxidative stress and the role of antioxidant supplements

A growing body of research has focused on the use of antioxidant supplements (Chhabria et al., 2022; Ilango et al., 2022) in the treatment of endocrine problems and the use of alternative therapies in conjunction with traditional treatment. An increase in OS has been linked to an increased risk of cellular damage and disease. When tissue damage has already occurred, antioxidant therapy may not stop it from progressing because other causes take over in the pathophysiology. Hormone replacement therapy and cardiac remodeling have been explored in relation to age-related disturbances in oxidant/antioxidant homeostasis (Szabó et al., 2021). The effects of supplementing the diet with concentrated red grape juice (RGJ), a polyphenol and vitamin E source, on neutrophil NADPH oxidase activity and other cardiovascular risk factors in hemodialysis patients were studied in clinical trials (Castilla et al., 2008). Both RGJ and vitamin E were able to lower the quantities of oxidized LDL in the plasma and the activity of *ex vivo* neutrophil NADPH oxidase. When taken in tandem, these effects were amplified even further. Because of this action, the consumption of RGJ may have a beneficial effect on cardiovascular health. Results show that natural antioxidants may act as NADPH oxidase inhibitory agents. However, there is a lack of understanding of the underlying mechanisms in this important area. In the next sections, the positive effects of hormones with antioxidative properties and the protective effects of antioxidants against hormone-induced oxidative stress are discussed in greater detail.

4.1 Thyroid hormones

Thyroid hormones regulate oxidative metabolism and play a crucial role in ROS generation, and are linked to the oxidative and antioxidative condition of an organism (Chainy & Sahoo, 2020; Sahoo & Chainy, 2007; Sahoo et al., 2019; Sahoo, Roy, & Chainy, 2008; Venditti & di Meo, 2006). Basal metabolic activity and oxygen metabolism are both considered to be normal when the thyroid condition of the body is considered to be normal. Thyroid hormone production can be high (hyperthyroidism) or low (hypothyroidism) for various pathophysiological reasons, thus affecting the basal metabolic activity and oxygen utilization.

4.1.1 Hypothyroidism

Hypothyroidism is thought to cause a decrease in oxidant generation because of its effect on metabolism and to help to counteract the effects of OS. In contrast, there is evidence connecting hypothyroidism with OS, however, the data demonstrating the relationship between thyroid hormone levels and OS are inconsistent. For instance, there is a lack of consensus among experts on the oxidative status of the plasma of hypothyroid females (Velayeti, Mansourian, Mojerloo, & Marjani, 2016). In the plasma of hypothyroid women, some authors have shown evidence of an increase in OS parameters, whereas other authors have failed to document such an increase (Velayeti et al., 2016). MDA levels are elevated in patients with primary hypothyroidism (Chakrabarti, Ghosh, Banerjee, Mukherjee, & Chowdhury, 2016). However, this stress signal is significantly diminished after L-thyroxine therapy (Chakrabarti et al., 2016). It has been reported that peroxidation of lipids increases with dyslipidemia and is a constant characteristic of hypothyroid individuals (Liberopoulos & Elisaf, 2002). In hypothyroidism, elevated TSH levels result in increased fat mobilization and elevated levels of free fatty acids (Sankha, Kumar, Madhuri, & Kumar, 2021). Reduced levels of thyroid hormone lead to increased levels of polyunsaturated fatty acids, which are more susceptible to oxidative damage (Hoch, 1988).

The hypo-metabolic state of hypothyroidism changes the oxidant production and the testicular antioxidant defense mechanism. Testicular AODS during development and maturation affected by persistent and transitory hypothyroidism was studied in rats (Chainy & Sahoo, 2020; Sahoo et al., 2019; Sahoo & Roy, 2012; Sahoo, Roy, Bhanja, & Chainy, 2008; Sahoo, Roy, & Chainy, 2006). The levels of hydrogen peroxide and protein carbonyl contents remain higher in the crude homogenate (Choudhury, Chainy, & Mishro, 2003) of testis from hypothyroid rats, while OS markers such as MDA levels decrease in hypothyroid rat testes (Mogulkoc, Baltaci, Aydin, Oztekin, & Sivrikaya, 2005). In addition, the testis contains higher levels of mitochondrial LPx and protein carbonylation during prolonged hypothyroidism (Choudhury et al., 2003). In hypothyroid rat testis, the level of oxidative damage indicated by an increase in mitochondrial membrane protein carbonylation was also noted (Chattopadhyay, Sahoo, Roy, Samanta, & Chainy, 2010). Significantly elevated protein carbonylation in hypothyroid immature rat testis is also indicative of the prevalence of OS in hypothyroidism (Sahoo & Roy, 2012). El-Kashlan, Nooh, Hassan, and Rizk (2015) have also observed that hypothyroidism causes an increase in testicular MDA and nitric oxide (NO) (El-Kashlan et al., 2015).

In another study, Zamoner et al. (2008) observed that the oxygen consumption of testis in newborn rats with hypothyroidism was considerably lower than in rats with euthyroidism, although there was no difference in lipid peroxide levels (Zamoner et al., 2008). Even while ROS levels do not rise, their findings suggest that in a hypothyroid state, decreased antioxidant activity could contribute to OS (Zamoner et al., 2008). According to the research conducted by Chattopadhyay, Choudhury, Roy, Chainy, and Samanta (2010) and Chattopadhyay, Sahoo, et al. (2010), the elevated pro-oxidant level and decreased antioxidant capability of hypothyroid mitochondria made them sensitive to oxidative damage (Chattopadhyay, Choudhury, et al., 2010). In the membrane fraction, the degree of damage was more visible. This was demonstrated by a greater degree of oxidative damage to membrane lipids and proteins. While membrane proteins were more sensitive to carbonylation, matrix fraction exhibited thiol residue damage. Transient hypothyroidism reduced mitochondrial LPx, whereas persistent hypothyroidism increased mitochondrial LPx and protein carbonylation in the testis (Sahoo, Roy, Bhanja, & Chainy, 2008). Furthermore, it has been demonstrated that the LPx levels in the germ cells of transiently hypothyroid rats are higher (Sahoo et al., 2006). Hypothyroidism and OS are linked, most likely due to the inability of the endogenous antioxidant system to protect cells from the accumulation of free radicals and consequent oxidative damage (Kochman, Jakubczyk, Bargiel, & Janda-Milczarek, 2021; Mancini et al., 2016).

In rats with hypothyroidism, testicular reduced glutathione (GSH) levels are lower than in control animals (Choudhury et al., 2003; El-Kashlan et al., 2015; Mogulkoc, Baltaci, Aydin, et al., 2005; Sahoo, Roy, Bhanja, & Chainy, 2008), and oxidized glutathione (GSSG) level remains elevated due to which a decrease in the ratio of reduced glutathione to the oxidized state of glutathione (GSH/GSSG) and an increased in oxidative stress index (OSI), the ratio between GSSG and total GSH ($OSI = 100\times (2 \times GSSG)/\text{total } GSH$) (Chainy & Sahoo, 2020) was marked indicating disruption of redox state and induction of OS (Choudhury et al., 2003; Sahoo & Roy, 2012). Furthermore, germ cells of rats with transient hypothyroidism contain less GSH (Sahoo et al., 2006). A lesser amount of glutathione and ascorbate was found in the mitochondrial matrix of hypothyroid rat testis, as well (Chattopadhyay, Sahoo, et al., 2010). Hypothyroidism impairs antioxidant defenses because glutathione and ascorbate levels drop and this disrupts the balance of antioxidants in the mitochondria, leading to testicular dysfunction (Chattopadhyay, Sahoo, et al., 2010).

Hypothyroidism regulates testicular enzymatic and non-enzymatic defenses in a manner that is dependent not only on the length of time that the animals are treated with hypothyroidism but also on their age (Chainy & Sahoo, 2020; Choudhury et al., 2003; El-Kashlan et al., 2015; Sahoo & Chainy, 2007; Sahoo & Roy, 2012; Sahoo, Roy, Bhanja, & Chainy, 2008). In hypothyroid rats, testicular SOD, GR, GPx, CAT, and GST levels were reduced significantly (Choudhury et al., 2003; El-Kashlan et al., 2015; Sahoo, Roy, Bhanja, & Chainy, 2008). Further research reveals that the CAT and SOD activity in the germ cells of transiently hypothyroid rats are significantly reduced (Sahoo et al., 2006; Sahoo, Roy, Bhanja, & Chainy, 2008). In testicular mitochondria, which do not contain CAT, GPx is the primary enzyme responsible for removing H_2O_2 from the cell. A reduction in GPx activity impacts the metabolic pathway necessary for testosterone production, which requires protection against peroxidation (Chandra et al., 2000). The fact is further supported by the finding that hypothyroid rats had reduced testosterone levels in their serum (Sahoo & Roy, 2012; Sahoo, Roy, Bhanja, & Chainy, 2008). This compromised testicular antioxidant status contributes to poor growth and development by affecting spermatogenesis and steroidogenesis in rats before puberty. This is indicated by reduced germ cell number due to increased apoptosis (Sahoo, 2013), the complete absence of round spermatids, decreased seminiferous tubule diameter and decreased testosterone level. This altered testicular physiology is caused by hypothyroidism, which shows up as lower fertility in maturity, as seen by less total viable germ cells and sperm counts (Sahoo, Roy, Bhanja, & Chainy, 2008; Sahoo et al., 2006). The hormetic activity of the goitrogen 6-n-propyl-2-thiouracil (PTU) was also observed. Mitochondrial GSH: GSSG was elevated after 30 days of PTU administration but decreased after 90 days of PTU treatment, resulting in decreased and increased mitochondrial LPx levels after 30 days and 90 days of PTU treatment, respectively (Sahoo, Roy, Bhanja, & Chainy, 2008). ROS generated by the mitochondrial electron transport system may serve as critical stimulating factors for a cascade of events with advantages ranging from transient cytoprotective effects to increased lifespan (Tapia, 2006). These studies on oxidative stress indicate that in a state of hypothyroidism, mitochondria are susceptible to oxidative damage due to elevated pro-oxidant levels and low antioxidant capacity, which ultimately results in tissue damage and compromised normal physiological circumstances.

4.1.2 Hyperthyroidism

Hyperthyroidism can result from numerous pathophysiological diseases, such as Graves' disease or pituitary or thyroid gland tumors, which cause thyroid cells to secrete more hormones than normal (Gulcelik et al., 2006). The hyperthyroid state is accompanied by an increase in the ratio of prooxidants to antioxidants, which promotes the accumulation of ROS and ultimately results in OS in liver (Chattopadhyay, Sahoo, et al., 2010; Chattopadhyay, Sahoo, Subudhi, & Chainy, 2007; Das & Chainy, 2001; Subudhi & Chainy, 2010; Venditti & di Meo, 2006), testis (Chainy & Sahoo, 2020; Sahoo et al., 2019, 2005; Sahoo, Roy, & Chainy, 2008), brain (Das & Chainy, 2004), Kidney (Jena & Chainy, 2011; Jena, Nityananda Chainy, & Dandapat, 2012; Sahoo & Chainy, 2007) and heart (Mishra et al., 2019). This large body of research strongly proved that hyperthyroidism is associated with an increase in oxygen metabolism as well as the formation of free radicals, which in turn led to an increase in OS (lipid peroxidation) in numerous tissues of rats proportional to its severity (Chainy & Sahoo, 2020; Mogulkoc, Baltaci, Aydin, et al., 2005; Sahoo & Chainy, 2007; Venditti & di Meo, 2006).

In a previous investigation, hormetic cellular effects of thyroid hormone were identified (Sahoo et al., 2005). After 1 and 3 days of T_3 (3,5,3′-tri-iodo-l-thyronine) administration, the level of mitochondrial LPx in the testes of rats decreased, whereas it increased after 5 days of treatment. Additionally, it is significant in the context of mitohormesis (Tapia, 2006). The hormetic cellular effects of thyroid hormone were also shown by the observation that the ratio of glutathione (GSH) to glutathione (GSSG) was high in the postmitochondrial fraction of rat testes after 1 day of T_3 treatment but low after 3 days of T_3 administration (Sahoo et al., 2007).

4.1.3 Role of antioxidants in different thyroid states

Several antioxidants were evaluated for their positive effects on hypo- and hyperthyroid-state-induced testicular OS and male fertility in general (Chainy & Sahoo, 2020; Dobrzyńska, Baumgartner, & Anderson, 2004; El-Kashlan et al., 2015; Mogulkoc, Baltaci, Oztekin, Aydin, & Tuncer, 2005; Ourique et al., 2013; Sahoo, Roy, & Chainy, 2008). In addition, a study reveals that supplementation with antioxidants and the replacement of thyroid hormone may benefit treating hypothyroidism (Gerenova & Gadjeva, 2007). Oral supplementation with exogenous antioxidants such as vitamin E and curcumin helps reduce the elevated blood transaminase activity and altered hepatic histoarchitecture in rats with hypothyroidism (Subudhi & Chainy, 2010).

Date palm pollen extract (DPP) co-administration reduced increased levels of MDA and NO and testicular DNA damage and apoptosis in hypothyroid and hyperthyroid rat testis while increasing testicular activity of the antioxidant enzymes CAT, SOD, GPx, and GR in hypothyroid and hyperthyroid rats (El-Kashlan et al., 2015). Because of the high concentration of phenolic and flavonoid components, trace minerals, vitamins (A, C, E), and specific amino acids in DPP extract are thought to have powerful antioxidant properties (El-Kashlan et al., 2015; Hassan, 2011). A decrease in nitric oxide synthase activity has also been linked to the extract (El-Kashlan et al., 2015). Supplementation with vitamin E and curcumin restored the elevated transaminase activity, deformed histoarchitecture, and downregulated antioxidant defense genes observed in hyperthyroid rats (Subudhi & Chainy, 2010). A study by Mishra et al. (2019) showed that antioxidants such as curcumin and vitamin E might reduce OS in the rat heart via modulating Nrf2 and Keap1 function (Mishra et al., 2019) under altered thyroid states. Because of their proven anti-inflammatory, anti-carcinogenic, and antioxidant capabilities, vitamin E and curcumin have been utilized as potential therapeutic agents (Zingg, 2007). Curcumin, the most potent turmeric antioxidant, works by blocking the Fenton reaction (Fig. 1), which results in the oxidation of Fe^{2+} to Fe^{3+} by inhibiting the production of superoxide anion and hydroxyl radicals (Pulla Reddy & Lokesh, 1996). Curcumin treatment of T_4-treated rats raises testicular SOD and CAT levels (Sahoo, Roy, & Chainy, 2008). The increased

GSH concentration in testes of curcumin-fed hyperthyroid rats (Sahoo, Roy, & Chainy, 2008) may result from the stimulated GSH production, which reduces OS. By increasing the transcription of the two Gcl genes, namely, Gclc and Gclm, for glutamate-cysteine ligase, the rate-limiting enzyme in glutathione synthesis, curcumin has been demonstrated to raise glutathione levels in the body (Dickinson et al., 2004). N-acetyl L-cysteine (NAC), a precursor of GSH, has been reported to minimize ROS-mediated DNA damage and increase sperm quality and motility (Lopes, Jurisicova, Sun, & Casper, 1998).

As a potent antioxidant and scavenger of free radicals, vitamin A also helps maintain thyroid hormone homeostasis (Nockels, Ewing, Phetteplace, Ritacco, & Mero, 1984). The link between thyroid hormones and vitamin A metabolism (Nockels et al., 1984) and its insufficiency is well-established. A lack of vitamin A can have various negative effects on the thyroid function of animals (Arthur, Beckett, & Mitchell, 1999). In addition to vitamin A, vitamins E and C are also considered to be among the most potent chain-breaking antioxidants due to their abilities to scavenge all three forms of free radicals, namely, superoxide, H_2O_2, and hydroxyl radicals, and thus prevent lipid peroxidation and raise sperm count (Agarwal, Nallella, Allamaneni, & Said, 2004). Patients suffering from hypothyroidism and hyperthyroidism were shown to have considerably lower levels of vitamins A, E, and C concentration when compared to the control group (Al-Rubae'i & Al-Musawi, 2011). Vitamin E inhibits the oxidation of iodide by quenching H_2O_2, the oxidizing agent needed for thyroid hormone biosynthesis (Zamora, Hidalgo, & Tappel, 1991). It has been observed that vitamin E can reduce the testicular OS caused by T_4, as shown by a decrease in the levels of LPx and PC (59). Vitamin E administration to rats that have been treated with T_4 leads to an increase in the testicular activity of both SOD and CAT. When vitamin E is administered to rats that have been treated with T_4, the GSH to GSSG ratio (GSH:GSSG) increases (Sahoo, Roy, & Chainy, 2008). Because of this, vitamin E can restore normal testicular physiology, which can be observed as an increase in both the total number of sperm and the number of viable sperm in hyperthyroid rats that have been treated with vitamin E (Sahoo, Roy, & Chainy, 2008).

Resveratrol (RSV), a natural polyphenol found in grapes and wine, plays a vital role as an antioxidant by scavenging superoxide, hydroxyl, and metal-induced radicals; hence, RSV protects cells against LPx in cell membranes and DNA damage produced by ROS, which is a result of ROS-induced OS (Ourique et al., 2013). In rats with hyperthyroidism, treatment with RSV results in decreased testicular hydroperoxide and TBARS, higher activity of CAT and GPx, and improved sperm motility (Ourique et al., 2013). However, it has been demonstrated that RSV has hormetic cellular effects. Although CAT expression was shown to be slightly but noticeably elevated at up to 100 μM RSV, with the greatest expression found at 50 μM RSV, it was shown that more than 100 μM RSV was hazardous, demonstrating the hormetic cellular impacts of RSV (Plauth et al., 2016). The ability of melatonin to neutralize free radicals, in addition to its direct antioxidant actions, helps to lower OS (Mogulkoc, Baltaci, Oztekin, et al., 2005). Pinealectomy reduces the amount of endogenous melatonin, leading to increased oxidative damage in the testis (Mogulkoc, Baltaci, Oztekin, et al., 2005).

Moreover, the addition of CAT and flavonoids such as quercetin (Q) and kaempferol (K), which have antioxidant characteristics, can protect human sperm from ROS-induced DNA damage, whereas excess T_3 and T_4 cause DNA damage in human sperm (Dobrzyńska et al., 2004). Carotenoids are naturally occurring pigments that play an important part in protecting cells by effectively scavenging free radicals (Dobrzyńska et al., 2004). Beta-carotene not only acts as an essential antioxidant that helps maintain the health of cells, but it also acts as a pool that can be transformed into vitamin A when required (Russell, 2004). Hyperthyroidism is linked to lower levels of beta-carotene when compared to the normal group (Chainy & Sahoo, 2020). MDA levels in hypothyroid rat testes are reduced by folic acid, which also functions as an antioxidant (Ibrahim, Tousson, Ali, & Mansour, 2011).

4.2 Mineralocorticoid (aldosterone)

Increased levels of aldosterone have a variety of harmful effects on the body, ranging from cardiovascular disease and metabolic disorders to high blood pressure, alteration in redox status, increased oxidative stress, inflammation, and fibrosis (Briet & Schiffrin, 2013; Rodrigues et al., 2021). Multiple clinical and animal research has conclusively proven that aldosterone is responsible for OS in subjects with chronic and stable heart failure (Kotlyar et al., 2006). According to the growing body of research, the nongenomic cardiovascular activities of aldosterone are thought to be produced by a wide variety of cellular pathways and mediated by various messenger systems, including reactive oxygen and nitrogen species. In rats, aldosterone/salt-induced hypertension results in renal damage and increased cortical ROS production (Nishiyama et al., 2004). Apoptosis in renal tubules is caused by an increase in renal OS (Patni et al., 2007). Aldosterone has been demonstrated to stimulate superoxide radical production in endothelial cells via activating Rec1 (Iwashima et al., 2008).

Moreover, adverse effects of aldosterone on kidney tubule cells include OS, oxidative DNA damage, and activation of the nuclear factor NF-κB (Queisser, Oteiza, Stopper, Oli, & Schupp, 2011). Aldosterone induces a rise in intracellular oxidant levels, particularly of superoxide anions. As a result, aldosterone-treated kidney cells had higher levels of the oxidized DNA alteration, 7,8-dihydro-8-oxo-guanine. Activation of NF-κB and DNA damage in response to aldosterone depends on the mineralocorticoid receptor. Because a redox-sensitive fluorogenic probe known as 2′,7′-dichlorodihydrofluorescein diacetate (DCFH2) interacts with numerous oxidants at varying reaction constants, it is an excellent indicator of redox alterations and general oxidative stress. An increase in OS was strongly induced by aldosterone, as evidenced by DCF fluorescence in LLC-PK1 and MDCK cells (Queisser et al., 2011).

4.2.1 Role of antioxidant treatment on aldosterone-induced changes in redox status

Several antioxidants, including tempol, N-acetyl cysteine (NAC), and α-lipoic acid (LA), have been shown to reduce aldosterone-induced oxidant generation, DNA damage, and NF-κB activation (Fig. 2). NAC provides the cysteine necessary for glutathione formation, and in addition to increasing other thiol pools, NAC can directly eliminate ROS and RNS (Zafarullah, Li, Sylvester, & Ahmad, 2003). LA functions as an antioxidant via many mechanisms, including glutathione reduction, free radical scavenging, and metal chelation (Han et al., 1997), whereas tempol primarily scavenges superoxide anions and, to a lesser extent, other radicals (Wilcox & Pearlman, 2008). Tempol is a redox-cycling (catalytic), metal-independent, and membrane-permeable antioxidant (Wilcox & Pearlman, 2008) and it is a SOD mimic because it stimulates $O_2^{\bullet-}$ metabolism at rates similar to SOD. In addition, tempol enhances the metabolism of a broad array of ROS and reactive nitrogen species (RNS), including hydroxyl radicals ($^{\bullet}OH$), and possesses CAT activity that limits additional production of $^{\bullet}OH$ and H_2O_2 by Fenton reactions (Queisser et al., 2011). Simultaneously incubating cells with the cell-permeable antioxidants LA,

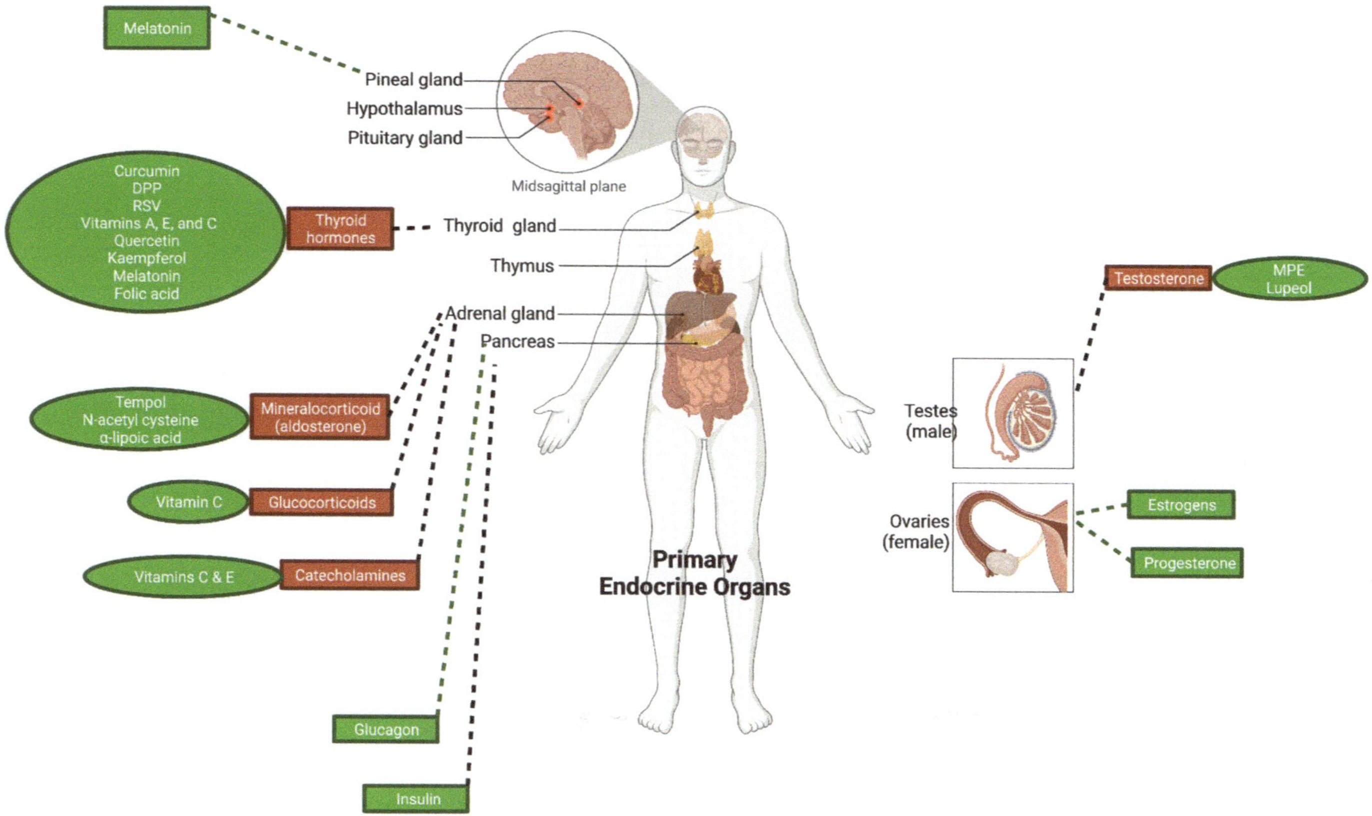

Fig. 2 Summary diagram of hormones causing oxidative stress or demonstrating antioxidant properties. The red shapes in this diagram reflect hormones that induce oxidative stress, whereas the green shapes represent hormones with antioxidant characteristics and antioxidants. The figure was created with BioRender.com.

NAC, and tempol, the rise in cell oxidants (DCF) was largely or completely inhibited. Antioxidants LA, NAC, and tempol reduced the aldosterone-induced increase in nuclear DNA binding by NF-κB. These results indicate that aldosterone promotes DNA damage and activates NF-κB in LLC-PK1 and MDCK cells by increasing the generation of ROS and RNS (Queisser et al., 2011).

4.3 Glucocorticoids

Glucocorticoids increase ROS generation, which may lead to various pathological conditions (Baičc et al., 2007). Glucocorticoid over secretion increases glucose levels, which then undergo glycation, increasing ROS formation and decreasing SOD, CAT, and GPx levels in the hippocampus, resulting in cognitive impairment (Sato, Takahashi, Sumitani, Takatsu, & Urano, 2010). Glucocorticoids have been linked to OS in the brain, lungs, and blood cells. Various inflammatory and autoimmune illnesses are treated with this hormone. As a result of an excess of the hormone, several health problems such as myopathy, osteoporosis, and diabetes were observed in patients. OS underlies each of the aforementioned physiological conditions. There is evidence that glucocorticoids, such as dexamethasone, can generate ROS directly or indirectly through endothelial nitric oxide synthase (eNOS) uncoupling by reducing the amount of tetrahydrofolate available (Simmons, Ungureanu-Longrois, Smith, Smith, & Kelly, 1996). It is possible that long-term corticosteroid use causes mitochondrial damage via OS, as evidenced by an increase in 8-hydroxy-2′-deoxyguanosine (8-OH-dG) from mitochondrial DNA (Mitsui et al., 2002). Excess glucocorticoids encourage the formation of free radicals. When administered to humans, corticosteroid therapy has been shown to raise levels of lipid peroxidation products like 4-hydroxynonenal (4-HNE) and a biomarker for nitrogen-free radical species such as 3-nitrotyrosine (NT) immunoreactivity (Camm et al., 2011). Glucocorticoid receptor expression and signaling have been altered by OS, which is likely one cause underlying corticosteroid insensitivity (Lewis, Ford, Rogers, & Britt, 2021). The expression of the active glucocorticoid receptor isoform GRα mediates the anti-inflammatory effects of corticosteroids and is reduced by OS. Furthermore, OS impairs glucocorticoid receptor DNA-binding activity as it affects the phosphorylation and activation of the glucocorticoid receptor, and it does this by modulating MAP kinases (Lewis et al., 2021).

4.3.1 Role of antioxidant treatment on glucocorticoid-induced redox status

Glucocorticoid-induced vascular dysfunction can be prevented and partially restored with antioxidant therapy. A combination of glucocorticoid and antioxidant therapy was considered safer for the developing brain in premature newborns (Camm et al., 2011). Overproduction of ROS by glucocorticoids (GC) leads to alterations in nitric oxide availability of vascular endothelium, resulting in vascular problems in patients with GC excess. Vitamin C (Fig. 2) delivery improved the lowered responses in forearm blood flow to reactive hyperemia caused by GC therapy, pointing to an overproduction of ROS in patients with too much GC (Iuchi et al., 2003).

4.4 Catecholamines

Catecholamines are the principal neurotransmitters that mediate several central nervous system processes, including motor control, cognition, emotion, memory processing, and endocrine regulation (Perlman & Chalfie, 1977). An increased catecholamine level is linked to an increased generation of ROS (Behonick, Novak, Nealley, & Baskin, 2001). Catechol neurotransmitters (dopamine, norepinephrine, epinephrine) and their precursor (L-DOPA) are highly reactive to ROS. Separation of free cytosolic dopamine protons from their respective hydroxyl groups increases dopamine oxidation to dopamine *o*-quinone. By one-electron oxidation, dopamine is oxidized to form dopamine *o*-semiquinone radical which is then oxidized to dopamine *o*-quinone by reducing two oxygen molecules to form superoxide radicals (Muñoz, Huenchuguala, Paris, & Segura-Aguilar, 2012) (Fig. 3). Dopamine o-semiquinone radicals react with one another to produce dopamine and dopamine o-quinone. Tyrosinase or other oxidative enzymes, including lactoperoxidase, cytochrome P450 forms, prostaglandin H synthase, xanthine oxidase, and dopamine β-monooxygenase, can also catalyze dopamine oxidation (Muñoz et al., 2012; Segura-Aguilar, Metodiewa, & Welch, 1998; Thompson, Capdevila, & Strobel, 2000). In the cytosol, dopamine o-quinone is not stable at physiological pH, and its amino chain cyclizes to produce aminochrome, which is produced when dopamine is oxidized by tyrosinase. Aminochrome is neurotoxic due to its ability to induce OS, endoplasmic reticulum stress, mitochondrial dysfunction, autophagy dysfunction, proteasomal malfunction, and neuroinflammation (Segura-Aguilar, 2019). As a result, under OS, dopamine

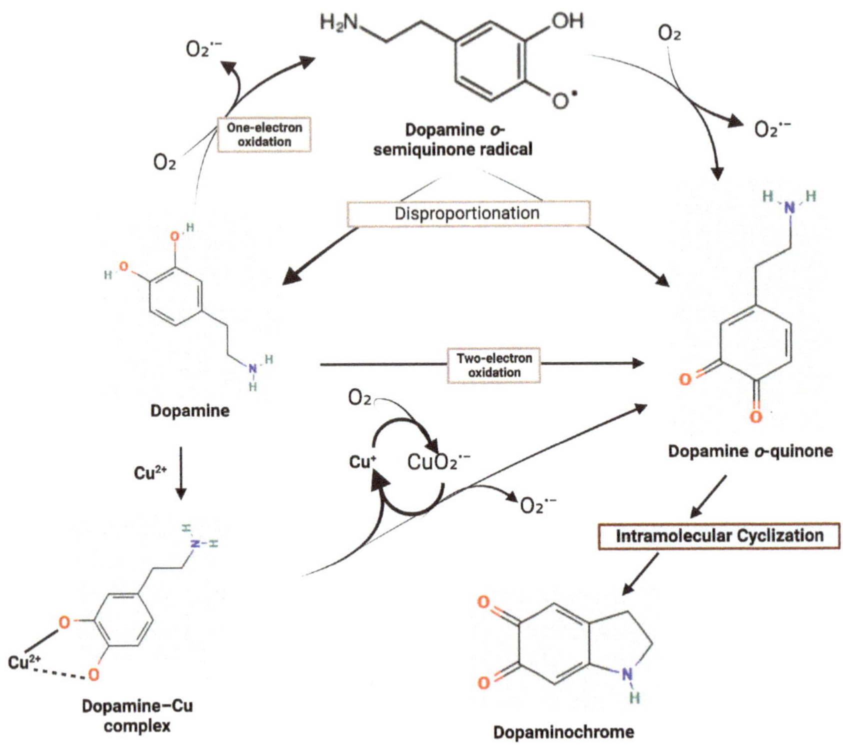

Fig. 3 Oxidation of dopamine. The figure was created with BioRender.com.

is oxidized in the brain to semiquinone, quinone, and dopaminochrome, which have neurotoxic effects. Also, auto-oxidation of dopamine produces superoxide.

At a physiologically attainable molar ratio of glutamate to catecholamines, glutamate inhibits catecholamine oxidation and hydroxyl radicals generated by catecholamine oxidation, quinoprotein, as well as DNA damage. The capability of glutamate to shield catecholamines from oxidation is due to its ability to inhibit copper redox activity by chelating the metal. Elements such as copper, manganese, and iron can also oxidize dopamine (Muñoz et al., 2012). Several enzymes, such as CuZn-superoxide dismutase, cytochrome oxidase, dopamine-β-hydroxylase, lysine oxidase, and ceruloplasmin, require copper as a cofactor in order to function properly. Because of the capacity of copper to cycle through redox states, it is involved in free radical processes as well. Through the Fenton reaction, the reduced form of copper (Cu^{+}) catalyzes the generation of hydroxyl radicals in the presence of hydrogen peroxide (Halliwell & Gutteridge, 2015). Complexes between copper and dopamine have been identified (Paris et al., 2001). By reacting free Cu^{2+} with the Cu^{2+}-catechol complex, copper catalyzes the oxidation of catechol to o-quinone (Fig. 3) by promoting the transfer of electrons from catechol to reduce Cu^{2+} to Cu^{+} and the production of $Cu(I)O_2^{+}$, which efficiently oxidizes catechol alone or when complexed with copper (Paris et al., 2001). Catecholamines in high dosages induce cardiac dysfunction, which is mediated by the generation of highly hazardous oxidation products, including free radicals and aminochromes (Dhalla, Adameova, & Kaur, 2010; Tappia et al., 2001).

4.4.1 Role of antioxidants on catecholamine-induced changes in redox status

Catecholamine oxidation products have also been linked to catecholamine-induced cardiotoxicity because vitamin E, an antioxidant, protects against the toxic effects of catecholamines (Singal, Khaper, Palace, & Kumar, 1998; Tappia et al., 2001). Catecholamine oxidation is also slowed down by ascorbate (Dillon, Root-Bernstein, & Lieder, 2004). The administration of a high dose of the synthetic catecholamine, isoproterenol, to rats caused an increase in malondialdehyde and the formation of conjugated dienes alongside a decrease in the glutathione redox ratio (GSH/GSSG ratio) in the heart. All of these catecholamine-induced alterations in the heart were mitigated by administering vitamin E (Tappia et al., 2001).

Cardiovascular complications have been linked to the catecholamine excess found in tumors derived from chromaffin cells, pheochromocytoma, and functional paraganglioma (PHEO) (Turková, Petrák, Škrha, Widimský, & Zelinka, 2013; Zelinka et al., 2012). Increased catecholamine synthesis in PHEO is associated with lower vitamin C levels and higher MDA levels, suggesting that PHEO is encountering an increase in OS. When PHEO is removed, vitamin C levels increase, and MDA levels drop significantly, but SOD activity remains unchanged (Turková et al., 2013).

4.5 Testosterone

The primary androgen in circulation, testosterone, is anabolic and known to raise metabolic rate; thus, OS is likely to increase due to higher testosterone levels. There is some debate about the effect of testosterone in promoting OS, though some research suggests that testosterone induces OS in the testes, muscles, and the human placenta (Aydilek, Aksakal, & Karakilçik, 2004; Chainy, Samantaray, & Samanta, 1997; Zhu, Bonet, & Knopp, 1997), while other research suggests that testosterone has antioxidant capabilities in prostate and nervous tissue (Ahlbom, Prins, & Ceccatelli, 2001; Guzmán et al., 2005; Tam, Gao, Leung, & Ho, 2003). However, more research has demonstrated that the connection between testosterone and OS may be more nuanced than was previously believed. This is because the role of testosterone-inducing OS or acting as an antioxidant agent is tissue, species, gender, and physiological condition specific. For example, it has been demonstrated that testosterone decreases the activity of three major AOEs such as CAT, SOD, and GPx in testes (Chainy et al., 1997) and increases the activity of CAT in the brain (Ahlbom et al., 2001), and decreases the activity of pro-oxidant enzymes, NADPH-oxidase in prostate (Tam et al., 2003), and reduces lipid peroxidation in rat brain (Guzmán et al., 2005) tissues.

By raising ROS anabolism and decreasing antioxidant detoxification, androgen deprivation causes OS in the rat ventral prostate. Castration induces 8-OH-dG and 4-HNE protein adducts in the regressing epithelium, indicating oxidative damage. NAD(P)H oxidase expression was dramatically raised, while the expression of important AOEs such as SOD2, GPx1, thioredoxin, and peroxiredoxin 5 was significantly reduced after castration (Tam et al., 2003). Ventral prostate epithelia of castrates had oxidative damage that was partially repaired by testosterone replacement therapy. In castrated rats, testosterone replacement induced a compensatory increase in the expression of CAT, GR, γ-glutamyl transpeptidase, and glutathione synthetase in the regenerating ventral prostate tissue. Also, testosterone replacement led to a partial reduction in the expression of NAD(P)H oxidases, but it led to a restoration of the expression of SOD2, GPx1, thioredoxin, and peroxiredoxin 5 in castrated rats (Tam et al., 2003). When androgens were present at physiological levels in LNCaP human carcinoma cells, they induced OS through increased mitochondrial activity, changed intracellular GSH levels, and increased γ-glutamyl transpeptidase activity (Ripple, Henry, Rago, & Wilding, 1997).

Testosterone administration also reduces SOD, CAT, GR, and GST levels in prostate glands in mice (Prasad, Kalra, Singh, & Shukla, 2008). However, androgens reduced the activity of cardiac AOEs in rats but increased the expression of Cu/ZnSOD in the aorta following orchidectomy (Chaves et al., 2006). Whereas orchidectomy lowered rat left ventricular AOE activities, testosterone replacement corrected the deficiency in tissue antioxidant capacity and significantly improved cardiac function recovery after ischemia/reperfusion *in vitro* (Borst, Quindry, Yarrow, Conover, & Powers, 2010). In contrast, castration of male rats resulted in a decrease in SOD, CAT, and GPx in the left ventricle, as well as reduced GSH and protein-thiol groups and an increase in lipid peroxidation, and greater NT concentrations in the left ventricle of the heart, according to Kłapcińska et al. (Kłapcińska et al., 2008). The left ventricular tissue concentrations of α- and γ-tocopherols showed a compensating response to increased OS produced by gonadectomy. After androgen replacement, serum testosterone levels returned to their normal range; however, the antioxidant status of left ventricular tissue declined. While a high dose of testosterone treatment alone diminished SOD, CAT, GPx, and GR activities in the left cardiac ventricle, it reversed the positive effect of endurance training on myocardial SOD and CAT activities and increased myocardial lipid peroxidation in adolescent male Wistar rats (Sadowska-Krępa et al., 2011). Based on these findings, Sadowska-Krępa et al. (2011) noted that adolescent anabolic androgen usage, either alone or in combination with endurance training, may increase the OS-related threat to cardiac health (Sadowska-Krępa et al., 2011). Moreover, it has been shown that testosterone increases ROS generation and OS in cardiomyocytes by signaling through the androgen receptor (AR) via genomic and nongenomic pathways.

OS has also been linked to testosterone through several indirect observations reported in human plasma (Chainy & Sahoo, 2020; Darbandi et al., 2018). Compared to non-alcoholic individuals, chronic alcoholics had lower testosterone, LH, FSH, AOEs, glutathione, and higher protein carbonyl and TBARS levels (Chainy & Sahoo, 2020). The same outcomes were found when testosterone was administered to adult rats, and the main AOEs (SOD, CAT, and GPx) were reduced (Chainy et al., 1997). However, after castration and testosterone supplementation in castrated rats, Azevedo, Lacava, Miyasaka, Chaves, and Curi (2001) found no difference in macrophage levels of SOD, CAT, or GPx (Azevedo et al., 2001). Furthermore, Adler et al. (2012) found that testosterone administration of human neutrophil granulocytes *in vitro* greatly increased MnSOD expression, demonstrating transcriptional control of AOEs by testosterone (Adler et al., 2012). Therefore, it would appear that the pro-oxidant effects of testosterone are tissue-specific. Additionally, the findings suggest that the role of testosterone as either an agent to induce OS or as an antioxidant agent is dependent not only on the type of tissue but also on the species, and the physiological conditions. Age-related non-communicable chronic diseases like obesity, metabolic syndrome, type 2 diabetes, and malignancy are all linked to male hypogonadism, defined by low testosterone levels in older men. Hormone replacement treatment, dietary and lifestyle changes, adhering to an optimal BMI, and dietary antioxidant supplements and/or phytomedicines may be the best options for treating these patients (Leisegang et al., 2021).

4.5.1 Role of antioxidants on testosterone-induced changes in redox status

Mango pulp extract (MPE) and lupeol have been examined for their antioxidant properties in prostate of male Swiss albino mice exposed to testosterone-induced OS (Fig. 2) (Prasad et al., 2008). In rats, oral administration of lupeol improved the antioxidant status of liver and reversed the tissue redox system changes caused by cadmium exposure (Sunitha, Nagaraj, & Varalakshmi, 2001). Dimethylolbutanoic acid-induced OS in the liver is efficiently reduced by supplementation with lupeol/MPE, which has been proven to restore antioxidant enzyme activity and reduce lipid peroxidation (Prasad, Kalra, & Shukla, 2007). The antilithiatic and free radical scavenging characteristics of mango pulp contribute to its positive effects on human health. These properties help lower lipid peroxidation and improve the activity of antioxidant enzymes (SOD and CAT) against isoproterenol, as reported in kidney and heart of rats (Bafna & Balaraman, 2005). Because of the OS that testosterone causes, there was a significant decrease in the levels of antioxidant enzymes and an increase in the rate of lipid peroxidation after it was administered. However, pretreatment with both MPE and lupeol reduced the frequency of testosterone-induced lipid peroxidation and improved the level of antioxidant enzymes SOD, CAT, GR, and GST in prostate tissue (Prasad et al., 2008).

5. Hormones exhibiting antioxidant properties

5.1 Estrogens

According to extensive research, estradiol has been proven to have a greater impact on the oxidant-antioxidant balance in numerous tissues. Due to their lipophilicity, estrogens concentrate in lipid-rich sections of cells, such as cellular membranes, and influence membrane fluidity as a result (Liang et al., 2001) and thus, serving as highly localized antioxidants *in vivo* (Prokai, Prokai-Tatrai, Perjési, & Simpkins, 2005).

A synthetic derivative of 17β-estradiol presented the mechanistic aspects of $^{\bullet}OH$ scavenging with increased potency to inhibit lipid peroxidation (Prokai-Tatrai, Perjesi, Rivera-Portalatin, Simpkins, & Prokai, 2008). In this regard, the molecular structure of estrogens provides certain advantages. The free phenolic hydroxyl group on the A-ring of all estrogens confers antioxidant properties and this is the only structural factor that influences their ability to scavenge free radicals (Prokai-Tatrai et al., 2008). It most likely entails fast transferring the hydrogen atom of the phenolic OH to radicals to stop free radical chain processes such as lipid peroxidation. The phenolic antioxidant can be recycled by reducing phenoxyl radicals by intracellular reductants (such as ascorbate and thiols) and enzymes (Kagan & Tyurina, 1998).

In ovariectomized mice, the liver and brain mitochondria produce more H_2O_2 (Viña et al., 2005). Many AOEs, such as GPx, CAT, and SOD, were significantly reduced in the hearts of ovariectomized mice (Muñoz-Castañeda et al., 2005). It has been demonstrated that E2 upregulates AOEs (Bellanti et al., 2013). When estrogen was replaced, these declines were reversed. Supplementation with E2 reversibly decreased the life span of ovariectomized mice (Bokov, Ko, & Richardson, 2009). A direct correlation between E2 levels and plasma antioxidant capacity during the menstrual cycle indicates that E2 is an antioxidant (Michos, Kiortsis, Evangelou, & Karkabounas, 2006). After menopause, lipid hydroperoxides increased, and GPx reduced in plasma in women; however, treatment with E2 reversed these effects. A strong correlation between erythrocyte GPx activity and variations in estradiol levels was reported (Massafra et al., 2000). Noticeably, the antioxidant effect of estrogen has been regarded as the primary mechanism for this hormone to protect skeletal and cardiac muscles (Persky et al., 2000), uterus (Diaz-Flores, Baiza-Gutman, Pedrón, & Hicks, 1999), and liver (Huh, Shin, Choi, & Lee, 1994) from damage.

Furthermore, when ovariectomized apoE-deficient mice were treated with E2, atherosclerosis severity was significantly decreased (Wing et al., 2009). Importantly, after E2 treatment, $O_2^{\bullet-}$ and expression of NADPH oxidase (Nox) were reduced, whereas Cu/ZnSOD and MnSOD were increased. This demonstrates that continuous oral estrogen treatment positively affects OS and atherosclerosis in apoE-deficient animals. Estradiol treatment increased the amounts of AOEs in liver of aging rats (Hamden, Carreau, Ellouz, Masmoudi, & el Feki, 2007). Estradiol treatment *in vitro* increased MnSOD gene expression in human neutrophil granulocytes, as Adler et al. demonstrated (Adler et al., 2012). In cows, also during the mid-luteal phase, when follicular E2 synthesis is at its highest, the expression of mRNA encoding SOD1 and SOD2 in the granulosa cells of dominant follicles is elevated (Kaïs, Al-Gubory, Bolifraud, & Garrel, 2008). 17β-estradiol was shown to increase MnSOD and EC-SOD expression in cultured rat vascular smooth muscle cells and human circulating monocytes, but it did not affect Cu/ZnSOD, GPx, or CAT (Strehlow et al., 2003). Therefore, estrogen may have a role in vasoprotection due to its anti-oxidative properties and promoting the expression and activity of MnSOD and EC-SOD (Strehlow et al., 2003). Estrogens have been shown to influence the expression of the AOE gene in skeletal muscle likewise. Baltgalvis, Greising, Warren, and Lowe (2010) evaluated 84 antioxidant genes in the skeletal muscles of ovariectomized mice 48h following E2 or placebo treatment. In ovariectomized mice administered with E2, the levels of GPx3, GPx2, GPx1, Nox4, and Txnip elevated in the soleus and EDL muscles (Baltgalvis et al., 2010). The antioxidative properties of estrogens may play a role in their ability to protect the cardiovascular system (Mendelsohn & Karas, 1999). Estrogen regulates target gene transcription via estrogen receptors (ERs), specifically ERα and ERβ. AOE and UCP expression was affected by E2 in a way that depended on the ERα/ERβ ratio (Miró et al., 2011). The antioxidant activity is crucial to the neuroprotective, vasoprotective, and cardioprotective effects of estrogens.

Hydroxysteroid dehydrogenases (HSDs) dominate peripheral steroid metabolism and their reactions are crucial for assessing steroid bioactivity. Glucocorticoids and mineralocorticoids metabolize mostly at C-11, progestins at C-20, androgens and estrogens at C-17, and 5-reduced steroids at C-3. Functionally, the HSDs can be categorized as true dehydrogenases, which convert hydroxysteroids to ketosteroids, and reductases, which convert ketosteroids to hydroxysteroids. Reductive 17β-HSD1 almost entirely converts estrone to estradiol (Luu-The, Zhang, Poirier, & Labrie, 1995), whereas oxidative 17β-HSD2 converts over 95% of estradiol to estrone (Wu et al., 1993). By employing nicotinamide cofactors NADPH/$NADP^+$ and NADH/NAD^+, HSD reactions can switch between converting active hydroxysteroids and inactive ketosteroids, thus changing the redox state (Jin & Lin, 1999; Talalay, Hurlock, & Williams-Ashman, 1958) as follows:

$$\text{Estrone} + \text{NADPH} = \text{Estradiol} + \text{NADP}^+ \ (\text{reductive Type 1 } 17\beta - \text{HSD})$$
$$\text{Estradiol} + \text{NAD} = \text{Estrone} + \text{NADH} \ (\text{oxidative Type 2/4 } 17\beta - \text{HSD})$$
$$\text{NET}: \text{NADPH} + \text{NAD} = \text{NADH} + \text{NADP}^+ \ (\text{Transhydrogenation})$$

Several endpoints, such as dopamine transporters and release, insulin sensitivity, tumor cell growth *in vitro*, and multiple inflammatory processes, are influenced by estrogen and progestogen hormesis (Strom, Theodorsson, & Theodorsson, 2011). Hormetic effects of female sex hormones on inflammation have been demonstrated. Most studies show that low concentrations of hormones are pro-inflammatory and high concentrations are anti-inflammatory, as seen with the effects of estrogens on the pro-inflammatory cytokine IL-1 (Polan, Kuo, Loukides, & Bottomly, 1990). The effects of equine estrogens on insulin sensitivity in postmenopausal women were studied, and it was demonstrated that a daily dose of 0.625 mg increased insulin sensitivity, whereas a dose of 1.25 mg lowered it. In general, it seems that estrogen improves insulin sensitivity at moderate doses, but this effect may be diminished at greater doses (Lindheim et al., 1993).

5.2 Progesterone

Progesterone, unlike estrogens, does not possess the molecular structure of an antioxidant; yet high levels of the hormone appear to prevent oxidative damage. It was observed that the antioxidant effect of progesterone greatly boosted brain function. The neuroprotective effects of progesterone are supported by several studies that show that the hormone protects against a variety of acute central nervous system injuries, including acute brain trauma, stroke, anoxic brain injury, and spinal cord injury; the hormone appears to exert its neuroprotective effects by protecting or rebuilding the blood–brain barrier; decreasing the development of cerebral edema; downregulating the inflammation cascade; limiting cellular necrosis and apoptosis; and also via an antioxidant effect (Stein, 2008). Progesterone administration raised the activities of GPx and GR, but it did not influence the activities of CAT and GST, as measured by the activities of AOEs in subcellular fractions of whole-brain homogenates collected from ovariectomized female rats (Pajović, Saičić, Spasić, Petrović, & Martinović, 1999). A dose-dependent reduction in lipid peroxidation was seen when progesterone was administered in several *in vitro* free radical-generating systems (Shimamura et al., 1995; Sugino et al., 1996). In a model of cerebral ischemia, progesterone administration decreased lipid peroxidation in the brain, regardless of whether the treatment was acute or long-term (Cai et al., 2008; Stein, 2008). Similarly, treatment with progesterone lowered MDA levels in the striatum and the hippocampus (Ozacmak & Sayan, 2009).

Moreover, He, Yang, Zhai, Shao, and Li (2011) demonstrated that long-term progesterone administration improved learning and memory in ovariectomized rats by increasing SOD activity and decreasing MDA levels in the hippocampus (He et al., 2011). In the brain, a plausible mechanism involves the reduction of progesterone to 5α-dihydro-progesterone, which is then converted to allopregnanolone. Allopregnanolone can restore the intracellular redox state, reducing ROS levels, lipid peroxidation, peroxide-induced cell death, and NF-κB activation (Zampieri et al., 2009). Considering the research conducted by Zampieri et al., 2009, it is possible to presume that allopregnanolone may reduce intracellular concentrations of ROS by increasing CAT activity, thereby restoring the intracellular redox state, as evidenced by reduced LPx and almost complete reversal of peroxide-induced apoptosis in fibroblasts from patients having an autosomal recessive neurodegenerative disorder, Niemann-Pick C disease (NPC) (Zampieri et al., 2009). Furthermore, *in vitro* studies in isolated human neutrophil granulocytes treated with progesterone enhances myeloperoxidase expression (Adler et al., 2012). These findings imply that progesterone lowers lipid peroxidation and OS, most likely via reducing free radical production and boosting endogenous free radical scavenging systems.

In humans, all steroid hormones are metabolized by the Aldo-keto reductases AKR1C1-AKR1C4 and AKR1D1. Steroid nucleus and side-chain reductions at C3, C5, C17, and C20 are catalyzed by these enzymes in a NADPH-dependent method (Rižner & Penning, 2014). AKR1C1 is a 20-ketosteroid reductase that converts progesterone to 20α-hydroxyprogesterone, rendering progesterone inactive. AKR1C1 functions as a 20-ketosteroid reductase to reduce progesterone to 20α-hydroxyprogesterone. AKR1C2 functions as a 3-ketosteroid reductase to reduce 5α-dihydrotestosterone to 3α-Adiol (5α-androstane-$3\alpha,17\beta$-diol), while AKR1C3 as a 17-ketosteroid reductase reduces Δ^4-androstene-3,17-dione to testosterone (Penning, Wangtrakuldee, & Auchus, 2019). Human AKR1C enzymes are pluripotent and utilize LPx byproducts such as 4-hydroxynonenal and 4-oxo-2-nonenal as additional substrates (Burczynski, Sridhar, Palackal, & Penning, 2001). It is not unexpected that, as part of the OS defense system, they are part of the antioxidant response gene battery and are controlled by the Keap1-Nrf2 system via AREs in their gene promoters (Tebay et al., 2015). The AKR1C genes are upregulated by electrophiles, ROS, and the Nrf2 activator R-sulforaphane, suggesting that steroid metabolism may be redox-controlled at the gene expression level and impacted by OS (Penning et al., 2019).

5.3 Glucagon

Glucagon plays a part in activating and regulating antioxidant systems, which are responsible for shielding organisms from the damaging effects of OS. Glucagon increased GSH efflux in cultured rat hepatocytes and perfused rat livers (Lu et al., 1990). This hormonal modulation of GSH efflux was the determining factor in the mobilization of the large hepatic reservoir of GSH for use outside of the liver (for example, by providing precursor cysteine through the hydrolysis of extracellular GSH or possibly through intact GSH uptake). In addition, glucagon has the potential to modulate the expression and activity of hepatic iNOS both *in vivo* and *in vitro* in response to pro-inflammatory stimuli. Glucagon effectively inhibited NO production in isolated rat hepatocytes that had been activated with interleukin-1 *in vitro* (Harbrecht, Wirant, Kim, & Billiar, 1996). *In vivo* treatment of rats with lipopolysaccharide (LPS) generated iNOS mRNA, iNOS protein, and plasma levels of nitrite/nitrate that were all considerably lowered by glucagon (Harbrecht, Perpetua, Fulmer, & Zhang, 2004). Alpha cells from diabetic and non-diabetic mice have high CAT expression levels (Bloch, Shichman, Vorobeychik, Bloch, & Vardi, 2007).

Insect antioxidative mechanisms were activated by mammalian glucagon (Alquicer, Kodrík, Krishnan, Večeřa, & Socha, 2009). It was found that injecting porcine glucagon into adult firebugs (*Pyrrhocoris apterus*) activated an antioxidant response when OS was elicited by paraquat, a diquaternary derivative of 4,4′-bipyridyl (Alquicer et al., 2009). Glucagon induces the antioxidant response by raising GSH, decreasing protein carbonyl levels in the hemolymph, and decreasing protein carbonyl and NT levels in the central nervous system. Glucagon partially decreased OS indicators induced by this redox cycling agent and OS or when co-injected with paraquat. These few pieces of evidence point to a close connection between antioxidant mechanisms and glucagon, which must be investigated further.

5.4 Insulin

In non-insulin-dependent diabetes mellitus (NIDDM), also known as type 2 diabetes, which is characterized by insulin resistance and relative, rather than absolute, insulin deficiency, there have been several reports of reduced plasma/serum total antioxidant status, free radical scavenging activity, ascorbate, and vitamin E levels, along with reduced activities of SOD, CAT, and GPx (Aguirre et al., 1998; Asmat, Abad, & Ismail, 2016; Ceriello et al., 1998; Maxwell et al., 1997; Paolisso et al., 1994). The expression of genes like heme oxygenase (HO) and HSP72, which are associated with providing cellular defense against OS, was significantly decreased in the skeletal muscles of patients with type 2 diabetes. This finding was closely associated with whole-body insulin sensitivity and muscle oxidative capacity (Bruce, Carey, Hawley, & Febbraio, 2003). During pregnancy, there was also a similar connection between GPx activity and markers of insulin resistance (Chen, Scholl, Leskiw, Donaldson, & Stein, 2003). Obesity is a significant contributor to increased OS, one of the primary causes of insulin resistance in men. A decrease in total antioxidant capacity and an increase in MDA were linked to insulin resistance and moderate obesity in males (Jia, Liu, Tian, Wang, & Lu, 2019). Elevated levels of oxidative markers, including TBARS, 8-OHdG, 4-HNE, 8-epi-prostaglandin F2a, hydroperoxides, and oxidized DNA bases, were detected in the serum, plasma, white blood cells, and pancreas biopsies of patients with type 2 diabetes (Bigagli & Lodovici, 2019). In both type 1 and type 2 diabetic patients, circulating and urine levels of 8-epi-prostaglandin F2a, an LPx product, were elevated (Bigagli & Lodovici, 2019; Ito, Sono, & Ito, 2019). Interestingly, exercise and the associated production of ROS are linked to improvements in insulin sensitivity and antioxidant defense through mitohormesis (Ristow et al., 2009; Tapia, 2006). By increasing mitochondrial formation of ROS in skeletal muscle, exercise reduces insulin resistance by increasing the expression of insulin sensitivity inducers such as *peroxisome proliferator-activated receptors-γ* (*PPARγ*) and transcriptional coactivators *PGC1α* and *PGC1β*, along with SOD 1 and 2, and GPx1. By reducing exercise-dependent ROS production with antioxidant supplements, the health benefits of exercise are nullified, and exercise does not improve insulin sensitivity or antioxidant defense (Ristow et al., 2009).

In hyperglycemic situations, a non-enzymatic process produces advanced glycation end products (AGEs), which have detrimental effects by generating ROS, secreting aberrant proteins or growth factors, and secreting inflammatory cytokines (Khan, Ooi, Parvus, Valdez, & Tsin, 2019). During the production of AGEs, including the autoxidation of glucose, Schiff bases, and Amadori adducts, reactive oxygen species (ROS) are formed. The UPS 20S proteasome degrades unstable, oxidized, and short-lived proteins. As the bulky form of AGEs inhibits access to the proteasomal core, proteolytic activity is decreased, which increases protein modification through an increase in oxidized and degraded proteins. In addition to the proteasome, other enzymes of the cellular repair processes are impacted by high amounts of ROS, resulting in a decrease in the antioxidative and repair capacity of the cells (Ott et al., 2014). Under imbalanced OS, an increase in ROS stimulates the oxidation of lipids and glucose, which in turn speeds up the synthesis of AGEs *in vivo* and perpetuates the vicious cycle of AGE/ROS (Chen, Lin, Bu, & Zhang, 2018). The avoidance of AGE production by limiting dietary glucose intake, blocking cellular AGE–RAGE (AGE receptors) interaction and enhancing AGE clearance may be the most effective means of preventing the accumulation of AGEs (Ott et al., 2014).

As a powerful inducer of the antioxidant defense system, insulin functions independently. By blocking the oxidative-stress-induced pro-apoptotic action of Bax, Bcl-2 protein inhibits ROS escape (Orzechowski, 2003) and by influencing the expression of Bcl-2 protein via protein kinase B (PKB), insulin increases the intracellular antioxidant state. AOEs, including alpha-class GSTs, microsomal epoxide hydrolase, and γ-glutamylcysteine ligase catalytic subunit, were upregulated in the presence of insulin in various *in vitro* studies (Kim, Abdelmegeed, & Novak, 2006; Kim, Woodcroft, Kim, & Novak, 2003; Kim, Woodcroft, & Novak, 2003; Langston, Circu, & Aw, 2008). Insulin-mediated AODS upregulation was carried out by PI3K/Akt/p70 S6 kinase signaling, thereby demonstrating that low insulin levels, such as those observed in diabetes, may increase the possibility that xenobiotics and OS will damage hepatocytes. In another study, insulin prevented a decrease in neuronal viability, decreased necrotic and apoptotic cell death, inhibited ascorbate/Fe^{2+}-mediated lipid and protein oxidation, and thus reduced neuronal OS in cultured cortical neurons (Duarte, Santos, Oliveira, & Rego, 2005). A study in cortical neurons demonstrated that insulin prevented the decline in intracellular uric acid and GSH/GSSG levels caused by OS and increased GR and reduced GPx activity. Phosphatidylinositol-3-kinase (PI-3K) or mitogen-induced extracellular kinase (MEK) inhibition abolished the effect of insulin on uric acid and GSH/GSSG, suggesting that insulin-mediated signaling pathways were activated. These findings show that insulin protects neurons through an antioxidant mechanism that modifies the GSH redox cycle. In both *in vitro* and *in vivo* studies, insulin administration can progressively and transiently boost a battery of AOEs by activating a critical transcription factor, nuclear factor erythroid 2 p45-related factor 2 (Nrf2) (Wang et al., 2012). Insulin modulates redox equilibrium and strengthens antioxidant defenses via the ERK-Akt-Nrf2 pathway (Wang et al., 2012). Insulin signaling and the cellular oxidant: antioxidant balance have a close connection, and insulin itself has a role to play as an antioxidant. Insulin facilitated the movement of Nrf2 into the nucleus and activated the expression of subsequent antioxidant proteins (Fig. 1). Through activation of Nrf2 transcription factor, insulin treatment has been shown to sequentially and transiently enhance a battery of antioxidant enzymes *in vitro* and *in vivo*. Such insulin-induced, Nrf2-mediated augmentation of antioxidant defense may need sequential activation of extracellular signal-regulated kinases (ERK) and protein kinase B (Akt) (Wang et al., 2012).

5.5 Melatonin

Melatonin (N-acetyl-5-methoxytryptamine) is an extremely effective direct free radical scavenger serving as a multifunctional antioxidant. Melatonin, in contrast to the vast majority of small-molecule biological antioxidants like ascorbic acid, α-tocopherol, and lipoic acid, does not go through redox cycling and does not promote oxidation. Instead, it removes the free electron from the system while undergoing chemical rearrangement, and as a result, it is sometimes referred to as a suicidal or terminal antioxidant. Each of these rearrangement products is an effective antioxidant in its way. On the other hand, melatonin was found in the central nervous system 10 min after its peripheral delivery, in contrast to vitamins E and C, which are not particularly effective at crossing the blood–brain barrier (Menendez-Pelaez et al., 1993). Due to its high lipophilicity and hydrophilicity, melatonin readily permeates cell membranes (Amini et al., 2021). According to nearly all studies that compared the antioxidant properties of melatonin and other well-known antioxidants, the antioxidant properties of melatonin were shown to be equal to, or even more effective than, those of other antioxidants (Amini et al., 2021; Hacışevki & Baba, 2018; Hardeland & Pandi-Perumal, 2005; Korkmaz et al., 2009; Reiter et al., 2016, 2009; Tarocco et al., 2019).

As a direct free radical scavenger, melatonin is highly effective because of its unique capacity to perform a cascade of scavenging actions (Allegra et al., 2003). Melatonin is an electron-rich molecule that can interact with free radicals through various processes, resulting in various stable products that can be eliminated through urine (Ma, Idle, Krausz, & Gonzalez, 2005). Since secondary and tertiary metabolites, such as cyclic 3-hydroxymelatonin (c3OHM), N^1-acetyl-N^2-formyl-5-methoxykynuramine (AFMK), and N^1-acetyl-5-methoxykynuramine (AMK), are also effective free radical scavengers, the melatonin antioxidant mechanism involves a cascade of free radical scavenger (Galano, Medina, Tan, & Reiter, 2015). While c3OHM is

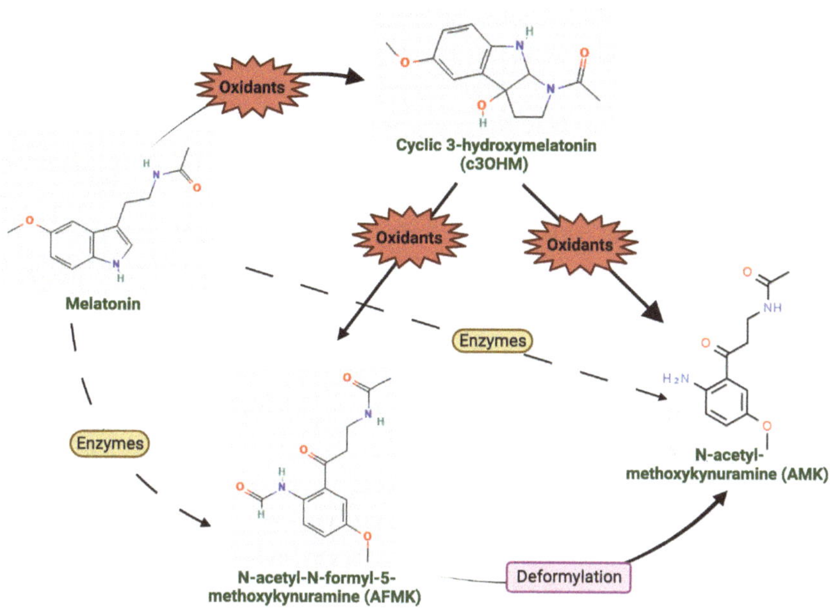

Fig. 4 Melatonin and its secondary and tertiary metabolites as effective free radical scavengers. The figure was created with BioRender.com.

the major byproduct of melatonin interactions with oxidants, AFMK and AMK are produced from c3OHM through further oxidation (Fig. 4). It has been demonstrated that, at least *in vitro*, AFMK can be deformylated to AMK by CAT (Galano & Reiter, 2018; Semak, Korik, Antonova, Wortsman, & Slominski, 2008). Melatonin can be oxidized to AFMK enzymatically or through chemical pathways by myeloperoxidase, cytochrome c (Cyt c), hemoglobin, ROS, and UV radiation (Galano & Reiter, 2018; Semak et al., 2008).

Melatonin has been demonstrated to promote the repair of oxidized DNA. This is likely attributed to the ability of melatonin to convert guanosine radical to guanosine via electron transfer (Galano & Reiter, 2018). Melatonin inhibited the synthesis of 8-hydroxy-2′-deoxyguanosine, a damaged DNA product, 60–70 times more effective than traditional antioxidants like ascorbate and α-tocopherol (Johns & Platts, 2014). Since melatonin safeguards lipid and protein molecules and DNA from oxidative damage, it must be widely distributed within cells (Hacışevki & Baba, 2018). Because of its inhibitory effects on lipid peroxidation, melatonin is well-known as a cell and organelle membrane stabilizer or protector. Free radical scavengers such as melatonin and its metabolites stop the initiation and spread of lipid peroxidation (Zhang & Zhang, 2014). In addition to scavenging peroxyl radicals, melatonin and its metabolites, AFMK and AMK, also directly scavenge the alkoxyl radical, a consequence of the transition metal-catalyzed breakdown of lipid peroxides (Drăgoi & Nicolae, 2018; Zhang & Zhang, 2014).

As an indirect antioxidant, melatonin can promote AOEs under baselines and experimentally induced settings. It has been reported to boost the activities of AOEs such as SOD, CAT, GPx (Fischer, Kleszczyński, Hardkop, Kruse, & Zillikens, 2013), GR (Fischer et al., 2013; Rodriguez et al., 2004), and G6PD (Hardeland, 2005). Melatonin also stimulates γ-glutamyl cysteine synthetase (the rate-limiting enzyme in GSH production) and GR (the enzyme converting GSSG to GSH), raising intracellular antioxidant GSH levels (Urata et al., 1999; Winiarska, Fraczyk, Malinska, Drozak, & Bryla, 2006). In addition, melatonin improves the efficiency of the mitochondrial electron transport chain, reducing electron leakage and free radical production (Reiter et al., 2018). Intriguingly, melatonin is a negative modulator of various pro-oxidant enzymes, including 5-lipoxygenase and 12-lipoxygenases as well as NO synthase (Hardeland & Pandi-Perumal, 2005), hence reducing free radical formation. Membrane or nuclear receptors are likely responsible for mediating the indirect antioxidant activities of melatonin, even though direct free radical scavenging characteristics of melatonin are not dependent on any receptor (Reiter, Tan, Terron, Flores, & Czarnocki, 2007). Through its interaction with calmodulin, melatonin is able to control the activities of antioxidant enzymes. This, in turn, suppresses the downstream processes that are responsible for the inactivation of the nuclear RORα melatonin receptor. This nuclear transcription factor eventually downregulates the expression of antioxidant enzymes activated by NF-κB. Consequently, the increase in antioxidant enzyme activity generated by melatonin is mediated by suppressing the RORα pathway (Tomás-Zapico & Coto-Montes, 2005).

Moreover, melatonin has been shown to prevent OS/nitrosative stress-induced mitochondrial dysfunction and inhibit the lethal effects of septic shock or ischemia/reperfusion injury in experimental models of neurodegenerative diseases such as Alzheimer's disease, Huntington's disease, and Parkinson's disease by maintaining complex respiratory activities, electron transport chain, and ATP production in mitochondria (Srinivasan, Spence, Pandi-Perumal, Brown, & Cardinali, 2011). When the blood supply to an organ is temporarily interrupted, melatonin can protect cells against cellular damage, not just in the brain but in other tissues, like the heart (Reiter & Tan, 2003). Due to its capacity to quickly cross the placenta, melatonin protects the fetus, maternal tissues, and placenta from oxidative damage during the reproductive process, demonstrating its high flexibility (Reiter et al., 2016, 2009). In conclusion, melatonin reduces OS and increases antioxidant capacity through several mechanisms.

6. Conclusion

Assessing the hormonal effects and function of antioxidants could lead to the discovery of therapeutic targets for illnesses associated with hormonal imbalance. Melatonin, glucagon, insulin, estrogens, and progesterone are hormones with antioxidant properties, whereas aldosterone, glucocorticoids, thyroid hormones, and catecholamines cause OS and disruption of redox status. Inducing OS or serving as an antioxidant is a tissue-, species-, and physiological-specific testosterone function. Various pathophysiological disorders are triggered by a decreased antioxidant status in tissues caused by hormonal imbalance. While antioxidants such as tempol, N-acetyl cysteine, and α-lipoic acid decrease OS induced by aldosterone, Vitamin C decreases OS caused by glucocorticoids, catecholamines, and thyroid hormones. Vitamin E mitigated the OS induced by thyroid hormone imbalance and catecholamines. Several antioxidants, such as vitamin A, folic acid, curcumin, date palm pollen extract, resveratrol, flavonoids, and melatonin, protect against the OS induced by thyroid hormonal abnormalities. Understanding how antioxidants and hormones work in conjunction to control inflammation, pathophysiology, and aging could lead to new treatments that promote healthy aging and longer life spans.

Acknowledgments

The authors acknowledge the Department of Veterinary Clinical Sciences (VCS) Core Lab for BioRender program support.

References

Adler, I., Tulassay, Z., Stark, J., Marczell, I., Nagy-Repas, P., Varbiro, S., et al. (2012). The effect of certain steroid hormones on the expression of genes involved in the metabolism of free radicals. *Gynecological Endocrinology*, *28*(11), 912–916. https://doi.org/10.3109/09513590.2012.683067.

Agarwal, A., Nallella, K. P., Allamaneni, S. S. R., & Said, T. M. (2004). Role of antioxidants in treatment of male infertility: An overview of the literature. *Reproductive Biomedicine Online*, *8*(6), 616–627. https://doi.org/10.1016/S1472-6483(10)61641-0.

Aguirre, F., Martin, I., Grinspon, D., Ruiz, M., Hager, A., de Paoli, T., et al. (1998). Oxidative damage, plasma antioxidant capacity, and glucemic control in elderly NIDDM patients. *Free Radical Biology & Medicine, 24*(4), 580–585. https://doi.org/10.1016/S0891-5849(97)00293-1.

Ahlbom, E., Prins, G. S., & Ceccatelli, S. (2001). Testosterone protects cerebellar granule cells from oxidative stress-induced cell death through a receptor mediated mechanism. *Brain Research, 892*(2), 255–262. https://doi.org/10.1016/S0006-8993(00)03155-3.

Allegra, M., Reiter, R. J., Tan, D. X., Gentile, C., Tesoriere, L., & Livrea, M. A. (2003). The chemistry of melatonin's interaction with reactive species. *Journal of Pineal Research, 34*(1), 1–10. https://doi.org/10.1034/J.1600-079X.2003.02112.X.

Alquicer, G., Kodrík, D., Krishnan, N., Večeřa, J., & Socha, R. (2009). Activation of insect anti-oxidative mechanisms by mammalian glucagon. *Comparative Biochemistry and Physiology. Part B, Biochemistry & Molecular Biology, 152*(3), 226–233. https://doi.org/10.1016/J.CBPB.2008.11.007.

Al-Rubae'i, S. H. N., & Al-Musawi, A. K. (2011). An evaluation of antioxidants and oxidative stress in Iraqi patients with thyroid gland dysfunction. *African Journal of Biochemistry Research, 5*(7), 188–196.

Amini, H., Rezabakhsh, A., Heidarzadeh, M., Hassanpour, M., Hashemzadeh, S., Ghaderi, S., et al. (2021). An examination of the putative role of melatonin in exosome biogenesis. *Frontiers in Cell and Development Biology, 9*, 1396. https://doi.org/10.3389/FCELL.2021.686551/BIBTEX.

Arthur, J. R., Beckett, G. J., & Mitchell, J. H. (1999). The interactions between selenium and iodine deficiencies in man and animals. *Nutrition Research Reviews, 12*(1), 55–73. https://doi.org/10.1079/095442299108728910.

Asmat, U., Abad, K., & Ismail, K. (2016). Diabetes mellitus and oxidative stress—A concise review. *Saudi Pharmaceutical Journal, 24*(5), 547–553. https://doi.org/10.1016/J.JSPS.2015.03.013.

Aydilek, N., Aksakal, M., & Karakilçik, A. Z. (2004). Effects of testosterone and vitamin E on the antioxidant system in rabbit testis. *Andrologia, 36*(5), 277–281. https://doi.org/10.1111/J.1439-0272.2004.00618.X.

Azevedo, R. B., Lacava, Z. G. M., Miyasaka, C. K., Chaves, S. B., & Curi, R. (2001). Regulation of antioxidant enzyme activities in male and female rat macrophages by sex steroids. *Brazilian Journal of Medical and Biological Research = Revista Brasileira de Pesquisas Medicas e Biologicas, 34*(5), 683–687. https://doi.org/10.1590/S0100-879X2001000500018.

Bafna, P. A., & Balaraman, R. (2005). Antioxidant activity of DHC-1, an herbal formulation, in experimentally-induced cardiac and renal damage. *Phytotherapy Research: PTR, 19*(3), 216–221. https://doi.org/10.1002/PTR.1659.

Baićc, J., Bjelakovič, G., Pavlovič, D., Kocić, G., Jevtovič, T., Stojanovič, I., et al. (2007). Glucocorticoids and oxidative stress. *Journal of Basic and Clinical Physiology and Pharmacology, 18*(2), 115–128. https://doi.org/10.1515/JBCPP.2007.18.2.115/MACHINEREADABLECITATION/RIS.

Baltgalvis, K. A., Greising, S. M., Warren, G. L., & Lowe, D. A. (2010). Estrogen regulates estrogen receptors and antioxidant gene expression in mouse skeletal muscle. *PLoS One, 5*(4), e10164. https://doi.org/10.1371/JOURNAL.PONE.0010164.

Barregard, L., Møller, P., Henriksen, T., Mistry, V., Koppen, G., Rossner, P., et al. (2013). Human and methodological sources of variability in the measurement of urinary 8-oxo-7,8-dihydro-2′-deoxyguanosine. *Antioxidants & Redox Signaling, 18*(18), 2377. https://doi.org/10.1089/ARS.2012.4714.

Behonick, G. S., Novak, M. J., Nealley, E. W., & Baskin, S. I. (2001). Toxicology update: The cardiotoxicity of the oxidative stress metabolites of catecholamines (aminochromes). *Journal of Applied Toxicology: JAT, 21*(Suppl 1), S15–S22. https://doi.org/10.1002/JAT.793.

Bellanti, F., Matteo, M., Rollo, T., de Rosario, F., Greco, P., Vendemiale, G., et al. (2013). Sex hormones modulate circulating antioxidant enzymes: Impact of estrogen therapy. *Redox Biology*, *1*(1), 340–346. https://doi.org/10.1016/J.REDOX.2013.05.003.

Bigagli, E., & Lodovici, M. (2019). Circulating oxidative stress biomarkers in clinical studies on type 2 diabetes and its complications. *Oxidative Medicine and Cellular Longevity*, *2019*. https://doi.org/10.1155/2019/5953685.

Bloch, K., Shichman, E., Vorobeychik, M., Bloch, D., & Vardi, P. (2007). Catalase expression in pancreatic alpha cells of diabetic and non-diabetic mice. *Histochemistry and Cell Biology*, *127*(2), 227–232. https://doi.org/10.1007/S00418-006-0248-4.

Bokov, A. F., Ko, D., & Richardson, A. (2009). The effect of gonadectomy and estradiol on sensitivity to oxidative stress. *Endocrine Research*, *34*(1–2), 43–58. https://doi.org/10.1080/07435800902913600.

Borst, S. E., Quindry, J. C., Yarrow, J. F., Conover, C. F., & Powers, S. K. (2010). Testosterone administration induces protection against global myocardial ischemia. *Hormone and Metabolic Research = Hormon- Und Stoffwechselforschung = Hormones et Metabolisme*, *42*(2), 122–129. https://doi.org/10.1055/S-0029-1241843.

Briet, M., & Schiffrin, E. L. (2013). Vascular actions of aldosterone. *Journal of Vascular Research*, *50*(2), 89–99. https://doi.org/10.1159/000345243.

Bruce, C. R., Carey, A. L., Hawley, J. A., & Febbraio, M. A. (2003). Intramuscular heat shock protein 72 and heme oxygenase-1 mRNA are reduced in patients with type 2 diabetes: Evidence that insulin resistance is associated with a disturbed antioxidant defense mechanism. *Diabetes*, *52*(9), 2338–2345. https://doi.org/10.2337/DIABETES.52.9.2338.

Burczynski, M. E., Sridhar, G. R., Palackal, N. T., & Penning, T. M. (2001). The reactive oxygen species- and Michael acceptor-inducible human aldo-keto reductase AKR1C1 reduces the alpha,beta-unsaturated aldehyde 4-hydroxy-2-nonenal to 1,4-dihydroxy-2-nonene. *The Journal of Biological Chemistry*, *276*(4), 2890–2897. https://doi.org/10.1074/JBC.M006655200.

Cai, W., Zhu, Y., Furuya, K., Li, Z., Sokabe, M., & Chen, L. (2008). Two different molecular mechanisms underlying progesterone neuroprotection against ischemic brain damage. *Neuropharmacology*, *55*(2), 127–138. https://doi.org/10.1016/J.NEUROPHARM.2008.04.023.

Camm, E. J., Tijsseling, D., Richter, H. G., Adler, A., Hansell, J. A., Derks, J. B., et al. (2011). Oxidative stress in the developing brain: Effects of postnatal glucocorticoid therapy and antioxidants in the rat. *PLoS One*, *6*(6), 21142. https://doi.org/10.1371/JOURNAL.PONE.0021142.

Castilla, P., Dávalos, A., Teruel, J. L., Cerrato, F., Fernández-Lucas, M., Merino, J. L., et al. (2008). Comparative effects of dietary supplementation with red grape juice and vitamin E on production of superoxide by circulating neutrophil NADPH oxidase in hemodialysis patients. *The American Journal of Clinical Nutrition*, *87*(4), 1053–1061. https://doi.org/10.1093/AJCN/87.4.1053.

Ceriello, A., Bortolotti, N., Crescentini, A., Motz, E., Lizzio, S., Russo, A., et al. (1998). Antioxidant defences are reduced during the oral glucose tolerance test in normal and non-insulin-dependent diabetic subjects. *European Journal of Clinical Investigation*, *28*(4), 329–333. https://doi.org/10.1046/J.1365-2362.1998.00295.X.

Chainy, G. B. N., & Sahoo, D. K. (2020). Hormones and oxidative stress: An overview. *Free Radical Research*, *54*(1). https://doi.org/10.1080/10715762.2019.1702656.

Chainy, G. B. N., Samantaray, S., & Samanta, L. (1997). Testosterone-induced changes in testicular antioxidant system. *Andrologia*, *29*(6), 343–349. https://doi.org/10.1111/J.1439-0272.1997.TB00328.X.

Chakrabarti, S. K., Ghosh, S., Banerjee, S., Mukherjee, S., & Chowdhury, S. (2016). Oxidative stress in hypothyroid patients and the role of antioxidant supplementation. *Indian Journal of Endocrinology and Metabolism*, *20*(5), 674–678. https://doi.org/10.4103/2230-8210.190555.

Chandra, R., Aneja, R., Rewal, C., Konduri, R., Dass, S. K., & Agarwal, S. (2000). An opium alkaloid-papaverine ameliorates ethanol-induced hepatotoxicity: Diminution of oxidative stress. *Indian Journal of Clinical Biochemistry: IJCB*, *15*(2), 155–160. https://doi.org/10.1007/BF02883745.

Chattopadhyay, S., Choudhury, S., Roy, A., Chainy, G. B. N., & Samanta, L. (2010). T3 fails to restore mitochondrial thiol redox status altered by experimental hypothyroidism in rat testis. *General and Comparative Endocrinology*, *169*(1), 39–47. https://doi.org/10.1016/J.YGCEN.2010.07.014.

Chattopadhyay, S., Sahoo, D. K., Roy, A., Samanta, L., & Chainy, G. B. N. (2010). Thiol redox status critically influences mitochondrial response to thyroid hormone-induced hepatic oxidative injury: A temporal analysis. *Cell Biochemistry and Function*, *28*(2), 126–134. https://doi.org/10.1002/CBF.1631.

Chattopadhyay, S., Sahoo, D. K., Subudhi, U., & Chainy, G. B. N. (2007). Differential expression profiles of antioxidant enzymes and glutathione redox status in hyperthyroid rats: A temporal analysis. *Comparative Biochemistry and Physiology Part C: Toxicology & Pharmacology*, *146*(3), 383–391. https://doi.org/10.1016/J.CBPC.2007.04.010.

Chaves, E. A., Pereira-Junior, P. P., Fortunato, R. S., Masuda, M. O., de Carvalho, A. C. C., de Carvalho, D. P., et al. (2006). Nandrolone decanoate impairs exercise-induced cardioprotection: Role of antioxidant enzymes. *The Journal of Steroid Biochemistry and Molecular Biology*, *99*(4–5), 223–230. https://doi.org/10.1016/J.JSBMB.2006.01.004.

Chen, S. K., Hsu, C. H., Tsai, M. L., Chen, R. H., & Drummen, G. P. C. (2013). Inhibition of oxidative stress by low-molecular-weight polysaccharides with various functional groups in skin fibroblasts. *International Journal of Molecular Sciences*, *14*(10), 19399. https://doi.org/10.3390/IJMS141019399.

Chen, J. H., Lin, X., Bu, C., & Zhang, X. (2018). Role of advanced glycation end products in mobility and considerations in possible dietary and nutritional intervention strategies. *Nutrition & Metabolism*, *15*(1), 1–18. https://doi.org/10.1186/S12986-018-0306-7.

Chen, X., Scholl, T. O., Leskiw, M. J., Donaldson, M. R., & Stein, T. P. (2003). Association of glutathione peroxidase activity with insulin resistance and dietary fat intake during normal pregnancy. *The Journal of Clinical Endocrinology & Metabolism*, *88*(12), 5963–5968. https://doi.org/10.1210/JC.2003-030544.

Chew, L. Y., Zhang, H., He, J., & Yu, F. (2021). The Nrf2-Keap1 pathway is activated by steroid hormone signaling to govern neuronal remodeling. *Cell Reports*, *36*(5). https://doi.org/10.1016/J.CELREP.2021.109466.

Chhabria, S., Mathur, S., Vadakan, S., Sahoo, D. K., Mishra, P., & Paital, B. (2022). A review on phytochemical and pharmacological facets of tropical ethnomedicinal plants as reformed DPP-IV inhibitors to regulate incretin activity. *Frontiers in Endocrinology*, *13*, 1027237. https://doi.org/10.3389/fendo.2022.1027237.

Choudhury, S., Chainy, G. B. N., & Mishro, M. M. (2003). Experimentally induced hypo- and hyper-thyroidism influence on the antioxidant defence system in adult rat testis. *Andrologia*, *35*(3), 131–140. https://doi.org/10.1046/J.1439-0272.2003.00548.X.

Cullinan, S. B., Gordan, J. D., Jin, J., Harper, J. W., & Diehl, J. A. (2004). The Keap1-BTB protein is an adaptor that bridges Nrf2 to a Cul3-based E3 Ligase: Oxidative stress sensing by a Cul3-Keap1 ligase. *Molecular and Cellular Biology*, *24*(19), 8477–8486. https://doi.org/10.1128/MCB.24.19.8477-8486.2004/ASSET/CD1AEB31-DD00-4CD7-9D3E-19688F6DB403/ASSETS/GRAPHIC/ZMB0190444350007.JPEG.

Darbandi, M., Darbandi, S., Agarwal, A., Sengupta, P., Durairajanayagam, D., Henkel, R., et al. (2018). Reactive oxygen species and male reproductive hormones. *Reproductive Biology and Endocrinology: RB&E*, *16*(1). https://doi.org/10.1186/S12958-018-0406-2.

Das, K., & Chainy, G. B. N. (2001). Modulation of rat liver mitochondrial antioxidant defence system by thyroid hormone. *Biochimica et Biophysica Acta*, *1537*(1), 1–13. https://doi.org/10.1016/S0925-4439(01)00048-5.

Das, K., & Chainy, G. B. N. (2004). Thyroid hormone influences antioxidant defense system in adult rat brain. *Neurochemical Research*, *29*(9), 1755–1766. https://doi.org/10.1023/B:NERE.0000035812.58200.A9.

Dhalla, N. S., Adameova, A., & Kaur, M. (2010). Role of catecholamine oxidation in sudden cardiac death. *Fundamental & Clinical Pharmacology*, *24*(5), 539–546. https://doi.org/10.1111/J.1472-8206.2010.00836.X.

Diaz-Flores, M., Baiza-Gutman, L. A., Pedrón, N. N., & Hicks, J. J. (1999). Uterine glutathione reductase activity: Modulation by estrogens and progesterone. *Life Sciences*, *65*(23), 2481–2488. https://doi.org/10.1016/S0024-3205(99)00514-7.

Dickinson, D. A., Levonen, A. L., Moellering, D. R., Arnold, E. K., Zhang, H., Darley-Usmar, V. M., et al. (2004). Human glutamate cysteine ligase gene regulation through the electrophile response element. *Free Radical Biology & Medicine*, *37*(8), 1152–1159. https://doi.org/10.1016/J.FREERADBIOMED.2004.06.011.

Dillon, P. F., Root-Bernstein, R. S., & Lieder, C. M. (2004). Antioxidant-independent ascorbate enhancement of catecholamine-induced contractions of vascular smooth muscle. *American Journal of Physiology - Heart and Circulatory Physiology*, *286*(6), 55–56. https://doi.org/10.1152/AJPHEART.00968.2003/ASSET/IMAGES/LARGE/ZH40060431040011.JPEG.

Dobrzyńska, M. M., Baumgartner, A., & Anderson, D. (2004). Antioxidants modulate thyroid hormone- and noradrenaline-induced DNA damage in human sperm. *Mutagenesis*, *19*(4), 325–330. https://doi.org/10.1093/MUTAGE/GEH037.

Drăgoi, C. M., & Nicolae, A. C. (2018). *Melatonin - Molecular Biology, Clinical and Pharmaceutical Approaches*. Intechopen. https://doi.org/10.5772/INTECHOPEN.74993.

Duarte, A. I., Santos, M. S., Oliveira, C. R., & Rego, A. C. (2005). Insulin neuroprotection against oxidative stress in cortical neurons—Involvement of uric acid and glutathione antioxidant defenses. *Free Radical Biology & Medicine*, *39*(7), 876–889. https://doi.org/10.1016/J.FREERADBIOMED.2005.05.002.

Dziezak, J. D. (1986). Preservatives: Antioxidants. The ultimate answer to oxidation. *Food Technology (Chicago)*, *40*(9), 94–102.

El-Kashlan, A. M., Nooh, M. M., Hassan, W. A., & Rizk, S. M. (2015). Therapeutic potential of date palm pollen for testicular dysfunction induced by thyroid disorders in male rats. *PLoS One*, *10*(10), e0139493. https://doi.org/10.1371/JOURNAL.PONE.0139493.

Fessel, J. P., Porter, N. A., Moore, K. P., Sheller, J. R., & Roberts, L. J. (2002). Discovery of lipid peroxidation products formed in vivo with a substituted tetrahydrofuran ring (isofurans) that are favored by increased oxygen tension. *Proceedings of the National Academy of Sciences of the United States of America*, *99*(26), 16713–16718. https://doi.org/10.1073/PNAS.252649099.

Fischer, T. W., Kleszczyński, K., Hardkop, L. H., Kruse, N., & Zillikens, D. (2013). Melatonin enhances antioxidative enzyme gene expression (CAT, GPx, SOD), prevents their UVR-induced depletion, and protects against the formation of DNA damage (8-hydroxy-2'-deoxyguanosine) in ex vivo human skin. *Journal of Pineal Research*, *54*(3), 303–312. https://doi.org/10.1111/JPI.12018.

Galano, A., Medina, M. E., Tan, D. X., & Reiter, R. J. (2015). Melatonin and its metabolites as copper chelating agents and their role in inhibiting oxidative stress: A physicochemical analysis. *Journal of Pineal Research*, *58*(1), 107–116. https://doi.org/10.1111/JPI.12196.

Galano, A., & Reiter, R. J. (2018). Melatonin and its metabolites vs oxidative stress: From individual actions to collective protection. *Journal of Pineal Research*, *65*(1), e12514. https://doi.org/10.1111/JPI.12514.

Gerenova, J., & Gadjeva, V. (2007). Oxidative stress and antioxidant enzyme activities in patients with Hashimoto's thyroiditis. *Comparative Clinical Pathology*, *16*(4), 259–264. https://doi.org/10.1007/s00580-007-0689-8.

Griffith, O. W. (1980). Determination of glutathione and glutathione disulfide using glutathione reductase and 2-vinylpyridine. *Analytical Biochemistry*, *106*(1), 207–212. https://doi.org/10.1016/0003-2697(80)90139-6.

Gulcelik, M. A., Gulcelik, N. E., Dinc, S., Kuru, B., Camlibel, M., & Alagol, H. (2006). The incidence of hyperthyroidism in patients with thyroid cancer in an area of iodine deficiency. *Journal of Surgical Oncology*, *94*(1), 35–39. https://doi.org/10.1002/JSO.20508.

Güntherberg, H., & Rost, J. (1966). The true oxidized glutathione content of red blood cells obtained by new enzymic and paper chromatographic methods. *Analytical Biochemistry*, *15*(2), 205–210. https://doi.org/10.1016/0003-2697(66)90025-X.

Gutteridge, J. M. C., & Halliwell, B. (1990). The measurement and mechanism of lipid peroxidation in biological systems. *Trends in Biochemical Sciences*, *15*(4), 129–135. https://doi.org/10.1016/0968-0004(90)90206-Q.

Guzmán, D. C., Mejía, G. B., Vázquez, I. E., García, E. H., del Angel, D. S., & Olguín, H. J. (2005). Effect of testosterone and steroids homologues on indolamines and lipid peroxidation in rat brain. *The Journal of Steroid Biochemistry and Molecular Biology*, *94*(4), 369–373. https://doi.org/10.1016/J.JSBMB.2004.11.006.

Hacışevki, A., & Baba, B. (2018). An overview of melatonin as an antioxidant molecule: A biochemical approach. *Melatonin - Molecular Biology, Clinical and Pharmaceutical Approaches*. https://doi.org/10.5772/INTECHOPEN.79421.

Halliwell, B. (2000). Lipid peroxidation, antioxidants and cardiovascular disease: How should we move forward? *Cardiovascular Research*, *47*(3), 410–418. https://doi.org/10.1016/S0008-6363(00)00097-3.

Halliwell, B., & Gutteridge, J. (2015). *Free radicals in biology and medicine*. https://books.google.com/books?hl=en&lr=&id=3DlKCgAAQBAJ&oi=fnd&pg=PP1&dq=Halliwell+B,+Gutteridge+JM.+Free+radicals+in+biology+and+medicine.+Oxford+University+Press,+USA%3B+2015.&ots=bpmFc-uriS&sig=RwMrUjKX-MnYZadf7dvxsXGqbys.

Hamden, K., Carreau, S., Ellouz, F., Masmoudi, H., & el Feki, A. (2007). Protective effect of 17beta-estradiol on oxidative stress and liver dysfunction in aged male rats. *Journal of Physiology and Biochemistry*, *63*(3), 195–201. https://doi.org/10.1007/BF03165782.

Han, D., Handelman, G., Marcocci, L., Sen, C. K., Roy, S., Kobuchi, H., et al. (1997). Lipoic acid increases de novo synthesis of cellular glutathione by improving cystine utilization. *BioFactors (Oxford, England)*, *6*(3), 321–338. https://doi.org/10.1002/BIOF.5520060303.

Harbrecht, B. G., Perpetua, M., Fulmer, M., & Zhang, B. (2004). Glucagon regulates hepatic inducible nitric oxide synthesis in vivo. *Shock (Augusta, Ga.)*, *22*(2), 157–162. https://doi.org/10.1097/01.SHK.0000131579.22409.33.

Harbrecht, B. G., Wirant, E. M., Kim, Y. M., & Billiar, T. R. (1996). Glucagon inhibits hepatocyte nitric oxide synthesis. *Archives of Surgery (Chicago, Ill.: 1960)*, *131*(12), 1266–1272. https://doi.org/10.1001/ARCHSURG.1996.01430240020002.

Hardeland, R. (2005). Antioxidative protection by melatonin. *Endocrine*, *27*(2), 119–130. https://doi.org/10.1385/ENDO:27:2:119.

Hardeland, R., & Pandi-Perumal, S. R. (2005). Melatonin, a potent agent in antioxidative defense: Actions as a natural food constituent, gastrointestinal factor, drug and prodrug. *Nutrition and Metabolism*, *2*(1), 1–15. https://doi.org/10.1186/1743-7075-2-22.

Harris, E. D. (1992). Regulation of antioxidant enzymes. *FASEB Journal: Official Publication of the Federation of American Societies for Experimental Biology*, *6*(9), 2675–2683. https://doi.org/10.1096/FASEBJ.6.9.1612291.

Hassan, H. M. M. (2011). Chemical Composition and nutritional value of palm pollen grains. *Global Journal of Biotechnology & Biochemistry*, *6*(1), 1–07.

He, L., Yang, H., Zhai, L. D., Shao, H., & Li, Y. S. (2011). A preliminary study on progesterone antioxidation in promoting learning and memory of young ovariectomized mice. *Archives of Medical Science. AMS*, 7(3), 397. https://doi.org/10.5114/AOMS.2011.23402.

Hoch, F. L. (1988). Lipids and thyroid hormones. *Progress in Lipid Research*, *27*(3), 199–270. https://doi.org/10.1016/0163-7827(88)90013-6.

Huh, K., Shin, U. S., Choi, J. W., & Lee, S. I. (1994). Effect of sex hormones on lipid peroxidation in rat liver. *Archives of Pharmacal Research*, *17*(2), 109–114. https://doi.org/10.1007/BF02974233.

Ibrahim, W., Tousson, E., Ali, E. M. M., & Mansour, M. A. (2011). Folic acid alleviates oxidative stress and hyperhomocysteinemia involved in testicular dysfunction of hypothyroid rats. *General and Comparative Endocrinology*, *174*(2), 143–149. https://doi.org/10.1016/J.YGCEN.2011.08.012.

Ilango, S., Sahoo, D. K., Paital, B., Kathirvel, K., Gabriel, J. I., Subramaniam, K., et al. (2022). A review on Annona muricata and its anticancer activity. *Cancers*, *14*(18), 4539. https://doi.org/10.3390/CANCERS14184539.

Ito, F., Sono, Y., & Ito, T. (2019). Measurement and clinical significance of lipid peroxidation as a biomarker of oxidative stress: Oxidative stress in diabetes, atherosclerosis, and chronic inflammation. *Antioxidants*, *8*(3). https://doi.org/10.3390/ANTIOX8030072.

Iuchi, T., Akaike, M., Mitsui, T., Ohshima, Y., Shintani, Y., Azuma, H., et al. (2003). Glucocorticoid excess induces superoxide production in vascular endothelial cells and elicits vascular endothelial dysfunction. *Circulation Research*, *92*(1), 81–87. https://doi.org/10.1161/01.RES.0000050588.35034.3C.

Iwashima, F., Yoshimoto, T., Minami, I., Sakurada, M., Hirono, Y., & Hirata, Y. (2008). Aldosterone induces superoxide generation via Rac1 activation in endothelial cells. *Endocrinology*, *149*(3), 1009–1014. https://doi.org/10.1210/EN.2007-0864.

Jena, S., & Chainy, G. B. N. (2011). Regulation of expression of antioxidant enzymes by vitamin E and curcumin in L-thyroxine-induced oxidative stress in rat renal cortex. *Molecular Biology Reports*, *38*(2), 1047–1054. https://doi.org/10.1007/S11033-010-0201-4.

Jena, S., Nityananda Chainy, G. B., & Dandapat, J. (2012). Expression of antioxidant genes in renal cortex of PTU-induced hypothyroid rats: Effect of vitamin E and curcumin. *Molecular Biology Reports*, *39*(2), 1193–1203. https://doi.org/10.1007/S11033-011-0849-4.

Jia, X. J., Liu, L. X., Tian, Y. M., Wang, R., & Lu, Q. (2019). The correlation between oxidative stress level and intra-abdominal fat in obese males. *Medicine*, *98*(7). https://doi.org/10.1097/MD.0000000000014469.

Jin, J. Z., & Lin, S. X. (1999). Human estrogenic 17β-hydroxysteroid dehydrogenase: Predominance of estrone reduction and its induction by NADPH. *Biochemical and Biophysical Research Communications*, *259*(2), 489–493. https://doi.org/10.1006/BBRC.1999.0704.

Johns, J. R., & Platts, J. A. (2014). Theoretical insight into the antioxidant properties of melatonin and derivatives. *Organic & Biomolecular Chemistry*, *12*(39), 7820–7827. https://doi.org/10.1039/C4OB01396D.

Kagan, V. E., & Tyurina, Y. Y. (1998). Recycling and redox cycling of phenolic antioxidants. *Annals of the New York Academy of Sciences*, *854*, 425–434. https://doi.org/10.1111/J.1749-6632.1998.TB09921.X.

Kaïs, K., Al-Gubory, H., Bolifraud, P., & Garrel, C. (2008). Regulation of key antioxidant enzymatic systems in the sheep endometrium by ovarian steroids. *Endocrinology*, *149*(9), 4428–4434. https://doi.org/10.1210/en.2008-0187.

Kensler, T. W., Wakabayashi, N., & Biswal, S. (2009). Cell survival responses to environmental stresses via the Kealp1-NRf2 pathway. *Annual Review of Pharmacology and Toxicology*, *47*, 89–116. https://doi.org/10.1146/ANNUREV.PHARMTOX.46.120604.141046.

Khan, R., Ooi, X. Y., Parvus, M., Valdez, L., & Tsin, A. (2019). Advanced glycation end products: Formation, role in diabetic complications, and potential in clinical applications. *The Eye and Foot in Diabetes*. https://doi.org/10.5772/INTECHOPEN.89408.

Kim, S. K., Abdelmegeed, M. A., & Novak, R. F. (2006). Identification of the insulin signaling cascade in the regulation of alpha-class glutathione S-transferase expression in primary cultured rat hepatocytes. *The Journal of Pharmacology and Experimental Therapeutics*, *316*(3), 1255–1261. https://doi.org/10.1124/JPET.105.096065.

Kim, S. K., Woodcroft, K. J., Kim, S. G., & Novak, R. F. (2003). Insulin and glucagon signaling in regulation of microsomal epoxide hydrolase expression in primary cultured rat hepatocytes. *Drug Metabolism and Disposition: The Biological Fate of Chemicals*, *31*(10), 1260–1268. https://doi.org/10.1124/DMD.31.10.1260.

Kim, S. K., Woodcroft, K. J., & Novak, R. F. (2003). Insulin and glucagon regulation of glutathione S-transferase expression in primary cultured rat hepatocytes. *The Journal of Pharmacology and Experimental Therapeutics*, *305*(1), 353–361. https://doi.org/10.1124/JPET.102.045153.

Kłapcińska, B., Jagsz, S., Sadowska-Krępa, E., Górski, J., Kempa, K., & Langfort, J. (2008). Effects of castration and testosterone replacement on the antioxidant defense system in rat left ventricle. *The Journal of Physiological Sciences*, *58*(3), 173–177. https://doi.org/10.2170/PHYSIOLSCI.RP002208.

Kochman, J., Jakubczyk, K., Bargiel, P., & Janda-Milczarek, K. (2021). The influence of oxidative stress on thyroid diseases. *Antioxidants*, *10*(9), 1442. https://doi.org/10.3390/ANTIOX10091442.

Korkmaz, A., Reiter, R. J., Topal, T., Manchester, L. C., Oter, S., & Tan, D. X. (2009). Melatonin: An established antioxidant worthy of use in clinical trials. *Molecular Medicine*, *15*(1–2), 43. https://doi.org/10.2119/MOLMED.2008.00117.

Kotlyar, E., Vita, J. A., Winter, M. R., Awtry, E. H., Siwik, D. A., Keaney, J. F., et al. (2006). The relationship between aldosterone, oxidative stress, and inflammation in chronic, stable human heart failure. *Journal of Cardiac Failure*, *12*(2), 122–127. https://doi.org/10.1016/J.CARDFAIL.2005.08.005.

Langston, W., Circu, M. L., & Aw, T. Y. (2008). Insulin stimulation of γ-glutamylcysteine ligase catalytic subunit expression increases endothelial GSH during oxidative stress: Influence of low glucose. *Free Radical Biology & Medicine*, *45*(11), 1591. https://doi.org/10.1016/J.FREERADBIOMED.2008.09.013.

Leisegang, K., Roychoudhury, S., Slama, P., & Finelli, R. (2021). The mechanisms and management of age-related oxidative stress in male hypogonadism associated with non-communicable chronic disease. *Antioxidants*, *10*(11). https://doi.org/10.3390/ANTIOX10111834/S1.

Lewis, B. W., Ford, M. L., Rogers, L. K., & Britt, R. D. (2021). Oxidative stress promotes corticosteroid insensitivity in asthma and COPD. *Antioxidants*, *10*(9), 1335. https://doi.org/10.3390/ANTIOX10091335.

Li, X., Yang, T., & Sun, Z. (2019). Hormesis in health and chronic diseases. *Trends in Endocrinology and Metabolism: TEM*, *30*(12), 944. https://doi.org/10.1016/J.TEM.2019.08.007.

Li, W., Yu, S., Liu, T., Kim, J. H., Blank, V., Li, H., et al. (2008). Heterodimerization with small Maf proteins enhances nuclear retention of Nrf2 via masking the NESzip motif. *Biochimica et Biophysica Acta*, *1783*(10), 1847–1856. https://doi.org/10.1016/J.BBAMCR.2008.05.024.

Liang, Y., Belford, S., Tang, F., Prokai, L., Simpkins, J. W., & Hughes, J. A. (2001). Membrane fluidity effects of estratrienes. *Brain Research Bulletin*, *54*(6), 661–668. https://doi.org/10.1016/S0361-9230(01)00483-X.

Liberopoulos, E., & Elisaf, M. (2002). Dyslipidemia in patients with thyroid disorders. *Hormones (Athens, Greece)*, *1*(4), 218–223. https://doi.org/10.14310/HORM.2002.1170.

Lindheim, S. R., Presser, S. C., Ditkoff, E. C., Vijod, M. A., Stanczyk, F. Z., & Lobo, R. A. (1993). A possible bimodal effect of estrogen on insulin sensitivity in postmenopausal women and the attenuating effect of added progestin. *Fertility and Sterility*, *60*(4), 664–667. https://doi.org/10.1016/S0015-0282(16)56218-9.

Loft, S., Danielsen, P., Løhr, M., Jantzen, K., Hemmingsen, J. G., Roursgaard, M., et al. (2012). Urinary excretion of 8-oxo-7,8-dihydroguanine as biomarker of oxidative damage to DNA. *Archives of Biochemistry and Biophysics*, *518*(2), 142–150. https://doi.org/10.1016/J.ABB.2011.12.026.

Lopes, S., Jurisicova, A., Sun, J. G., & Casper, R. F. (1998). Reactive oxygen species: Potential cause for DNA fragmentation in human spermatozoa. *Human Reproduction (Oxford, England)*, *13*(4), 896–900. https://doi.org/10.1093/HUMREP/13.4.896.

Lu, S. C., Garcia-Ruiz, C., Kuhlenkamp, J., Ookhtens, M., Salas-Prato, M., & Kaplowitz, N. (1990). Hormonal regulation of glutathione efflux. *The Journal of Biological Chemistry*, *265*(27), 16088–16095. https://doi.org/10.1016/S0021-9258(17)46192-8.

Luu-The, V., Zhang, Y., Poirier, D., & Labrie, F. (1995). Characteristics of human types 1, 2 and 3 17 beta-hydroxysteroid dehydrogenase activities: Oxidation/reduction and inhibition. *The Journal of Steroid Biochemistry and Molecular Biology*, *55*(5–6), 581–587. https://doi.org/10.1016/0960-0760(95)00209-X.

Ma, X., Idle, J. R., Krausz, K. W., & Gonzalez, F. J. (2005). Metabolism of melatonin by human cytochromes p450. *Drug Metabolism and Disposition: The Biological Fate of Chemicals*, *33*(4), 489–494. https://doi.org/10.1124/DMD.104.002410.

Mancini, A., di Segni, C., Raimondo, S., Olivieri, G., Silvestrini, A., Meucci, E., et al. (2016). Thyroid hormones, oxidative stress, and inflammation. *Mediators of Inflammation*, *2016*. https://doi.org/10.1155/2016/6757154.

Marnett, L. J. (2002). Oxy radicals, lipid peroxidation and DNA damage. *Toxicology*, *181–182*, 219–222. https://doi.org/10.1016/S0300-483X(02)00448-1.

Massafra, C., Gioia, D., de Felice, C., Picciolini, E., de Leo, V., Bonifazi, M., et al. (2000). Effects of estrogens and androgens on erythrocyte antioxidant superoxide dismutase, catalase and glutathione peroxidase activities during the menstrual cycle. *The Journal of Endocrinology*, *167*(3), 447–452. https://doi.org/10.1677/JOE.0.1670447.

Maxwell, S. R. J., Thomason, H., Sandler, D., LeGuen, C., Baxter, M. A., Thorpe, G. H. G., et al. (1997). Poor glycaemic control is associated with reduced serum free radical scavenging (antioxidant) activity in non-insulin-dependent diabetes mellitus. *Annals of Clinical Biochemistry*, *34*(*Pt 6*), 638–644. https://doi.org/10.1177/000456329703400607.

McGill, M. R., & Jaeschke, H. (2015). A direct comparison of methods used to measure oxidized glutathione in biological samples: 2-vinylpyridine and N-ethylmaleimide. *Toxicology Mechanisms and Methods*, *25*(8), 589. https://doi.org/10.3109/15376516.2015.1094844.

Mendelsohn, M. E., & Karas, R. H. (1999). The protective effects of estrogen on the cardiovascular system. *The New England Journal of Medicine*, *340*(23), 1801–1811. https://doi.org/10.1056/NEJM199906103402306.

Menendez-Pelaez, A., Poeggeler, B., Reiter, R. J., Barlow-Walden, L., Pablos, M. I., & Tan, D.-X. (1993). Nuclear localization of melatonin in different mammalian tissues: immunocytochemical and radioimmunoassay evidence. *Journal of Cellular Biochemistry*, *53*(4), 373–382. https://doi.org/10.1002/JCB.240530415.

Michos, C., Kiortsis, D., Evangelou, A., & Karkabounas, S. (2006). Antioxidant protection during the menstrual cycle: The effects of estradiol on ascorbic-dehydroascorbic acid plasma levels and total antioxidant plasma status in eumenorrhoic women during the menstrual cycle. *Acta Obstetricia et Gynecologica Scandinavica*, *85*(8), 960–965. https://doi.org/10.1080/00016340500432812.

Miró, A. M., Sastre-Serra, J., Pons, D. G., Valle, A., Roca, P., & Oliver, J. (2011). 17β-Estradiol regulates oxidative stress in prostate cancer cell lines according to ERalpha/ERbeta ratio. *The Journal of Steroid Biochemistry and Molecular Biology*, *123*(3–5), 133–139. https://doi.org/10.1016/J.JSBMB.2010.12.004.

Mishra, P., Paital, B., Jena, S., Swain, S. S., Kumar, S., Yadav, M. K., et al. (2019). Possible activation of NRF2 by Vitamin E/Curcumin against altered thyroid hormone induced oxidative stress via NFκB/AKT/mTOR/KEAP1 signalling in rat heart. *Scientific Reports*, *9*(1). https://doi.org/10.1038/S41598-019-43320-5.

Mitsui, T., Umaki, Y., Nagasawa, M., Akaike, M., Aki, K., Azuma, H., et al. (2002). Mitochondrial damage in patients with long-term corticosteroid therapy: Development of oculoskeletal symptoms similar to mitochondrial disease. *Acta Neuropathologica*, *104*(3), 260–266. https://doi.org/10.1007/S00401-002-0553-5.

Mogulkoc, R., Baltaci, A. K., Aydin, L., Oztekin, E., & Sivrikaya, A. (2005). The effect of thyroxine administration on lipid peroxidation in different tissues of rats with hypothyroidism. *Acta Physiologica Hungarica*, *92*(1), 39–46. https://doi.org/10.1556/APHYSIOL.92.2005.1.6.

Mogulkoc, R., Baltaci, A. K., Oztekin, E., Aydin, L., & Tuncer, I. (2005). Hyperthyroidism causes lipid peroxidation in kidney and testis tissues of rats: Protective role of melatonin. *Neuro Endocrinology Letters*, *26*(6), 806–810. https://europepmc.org/article/med/16380687.

Montuschi, P., Barnes, P. J., & Roberts, L. J. (2004). Isoprostanes: Markers and mediators of oxidative stress. *The FASEB Journal*, *18*(15), 1791–1800. https://doi.org/10.1096/FJ.04-2330REV.

Muñoz, P., Huenchuguala, S., Paris, I., & Segura-Aguilar, J. (2012). Dopamine oxidation and autophagy. *Parkinson's Disease*, *2012*. https://doi.org/10.1155/2012/920953.

Muñoz-Castañeda, J. R., Montilla, P., Muñoz, M. C., Bujalance, I., Muntané, J., & Túnez, I. (2005). Effect of 17-beta-estradiol administration during adriamycin-induced cardiomyopathy in ovariectomized rat. *European Journal of Pharmacology*, *523*(1–3), 86–92. https://doi.org/10.1016/J.EJPHAR.2005.08.056.

Nguyen, T., Yang, C. S., & Pickett, C. B. (2004). The pathways and molecular mechanisms regulating Nrf2 activation in response to chemical stress. *Free Radical Biology and Medicine*, *37*(4), 433–441. https://doi.org/10.1016/j.freeradbiomed.2004.04.033.

Nishiyama, A., Yao, L., Nagai, Y., Miyata, K., Yoshizumi, M., Kagami, S., et al. (2004). Possible contributions of reactive oxygen species and mitogen-activated protein kinase to renal injury in aldosterone/salt-induced hypertensive rats. *Hypertension (Dallas, Tex.: 1979)*, *43*(4), 841–848. https://doi.org/10.1161/01.HYP.0000118519.66430.22.

Nockels, C. F., Ewing, D. L., Phetteplace, H., Ritacco, K. A., & Mero, K. N. (1984). Hypothyroidism: An early sign of vitamin A deficiency in chickens. *The Journal of Nutrition*, *114*(9), 1733–1736. https://doi.org/10.1093/JN/114.9.1733.

Orzechowski, A. (2003). Justification for antioxidant preconditioning (or how to protect insulin-mediated actions under oxidative stress). *Journal of Biosciences*, *28*(1), 39–49. https://doi.org/10.1007/BF02970130.

Ott, C., Jacobs, K., Haucke, E., Navarrete Santos, A., Grune, T., & Simm, A. (2014). Role of advanced glycation end products in cellular signaling. *Redox Biology*, *2*(1), 411. https://doi.org/10.1016/J.REDOX.2013.12.016.

Ourique, G. M., Finamor, I. A., Saccol, E. M. H., Riffel, A. P. K., Pês, T. S., Gutierrez, K., et al. (2013). Resveratrol improves sperm motility, prevents lipid peroxidation and enhances antioxidant defences in the testes of hyperthyroid rats. *Reproductive Toxicology (Elmsford, N.Y.)*, *37*, 31–39. https://doi.org/10.1016/J.REPROTOX.2013.01.006.

Ozacmak, V. H., & Sayan, H. (2009). The effects of 17beta estradiol, 17alpha estradiol and progesterone on oxidative stress biomarkers in ovariectomized female rat brain subjected to global cerebral ischemia. *Physiological Research*, *58*(6), 909–912. https://doi.org/10.33549/PHYSIOLRES.931647.

Pajović, S. B., Saičić, Z. S., Spasić, M. B., Petrović, V. M., & Martinović, J. V. (1999). Effects of progesterone and estradiol benzoate on glutathione dependent antioxidant enzyme activities in the brain of female rats. *General Physiology and Biophysics*, *18*(1), 35–44. https://europepmc.org/article/med/10378119.

Paolisso, G., D'Amore, A., Balbi, V., Volpe, C., Galzerano, D., Giugliano, D., et al. (1994). Plasma vitamin C affects glucose homeostasis in healthy subjects and in non-insulin-dependent diabetics. *The American Journal of Physiology*, *266*(2 Pt 1). https://doi.org/10.1152/AJPENDO.1994.266.2.E261.

Paris, I., Dagnino-Subiabre, A., Marcelain, K., Bennett, L. B., Caviedes, P., Caviedes, R., et al. (2001). Copper neurotoxicity is dependent on dopamine-mediated copper uptake and one-electron reduction of aminochrome in a rat substantia nigra neuronal cell line. *Journal of Neurochemistry*, 77(2), 519–529. https://doi.org/10.1046/J.1471-4159.2001.00243.X.

Patni, H., Mathew, J. T., Luan, L., Franki, N., Chander, P. N., & Singhal, P. C. (2007). Aldosterone promotes proximal tubular cell apoptosis: Role of oxidative stress. *American Journal of Physiology. Renal Physiology*, *293*(4). https://doi.org/10.1152/AJPRENAL.00147.2007.

Penning, T. M., Wangtrakuldee, P., & Auchus, R. J. (2019). Structural and functional biology of aldo-keto reductase steroid-transforming enzymes. *Endocrine Reviews*, *40*(2), 447–475. https://doi.org/10.1210/ER.2018-00089.

Perlman, R. L., & Chalfie, M. (1977). Catecholamine release from the adrenal medulla. *Clinics in Endocrinology and Metabolism*, *6*(3), 551–576. https://doi.org/10.1016/S0300-595X(77)80071-6.

Persky, A. M., Green, P. S., Stubley, L., Howell, C. O., Zaulyanov, L., Brazeau, G. A., et al. (2000). Protective effect of estrogens against oxidative damage to heart and skeletal muscle in vivo and in vitro. *Proceedings of the Society for Experimental Biology and Medicine. Society for Experimental Biology and Medicine (New York, N.Y.)*, *223*(1), 59–66. https://doi.org/10.1046/J.1525-1373.2000.22308.X.

Plastaras, J. P., Dedon, P. C., & Marnett, L. J. (2002). Effects of DNA structure on oxopropenylation by the endogenous mutagens malondialdehyde and base propenal. *Biochemistry*, *41*(15), 5033–5042. https://doi.org/10.1021/BI0113059/ASSET/IMAGES/LARGE/BI0113059F00011.JPEG.

Plauth, A., Geikowski, A., Cichon, S., Wowro, S. J., Liedgens, L., Rousseau, M., et al. (2016). Hormetic shifting of redox environment by pro-oxidative resveratrol protects cells against stress. *Free Radical Biology and Medicine*, *99*, 608–622. https://doi.org/10.1016/J.FREERADBIOMED.2016.08.006.

Polan, M. L., Kuo, A., Loukides, J., & Bottomly, K. (1990). Cultured human luteal peripheral monocytes secrete increased levels of interleukin-1. *Journal of Clinical Endocrinology and Metabolism*, *70*(2), 480–484. https://doi.org/10.1210/jcem-70-2-480.

Prasad, S., Kalra, N., & Shukla, Y. (2007). Hepatoprotective effects of lupeol and mango pulp extract of carcinogen induced alteration in Swiss albino mice. *Molecular Nutrition & Food Research*, *51*(3), 352–359. https://doi.org/10.1002/MNFR.200600113.

Prasad, S., Kalra, N., Singh, M., & Shukla, Y. (2008). Protective effects of lupeol and mango extract against androgen induced oxidative stress in Swiss albino mice. *Asian Journal of Andrology*, *10*(2), 313–318. https://doi.org/10.1111/J.1745-7262.2008.00313.X.

Prokai, L., Prokai-Tatrai, K., Perjési, P., & Simpkins, J. W. (2005). Mechanistic insights into the direct antioxidant effects of estrogens. *Drug Development Research, 66*(2), 118–125. https://doi.org/10.1002/DDR.20050.

Prokai-Tatrai, K., Perjesi, P., Rivera-Portalatin, N. M., Simpkins, J. W., & Prokai, L. (2008). Mechanistic investigations on the antioxidant action of a neuroprotective estrogen derivative. *Steroids, 73*(3), 280. https://doi.org/10.1016/J.STEROIDS.2007.10.011.

Pulla Reddy, A. C., & Lokesh, B. R. (1996). Effect of curcumin and eugenol on iron-induced hepatic toxicity in rats. *Toxicology, 107*(1), 39–45. https://doi.org/10.1016/0300-483X(95)03199-P.

Queisser, N., Oteiza, P. I., Stopper, H., Oli, R. G., & Schupp, N. (2011). Aldosterone induces oxidative stress, oxidative DNA damage and NF-κB-activation in kidney tubule cells. *Molecular Carcinogenesis, 50*(2), 123–135. https://doi.org/10.1002/MC.20710.

Reiter, R. J., Mayo, J. C., Tan, D. X., Sainz, R. M., Alatorre-Jimenez, M., & Qin, L. (2016). Melatonin as an antioxidant: Under promises but over delivers. *Journal of Pineal Research, 61*(3), 253–278. https://doi.org/10.1111/JPI.12360.

Reiter, R. J., & Tan, D. X. (2003). Melatonin: A novel protective agent against oxidative injury of the ischemic/reperfused heart. *Cardiovascular Research, 58*(1), 10–19. https://doi.org/10.1016/S0008-6363(02)00827-1/2/58-1-10-FIG3.GIF.

Reiter, R. J., Tan, D. X., Manchester, L. C., Paredes, S. D., Mayo, J. C., & Sainz, R. M. (2009). Melatonin and reproduction revisited. *Biology of Reproduction, 81*(3), 445–456. https://doi.org/10.1095/BIOLREPROD.108.075655.

Reiter, R. J., Tan, D. X., Rosales-Corral, S., Galano, A., Zhou, X. J., & Xu, B. (2018). Mitochondria: Central organelles for melatonin's antioxidant and anti-aging actions. *Molecules, 23*(2), 509. https://doi.org/10.3390/MOLECULES23020509.

Reiter, R. J., Tan, D.-X., Terron, M. P., Flores, L. J., & Czarnocki, Z. (2007). Melatonin and its metabolites: New findings regarding their production and their radical scavenging actions. *Acta Biochimica Polonica, 54*(1), 1–9. www.actabp.pl.

Ripple, M. O., Henry, W. F., Rago, R. P., & Wilding, G. (1997). Prooxidant-antioxidant shift induced by androgen treatment of human prostate carcinoma cells. *Journal of the National Cancer Institute, 89*(1), 40–48. https://doi.org/10.1093/JNCI/89.1.40.

Ristow, M., Zarse, K., Oberbach, A., Klöting, N., Birringer, M., Kiehntopf, M., et al. (2009). Antioxidants prevent health-promoting effects of physical exercise in humans. *Proceedings of the National Academy of Sciences of the United States of America, 106*(21), 8665. https://doi.org/10.1073/PNAS.0903485106.

Rižner, T. L., & Penning, T. M. (2014). Role of aldo-keto reductase family 1 (AKR1) enzymes in human steroid metabolism. *Steroids, 79*, 49–63. https://doi.org/10.1016/J.STEROIDS.2013.10.012.

Roberts, L. J., & Morrow, J. D. (2000). Measurement of F(2)-isoprostanes as an index of oxidative stress in vivo. *Free Radical Biology & Medicine, 28*(4), 505–513. https://doi.org/10.1016/S0891-5849(99)00264-6.

Rodrigues, D., Costa, T. J., Silva, J. F., de Oliveira Neto, J. T., Alves, J. V., Fedoce, A. G., et al. (2021). Aldosterone negatively regulates Nrf2 activity: An additional mechanism contributing to oxidative stress and vascular dysfunction by aldosterone. *International Journal of Molecular Sciences, 22*(11). https://doi.org/10.3390/IJMS22116154.

Rodriguez, C., Mayo, J. C., Sainz, R. M., Antolín, I., Herrera, F., Martín, V., et al. (2004). Regulation of antioxidant enzymes: A significant role for melatonin. *Journal of Pineal Research, 36*(1), 1–9. https://doi.org/10.1046/J.1600-079X.2003.00092.X.

Russell, R. M. (2004). The enigma of beta-carotene in carcinogenesis: What can be learned from animal studies. *The Journal of Nutrition, 134*(1). https://doi.org/10.1093/JN/134.1.262S.

Sadowska-Krępa, E., Kłapcińska, B., Jagsz, S., Sobczak, A., Chrapusta, S. J., Chalimoniuk, M., et al. (2011). High-dose testosterone propionate treatment reverses the effects of endurance training on myocardial antioxidant defenses in adolescent male rats. *Cardiovascular Toxicology*, *11*(2), 118. https://doi.org/10.1007/S12012-011-9105-3.

Sahoo, D. K. (2013). Increased germ cell apoptosis during testicular development and maturation by experimentally induced transient and persistent hypothyroidism. *Webmedcentral*, *4*(12), 209. https://doi.org/10.9754/JOURNAL.WPLUS.2013.00209.

Sahoo, D. K., Borcherding, D. C., Chandra, L., Jergens, A. E., Atherly, T., Bourgois-Mochel, A., et al. (2022). Differential transcriptomic profiles following stimulation with lipopolysaccharide in intestinal organoids from dogs with inflammatory bowel disease and intestinal mast cell tumor. *Cancers*, *14*(14), 3525. https://doi.org/10.3390/CANCERS14143525.

Sahoo, D. K., & Chainy, G. B. N. (2007). Tissue specific response of antioxidant defence systems of rat to experimentally-induced hyperthyroidism. *National Academy Science Letters*, *30*(7–8).

Sahoo, D. K., Jena, S., & Chainy, G. B. N. (2019). Thyroid dysfunction and testicular redox status: An intriguing association. *Oxidants, Antioxidants, and Impact of the Oxidative Status in Male Reproduction*, *149–170*. https://doi.org/10.1016/B978-0-12-812501-4.00015-8.

Sahoo, D. K., & Roy, A. (2012). Compromised rat testicular antioxidant defence system by hypothyroidism before puberty. *International Journal of Endocrinology*, *2012*. https://doi.org/10.1155/2012/637825.

Sahoo, D. K., Roy, A., Bhanja, S., & Chainy, G. B. N. (2005). Experimental hyperthyroidism-induced oxidative stress and impairment of antioxidant defence system in rat testis. *Indian Journal of Experimental Biology*, *43*(11).

Sahoo, D. K., Roy, A., Bhanja, S., & Chainy, G. B. N. (2008). Hypothyroidism impairs antioxidant defence system and testicular physiology during development and maturation. *General and Comparative Endocrinology*, *156*(1). https://doi.org/10.1016/j.ygcen.2007.11.007.

Sahoo, D. K., Roy, A., & Chainy, G. B. N. (2006). PTU-induced neonatal hypothyroidism modulates antioxidative status and population of rat testicular germ cells. *National Academy Science Letters*, *29*(3–4).

Sahoo, D. K., Roy, A., & Chainy, G. B. N. (2008). Protective effects of vitamin E and curcumin on L-thyroxine-induced rat testicular oxidative stress. *Chemico-Biological Interactions*, *176*(2–3), 121–128. https://doi.org/10.1016/J.CBI.2008.07.009.

Sahoo, D. K., Roy, A., Chattopadhyay, S., & Chainy, G. B. N. (2007). Effect of T3 treatment on glutathione redox pool and its metabolizing enzymes in mitochondrial and post-mitochondrial fractions of adult rat testes. *Indian Journal of Experimental Biology*, *45*(4).

Sankha, S., Kumar, Y., Madhuri, A., & Kumar, M. (2021). Antioxidant status and oxidative stress in hypothyroidism. *Journal of Datta Meghe Institute of Medical Sciences University*, *16*(3), 508. https://go.gale.com/ps/i.do?p=AONE&sw=w&issn=09743901&v=2.1&it=r&id=GALE%7CA697888895&sid=googleScholar&linkaccess=fulltext.

Sato, H., Takahashi, T., Sumitani, K., Takatsu, H., & Urano, S. (2010). Glucocorticoid generates ROS to induce oxidative injury in the hippocampus, leading to impairment of cognitive function of rats. *Journal of Clinical Biochemistry and Nutrition*, *47*(3), 224–232. https://doi.org/10.3164/JCBN.10-58.

Segura-Aguilar, J. (2019). On the role of aminochrome in mitochondrial dysfunction and endoplasmic reticulum stress in Parkinson's disease. *Frontiers in Neuroscience*, *13*, 271. https://doi.org/10.3389/FNINS.2019.00271/BIBTEX.

Segura-Aguilar, J., Metodiewa, D., & Welch, C. J. (1998). Metabolic activation of dopamine o-quinones to o-semiquinones by NADPH cytochrome P450 reductase may play an important role in oxidative stress and apoptotic effects. *Biochimica et Biophysica Acta (BBA) - General Subjects, 1381*(1), 1–6. https://doi.org/10.1016/S0304-4165(98)00036-1.

Semak, I., Korik, E., Antonova, M., Wortsman, J., & Slominski, A. (2008). Metabolism of melatonin by cytochrome P450s in rat liver mitochondria and microsomes. *Journal of Pineal Research, 45*(4), 515–523. https://doi.org/10.1111/J.1600-079X.2008.00630.X.

Shimamura, K., Sukino, N., Yoshida, Y., Nakamura, Y., Ogino, K., & Kato, H. (1995). Changes in lipid peroxide and antioxidant enzyme activities in corpora lutea during pseudopregnancy in rats. *Journal of Reproduction and Fertility, 105*(2), 253–257. https://doi.org/10.1530/JRF.0.1050253.

Sies, H. (1991). Oxidative stress: From basic research to clinical application. *The American Journal of Medicine, 91*(3C). https://doi.org/10.1016/0002-9343(91)90281-2.

Simmons, W. W., Ungureanu-Longrois, D., Smith, G. K., Smith, T. W., & Kelly, R. A. (1996). Glucocorticoids regulate inducible nitric oxide synthase by inhibiting tetrahydrobiopterin synthesis and L-arginine transport. *The Journal of Biological Chemistry, 271*(39), 23928–23937. https://doi.org/10.1074/JBC.271.39.23928.

Singal, P. K., Khaper, N., Palace, V., & Kumar, D. (1998). The role of oxidative stress in the genesis of heart disease. *Cardiovascular Research, 40*(3), 426–432. https://doi.org/10.1016/S0008-6363(98)00244-2/2/40-3-426-FIG1.GIF.

Srinivasan, V., Spence, D. W., Pandi-Perumal, S. R., Brown, G. M., & Cardinali, D. P. (2011). Melatonin in mitochondrial dysfunction and related disorders. *International Journal of Alzheimer's Disease, 2011*. https://doi.org/10.4061/2011/326320.

Stein, D. G. (2008). Progesterone exerts neuroprotective effects after brain injury. *Brain Research Reviews, 57*(2), 386–397. https://doi.org/10.1016/J.BRAINRESREV.2007.06.012.

Strehlow, K., Rotter, S., Wassmann, S., Adam, O., Grohé, C., Laufs, K., et al. (2003). Modulation of antioxidant enzyme expression and function by estrogen. *Circulation Research, 93*(2), 170–177. https://doi.org/10.1161/01.RES.0000082334.17947.11.

Strom, J. O., Theodorsson, A., & Theodorsson, E. (2011). Hormesis and female sex hormones. *Pharmaceuticals, 4*(5), 726. https://doi.org/10.3390/PH4050726.

Subudhi, U., & Chainy, G. B. N. (2010). Expression of hepatic antioxidant genes in l-thyroxine-induced hyperthyroid rats: Regulation by vitamin E and curcumin. *Chemico-Biological Interactions, 183*(2), 304–316. https://doi.org/10.1016/J.CBI.2009.11.004.

Sugino, N., Shimamura, K., Tamura, H., Ono, M., Nakamura, Y., Ogino, K., et al. (1996). Progesterone inhibits superoxide radical production by mononuclear phagocytes in pseudopregnant rats. *Endocrinology, 137*(2), 749–754. https://doi.org/10.1210/ENDO.137.2.8593826.

Sun, Z., Zhang, S., Chan, J. Y., & Zhang, D. D. (2007). Keap1 controls postinduction repression of the Nrf2-mediated antioxidant response by escorting nuclear export of Nrf2. *Molecular and Cellular Biology, 27*(18), 6334–6349. https://doi.org/10.1128/MCB.00630-07/ASSET/3180B987-0772-47EA-A237-1FFC2D136780/ASSETS/GRAPHIC/ZMB0180769920009.JPEG.

Sunitha, S., Nagaraj, M., & Varalakshmi, P. (2001). Hepatoprotective effect of lupeol and lupeol linoleate on tissue antioxidant defence system in cadmium-induced hepatotoxicity in rats. *Fitoterapia, 72*(5), 516–523. https://doi.org/10.1016/S0367-326X(01)00259-3.

Szabó, R., Hoffmann, A., Börzsei, D., Kupai, K., Veszelka, M., Berkó, A. M., et al. (2021). Hormone replacement therapy and aging: A potential therapeutic approach for age-related oxidative stress and cardiac remodeling. *Oxidative Medicine and Cellular Longevity, 2021*. https://doi.org/10.1155/2021/8364297.

Szwarc, M. M., Kommagani, R., Lessey, B. A., & Lydon, J. P. (2014). The p160/steroid receptor coactivator family: Potent arbiters of uterine physiology and dysfunction. *Biology of Reproduction*, *91*(5), 122–123. https://doi.org/10.1095/BIOLREPROD.114.125021.

Talalay, P., Hurlock, B., & Williams-Ashman, H. G. (1958). On a coenzymatic function of estradiol-17beta. *Proceedings of the National Academy of Sciences of the United States of America*, *44*(9), 862–884. https://doi.org/10.1073/PNAS.44.9.862.

Tam, N. N. C., Gao, Y., Leung, Y. K., & Ho, S. M. (2003). Androgenic regulation of oxidative stress in the rat prostate: Involvement of NAD(P)H oxidases and antioxidant defense machinery during prostatic involution and regrowth. *The American Journal of Pathology*, *163*(6), 2513–2522. https://doi.org/10.1016/S0002-9440(10)63606-1.

Tapia, P. C. (2006). Sublethal mitochondrial stress with an attendant stoichiometric augmentation of reactive oxygen species may precipitate many of the beneficial alterations in cellular physiology produced by caloric restriction, intermittent fasting, exercise and dietary phytonutrients: "Mitohormesis" for health and vitality. *Medical Hypotheses*, *66*(4), 832–843. https://doi.org/10.1016/J.MEHY.2005.09.009.

Tappia, P. S., Hata, T., Hozaima, L., Sandhu, M. S., Panagia, V., & Dhalla, N. S. (2001). Role of oxidative stress in catecholamine-induced changes in cardiac sarcolemmal Ca2+ transport. *Archives of Biochemistry and Biophysics*, *387*(1), 85–92. https://doi.org/10.1006/ABBI.2000.2234.

Tarocco, A., Caroccia, N., Morciano, G., Wieckowski, M. R., Ancora, G., Garani, G., et al. (2019). Melatonin as a master regulator of cell death and inflammation: Molecular mechanisms and clinical implications for newborn care. *Cell Death & Disease*, *10*(4), 1–12. https://doi.org/10.1038/s41419-019-1556-7.

Tebay, L. E., Robertson, H., Durant, S. T., Vitale, S. R., Penning, T. M., Dinkova-Kostova, A. T., et al. (2015). Mechanisms of activation of the transcription factor Nrf2 by redox stressors, nutrient cues, and energy status and the pathways through which it attenuates degenerative disease. *Free Radical Biology & Medicine*, *88*(Pt B), 108–146. https://doi.org/10.1016/J.FREERADBIOMED.2015.06.021.

Thompson, C. M., Capdevila, J. H., & Strobel, H. W. (2000). Recombinant cytochrome P450 2D18 metabolism of dopamine and arachidonic acid. *Journal of Pharmacology and Experimental Therapeutics*, *294*(3), 1120–1130. https://jpet.aspetjournals.org/content/294/3/1120.short.

Tietze, F. (1969). Enzymic method for quantitative determination of nanogram amounts of total and oxidized glutathione: Applications to mammalian blood and other tissues. *Analytical Biochemistry*, *27*(3), 502–522. https://doi.org/10.1016/0003-2697(69)90064-5.

Tomás-Zapico, C., & Coto-Montes, A. (2005). A proposed mechanism to explain the stimulatory effect of melatonin on antioxidative enzymes. *Journal of Pineal Research*, *39*(2), 99–104. https://doi.org/10.1111/J.1600-079X.2005.00248.X.

Turková, H., Petrák, O., Škrha, J., Widimský, J., & Zelinka, T. (2013). Pheochromocytoma and markers of oxidative stress. *Physiological Research*, *62*(3), 331–335. https://doi.org/10.33549/PHYSIOLRES.932449.

Urata, Y., Honma, S., Goto, S., Todoroki, S., Iida, T., Cho, S., et al. (1999). Melatonin induces gamma-glutamylcysteine synthetase mediated by activator protein-1 in human vascular endothelial cells. *Free Radical Biology & Medicine*, *27*(7–8), 838–847. https://doi.org/10.1016/S0891-5849(99)00131-8.

Vargatu, I. (2016). Williams textbook of endocrinology. *Acta Endocrinologica (Bucharest)*, *12*(1), 113. https://doi.org/10.4183/AEB.2016.113.

Velayeti, J., Mansourian, A., Mojerloo, M., & Marjani, A. (2016). Evaluation of oxidative stress and thyroid hormone status in hemodialysis patients in Gorgan. *Indian Journal of Endocrinology and Metabolism*, *20*(3), 348. https://doi.org/10.4103/2230-8210.179986.

Venditti, P., & di Meo, S. (2006). Thyroid hormone-induced oxidative stress. *Cellular and Molecular Life Sciences: CMLS*, *63*(4), 414–434. https://doi.org/10.1007/S00018-005-5457-9.

Viña, J., Borrás, C., Gambini, J., Sastre, J., Pallardó, F., & v. (2005). Why females live longer than males: Control of longevity by sex hormones. *Science of Aging Knowledge Environment: SAGE KE, 2005*(23). https://doi.org/10.1126/SAGEKE.2005.23.PE17.
Wang, X., Wu, H., Chen, H., Liu, R., Liu, J., Zhang, T., et al. (2012). Does insulin bolster antioxidant defenses via the extracellular signal-regulated kinases-protein kinase B-nuclear factor erythroid 2 p45-related factor 2 pathway? *Antioxidants & Redox Signaling, 16*(10), 1061–1070. https://doi.org/10.1089/ARS.2011.4460.
Wilcox, C. S., & Pearlman, A. (2008). Chemistry and antihypertensive effects of tempol and other nitroxides. *Pharmacological Reviews, 60*(4), 418–469. https://doi.org/10.1124/PR.108.000240.
Wing, L. Y. C., Chen, Y. C., Shih, Y. Y., Cheng, J. C., Lin, Y. J., & Jiang, M. J. (2009). Effects of oral estrogen on aortic ROS-generating and -scavenging enzymes and atherosclerosis in apoE-deficient mice. *Experimental Biology and Medicine (Maywood, N.J.), 234*(9), 1037–1046. https://doi.org/10.3181/0811-RM-332.
Winiarska, K., Fraczyk, T., Malinska, D., Drozak, J., & Bryla, J. (2006). Melatonin attenuates diabetes-induced oxidative stress in rabbits. *Journal of Pineal Research, 40*(2), 168–176. https://doi.org/10.1111/J.1600-079X.2005.00295.X.
Wu, L., Einstein, M., Geissler, W. M., Chan, H. K., Elliston, K. O., & Andersson, S. (1993). Expression cloning and characterization of human 17 beta-hydroxysteroid dehydrogenase type 2, a microsomal enzyme possessing 20 alpha-hydroxysteroid dehydrogenase activity. *Journal of Biological Chemistry, 268*(17), 12964–12969. https://doi.org/10.1016/S0021-9258(18)31480-7.
Zafarullah, M., Li, W. Q., Sylvester, J., & Ahmad, M. (2003). Molecular mechanisms of N-acetylcysteine actions. *Cellular and Molecular Life Sciences: CMLS, 60*(1), 6–20. https://doi.org/10.1007/S000180300001.
Zamoner, A., Barreto, K. P., Wilhelm Filho, D., Sell, F., Mara Woehl, V., Guma, F. C. R., et al. (2008). Propylthiouracil-induced congenital hypothyroidism upregulates vimentin phosphorylation and depletes antioxidant defenses in immature rat testis. *Journal of Molecular Endocrinology, 40*(3), 125–135. https://doi.org/10.1677/JME-07-0089.
Zamora, R., Hidalgo, F. J., & Tappel, A. L. (1991). Comparative antioxidant effectiveness of dietary beta-carotene, vitamin E, selenium and coenzyme Q10 in rat erythrocytes and plasma. *The Journal of Nutrition, 121*(1), 50–56. https://doi.org/10.1093/JN/121.1.50.
Zampieri, S., Mellon, S. H., Butters, T. D., Nevyjel, M., Covey, D. F., Bembi, B., et al. (2009). Oxidative stress in NPC1 deficient cells: Protective effect of allopregnanolone. *Journal of Cellular and Molecular Medicine, 13*(9b), 3786. https://doi.org/10.1111/J.1582-4934.2008.00493.X.
Zelinka, T., Petrák, O., Turková, H., Holaj, R., Štrauch, B., Kršek, M., et al. (2012). High incidence of cardiovascular complications in pheochromocytoma. *Hormone and Metabolic Research, 44*(5), 379–384. https://doi.org/10.1055/S-0032-1306294.
Zhang, D. D., Lo, S. C., Sun, Z., Habib, G. M., Lieberman, M. W., & Hannink, M. (2005). Ubiquitination of Keap1, a BTB-Kelch substrate adaptor protein for Cul3, targets Keap1 for degradation by a proteasome-independent pathway. *Journal of Biological Chemistry, 280*(34), 30091–30099. https://doi.org/10.1074/JBC.M501279200.
Zhang, H. M., & Zhang, Y. (2014). Melatonin: A well-documented antioxidant with conditional pro-oxidant actions. *Journal of Pineal Research, 57*(2), 131–146. https://doi.org/10.1111/JPI.12162.
Zhu, X., Bonet, B., & Knopp, R. H. (1997). 17beta-Estradiol, progesterone, and testosterone inversely modulate low-density lipoprotein oxidation and cytotoxicity in cultured placental trophoblast and macrophages. *American Journal of Obstetrics and Gynecology, 177*(1), 196–209. https://doi.org/10.1016/S0002-9378(97)70462-9.
Zingg, J. M. (2007). Vitamin E: An overview of major research directions. *Molecular Aspects of Medicine, 28*(5–6), 400–422. https://doi.org/10.1016/J.MAM.2007.05.004.

CHAPTER NINE

Ascorbic acid as antioxidant

Agnieszka Gęgotek and Elżbieta Skrzydlewska*
Department of Inorganic and Analytical Chemistry, Medical University of Bialystok, Bialystok, Poland
*Corresponding author: e-mail address: skrzydle@amb.edu.pl

Contents

Abstract

Ascorbic acid, as one of the basic exogenous vitamins, is known for its tremendous antioxidant properties. This review has been prepared to show not only the importance of ascorbic acid as a free radical scavenger, but also to summarize its antioxidant action based on other mechanisms, including activation of intracellular antioxidant systems. Ascorbic acid interacts with small molecule antioxidants, including tocopherol, glutathione and thioredoxin, but also can stimulate the biosynthesis and activation of antioxidant enzymes, such as superoxide dismutase, catalase or glutathione peroxidase. Moreover, ascorbic acid promotes the activity of several transcription factors (Nrf2, Ref-1, AP-1), which enables the expression of genes encoding antioxidant proteins. Additionally, it supports the action of other exogenous antioxidants, mainly polyphenols. In this connection, both DNA, protein and lipids are protected against oxidation. Although ascorbic acid has strong antioxidant properties, it can also have pro-oxidant effects in the presence of free transition metals. However, its role in prevention of DNA mutation and cellular apoptosis, especially in relation to cancer cells is controversial.

Vitamins and Hormones, Volume 121
ISSN 0083-6729
https://doi.org/10.1016/bs.vh.2022.10.008

1. Introduction

Ascorbic acid, commonly known as vitamin C, is one of the basic and best-known compounds necessary for the proper functioning of the human body. It was described and isolated for the first time in 1928 by the Hungarian biochemist Albert Szent-Györgyi, who was awarded the Nobel Prize in 1937. The name, ascorbic acid, refers to scurvy (scorbutus), because the deficiency of this compound was initially associated solely with the development of this disease (Moss, 1989).

Currently, it is known that ascorbic acid is necessary for the proper functioning of the human body. It contributes to many processes, including strengthening and sealing blood vessels, regulation of microbial absorption by leukocytes, lowering the level of cholesterol, as well as acceleration of wound healing (May & Harrison, 2013; Ströhle & Hahn, 2009). Other important functions of ascorbic acid include the regulation of collagen production, slowing down the aging process of the skin, and lowering blood pressure (Crisan, Roman, Crisan, Scharffetter-Kochanek, & Badea, 2015; Juraschek, Guallar, Appel, & Miller, 2012). Therefore, ascorbic acid deficiency is associated with numerous disorders, including general weakness, fatigue, muscle and joint pain, lack of appetite lowering immunity, tendency to bruise swollen and bleeding gums. Moreover, chronic ascorbic acid deficiency may also contribute to the development of neoplastic changes and atherosclerosis (Ang, Pullar, Currie, & Vissers, 2018; Langlois, Duprez, Delanghe, De Buyzere, & Clement, 2001). On the other hand, excessive consumption of this compound can be dangerous. It has been reported that ascorbic acid overdosage causes gastrointestinal disturbances, including abdominal pain, nausea, vomiting and diarrhea, and skin rash (Kodentsova, 2014). Moreover, very high doses of ascorbic acid may also contribute to acidification of the urine and, consequently, to the formation of kidney stones (Lamarche, Nair, Peguero, & Courville, 2011). However, ascorbic acid is water-soluble that means it is easily excreted with sweat and urine. Consequently, it is not easy to overdose on this broad-spectrum compound.

2. Chemical structure and bioavailability

Ascorbic acid is an organic compound belonging to the group of unsaturated polyhydroxy alcohols. It is a water-soluble ketolactone, whose

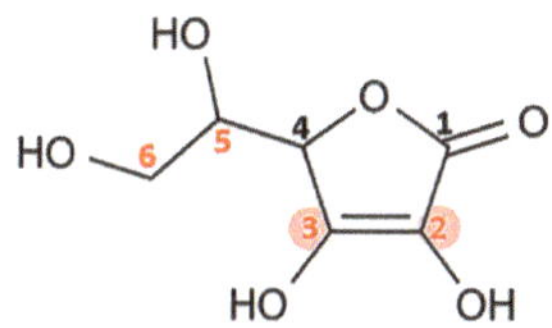

Fig. 1 Chemical structure of ascorbic acid with the indication of the carbon atoms important for the antioxidant properties: C2, C3, C5 and C6 with hydroxyl groups and C2 and C3 with double bonds.

center is formed by a five-membered carbon ring (Fig. 1). Strong reducing properties characterizing ascorbic acid result from the presence of double bonds at the C2 and C3 carbons, as well as four hydroxyl groups in positions C2, C3, C5 and C6. Moreover, due to the proximity of the hydroxyl and carbonyl groups, ascorbic acid is an ideal hydrogen or electron donor that makes it the cofactor of many enzymatic reactions in living organisms (Davey et al., 2000).

Ascorbic acid occurs naturally in the L configuration and it this form that is biologically active. D-ascorbic acid has similar physicochemical properties, but does not act as an enzymatic cofactor and because it is not considered to be a vitamin (Organisciak, Bicknell, & Darrow, 1992). On the other hand, the oxidized form of ascorbic acid, called dehydroascorbic acid, only in the case of L configuration is biologically active (Forman et al., 2020; Hildebrandt, 2011).

In most mammals, ascorbic acid is essential for proper functioning and can be produced in a multistep pathway from glucose. However, in humans, synthesis does not occur due to the lack of L-gulonolactone oxidase. Hence, ascorbic acid must be supplied in the diet (Nishikimi & Yagi, 1991). Citris is the main and most abundant source of ascorbic acid. However, analytical studies show a number of dietary fruits and vegetables, as well as meat can derive this compound (Mazurek & Pankiewicz, 2012; Valente, Albuquerque, Sanches-Silva, & Costa, 2011; Vanderslice, Higgs, Hayes, & Block, 1990). Table 1 presents the content of ascorbic acid in some of the most essential foods.

Ascorbic acid, as a nutrient, is absorbed from the lumen of the intestine mainly by the enterocytes and circulates in the blood throughout the body, enabling it to be taken up by all cells (Levine et al., 1996). Another way of getting ascorbic acid into the body is by its transdermal application (Pinnell et al., 2001). Ascorbic acid can also be reabsorbed by the cells of the renal tubular epithelium into the blood filtered by the kidneys (Rose, 1988).

Table 1 The amount of ascorbic acid in basic food products with its greatest amount (Douglas, Hemilä, Chalker, D'Souza, & Treacy, 2004; García-Closas et al., 2004; Hill, 2018)

Products	Amount [mg] per 100 g
Kakadu plum	5300
Acerola cherries	1600–1700
Wild rose	250–800
Blackcurrant	150–300
Guava	230
Peppers	125–200
Brussels	65–145
Broccoli	65–100
Liver	22–30
Cod	2
Trout	1
Cow's milk	1

As ascorbic acid is soluble in water its transfer through the lipid plasma membrane is hampered. Simple diffusion of ascorbic acid plays only a small role in its transport across the membranes and it is assisted by specific transport systems. The best known mechanisms of ascorbic acid transport are: (I) diffusion through transmembrane channels; (II) facilitated diffusion through exocytosis in secretory vesicles; (III) transmission by glucose-sensitive and -insensitive transporters and (IV) secondary active transport through the sodium-dependent transporters SVCT1/2 (MacDonald, Thumser, & Sharp, 2002; Wilson, 2005). Therefore, the bioavailability of the ascorbic acid to all cellular processes, where it is needed, is suitably provided when a balanced diet is consumed.

3. Antioxidant properties

3.1 Suppressing the generation of free radicals

Ascorbic acid is one of the basic low-molecular antioxidants functioning in the human body. It takes part in the regulation reactive oxygen species

(ROS) levels and the effectiveness of other antioxidants. Ascorbic acid regulates the level of ROS already at the stage of their formation. The main sources of ROS are the mitochondrial respiratory chain and specific enzymes, such as NADPH oxidases (NOXs) or xanthine oxidase (XO). Ascorbic acid has been shown to modify both of the above systems (Fig. 2). XO is an enzyme that generates ROS under the oxidation of hypoxanthine to xanthine and then to uric acid. Both of these reactions are necessary for the functioning of the organism, but the result of this process is the production of hydrogen peroxide and superoxide generation (Hille & Nishino, 1995). It is known that XO can also directly oxidize ascorbic acid (Nishikimi, 1975), however, ascorbic acid supplementation significantly protects an organism against XO hyperactivity (Francis, Anderson, Jenkinson, & Parke, 1989). This antioxidant has no effect on the activity of XO under physiological conditions, as found with skin cells (fibroblasts and keratinocytes). However, in cells under stress caused by, e.g., UV radiation or hydrogen peroxide, the XO hyperactivation is inhibited by treatment with ascorbic acid (Gęgotek, Ambrożewicz, Jastrząb, Jarocka-Karpowicz, & Skrzydlewska, 2019; Gęgotek et al., 2017). Decreases in XO activity induced by ascorbic acid may be useful in preventing or reducing reperfusion injuries and stimulating neutrophils (Dwenger, Funck, Lueken, Schweizer, & Lehmann, 1992).

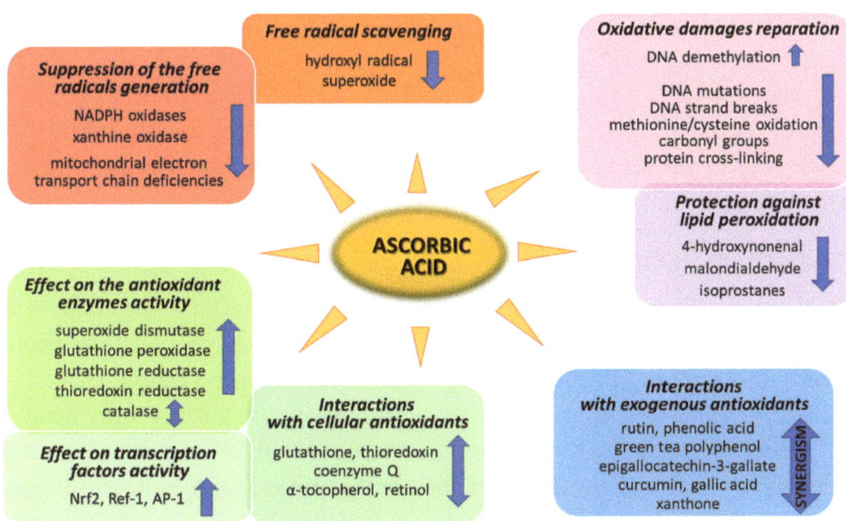

Fig. 2 Pathways of the antioxidant action of ascorbic acid.

NOXs is a group of transmembrane enzymes present in many t cells that, similarly to XO, generate superoxide or hydrogen peroxide as signaling molecules. The controls of these ROS are critical for proper functioning (Brandes, Weissmann, & Schröder, 2014). However, NOX hyperactivity inevitably leads to oxidative stress that can be prevented by ascorbic acid (Cobley, McHardy, Morton, Nikolaidis, & Close, 2015). As described in the case of XO in skin cells, NOX activity is not affected by ascorbic acid in healthy cells. Only strong stress inducers, such as hydrogen peroxide or UVB irradiation, activate NOX strongly enough, that ascorbic acid starts to inhibit this enzyme (Gęgotek, Ambrożewicz, et al., 2019; Gęgotek et al., 2017). Other data showing that ascorbic acid can inhibit NOX in microvascular endothelial cells, suggest that this vitamin may reduce the development of sepsis (Wilson, 2009; Wu, Schuster, Tyml, & Wilson, 2007). However, ascorbic acid shows NOX activating properties in embryonic stem cells, where the NOX-induced increase of ROS levels increases cardiomyogenesis (Bartsch et al., 2011).

Also, mitochondria play an important function in ascorbic acid antioxidant action. On the one hand, due to the Complexes II and III activity, mitochondria regenerate ascorbic acid from its oxidized form, thus maintaining redox status in both the mitochondrial matrix and in the cytoplasm (Li, Cobb, & May, 2002). On the other hand, ascorbic acid favors the sealing of the mitochondrial electron transport chain (Martinovich, Golubeva, Martinovich, & Cherenkevich, 2012) and reduction of superoxide production especially in cells with electron transport chain deficiencies (Sharma & Morgan, 2001). However, the described effect raises concerns as to whether the influence of ascorbic acid on mitochondrial processes does not reduce the elimination of damaged cells by apoptosis and thus promotes carcinogenesis (Bakalova, Zhelev, Miller, Aoki, & Higashi, 2020).

3.2 Ascorbic acid interaction with cellular antioxidants

The first line of natural cell protection against uncontrolled overproduction of ROS is created by low molecular weight antioxidants, such as glutathione (GSH), thioredoxin (Trx), coenzyme Q, α-tocopherol and retinol. Ascorbic acid also belongs to this group, but its antioxidant effect is not only limited to ROS elimination, but also to the interaction with the other mentioned molecules. This small molecule in the center of oxidation–reduction reaction cycle consisting also of tocopherol and GSH/Trx (Fig. 3). Because some of these compounds (GSH/Trx) are not soluble in lipids, they can

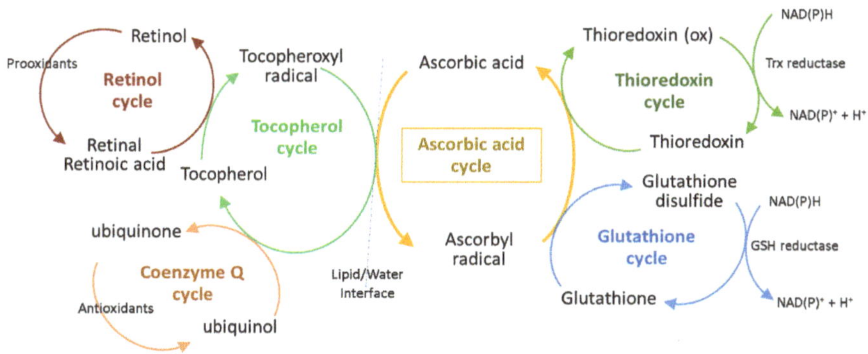

Fig. 3 The scheme of ascorbic acid interaction with glutathione (GSH), thioredoxin (Trx), coenzyme Q, α-tocopherol and retinol (Li, Qu, & May, 2001; Packer, Weber, & Rimbach, 2001; Peake, 2003; Young & Gregoriadis, 1996).

act only in cytoplasm and cannot protect the cell membrane against ROS. Also, ascorbic acid is a hydrophilic molecule, however, it can react over lipid/water interface with tocopherol and its derivatives. Under oxidative conditions tocopherol neutralizes free radicals, which attack cell membrane components, and itself becomes an oxidized form (tocopheroxyl radical). Ascorbic acid from cytoplasm restores the reduced form of tocopherol in the lipid fraction, so that it can further protect cell membranes. This reaction leads to the oxidation of the ascorbic acid to the ascorbyl radical, which is reduced in the cytoplasm by thiol group of GSH or Trx (Li et al., 2001; Packer et al., 2001; Peake, 2003). Furthermore, reduction of tocopherol by ascorbic acid allows it to interact with lipophilic retinol and control its reduced pool (Sünder & Flachowsky, 2001; Young & Gregoriadis, 1996) which provides for maintaining the continuity of the membranes.

Another lipophilic compound—coenzyme Q, which interacts with the tocopheroxyl radical, regenerating its antioxidant form, can also participate in these cycles (Beyer, 1994). Moreover, a lot of data suggest that coenzyme Q antioxidant properties go hand in hand and even complement the attributes of ascorbic acid (Kobori et al., 2014; Nemati, Shahir, Harakinezhad, & Lotfalhian, 2017; Pavlović et al., 2001; Shargorodsky, Debby, Matas, & Zimlichman, 2010). However, so far it is not possible to accurately describe their relationship.

3.3 ROS scavenging by ascorbic acid

Every living organism constantly generates reactive oxygen and nitrogen species (ROS/RNS) which participate in a variety of physiological activities.

However, their level increases significantly in pathological conditions as a result of dysfunction of pro-oxidative systems of cells/biological fluids. In both cases, the ROS should be counterbalanced by an effective natural antioxidant system. Disruption of this balance leads to oxidative stress and, as a consequence, the possibility of ROS interactions with the endogenous components of the organism, such as nucleic acids, proteins, lipids and small molecules, results in irreversible oxidative damage to cells and their components. The living cell's antioxidant system has three lines of defense: (I) free radical scavenging; (II) biosynthesis and activation of antioxidant enzymes; and (III) repair of oxidative damage. Ascorbic acid has been shown as a molecule involved in all these stages (Fig. 2). Its antioxidant properties as a free radicals scavenger are related to its ability to form a stabilized radical (Fig. 4). This allows ascorbic acid to react with more reactive molecules, including hydroxyl radical or superoxide, that prevent their interaction with biomolecules that are important for proper cells functioning (Tariq, 2007). On the other hand, strong antioxidant properties of ascorbic acid can induce transformation of Fe^{3+} into Fe^{2+}. However, the ascorbate–Fe^{2+} chelate may catalyze the ROS generation via Fenton reaction (Tariq, 2007).

3.4 Effect of ascorbic acid on the main antioxidant enzymes activity

To prevent oxidative stress or reduce its destructive effect on cellular compartments, ascorbic acid significantly stimulates the antioxidant system to remove free radicals, not only at the level of low molecular weight antioxidants, but by acting on antioxidant enzymes (Fig. 2). An example of such an enzyme is superoxide dismutase (Cu,Zn-SOD; SOD), which is responsible for superoxide conversion to less cytotoxic hydrogen peroxide. There are no conclusive data about effect of ascorbic acid on SOD activity in cells under physiologic conditions. However, a series of data suggest an essential

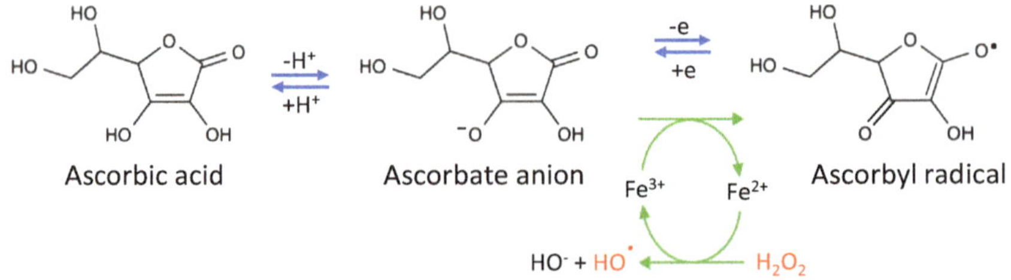

Fig. 4 The scheme of ascorbic acid reaction under oxidative conditions.

function of this vitamin on SOD activity under oxidative stress. Regardless of whether the oxidative stress was caused by heavy metals, UV radiation, brain damage or depression, in all cases the supplementation with ascorbic acid significantly increased the activity of SOD in in vitro cultured cells, as well as plasma of animals (Erdogan, Erdogan, Celik, & Unlu, 2005; Gęgotek, Ambrożewicz, et al., 2019; Gęgotek et al., 2017; Ishaq et al., 2013; May, Qu, & Meredith, 2012; Moretti et al., 2013). Moreover, as a result, this not only leads to a lowering of the level of superoxide but also protects lipids against oxidation observed as decreased level of lipid peroxidation products including malondialdehyde (MDA) (Erdogan et al., 2005; Gęgotek, Ambrożewicz, et al., 2019; Gęgotek et al., 2017). On the other hand, ascorbic acid in SOD depleted cells also reduces the superoxide level preventing oxidative stress (Saffi, Sonego, Varela, & Salvador, 2006; Tamari et al., 2013). However, other studies indicates that oral ascorbic acid supplementation does not significantly affects SOD activity in human plasma (Washio et al., 2008).

The hydrogen peroxide formed in the reaction catalyzed by SOD in the cells is decomposed into oxygen and water by another antioxidant enzyme—catalase (CAT). Ascorbic acid influences the activity of this next enzyme in a various ways depending on the type of cells. In the case of plant cells high concentration of ascorbic acid, being a signal about a high antioxidant potential of the cell, reduces the catalase activity and causes an increase in hydrogen peroxide level (Notulae Botanicae Horti Agrobotanici Cluj-Napoca et al., 2008), allowing this molecule to carrying out a signaling function. Similar effect of ascorbic acid on CAT activity is also observed in rapidly proliferating mammalian cells, especially cancer cells, e.g., gastric cancer cells, glioblastoma and carcinoma cells (Klingelhoeffer et al., 2012; Sun, Zheng, Li, Guo, & Wang, 2006). However, in the case of non-neoplastic cells, such as keratinocytes ascorbic acid significantly enhances CAT activity, thereby increasing the antioxidant potential of these cells which is important due to the peripheral location of keratinocytes in the skin (Gęgotek, Ambrożewicz, et al., 2019). In the same experiment, it has been shown that following UV-induced oxidative stress, ascorbic acid affects catalase activity in different skin cells. In UV irradiated keratinocytes ascorbic acid stimulates CAT activity and protects cells against UV-induced hydrogen peroxide overexpression, while this effect is not observed in UV irradiated skin fibroblasts (Gęgotek, Ambrożewicz, et al., 2019).

Ascorbic acid can act as an antioxidant also by influencing the GSH-based system. Many experiments show that ascorbic acid can prevent disturbances in this system induced by chemical factors or the development of diseases (Moretti et al., 2013; Muthuvel et al., 2006). However, there are two patterns of action of this compound: (I) ascorbic acid significantly increases the level of GSH without affecting or even reducing the activity of enzymes associated with this molecule (glutathione peroxidase, GSH-Px; glutathione reductase, GSSG-R) and (II) ascorbic acid action is based primarily on inducing GSH-Px and GSSG-R activity. It has been found that hepatoprotective and gastroprotective effects of ascorbic acid are based mainly on GSH-Px and GSSG-R activation (Adikwu & Deo, 2013; Koc, Imik, & Odabasoglu, 2008), while in the case of UV-irradiated skin cells ascorbic acid leads to GSH-Px and GSSG-R downregulation with a simultaneous increase in the GSH level (Gęgotek, Ambrożewicz, et al., 2019; Gęgotek et al., 2017). On the other hand, ascorbic acid has such a strong antioxidant effect that can replace GSH-based system observed as decrease in GSH-Px activity and GSH level in bovine semen (Hu et al., 2010).

Another important element for redox balance maintenance in cell is system based on Trx. So far it has been widely shown that oxidized form of ascorbic acid activates Trx reductase (TrxR), resulting in restoration of the reduced form of the ascorbic acid (May & Harrison, 2013). Additionally, it has been noted that ascorbic acid induction of TrxR activity is accompanied by stimulation of GSH-based antioxidant system (Krishnan, Kodrík, Kłudkiewicz, & Sehnal, 2009). In skin cells, ascorbic acid is able to enhance both the Trx level and TrxR activity under physiologic conditions, but also it can prevent UV-induced lowering of these parameters (Gęgotek, Ambrożewicz, et al., 2019). Moreover, ascorbic acid, also by the thioredoxin-interacting protein (TXNIP) stimulation, affects inflammasome functioning, thus reducing ROS production and expression of proinflammatory proteins (interleukins 1β, 18 and caspase-1) in human macrophages (Choe & Kim, 2017). On the other hand, the effect of ascorbic acid on TrxR activity, can be a promising tool in therapies against cancers which are described by TrxR hyperactivity (Pan et al., 2020).

3.5 Effect of ascorbic acid on transcription factors

Another aspect in ascorbic acid action as an antioxidant is its effect on gene expression resulting in the biosynthesis of antioxidant proteins. Among the most important transcriptional factors involved in cellular antioxidant response are Nrf2 (nuclear factor, erythroid 2-like 2), Ref-1 (redox effector factor 1) and AP-1 (activator protein 1) (Fig. 2). Nrf2 is a protein that is common in the cytoplasm regardless of the oxidation–reduction conditions. Under physiological conditions it is associated to its inhibitor Keap1. In oxidative conditions, Keap1 changes conformation and dissociates from Nrf2. The free form of Nrf2 transfers to the nucleus, where it heterodimerizes with sMaf protein and this protein complex can bind to the DNA in a sequence-specific manner—to the antioxidant response element called ARE and starts antioxidant proteins biosynthesis (Gęgotek & Skrzydlewska, 2015; Ma, 2013; Tonelli, Chio, & Tuveson, 2018). Ascorbic acid is known as an activator of Nrf2 factor, as well as the whole Keap1/Nrf2/ARE pathway, and its deficiency leads to the impaired Nrf2 action resulting in inflammation and apoptosis (Xu et al., 2016a, 2016b). It is especially visible in cells under chemically or physically induced oxidative stress (Gęgotek, Ambrożewicz, et al., 2019; Vineetha, Binu, Arathi, & Nair, 2018). This is most important in the case of cells that are constantly exposed to the harmful stressors, such as hepatocytes and keratinocytes (Gęgotek, Ambrożewicz, et al., 2019; Li et al., 2019; Wagner, Ernst, Iori, Desel, & Rimbach, 2010). In keratinocytes ascorbic acid, on the one hand, decreases the level of Nrf2 inhibitor—protein Keap1, and on the other, increases free Nrf2 expression, as well as its activators, including p62 and KAP1 (Gęgotek, Ambrożewicz, et al., 2019; Wagner et al., 2010). At the same time, ascorbic acid favors the Keap1 conformational changes induced by other antioxidants, like polyphenols, which, additionally, stimulates Nrf2 dissociation (Gęgotek, Jarocka-Karpowicz, & Skrzydlewska, 2020). In hepatocytes Nrf2 activation by ascorbic acid results in antioxidant enzymes' expression observed as decreased level of lipid hydroperoxides (Li et al., 2019). On the other hand, some data indicate that high ascorbic acid concentrations may lead to a disturbed Keap1/Nrf2/ARE pathway activation (Wagner et al., 2011). However, according to the dual role of Nrf2 in cancers, its activation by the ascorbic acid can be detrimental (Abbasi et al., 2021; Mostafavi-Pour, Ramezani, Keshavarzi, & Samadi, 2017).

Nrf2 is also known as a factor which activity is strongly related to the transcriptional action of NF-κB (nuclear factor kappa-light-chain-enhancer of activated B cells) (Wardyn, Ponsford, & Sanderson, 2015). Nrf2-induced expression of heme oxygenase 1 inhibits proinflammatory signaling operated by NF-κB (Li et al., 2008), while free Keap1, following Nrf2 dissociation, can additionally amplify this effect, through adducts created with IKKβ (inhibitor of nuclear factor kappa-B kinase subunit beta), which is a positive regulator of NF-κB (Tian et al., 2012). Therefore, ascorbic acid activating Nrf2 contributes significantly to lowering the activity of NF-κB. Moreover, ascorbic acid also directly influences the activity of this molecule by reducing its phosphorylation (Diomede et al., 2019), resulting in blocking the NF-κB/TNFα pathway (Son, Mo, Rhee, & Pyo, 2004). Unfortunately, there are divergent data on the effect of ascorbic acid on the levels of NF-κB in keratinocytes exposed to UV irradiation. Some of the results show that the level of NF-κB is lowered by the use of ascorbic acid (Gęgotek, Ambrożewicz, et al., 2019; Gegotek et al., 2016), while some show an increase in the level of NF-κB (Catani, Savini, Rossi, Melino, & Avigliano, 2005). However, all mentioned results independently show that ascorbic acid increases the survival of UV-irradiated keratinocytes.

The activities of Ref-1 and AP-1 factors are closely related to each other. Ref-1 is a Trx-dependent endonuclease that facilitates AP-1 DNA-binding activity (Xanthoudakis & Curran, 1992), while AP-1 is a heterodimer composed of proteins belonging to the c-Fos, c-Jun, ATF and Maf families and manifests transcription activity by gene expression regulation in response to a variety of stimuli, including cytokines, growth factors and oxidative stress (Hess, Angel, & Schorpp-Kistner, 2004). Only oxidized ascorbic acid indirectly affects Ref-1 activity according to the decrease in Trx level (Krishnan et al., 2009). Ascorbic acid has been found to suppress the action of AP-1 in epidermal keratinocytes, preventing maintenance of these cells when their DNA is oxidatively modified predisposing to the formation of cancer (Huang, Ma, & Dong, 1996). Moreover, in keratinocytes exposed to UV radiation, where AP-1 should be activated, supplementation with ascorbic acid induces a reduced level of the active positive component, c-Jun, and increased levels of fra-1 messenger that is an AP-1 inhibitor (Catani et al., 2005). Similar silencing of AP-1 activity by ascorbic acid is observed in respiratory epithelial cells resulting in decreased levels of signaling molecules, such as pro-inflammatory chemokines (Lee, Lim, & Kim, 2021).

4. Ascorbic acid and oxidative modifications

4.1 Oxidative damages reparation

One of the most dangerous damages in cells occurring under oxidative stress is oxidative DNA modifications, including single-strand breaks and oxidative base damage with 8-oxoguanine (8-oxoG) formation (Lu, Li, Gu, Wright, & Chang, 2001). Repair of both of these accumulated damages involves DNA polymerase β (Lan et al., 2004). There are no clear data on how ascorbic acid affects activity of DNA polymerases, however, the vitamin can favor DNA reparation by demethylation (Gehring, Reik, & Henikoff, 2009). It has been found that ascorbic acid can directly enhance DNA demethylation in a various cell types (Blaschke et al., 2013; Yin et al., 2013). On the other hand, ascorbic acid directly decreased DNA mutations induced by H_2O_2, including 8-oxoG (Lutsenko, Cárcamo, & Golde, 2002), as well as the extent of DNA strand breaks (He & Häder, 2002) (Fig. 2).

ROS also interact with amino acids and lead to the generation of a number of oxidatively modified proteins with impaired functions. The most common protein oxidative modifications are methionine/cysteine oxidation, carbonyl group formation, or cross-linking (Stadtman & Levine, 2006). Some of them are reversible and ascorbic acid as a donor of hydrogen or electron can reduce these proteins and restore their functions (Davey et al., 2000). However, many modifications are irreversible and, in addition, their proteolysis is not possible due to the disruptions in the proteasomes functioning under oxidative stress (Friguet, 2006). The protective properties of ascorbic acid are based on eliminating the action of many proteasome inhibitors (Perrone et al., 2009; Zou et al., 2006) and promoting the removal of oxidatively damaged proteins from the cytoplasm of cells (Fig. 2).

4.2 Prevention lipid peroxidation

The destructive activity of ROS also relates to lipids. Molecules included in the lipid bilayer as well as those non-membrane-forming lipids often undergo non-enzymatic or enzymatic oxidative reactions with the simultaneous formation of lipid peroxidation products. Lipid peroxidation products are reactive aldehydes [e.g., 4-hydroxynonenal (4-HNE) and malondialdehyde (MDA)] or cyclization products, including isoprostanes (Guéraud et al., 2010). These compounds are important signaling molecules in cellular functioning, however, their overproduction under oxidative stress disturbs the continuity of biological membranes, as well as the basic metabolic processes of cells (Ramana, Srivastava, & Singhal, 2013). It has been described that the reduced level of ascorbic acid in cells or plasma inevitably favors the formation of lipid peroxidation products in the whole organism (Marchlewicz et al., 2007; Ranjan, Swarup, Naresh, & Patra, 2005; Surapaneni & Venkataramana, 2007). It is clear that ascorbic acid is important molecule in lipid peroxidation prevention (Adikwu & Deo, 2013; Krishna Mohan & Venkataramana, 2007; Sönmez, Türk, & Yüce, 2005) (Fig. 2). It has been found that ascorbic acid significantly protects cells/organism against increase in 4-HNE and MDA level under stress conditions, including UV-induced oxidative stress in skin cells (Gęgotek, Ambrożewicz, et al., 2019; Gęgotek et al., 2017; Gęgotek, Jarocka-Karpowicz, & Skrzydlewska, 2020) or cornea (Chen et al., 2020), toxic effects of xenobiotics in liver cells (Akbulut et al., 2014; Hernández-Guerra et al., 2006; Raina, Baba, Verma, Sultana, & Singh, 2015; Ray, Roy, & Sengupta, 2006) and erythrocytes (Das et al., 2007), as well as oxidative stress related myonecrosis (Tonon et al., 2012). Ascorbic acid action on the level of isoprostanes, results in arresting the pro-inflammatory effect of these lipid peroxidation products. Ascorbic acid efficiently decreases the expression of isoprostanes in plasma of patients with a wide variety of diseases including diabetes or hypertension (Mason, Rasmussen, van Loon, Salmon, & Wadley, 2019; Sato et al., 2011). Therefore, this molecule through its antioxidant properties, prevents inflammation mediated by lipid peroxidation.

5. Ascorbic acid cooperation with other antioxidants

Despite of the fact that ascorbic acid itself has strong antioxidant properties, it can interact with other exogenous molecules and induce synergistic protective effects (Fig. 2). The compounds with which ascorbic acid interacts include substances used in pharmacotherapy of different diseases, e.g., antibiotics, such as ceftriaxone, gallic acid or xanthone (Abdel-Daim, 2016; Abdel-Daim & El-Ghoneimy, 2015; Naksuriya & Okonogi, 2015). However, the best-studied compounds that are synergistic with ascorbic acid belong to the group of natural polyphenols, especially curcumin, epigallocatechin-3-gallate, phenolic acid and green tea polyphenol (GTP) (Andarwulan et al., 2012; Enko & Gliszczyńska-Świgło, 2015; Li, Shao, Zhu, Zhou, & Xu, 2013; Li, Wu, & Tu, 2010; Luca et al., 2014; Naksuriya & Okonogi, 2015). In the case of the best known combination of ascorbic acid and polyphenol—rutin (often used in oral pharmaceuticals against fever and colds), synergistic action consists in the enhanced antioxidant effect of the particles by restoring their reduced forms. As it has been described following in vitro experiments on UV irradiated skin cells, ascorbic acid and rutin support cytoprotective properties of each other at every possible stage of activity (Gęgotek, Ambrożewicz, et al., 2019; Gęgotek, Domingues, & Skrzydlewska, 2020; Gęgotek, Jarocka-Karpowicz, & Skrzydlewska, 2019; Gęgotek, Jarocka-Karpowicz, & Skrzydlewska, 2020;

Hernández-Herrero & Frutos, 2015). Ascorbic acid, as mentioned before, has many pathways to penetrate through cell membranes, while the bioavailability of rutin is limited. Therefore, ascorbic acid promotes the penetration of rutin into cells, e.g., through bilitranslocase activation (Gęgotek, Ambrożewicz, et al., 2019). Ascorbic acid—induced enhanced level of rutin leads to the effective synergism of these antioxidants in: scavenging of free radicals, as well as activating the cellular antioxidant system (Gęgotek, Ambrożewicz, et al., 2019; Gęgotek, Jastrząb, Dobrzyńska, Biernacki, & Skrzydlewska, 2021). To enhance this antioxidant effect, ascorbic acid and rutin also cooperate in the effective activation of Nrf2, including development of adducts with Keap1 (Gęgotek, Ambrożewicz, et al., 2019; Gęgotek, Jarocka-Karpowicz, & Skrzydlewska, 2020), leading to the more successful protection against UV-induced changes in protein structure (Gęgotek, Jarocka-Karpowicz, & Skrzydlewska, 2019; Gęgotek, Jarocka-Karpowicz, & Skrzydlewska, 2020), as well as the entire proteomic profile (Gęgotek, Domingues, & Skrzydlewska, 2020).

6. Summary

As described in this review, ascorbic acid exhibits its antioxidant properties at various levels of the functioning of living cells: starting from the direct scavenging of free radicals, through the activation of intracellular antioxidant systems, up to supporting the action of other exogenous antioxidants. In this connection, both DNA, protein and lipids are protected against oxidation. Although ascorbic acid has strong antioxidant properties, it can also have pro-oxidant effects in the presence of free transition metals. Moreover, its role in prevention of DNA mutation and cellular apoptosis is controversial, especially in relation to cancer cells.

References

Abbasi, A., Mostafavi-Pour, Z., Amiri, A., Keshavarzi, F., Nejabat, N., Ramezani, F., et al. (2021). Chemoprevention of prostate cancer cells by vitamin C plus quercetin: Role of Nrf2 in inducing oxidative stress. *Nutrition and Cancer*, *73*(10), 2003–2013. https://doi.org/10.1080/01635581.2020.1819346.

Abdel-Daim, M. M. (2016). Synergistic protective role of ceftriaxone and ascorbic acid against subacute diazinon-induced nephrotoxicity in rats. *Cytotechnology*, *68*(2), 279–289. https://doi.org/10.1007/s10616-014-9779-z.

Abdel-Daim, M. M., & El-Ghoneimy, A. (2015). Synergistic protective effects of ceftriaxone and ascorbic acid against subacute deltamethrin-induced nephrotoxicity in rats. *Renal Failure*, *37*(2), 297–304. https://doi.org/10.3109/0886022X.2014.983017.

Adikwu, E., & Deo, O. (2013). Hepatoprotective effect of vitamin C (ascorbic acid). *The Journal of Pharmacy and Pharmacology*, *04*(01), 84–92. https://doi.org/10.4236/pp.2013.41012.

Akbulut, S., Elbe, H., Eris, C., Dogan, Z., Toprak, G., Otan, E., et al. (2014). Cytoprotective effects of amifostine, ascorbic acid and N-acetylcysteine against methotrexate-induced hepatotoxicity in rats. *World Journal of Gastroenterology*, *20*(29), 10158–10165. https://doi.org/10.3748/wjg.v20.i29.10158.

Andarwulan, N., Kurniasih, D., Apriady, R. A., Rahmat, H., Roto, A. V., & Bolling, B. W. (2012). Polyphenols, carotenoids, and ascorbic acid in underutilized medicinal vegetables. *Journal of Functional Foods*, *4*(1), 339–347. https://doi.org/10.1016/j.jff.2012.01.003.

Ang, A., Pullar, J. M., Currie, M. J., & Vissers, M. C. M. (2018). Vitamin C and immune cell function in inflammation and cancer. *Biochemical Society Transactions*, *46*(5), 1147–1159. https://doi.org/10.1042/BST20180169.

Bakalova, R., Zhelev, Z., Miller, T., Aoki, I., & Higashi, T. (2020). Vitamin C versus Cancer: Ascorbic acid radical and impairment of mitochondrial respiration? *Oxidative Medicine and Cellular Longevity*, *2020*. https://doi.org/10.1155/2020/1504048.

Bartsch, C., Bekhite, M. M., Wolheim, A., Richter, M., Ruhe, C., Wissuwa, B., et al. (2011). NADPH oxidase and eNOS control cardiomyogenesis in mouse embryonic stem cells on ascorbic acid treatment. *Free Radical Biology and Medicine*, *51*(2), 432–443. https://doi.org/10.1016/j.freeradbiomed.2011.04.029.

Beyer, R. E. (1994). The role of ascorbate in antioxidant protection of biomembranes: Interaction with vitamin E and coenzyme Q. *Journal of Bioenergetics and Biomembranes*, *26*(4), 349–358. https://doi.org/10.1007/BF00762775.

Blaschke, K., Ebata, K. T., Karimi, M. M., Zepeda-Martínez, J. A., Goyal, P., Mahapatra, S., et al. (2013). Vitamin C induces Tet-dependent DNA demethylation and a blastocyst-like state in ES cells. *Nature*, *500*(7461), 222–226. https://doi.org/10.1038/nature12362.

Brandes, R. P., Weissmann, N., & Schröder, K. (2014). Nox family NADPH oxidases: Molecular mechanisms of activation. *Free Radical Biology and Medicine*, *76*, 208–226. https://doi.org/10.1016/j.freeradbiomed.2014.07.046.

Catani, M. V., Savini, I., Rossi, A., Melino, G., & Avigliano, L. (2005). Biological role of vitamin C in keratinocytes. *Nutrition Reviews*, *63*(3), 81–90. https://doi.org/10.1111/j.1753-4887.2005.tb00125.x.

Chen, W., Guo, J., Guo, H., Kong, X., Bai, J., & Long, P. (2020). Protective effect of vitamin C against infancy rat corneal injury caused by acute UVB irradiation. *BioMed Research International*, *2020*. https://doi.org/10.1155/2020/8089273.

Choe, J. Y., & Kim, S. K. (2017). Quercetin and ascorbic acid suppress fructose-induced NLRP3 Inflammasome activation by blocking intracellular shuttling of TXNIP in human macrophage cell lines. *Inflammation*, *40*(3), 980–994. https://doi.org/10.1007/s10753-017-0542-4.

Cobley, J. N., McHardy, H., Morton, J. P., Nikolaidis, M. G., & Close, G. L. (2015). Influence of vitamin C and vitamin e on redox signaling: Implications for exercise adaptations. *Free Radical Biology and Medicine*, *84*, 65–76. https://doi.org/10.1016/j.freeradbiomed.2015.03.018.

Crisan, D., Roman, I., Crisan, M., Scharffetter-Kochanek, K., & Badea, R. (2015). The role of vitamin C in pushing back the boundaries of skin aging: An ultrasonographic approach. *Clinical, Cosmetic and Investigational Dermatology*, *8*, 463–470. https://doi.org/10.2147/CCID.S84903.

Das, K. K., Gupta, A. D., Dhundasi, S. A., Patil, A. M., Das, S. N., & Ambekar, J. G. (2007). Protective role of L-ascorbic acid on antioxidant defense system in erythrocytes of albino rats exposed to nickel sulfate. *Biometals*, *20*(2), 177–184. https://doi.org/10.1007/s10534-006-9025-z.

Davey, M. W., Van Montagu, M., Inz, D., Sanmartin, M., Kanellis, A., Smirnoff, N., et al. (2000). Plant L-ascorbic acid: Chemistry, function, metabolism, bioavailability and effects of processing. *Journal of the Science of Food and Agriculture*, *80*(7), 825–860. https://doi.org/10.1002/(SICI)1097-0010(20000515)80:7<825::AID-JSFA598>3.0.CO;2-6.

Diomede, F., Marconi, G. D., Guarnieri, S., D'Attilio, M., Cavalcanti, M. F. X. B., Mariggiò, M. A., et al. (2019). A novel role of ascorbic acid in anti-inflammatory pathway and ROS generation in HEMA treated dental pulp stem cells. *Materials*, *13*(1), 130. https://doi.org/10.3390/ma13010130.

Douglas, R., Hemilä, H., Chalker, E., D'Souza, R., & Treacy, B. (2004). Vitamin C for preventing and treating the common cold. In *Cochrane Database of Systematic Reviews* John Wiley & Sons, Ltd. https://doi.org/10.1002/14651858.cd000980.pub2.

Dwenger, A., Funck, M., Lueken, B., Schweizer, G., & Lehmann, U. (1992). Effect of ascorbic acid on neutrophil functions and hypoxanthine/xanthine oxidase-generated, oxygen-derived radicals. *Clinical Chemistry and Laboratory Medicine*, *30*(4), 187–192. https://doi.org/10.1515/cclm.1992.30.4.187.

Enko, J., & Gliszczyńska-Świgło, A. (2015). Influence of the interactions between tea (Camellia sinensis) extracts and ascorbic acid on their antioxidant activity: Analysis with interaction indexes and isobolograms. *Food Additives and Contaminants - Part A Chemistry, Analysis, Control, Exposure and Risk Assessment*, *32*(8), 1234–1242. https://doi.org/10.1080/19440049.2015.1049218.

Erdogan, Z., Erdogan, S., Celik, S., & Unlu, A. (2005). Effects of ascorbic acid on cadmium-induced oxidative stress and performance of broilers. *Biological Trace Element Research*, *104*(1), 19–31. https://doi.org/10.1385/bter:104:1:019.

Forman, K., Martínez, F., Cifuentes, M., Fernández, M., Bertinat, R., Torres, P., et al. (2020). Dehydroascorbic acid, the oxidized form of vitamin C, improves renal histology and function in old mice. *Journal of Cellular Physiology*, *235*(12), 9773–9784. https://doi.org/10.1002/jcp.29791.

Francis, A. J., Anderson, D., Jenkinson, P. C., & Parke, D. V. W. (1989). The protective effects of L-ascorbic acid and DL-α-tocopherol on cultured rat embryos treated with xanthine/xanthine oxidase. *Mutation Research, Fundamental and Molecular Mechanisms of Mutagenesis*, *214*(1), 137–145. https://doi.org/10.1016/0027-5107(89)90207-8.

Friguet, B. (2006, May 22). Oxidized protein degradation and repair in ageing and oxidative stress. *FEBS Letters*. No longer published by Elsevier https://doi.org/10.1016/j.febslet.2006.03.028.

García-Closas, R., Berenguer, A., Tormo, M. J., Sánchez, M. J., Quirós, J. R., Navarro, C., et al. (2004). Dietary sources of vitamin C, vitamin E and specific carotenoids in Spain. *British Journal of Nutrition*, *91*(6), 1005–1011. https://doi.org/10.1079/bjn20041130.

Gęgotek, A., Ambrożewicz, E., Jastrząb, A., Jarocka-Karpowicz, I., & Skrzydlewska, E. (2019). Rutin and ascorbic acid cooperation in antioxidant and antiapoptotic effect on human skin keratinocytes and fibroblasts exposed to UVA and UVB radiation. *Archives of Dermatological Research*, *311*(3), 203–219. https://doi.org/10.1007/s00403-019-01898-w.

Gęgotek, A., Bielawska, K., Biernacki, M., Zaręba, I., Surażyński, A., & Skrzydlewska, E. (2017). Comparison of protective effect of ascorbic acid on redox and endocannabinoid systems interactions in in vitro cultured human skin fibroblasts exposed to UV radiation and hydrogen peroxide. *Archives of Dermatological Research*, *309*(4), 285–303. https://doi.org/10.1007/s00403-017-1729-0.

Gegotek, A., Biernacki, M., Ambrozewicz, E., Surazyński, A., Wroński, A., & Skrzydlewska, E. (2016). The cross-talk between electrophiles, antioxidant defence and the endocannabinoid system in fibroblasts and keratinocytes after UVA and UVB irradiation. *Journal of Dermatological Science*, *81*(2), 107–117. https://doi.org/10.1016/j.jdermsci.2015.11.005.

Gęgotek, A., Domingues, P., & Skrzydlewska, E. (2020). Natural exogenous antioxidant defense against changes in human skin fibroblast proteome disturbed by UVA radiation. *Oxidative Medicine and Cellular Longevity*, *2020*, 3216415. https://doi.org/10.1155/2020/3216415.

Gęgotek, A., Jarocka-Karpowicz, I., & Skrzydlewska, E. (2019). Synergistic cytoprotective effects of rutin and ascorbic acid on the proteomic profile of 3D-cultured keratinocytes exposed to UVA or UVB radiation. *Nutrients*, *11*(11), 2672. https://doi.org/10.3390/nu11112672.

Gęgotek, A., Jarocka-Karpowicz, I., & Skrzydlewska, E. (2020). Cytoprotective effect of ascorbic acid and Rutin against oxidative changes in the proteome of skin fibroblasts cultured in a three-dimensional system. *Nutrients*, *12*(4), 1074. https://doi.org/10.3390/nu12041074.

Gęgotek, A., Jastrząb, A., Dobrzyńska, M., Biernacki, M., & Skrzydlewska, E. (2021). Exogenous antioxidants impact on UV-induced changes in membrane phospholipids and the effectiveness of the endocannabinoid system in human skin cells. *Antioxidants*, *10*(8), 1260. https://doi.org/10.3390/antiox10081260.

Gęgotek, A., & Skrzydlewska, E. (2015). The role of transcription factor Nrf2 in skin cells metabolism. *Archives of Dermatological Research*, *307*(5), 385–396. https://doi.org/10.1007/s00403-015-1554-2.

Gehring, M., Reik, W., & Henikoff, S. (2009). DNA demethylation by DNA repair. *Trends in Genetics*, *25*(2), 82–90. https://doi.org/10.1016/j.tig.2008.12.001.

Guéraud, F., Atalay, M., Bresgen, N., Cipak, A., Eckl, P. M., Huc, L., et al. (2010). Chemistry and biochemistry of lipid peroxidation products. *Free Radical Research*, *44*(10), 1098–1124. https://doi.org/10.3109/10715762.2010.498477.

He, Y. Y., & Häder, D. P. (2002). UV-B-induced formation of reactive oxygen species and oxidative damage of the cyanobacterium Anabaena sp.: Protective effects of ascorbic acid and N-acetyl-L-cysteine. *Journal of Photochemistry and Photobiology B: Biology*, *66*(2), 115–124. https://doi.org/10.1016/S1011-1344(02)00231-2.

Hernández-Guerra, M., García-Pagán, J. C., Turnes, J., Bellot, P., Deulofeu, R., Abraldes, J. G., et al. (2006). Ascorbic acid improves the intrahepatic endothelial dysfunction of patients with cirrhosis and portal hypertension. *Hepatology*, *43*(3), 485–491. https://doi.org/10.1002/hep.21080.

Hernández-Herrero, J. A., & Frutos, M. J. (2015). Influence of rutin and ascorbic acid in colour, plum anthocyanins and antioxidant capacity stability in model juices. *Food Chemistry*, *173*, 495–500. https://doi.org/10.1016/j.foodchem.2014.10.059.

Hess, J., Angel, P., & Schorpp-Kistner, M. (2004). AP-1 subunits: Quarrel and harmony among siblings. *Journal of Cell Science*, *117*(25), 5965–5973. https://doi.org/10.1242/jcs.01589.

Hildebrandt, T. M. (2011). Modulation of sulfide oxidation and toxicity in rat mitochondria by dehydroascorbic acid. *Biochimica et Biophysica Acta - Bioenergetics*, *1807*(9), 1206–1213. https://doi.org/10.1016/j.bbabio.2011.06.003.

Hill, C. H. (2018). *20 Foods That Are High in Vitamin C*. Retrieved March 12, 2022, from https://www.healthline.com/nutrition/vitamin-c-foods#TOC_TITLE_HDR_2.

Hille, R., & Nishino, T. (1995). Xanthine oxidase and xanthine dehydrogenase. *The FASEB Journal*, *9*(11), 995–1003. https://doi.org/10.1096/fasebj.9.11.7649415.

Hu, J. H., Tian, W. Q., Zhao, X. L., Zan, L. S., Wang, H., Li, Q. W., et al. (2010). The cryoprotective effects of ascorbic acid supplementation on bovine semen quality. *Animal Reproduction Science*, *121*(1–2), 72–77. https://doi.org/10.1016/j.anireprosci.2010.04.180.

Huang, C., Ma, W., & Dong, Z. (1996). Inhibitory effects of ascorbic acid on AP-1 activity and transformation of JB6 cells. *International Journal of Oncology*, *8*(2), 389–393. https://doi.org/10.3892/ijo.8.2.389.

Ishaq, G. M., Saidu, Y., Bilbis, L. S., Muhammad, S. A., Jinjir, N., & Shehu, B. B. (2013). Effects of α-tocopherol and ascorbic acid in the severity and management of traumatic brain injury in albino rats. *Journal of Neurosciences in Rural Practice*, *4*(3), 292–297. https://doi.org/10.4103/0976-3147.118784.

Juraschek, S. P., Guallar, E., Appel, L. J., & Miller, E. R. (2012). Effects of vitamin C supplementation on blood pressure: A meta-analysis of randomized controlled trials. *The American Journal of Clinical Nutrition*, *95*(5), 1079–1088. https://doi.org/10.3945/ajcn.111.027995.

Klingelhoeffer, C., Kämmerer, U., Koospal, M., Mühling, B., Schneider, M., Kapp, M., et al. (2012). Natural resistance to ascorbic acid induced oxidative stress is mainly mediated by catalase activity in human cancer cells and catalase-silencing sensitizes to oxidative stress. *BMC Complementary and Alternative Medicine*, *12*(1), 1–10. https://doi.org/10.1186/1472-6882-12-61.

Kobori, Y., Ota, S., Sato, R., Yagi, H., Soh, S., Arai, G., et al. (2014). Antioxidant cosupplementation therapy with vitamin C, vitamin E, and coenzyme Q10 in patients with oligoasthenozoospermia. *Archivio Italiano Di Urologia e Andrologia*, *86*(1), 1. https://doi.org/10.4081/aiua.2014.1.1.

Koc, M., Imik, H., & Odabasoglu, F. (2008). Gastroprotective and anti-oxidative properties of ascorbic acid on indomethacin-induced gastric injuries in rats. *Biological Trace Element Research*, *126*(1–3), 222–236. https://doi.org/10.1007/s12011-008-8205-9.

Kodentsova, V. M. (2014). Gradation in the level of vitamin consumption: Possible risk of excessive consumption. *Voprosy Pitaniia*, *83*(3), 41–51. Retrieved from https://europepmc.org/article/med/25300108.

Krishna Mohan, S., & Venkataramana, G. (2007). Status of lipid peroxidation, glutathione, ascorbic acid, vitamin E and antioxidant enzymes in patients with pregnancy - induced hypertension. *Indian Journal of Physiology and Pharmacology*, *51*(3), 284–288. Retrieved from https://europepmc.org/article/med/18341226.

Krishnan, N., Kodrík, D., Kłudkiewicz, B., & Sehnal, F. (2009). Glutathione-ascorbic acid redox cycle and thioredoxin reductase activity in the digestive tract of Leptinotarsa decemlineata (say). *Insect Biochemistry and Molecular Biology*, *39*(3), 180–188. https://doi.org/10.1016/j.ibmb.2008.11.001.

Lamarche, J., Nair, R., Peguero, A., & Courville, C. (2011). Vitamin C-induced oxalate nephropathy. *International Journal of Nephrology*, *2011*, 1–4. https://doi.org/10.4061/2011/146927.

Lan, L., Nakajima, S., Oohata, Y., Takao, M., Okano, S., Masutani, M., et al. (2004). In situ analysis of repair processes for oxidative DNA damage in mammalian cells. *Proceedings of the National Academy of Sciences of the United States of America*, *101*(38), 13738–13743. https://doi.org/10.1073/pnas.0406048101.

Langlois, M., Duprez, D., Delanghe, J., De Buyzere, M., & Clement, D. L. (2001). Serum vitamin C concentration is low in peripheral arterial disease and is associated with inflammation and severity of atherosclerosis. *Circulation*, *103*(14), 1863–1868. https://doi.org/10.1161/01.CIR.103.14.1863.

Lee, A. J., Lim, J. W., & Kim, H. (2021). Ascorbic acid suppresses house dust mite-induced expression of interleukin-8 in human respiratory epithelial cells. *Journal of Cancer Prevention*, *26*(1), 64–70. https://doi.org/10.15430/jcp.2021.26.1.64.

Levine, M., Conry-Cantilena, C., Wang, Y., Welch, R. W., Washko, P. W., Dhariwal, K. R., et al. (1996). Vitamin C pharmacokinetics in healthy volunteers: Evidence for a recommended dietary allowance. *Proceedings of the National Academy of Sciences of the United States of America*, *93*(8), 3704–3709. https://doi.org/10.1073/pnas.93.8.3704.

Li, X., Cobb, C. E., & May, J. M. (2002). Mitochondrial recycling of ascorbic acid from dehydroascorbic acid: Dependence on the electron transport chain. *Archives of Biochemistry and Biophysics*, *403*, 103–110. Academic Press https://doi.org/10.1016/S0003-9861(02)00205-9.

Li, Y., Darwish, W. S., Chen, Z., Hui, T., Wu, Y., Hirotaka, S., et al. (2019). Identification of lead-produced lipid hydroperoxides in human HepG2 cells and protection using rosmarinic and ascorbic acids with a reference to their regulatory roles on Nrf2-Keap1 antioxidant pathway. *Chemico-Biological Interactions, 314*, 108847. https://doi.org/10.1016/j.cbi.2019.108847.

Li, W., Khor, T. O., Xu, C., Shen, G., Jeong, W. S., Yu, S., et al. (2008). Activation of Nrf2-antioxidant signaling attenuates NFκB-inflammatory response and elicits apoptosis. *Biochemical Pharmacology, 76*(11), 1485–1489. https://doi.org/10.1016/j.bcp.2008.07.017.

Li, X., Qu, Z. C., & May, J. M. (2001). GSH is required to recycle ascorbic acid in cultured liver cell lines. *Antioxidants and Redox Signaling, 3*(6), 1089–1097. https://doi.org/10.1089/152308601317203594.

Li, L., Shao, J., Zhu, X., Zhou, G., & Xu, X. (2013). Effect of plant polyphenols and ascorbic acid on lipid oxidation, residual nitrite and N-nitrosamines formation in dry-cured sausage. *International Journal of Food Science & Technology, 48*(6), 1157–1164. https://doi.org/10.1111/ijfs.12069.

Li, W., Wu, J. X., & Tu, Y. Y. (2010). Synergistic effects of tea polyphenols and ascorbic acid on human lung adenocarcinoma SPC-A-1 cells. *Journal of Zhejiang University. Science. B, 11*(6), 458–464. https://doi.org/10.1631/jzus.B0900355.

Lu, A. L., Li, X., Gu, Y., Wright, P. M., & Chang, D. Y. (2001). Repair of oxidative DNA damage: Mechanisms and functions. *Cell Biochemistry and Biophysics, 35*(2), 141–170. https://doi.org/10.1385/CBB:35:2:141.

Luca, A., Alexa, T., Dondaş, A., Andron, G., Bădescu, M., Alexa, I. D., et al. (2014). Pain modulation by curcumin and ascorbic acid in mice. *Revista Medico-Chirurgicală ̆ a Societă ̆ţii de Medici Ş,i Naturalis̨,ti Din Iaş,i, 118*(2), 346–351. Retrieved from https://europepmc.org/article/med/25076698.

Lutsenko, E. A., Cárcamo, J. M., & Golde, D. W. (2002). Vitamin C prevents DNA mutation induced by oxidative stress. *Journal of Biological Chemistry, 277*(19), 16895–16899. https://doi.org/10.1074/jbc.M201151200.

Ma, Q. (2013). Role of Nrf2 in oxidative stress and toxicity. *Annual Review of Pharmacology and Toxicology, 53*(1), 401–426. https://doi.org/10.1146/annurev-pharmtox-011112-140320.

MacDonald, L., Thumser, A. E., & Sharp, P. (2002). Decreased expression of the vitamin C transporter SVCT1 by ascorbic acid in a human intestinal epithelial cell line. *British Journal of Nutrition, 87*(2), 97–100. https://doi.org/10.1079/bjn2001492.

Marchlewicz, M., Wiszniewska, B., Gonet, B., Baranowska-Bosiacka, I., Safranow, K., Kolasa, A., et al. (2007). Increased lipid peroxidation and ascorbic acid utilization in testis and epididymis of rats chronically exposed to lead. *Biometals, 20*(1), 13–19. https://doi.org/10.1007/s10534-006-9009-z.

Martinovich, G. G., Golubeva, E. N., Martinovich, I. V., & Cherenkevich, S. N. (2012). Redox regulation of calcium signaling in cancer cells by ascorbic acid involving the mitochondrial electron transport chain. *Journal of Biophysics, 2012*. https://doi.org/10.1155/2012/921653.

Mason, S. A., Rasmussen, B., van Loon, L. J. C., Salmon, J., & Wadley, G. D. (2019). Ascorbic acid supplementation improves postprandial glycaemic control and blood pressure in individuals with type 2 diabetes: Findings of a randomized cross-over trial. *Diabetes, Obesity and Metabolism, 21*(3), 674–682. https://doi.org/10.1111/dom.13571.

May, J. M., & Harrison, F. E. (2013). Role of vitamin C in the function of the vascular endothelium. *Antioxidants and Redox Signaling, 19*(17), 2068–2083. https://doi.org/10.1089/ars.2013.5205.

May, J. M., Qu, Z. C., & Meredith, M. E. (2012). Mechanisms of ascorbic acid stimulation of norepinephrine synthesis in neuronal cells. *Biochemical and Biophysical Research Communications, 426*(1), 148–152. https://doi.org/10.1016/j.bbrc.2012.08.054.

Mazurek, A., & Pankiewicz, U. (2012). Changes of dehydroascorbic acid content in relation to total content of vitamin C in selected fruits and vegetables. *Acta Scientiarum Polonorum. Hortorum Cultus*, *6*(11), 169–177. Retrieved from https://www.infona.pl//resource/bwmeta1.element.agro-7e5cbfa9-553c-409e-af59-ef8896c85835.

Moretti, M., Budni, J., Dos Santos, D. B., Antunes, A., Daufenbach, J. F., Manosso, L. M., et al. (2013). Protective effects of ascorbic acid on behavior and oxidative status of restraint-stressed mice. *Journal of Molecular Neuroscience*, *49*(1), 68–79. https://doi.org/10.1007/s12031-012-9892-4.

Moss, R. W. (1989). Free radical: Albert Szent-Gyorgyi and the Battle over vitamin C. *Journal of the History of Biology*, *22*(1). Retrieved from https://philpapers.org/rec/MOSFRA.

Mostafavi-Pour, Z., Ramezani, F., Keshavarzi, F., & Samadi, N. (2017). The role of quercetin and vitamin c in NRF2-dependent oxidative stress production in breast cancer cells. *Oncology Letters*, *13*(3), 1965–1973. https://doi.org/10.3892/ol.2017.5619.

Muthuvel, R., Venkataraman, P., Krishnamoorthy, G., Gunadharini, D. N., Kanagaraj, P., Jone Stanley, A., et al. (2006). Antioxidant effect of ascorbic acid on PCB (Aroclor 1254) induced oxidative stress in hypothalamus of albino rats. *Clinica Chimica Acta*, *365*(1–2), 297–303. https://doi.org/10.1016/j.cca.2005.09.006.

Naksuriya, O., & Okonogi, S. (2015). Comparison and combination effects on antioxidant power of curcumin with gallic acid, ascorbic acid, and xanthone. *Drug Discoveries & Therapeutics*, *9*(2), 136–141. https://doi.org/10.5582/ddt.2015.01013.

Nemati, M., Shahir, M., Harakinezhad, M., & Lotfalhian, H. (2017). Cold-induced ascites in broilers: Effects of vitamin C and coenzyme Q10. *Revista Brasileira de Ciência Avícola*, *19*(3), 537–544. https://doi.org/10.1590/1806-9061-2017-0463.

Nishikimi, M. (1975). Oxidation of ascorbic acid with superoxide anion generated by the xanthine-xanthine oxidase system. *Biochemical and Biophysical Research Communications*, *63*(2), 463–468. https://doi.org/10.1016/0006-291X(75)90710-X.

Nishikimi, M., & Yagi, K. (1991). Molecular basis for the deficiency in humans of gulonolactone oxidase, a key enzyme for ascorbic acid biosynthesis. *American Journal of Clinical Nutrition*, *54*(6 SUPPL), 1203S–1208S. https://doi.org/10.1093/ajcn/54.6.1203s.

Notulae Botanicae Horti Agrobotanici Cluj-Napoca, Dolatabadian, A., & Ali Mohammad Modarres Sanavy, S. (2008). Effect of the ascorbic acid, pyridoxine and hydrogen peroxide treatments on germination, catalase activity, protein and malondialdehyde content of three oil seeds. *Notulae Botanicae Horti Agrobotanici Cluj-Napoca*, *36*(2), 61–66. Retrieved from www.notulaebotanicae.ro.

Organisciak, D. T., Bicknell, I. R., & Darrow, R. M. (1992). The effects of l- and d-ascorbic acid administration on retinal tissue levels and light damage in rats. *Current Eye Research*, *11*(3), 231–241. https://doi.org/10.3109/02713689209001774.

Packer, L., Weber, S. U., & Rimbach, G. (2001). Molecular aspects of α-tocotrienol antioxidant action and cell signalling. *Journal of Nutrition*, *131*, 369S. American Institute of Nutrition https://doi.org/10.1093/jn/131.2.369s.

Pan, S., Yang, J., Ji, S., Li, T., Gao, S., Sun, C., et al. (2020). Cancer therapy by targeting thioredoxin reductase based on selenium-containing dynamic covalent bond. *CCS Chemistry*, *2*(3), 225–235. https://doi.org/10.31635/ccschem.020.202000124.

Pavlović, S. Z., Ognjanović, B. I., Štajn, A.Š., Źikić, R. V., Saičić, Z. S., & Petrović, V. M. (2001). The effect of coenzyme Q10 on blood ascorbic acid, vitamin E, and lipid peroxide in chronic cadmium intoxication. *Journal of Environmental Pathology, Toxicology and Oncology*, *20*(2), 133–140. https://doi.org/10.1615/jenvironpatholtoxicoloncol.v20.i2.80.

Peake, J. M. (2003). Vitamin C: Effects of exercise and requirements with training. *International Journal of Sport Nutrition and Exercise Metabolism*, *13*(2), 125–151. https://doi.org/10.1123/ijsnem.13.2.125.

Perrone, G., Hideshima, T., Ikeda, H., Okawa, Y., Calabrese, E., Gorgun, G., et al. (2009). Ascorbic acid inhibits antitumor activity of bortezomib in vivo. *Leukemia*, *23*(9), 1679–1686. https://doi.org/10.1038/leu.2009.83.

Pinnell, S. R., Yang, H., Omar, M., Riviere, N. M., DeBuys, H. V., Walker, L. C., et al. (2001). Topical L-ascorbic acid: Percutaneous absorption studies. *Dermatologic Surgery*, *27*(2), 137–142. https://doi.org/10.1046/j.1524-4725.2001.00264.x.

Raina, R., Baba, N. A., Verma, P. K., Sultana, M., & Singh, M. (2015). Hepatotoxicity induced by subchronic exposure of fluoride and chlorpyrifos in Wistar rats: Mitigating effect of ascorbic acid. *Biological Trace Element Research*, *166*(2), 157–162. https://doi.org/10.1007/s12011-015-0263-1.

Ramana, K. V., Srivastava, S., & Singhal, S. S. (2013). Lipid peroxidation products in human health and disease. *Oxidative Medicine and Cellular Longevity*, *2013*(583438), 583438. https://doi.org/10.1155/2013/583438.

Ranjan, R., Swarup, D., Naresh, R., & Patra, R. C. (2005). Enhanced erythrocytic lipid peroxides and reduced plasma ascorbic acid, and alteration in blood trace elements level in dairy cows with mastitis. *Veterinary Research Communications*, *29*(1), 27–34. https://doi.org/10.1023/B:VERC.0000046740.59694.5d.

Ray, S., Roy, K., & Sengupta, C. (2006). Cisplatin-induced lipid peroxidation and its inhibition with ascorbic acid. *Indian Journal of Pharmaceutical Sciences*, *68*(2), 199–204. https://doi.org/10.4103/0250-474x.25715.

Rose, R. C. (1988). Transport of ascorbic acid and other water-soluble vitamins. *BBA - Reviews on Biomembranes*. Elsevier https://doi.org/10.1016/0304-4157(88)90014-7.

Saffi, J., Sonego, L., Varela, Q. D., & Salvador, M. (2006). Antioxidant activity of L-ascorbic acid in wild-type and superoxide dismutase deficient strains of Saccharomyces cerevisiae. *Redox Report*, *11*(4), 179–184. https://doi.org/10.1179/135100006X116691.

Sato, K., Dohi, Y., Kojima, M., Miyagawa, K., Takase, H., Katada, E., et al. (2011). Effects of ascorbic acid on ambulatory blood pressure in elderly patients with refractory hypertension. *Arzneimittel-Forschung*, *56*(07), 535–540. https://doi.org/10.1055/s-0031-1296748.

Shargorodsky, M., Debby, O., Matas, Z., & Zimlichman, R. (2010). Effect of long-term treatment with antioxidants (vitamin C, vitamin E, coenzyme Q10 and selenium) on arterial compliance, humoral factors and inflammatory markers in patients with multiple cardiovascular risk factors. *Nutrition and Metabolism*, 7(1), 1–8. https://doi.org/10.1186/1743-7075-7-55.

Sharma, P., & Morgan, P. D. (2001). Ascorbate reduces superoxide production and improves mitochondrial respiratory chain function in human fibroblasts with electron transport chain deficiencies. *Mitochondrion*, *1*(2), 191–198. https://doi.org/10.1016/S1567-7249(01)00016-2.

Son, E. W., Mo, S. J., Rhee, D. K., & Pyo, S. (2004). Vitamin C blocks TNF-α-induced NF-κB activation and ICAM-1 expression in human neuroblastoma cells. *Archives of Pharmacal Research*, *27*(10), 1073–1079. https://doi.org/10.1007/BF02975434.

Sönmez, M., Türk, G., & Yüce, A. (2005). The effect of ascorbic acid supplementation on sperm quality, lipid peroxidation and testosterone levels of male Wistar rats. *Theriogenology*, *63*(7), 2063–2072. https://doi.org/10.1016/j.theriogenology.2004.10.003.

Stadtman, E. R., & Levine, R. L. (2006). Protein oxidation. *Annals of the New York Academy of Sciences*, *899*(1), 191–208. https://doi.org/10.1111/j.1749-6632.2000.tb06187.x.

Ströhle, A., & Hahn, A. (2009). Vitamin C und immunfunktion. *Medizinische Monatsschrift Fur Pharmazeuten*, *32*(2), 49–54. Retrieved from https://europepmc.org/article/med/19263912.

Sun, Y., Zheng, Q., Li, G., Guo, D., & Wang, Z. (2006). Mechanism of ascorbic acid-induced reversion against malignant phenotype in human gastric cancer cells. *Biomedical and Environmental Sciences*, *19*(5), 385–391.

Sünder, A., & Flachowsky, G. (2001). Influence of high vitamin E dosages on retinol and carotinoid concentration in body tissues and eggs of laying hens. *Archives of Animal Nutrition*, *55*(1), 43–52. https://doi.org/10.1080/17450390109386181.

Surapaneni, K., & Venkataramana, G. (2007). Status of lipid peroxidation, glutathione, ascorbic acid, vitamin E and antioxidant enzymes in patients with osteoarthritis. *Indian Journal of Medical Sciences*, *61*(1), 9–14. https://doi.org/10.4103/0019-5359.29592.

Tamari, Y., Nawata, H., Inoue, E., Yoshimura, A., Yoshii, H., Kashino, G., et al. (2013). Protective roles of ascorbic acid in oxidative stress induced by depletion of superoxide dismutase in vertebrate cells. *Free Radical Research*, *47*(1), 1–7. https://doi.org/10.3109/10715762.2012.734916.

Tariq, S. A. (2007). Role of ascorbic acid in scavenging free radicals and lead toxicity from biosystems. *Molecular Biotechnology*, *Vol. 37*, 62–65. https://doi.org/10.1007/s12033-007-0045-x.

Tian, H., Zhang, B. F., Di, J. H., Jiang, G., Chen, F. F., Li, H. Z., et al. (2012). Keap1: One stone kills three birds Nrf2, IKKβ and Bcl-2/Bcl-xL. *Cancer Letters*, *325*(1), 26–34. https://doi.org/10.1016/j.canlet.2012.06.007.

Tonelli, C., Chio, I. I. C., & Tuveson, D. A. (2018). Transcriptional regulation by Nrf2. *Antioxidants & Redox Signaling*, *29*(17), 1727–1745. https://doi.org/10.1089/ars.2017.7342.

Tonon, E., Ferretti, R., Shiratori, J. H., Santo Neto, H., Marques, M. J., & Minatel, E. (2012). Ascorbic acid protects the diaphragm muscle against myonecrosis in mdx mice. *Nutrition*, *28*(6), 686–690. https://doi.org/10.1016/j.nut.2011.09.013.

Valente, A., Albuquerque, T. G., Sanches-Silva, A., & Costa, H. S. (2011). Ascorbic acid content in exotic fruits: A contribution to produce quality data for food composition databases. *Food Research International*, *44*(7), 2237–2242. https://doi.org/10.1016/j.foodres.2011.02.012.

Vanderslice, J. T., Higgs, D. J., Hayes, J. M., & Block, G. (1990). Ascorbic acid and dehydroascorbic acid content of foods-as-eaten. *Journal of Food Composition and Analysis*, *3*(2), 105–118. https://doi.org/10.1016/0889-1575(90)90018-H.

Vineetha, R. C., Binu, P., Arathi, P., & Nair, R. H. (2018). L-ascorbic acid and α-tocopherol attenuate arsenic trioxide-induced toxicity in H9c2 cardiomyocytes by the activation of Nrf2 and Bcl2 transcription factors. *Toxicology Mechanisms and Methods*, *28*(5), 353–360. https://doi.org/10.1080/15376516.2017.1422578.

Wagner, A. E., Boesch-Saadatmandi, C., Breckwoldt, D., Schrader, C., Schmelzer, C., Döring, F., et al. (2011). Ascorbic acid partly antagonizes resveratrol mediated heme oxygenase-1 but not paraoxonase-1 induction in cultured hepatocytes—Role of the redox-regulated transcription factor Nrf2. *BMC Complementary and Alternative Medicine*, *11*(1), 1–8. https://doi.org/10.1186/1472-6882-11-1.

Wagner, A. E., Ernst, I., Iori, R., Desel, C., & Rimbach, G. (2010). Sulforaphane but not ascorbigen, indole-3-carbinole and ascorbic acid activates the transcription factor Nrf2 and induces phase-2 and antioxidant enzymes in human keratinocytes in culture. *Experimental Dermatology*, *19*(2), 137–144. https://doi.org/10.1111/j.1600-0625.2009.00928.x.

Wardyn, J. D., Ponsford, A. H., & Sanderson, C. M. (2015). Dissecting molecular cross-talk between Nrf2 and NF-κB response pathways. *Biochemical Society Transactions*, *43*(4), 621–626. https://doi.org/10.1042/BST20150014.

Washio, K., Inagaki, M., Tsuji, M., Morio, Y., Akiyama, S., Gotoh, H., et al. (2008). Oral vitamin C supplementation in hemodialysis patients and its effect on the plasma level of oxidized ascorbic acid and cu/Zn superoxide dismutase, an oxidative stress marker. *Nephron. Clinical Practice*, *109*(2), c49–c54. https://doi.org/10.1159/000137628.

Wilson, J. X. (2005). Regulation of vitamin C transport. *Annual Review of Nutrition*, *25*(1), 105–125. https://doi.org/10.1146/annurev.nutr.25.050304.092647.

Wilson, J. X. (2009). Mechanism of action of vitamin C in sepsis: Ascorbate modulates redox signaling in endothelium. *BioFactors*, *35*(1), 5–13. https://doi.org/10.1002/biof.7.
Wu, F., Schuster, D. P., Tyml, K., & Wilson, J. X. (2007). Ascorbate inhibits NADPH oxidase subunit p47phox expression in microvascular endothelial cells. *Free Radical Biology and Medicine*, *42*(1), 124–131. https://doi.org/10.1016/j.freeradbiomed.2006.10.033.
Xanthoudakis, S., & Curran, T. (1992). Identification and characterization of Ref-1, a nuclear protein that facilitates AP-1 DNA-binding activity. *The EMBO Journal*, *11*(2), 653–665. https://doi.org/10.1002/j.1460-2075.1992.tb05097.x.
Xu, H. J., Jiang, W. D., Feng, L., Liu, Y., Wu, P., Jiang, J., et al. (2016a). Dietary vitamin C deficiency depressed the gill physical barriers and immune barriers referring to Nrf2, apoptosis, MLCK, NF-κB and TOR signaling in grass carp (Ctenopharyngodon idella) under infection of Flavobacterium columnare. *Fish & Shellfish Immunology*, *58*, 177–192. https://doi.org/10.1016/j.fsi.2016.09.029.
Xu, H. J., Jiang, W. D., Feng, L., Liu, Y., Wu, P., Jiang, J., et al. (2016b). Dietary vitamin C deficiency depresses the growth, head kidney and spleen immunity and structural integrity by regulating NF-κB, TOR, Nrf2, apoptosis and MLCK signaling in young grass carp (Ctenopharyngodon idella). *Fish & Shellfish Immunology*, *52*, 111–138. https://doi.org/10.1016/j.fsi.2016.02.033.
Yin, R., Mao, S. Q., Zhao, B., Chong, Z., Yang, Y., Zhao, C., et al. (2013). Ascorbic acid enhances tet-mediated 5-methylcytosine oxidation and promotes DNA demethylation in mammals. *Journal of the American Chemical Society*, *135*(28), 10396–10403. https://doi.org/10.1021/ja4028346.
Young, A. M., & Gregoriadis, G. (1996). Photolysis of retinol in liposomes and its protection with tocopherol and Oxybenzone. *Photochemistry and Photobiology*, *63*(3), 344–352. https://doi.org/10.1111/j.1751-1097.1996.tb03037.x.
Zou, W., Yue, P., Lin, N., He, M., Zhou, Z., Lonial, S., et al. (2006). Vitamin C inactivates the proteasome inhibitor PS-341 in human cancer cells. *AACR*, *12*(1), 273–280. https://doi.org/10.1158/1078-0432.CCR-05-0503.

CHAPTER TEN

Free radicals, antioxidants, nuclear factor-E2-related factor-2 and liver damage

Erika Ramos-Tovar[a,*] and Pablo Muriel[b,*]
[a]Postgraduate Studies and Research Section, School of Higher Education in Medicine-IPN, Mexico City, Mexico
[b]Laboratory of Experimental Hepatology, Department of Pharmacology, Cinvestav-IPN, Mexico City, Mexico
*Corresponding authors: e-mail address: erikaramost@gmail.com; pmuriel@cinvestav.mx

Contents

Abstract

The liver performs various biochemical and molecular functions. Its location as a portal to blood arriving from the intestines makes it susceptible to several insults, leading to diverse pathologies, including alcoholic liver disease, viral infections, nonalcoholic steatohepatitis, and hepatocellular carcinoma, which are causes of death worldwide. Illuminating the molecular mechanism underlying hepatic injury will provide targets to develop new therapeutic strategies to fight liver maladies. In this regard, reactive oxygen species (ROS) are well-recognized mediators of liver damage. ROS induce nuclear factor-κB and the nucleotide-binding oligomerization domain (NOD)-like receptor protein 3 inflammasome, which are the main proinflammatory signaling pathways that upregulate several proinflammatory and profibrogenic mediators. Additionally,

Vitamins and Hormones, Volume 121
ISSN 0083-6729
https://doi.org/10.1016/bs.vh.2022.09.006

oxygen-derived free radicals induce hepatic stellate cell activation to produce exacerbated quantities of extracellular matrix proteins, leading to fibrosis, cirrhosis and eventually hepatocellular carcinoma. Exogenous and endogenous antioxidants counteract the harmful effects of ROS, preventing liver necroinflammation and fibrogenesis. Therefore, several researchers have demonstrated that the administration of antioxidants, mainly derived from plants, affords beneficial effects on the liver. Notably, nuclear factor-E2-related factor-2 (Nrf2) is a major factor against oxidative stress in the liver. Increasing evidence has demonstrated that Nrf2 plays an important role in liver necroinflammation and fibrogenesis via the induction of antioxidant response element genes. The use of Nrf2 inducers seems to be an interesting approach to prevent/attenuate hepatic disorders, particularly under conditions where ROS play a causative role.

1. Liver disease etiology

Liver disease includes a broad spectrum, including nonalcoholic liver disease (NAFLD), nonalcoholic steatohepatitis (NASH), hepatitis B virus (HBV), hepatitis C virus (HCV), cholestatic diseases and harmful alcohol consumption (11%), which may progress to fibrosis, cirrhosis, and hepatocellular carcinoma (HCC) (Chalasani et al., 2018; Younossi et al., 2016). Histopathological analysis of hepatic tissue samples shows four key features: steatosis, lobular inflammation, hepatocyte ballooning, and fibrosis (Forlano et al., 2020). NAFLD is characterized by $\geq$5% hepatic triglyceride accumulation in large and small vesicles in the absence of any secondary causes, such as significant alcohol consumption (Angulo, 2002; Chalasani et al., 2018; Karanjia et al., 2016; Younossi et al., 2016). In addition to hepatic steatosis, lobular inflammation is the main feature of NASH, with Kupffer cell aggregates, small mononuclear cell clusters, ballooning hepatocytes, and cells with flocculating cytoplasm that may contain Mallory-Denk bodies (Brunt, 2012; Castera, Friedrich-Rust, & Loomba, 2019). NASH is also characterized by encapsulated damaged tissue with collagen scars (Schuppan, Surabattula, & Wang, 2018).

NAFLD severity appears to increase with age, and a higher prevalence of hepatic steatosis is shown in Hispanics than in Caucasians (Browning et al., 2004). More men than women develop liver cirrhosis, and this terminal disease is responsible for more than 1 million deaths annually (Rowe, 2017). The global prevalence of NAFLD is approximately 25% of the adult population (Chalasani et al., 2018). The development of NASH occurs in 40% of patients with NAFLD, and approximately 40–50% of patients with NASH progress to hepatic fibrosis (Ekstedt et al., 2006; Younossi et al., 2016). Currently, cirrhosis of any etiology is the 11th most common cause of death globally, and liver cancer is the 16th leading cause of death (Younossi et al., 2019). A projection to 2030 in liver pathologies indicates that these diseases will continue to increase as a cause of death (Muriel, 2017).

The main causes of liver cirrhosis are NAFLD and alcoholic liver disease (ALD), which have surpassed viral hepatitis (Shah et al., 2019; Wong & Singal, 2020). Obesity, type 2 diabetes mellitus and increased consumption of fat, sucrose and fructose are important risk factors for NAFLD and hepatocellular carcinoma (HCC) (Asrani et al., 2019; Márquez-Quiroga et al., 2022; Muriel, López-Sánchez, & Ramos-Tovar, 2021).

2. Reactive oxygen species (ROS) in liver health and disease

Oxygen is one of the most prevalent elements on Earth and is essential for energy production by aerobic organisms. However, oxygen also initiates the generation of free radicals, producing oxidative stress and damaging the organism's cells and tissues. Oxidative stress is defined as a severe imbalance between ROS production (Table 1) and antioxidant defenses (Halliwell, 2007).

Free radicals are produced in the human body in both healthy and disease states. In the health state, they can arise as regulatory mechanisms or species for intercellular signaling. Several organelles, such as the mitochondria, peroxisomes, and endoplasmic reticulum, generate ROS during physiological processes (Begriche et al., 2013). Additionally, enzymes in the cytosol and membranes produce ROS, particularly hydrogen peroxide (Łukawska-Kuź ma, Podgórska, & Rychter, 2012). ROS modulate several specific signaling pathways that regulate many cellular physiological and pathological processes in diverse human diseases (Zhang et al., 2016). Free radicals can indiscriminately induce the oxidation of lipids and proteins and cause damage to

Table 1 Main reactive oxygen species.

Formula	**Name**
O_2	Superoxide anion
$^{\bullet}OH$	Hydroxyl radical
$RO^{\bullet}$	Alkoxyl radical
$ROO^{\bullet}$	Peroxyl radical
H_2O_2	Hydrogen peroxide
$HOO^{\bullet}$	Hydroperoxyl radical

the DNA chain (Thanan et al., 2014). ROS react with double bonds of polyunsaturated fatty acids, amplify oxidative damage and form derivative molecules, such as malondialdehyde (MDA) and 4-hydroxy-2-nonenal (4-HNE), two major bioactive toxic products and markers of lipid peroxidation (Ayala, Muñoz, & Argüelles, 2014). ROS can add double bonds of DNA bases and abstract an H atom from pyrimidine bases. Thus, oxidative base lesions persisting in DNA can cause mutations leading to cancer (Cooke et al., 2003; Peluso et al., 2020).

Oxidative stress is considered critical in various diseases, including liver pathologies (Muriel, 2009b). The liver performs many biochemical and metabolic processes. It is exposed to xenobiotics through communication with the gastrointestinal tract via the portal system, making it particularly susceptible to ROS. In response, antioxidant systems have been developed in the liver to maintain redox homeostasis; however, excessive ROS production can surpass antioxidant systems and jeopardize liver health (Rolo, Teodoro, & Palmeira, 2012; Wu & Cederbaum, 2009).

3. Antioxidants in liver health and disease

An antioxidant is any substance or action that delays or minimizes free radical exposure. Antioxidants can be synthesized or obtained from the diet (Halliwell, 2007). Fruits and vegetables are rich sources of a wide range of beneficial nutrients and nonnutrients, including antioxidants (Slavin & Lloyd, 2012). According to their mechanism of action, antioxidants can be classified as primary antioxidants, mainly chain breakers, able to scavenge radical species, and secondary antioxidants; singlet oxygen quenchers; peroxide decomposers, metal chelators; oxidative enzyme inhibitors, and UV radiation absorbers (Pisoschi & Pop, 2015). Cells can neutralize oxidative stress using endogenous antioxidants and systems specifically designed to counteract these reactive molecules. Antioxidant defense mechanisms generally control free radical production. Some baseline oxidative damage occurs in all individuals in a stable state (Evans & Halliwell, 1999). Endogenous antioxidant defenses are designed to counteract oxidative damage to proteins, lipids, DNA and other molecules (Gutteridge & Halliwell, 2018). The principal endogenous defenses can be classified as enzymatic and nonenzymatic antioxidants (Arauz, Ramos-Tovar, & Muriel, 2016) (Table 2). The composition of antioxidants varies between tissues and is affected by nutrition and cellular redox states.

Table 2 The most important antioxidant defenses of the human body.

Enzymatic antioxidants	Nonenzymatic antioxidants
Superoxide dismutase	Ascorbic acid
Catalase	α-Tocopherol
Glutathione peroxidase	β-Carotene
Thioredoxin	Glutathione
Peroxiredoxin	Retinol
Glutathione transferase	Uric acid

3.1 Importance of nuclear factor-E2-related factor-2 (Nrf2) in the liver

The first evidence of the Nrf2 signaling pathway was reported in 1991 by Rushmore et al., who found a discrete regulatory element that encoded phase II drug metabolism enzymes called the antioxidant response element (ARE) (Rushmore, Morton, & Pickett, 1991). Itoh et al. reported in 1997 that Nrf2 forms a heterodimer with a small Maf protein (sMaf), and the heterodimer is responsible for binding to the ARE to induce target genes. The binding of Nrf2 with the ARE was confirmed by mutations in the Nrf2 gene, which prevented the induction of the phase II genes (Itoh et al., 1997). The cytoplasmic protein repressor Kelch-like ECH-associated protein-1 (Keap-1) interacts with Cul3, an adaptor protein, and regulates Nrf2 function. Following oxidative stress and other stimuli, the sequestration complex dissociates, and free Nrf2 translocates to the nucleus, binds to the ARE, and induces the transcription of a battery of antioxidant genes, such as thioredoxin (TRX) (Krajka-Kuźniak, Paluszczak, & Baer-Dubowska, 2017). The Nrf2 signaling pathway acts as a first-line defense against cellular damage due to oxidative stress. It is considered the master regulator of the antioxidant response, exhibiting antiinflammatory and cytoprotective properties (Bellezza et al., 2018). Sex differences were found in the expression of xenobiotic genes in the human liver. The transcription of ARE in females induces higher levels of catalase, heme oxygenase 1 (HO-1), and superoxide dismutase 1 compared with males, as assessed by quantitative polymerase chain reaction (Liu et al., 2021). Because of its location, the liver is in charge of first-pass metabolism; therefore, it is exposed to various xenobiotics and is a major multifunctional organ responsible for detoxification and metabolism; consequently, ROS production is inevitable. Therefore, rapid and efficient antioxidant systems are present in the liver, such as the Nrf2 signaling pathway, to counteract the noxious effects of ROS (Ramos-Tovar & Muriel, 2020a, 2020b). A growing body of evidence indicates that the Nrf2 signaling pathway plays a complex, multicellular role during liver inflammation, fibrosis, hepatocarcinogenesis, and regeneration by inducing its target genes (Jaeschke, 2011; Ramos-Tovar & Muriel, 2020a).

3.2 The Nrf2 signaling pathway protects against inflammation and attenuates liver damage

In the liver, activation of the Nrf2 signaling pathway attenuates injury in diverse diseases, including chronic diseases such as viral hepatitis, ALD, NAFLD and NASH. HO-1, which is an essential antioxidant enzyme against oxidative stress by catalyzing the breakdown of prooxidant heme to carbon oxide (CO) and free iron (Bataille & Manautou, 2012; Tang, 2014), leads to the formation of bioactive metabolites, mainly biliverdin and bilirubin, which regulate immune responses (Wu et al., 2019). Additionally, HO-1 induction in rodent models of hepatic inflammation improved liver damage and downregulated proinflammatory cytokine levels (Sass, Barikbin, & Tiegs, 2012). Two of the most important endogenous antioxidants are reduced glutathione (GSH) and TRX, which are upregulated by the Nrf2 signaling pathway (Jaganjac et al., 2020; Xu et al., 2019). The TRX system comprises TRX, TRX reductase, nicotinamide adenine dinucleotide phosphate (NADPH) and TRX interacting protein (TXNIP) (Lu & Holmgren, 2014). Oxidized TRX is reduced by thioredoxin reductase, receiving electrons from NADPH (Fig. 1) (Lu & Holmgren, 2014). TRX contributes to maintaining cellular redox homeostasis and protein function (Bertini et al., 1999).

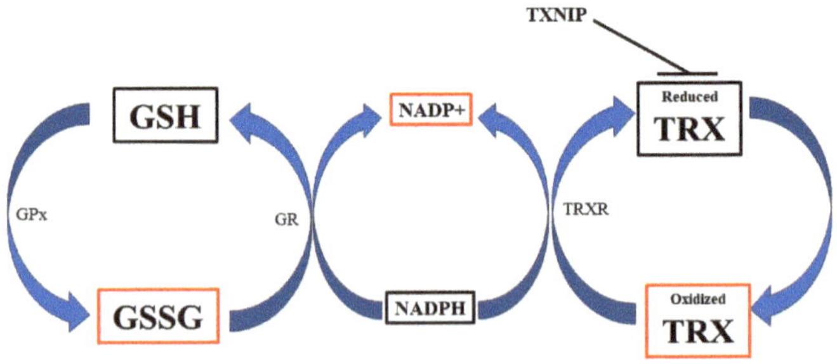

Fig. 1 Thioredoxin (TRX) and reduced glutathione (GSH) preserve the redox homeostasis of the cell. TRX is reduced by thioredoxin reductase (TRXR), which uses nicotinamide adenine dinucleotide phosphate (NADPH) as a source of electrons. NADPH reduces glutathione reductase (GR), producing GSH from the oxidized form (GSSG). GPx: glutathione peroxidase.

The nucleotide-binding oligomerization domain (NOD)-like receptor protein 3 (NLRP3) inflammasome plays a prominent proinflammatory role. It is the principal regulator of interleukin (IL)-1β maturation and release, which triggers the expression of proinflammatory genes by activating nuclear factor-κB (NF-κB), another proinflammatory signaling pathway involved in liver diseases (Muriel, 2009a). Notably, NLRP3 inflammasome activation induces severe liver inflammation and hepatic stellate cell (HSC) activation, which are the main factors responsible for collagen deposition and fibrosis (see below for a description of this process) (Tao et al., 2020; Wree, Eguchi, et al., 2014; Wree, McGeough, et al., 2014). In ROS overproduction, TXNIP, an inhibitor of TRX-1, separates from TRX-1 and activates the NLRP3 inflammasome, exacerbating the proinflammatory process and thereby triggering liver damage (Ramos-Tovar & Muriel, 2020a, 2020b). Notably, Nrf2 signaling pathway activation upregulates TRX-1 and downregulates the NLRP3 inflammasome in an NF-κB-dependent manner (Kabe et al., 2005). Therefore, the Nrf2 signaling pathway plays a significant antiinflammatory role, in addition to its essential antioxidant effect, by downregulating the NLRP3 inflammatory pathway (Fig. 2).

4. Oxidative stress, Nrf2 and liver fibrosis

Liver fibrosis, the abnormal and excessive deposition of scar tissue within the hepatic parenchyma, is a worldwide health problem (Muriel, 2017) that, if unattended, may progress to cirrhosis and HCC (Higashi, Friedman, & Hoshida, 2017). HSCs, when activated, are the main executors of fibrogenesis (Ignat et al., 2020; Roehlen, Crouchet, & Baumert, 2020). Therefore, activation of these cells is the limiting event that triggers extracellular matrix (ECM) protein overproduction that eventually leads to fibrosis (Ignat et al., 2020). Transforming growth factor-β (TGF-β) and other cytokines and factors, mainly produced by Kupffer cells, induce HSC activation (Tsai, Lin, & Huang, 2010). Furthermore, TGF-β is considered the most profibrogenic factor because it mediates the canonical Smad molecular signaling pathway, which comprises the carboxyl terminus phosphorylation of Smad2/3 (Guan et al., 2018; Liu et al., 2019; Xu et al., 2016). A growing body of evidence indicates that oxidative stress plays a significant causative role in fibrogenesis (Catanzaro et al., 2017; Gong & Yang, 2020; Ramos-Tovar & Muriel, 2020b). As previously described, the Nrf2-ARE signaling pathway upregulates several endogenous antioxidants and phase II detoxification enzymes. However, evidence indicates that

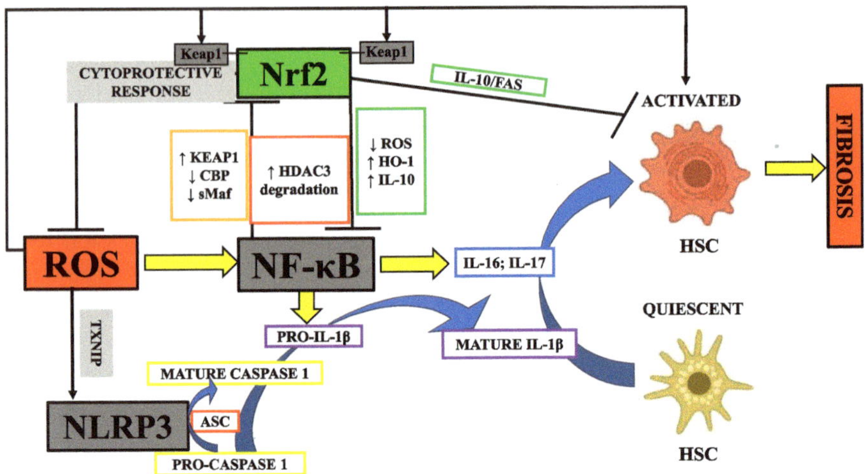

Fig. 2 Nuclear factor-E2-related factor-2 (Nrf2), nuclear factor-κB (NF-κB), and NOD-like receptor protein 3 (NLRP3) inflammasome crosstalk regulates liver fibrosis. Reactive oxygen species (ROS) can directly activate hepatic stellate cells (HSCs), producing large amounts of extracellular matrix proteins leading to liver fibrosis. Additionally, ROS induce the NF-κB signaling pathway to produce proinflammatory and profibrogenic cytokines, including interleukin (IL)-6 and IL-17, which are effective activators of HSCs, and release pro-IL-1β. ROS activate NLRP3 inflammasome via TXNIP. Next, apoptosis-associated speck-like protein containing a caspase recruitment domain (ASC) helps to produce mature caspase 1 from pro-caspase 1. Mature caspase 1 then transforms pro-IL-1β to mature IL-1β, which activates HSCs, promoting fibrosis. ROS are activators of the Nrf2 system by dissociating Kelch-like ECH-associated protein 1 (Keap1) from Nrf2, allowing it to translocate to the nucleus to upregulate target genes. Nrf2 and NF-κB can downregulate each other through several signaling molecules, as specified in the figure. Nrf2 may block the activation of HSCs by an IL-10/FAS-dependent mechanism. CBP: CREB-binding protein; sMaf: small Maf; HDAC3: histone deacetylase 3; HO-1: heme oxygenase-1.

the Nrf2-ARE signaling pathway may play a protective role in liver diseases, including fibrosis (Ramos-Tovar & Muriel, 2020a, 2020b). Few reports have explored the crosstalk between the Nrf2-ARE and TGF-β1/Smad3 signaling pathways. Prestigiacomo and Suter-Dick showed that Nrf2 mediates Smad7 elevation, leading to TGF-β inhibition (Prestigiacomo & Suter-Dick, 2018), but researchers must be encouraged to investigate this effect because of its implication in uncovering new therapeutic targets and markers of fibrogenesis. In this part of the chapter, we illuminate the molecular mechanisms involved in the effects of oxidative stress and the Nrf2-regulated antioxidant response on hepatic fibrogenesis.

4.1 Oxidative stress promotes liver fibrogenesis

Independent of the etiology, alcohol, viruses, obesity, or cholestasis, liver fibrosis is a wound-healing response characterized by the exacerbated production of ECM proteins (scar tissue and connective tissue), mainly type I collagen (Higashi et al., 2017; Ignat et al., 2020; Roehlen et al., 2020). If unattended, hepatic fibrosis progresses to cirrhosis, which affects 1% of the world population and is associated with elevated mortality risk (Higashi et al., 2017). Once activated by chronic hepatic injury, HSCs transdifferentiate into α-smooth muscle actin (α-SMA)-positive myofibroblast (MFB)-like cells with highly proliferative properties that produce large quantities of ECM proteins. Activation of HSCs is achieved by many paracrine factors, such as platelet-derived growth factor (PDGF), connective tissue growth factor (CTGF), TGF-β, and notably ROS(Ramos-Tovar & Muriel, 2020a, 2020b) (Fig. 3).

ROS endogenously produced by liver metabolism, particularly in mitochondria and CYP450 reactions, disrupt hepatic cell homeostasis and

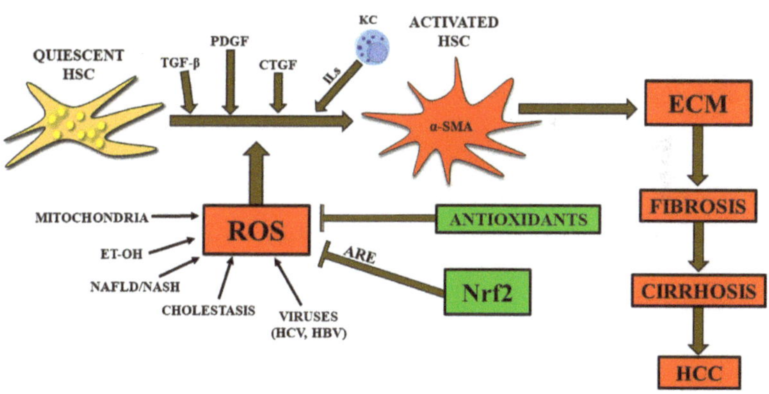

Fig. 3 Activation of hepatic stellate cells (HSCs) is the limiting step toward fibrosis. Quiescent HSCs can be activated by several factors and cytokines, including transforming growth factor-β (TGF-β), platelet-derived growth factor (PDGF), connective tissue growth factor (CTGF), and several interleukins (ILs) produced and released by Kupffer cells (KCs). Notably, reactive oxygen species (ROS), produced by mitochondrial metabolism, alcohol (ET-OH) consumption, nonalcoholic fatty liver disease (NAFLD)/nonalcoholic steatohepatitis (NASH), cholestasis, hepatic C virus (HCV), and hepatic B virus (HBV), effectively induce HSC activation. Once activated, HSCs lose lipid droplets, express α-smooth muscle actin (α-SMA) and produce large quantities of extracellular matrix (ECM) proteins, leading to fibrosis, cirrhosis, and ultimately hepatocellular carcinoma (HCC). Antioxidants and the nuclear factor-E2-related factor-2 (Nrf2)-antioxidant response element (ARE) signaling pathway decrease ROS, thereby inhibiting HSC activation and fibrogenesis.

presumably trigger fibrogenesis (Sanchez-Valle et al., 2012). Kupffer cells, the resident macrophages of the liver, respond to insults by producing ROS, which trigger HSC activation to produce exacerbated amounts of ECM proteins, leading to fibrosis and disruption of the hepatic parenchyma structure and function (Gong & Yang, 2020). Additionally, activated HSCs phagocytize apoptotic bodies to increase NADPH, leading to the formation of more free radicals, further increasing liver damage and fibrogenesis (Gong & Yang, 2020). Another mechanism by which ROS may induce fibrogenesis is by interacting with HSCs, producing DNA methylation, histone disruption, and gene silencing (Li et al., 2016). Therefore, ROS act as a profibrogenic mediator induced by alcohol, obesity (NAFLD/NASH), cholestasis or chronic viral infections or as a conjoint pathological mechanism, contributing to both the onset and progression of liver fibrosis (Gupta et al., 2014; Li et al., 2015; Luangmonkong et al., 2018). Consequently, attenuation of oxidative stress by several means has shown antifibrotic effects, as reported by researchers (Gebhardt, 2002; Parola & Robino, 2001; Tsai et al., 2010). The antioxidant and protective effects of Nrf2 can be induced by the genetic engineering of natural and synthetic compounds, leading to protection of the liver from various insults (Fig. 3) (Keum & Choi, 2014; Magesh, Chen, & Hu, 2012).

4.2 TGF-β/Smads signaling pathway of fibrosis

TGF-β is a well-recognized activator of HSCs and is considered the most profibrogenic factor in the liver. Smad proteins are important effectors of TGF-β-induced hepatic fibrogenesis. The canonical pathway involves TGF-β binding to type II TGF-β receptor (TBRII), activating its kinase activity, to phosphorylate TBRI, which phosphorylates Smad2 and Smad3 in their carboxyl terminus. Next, pSmadC2 or pSmadC3 binds to Smad4, forming a complex capable of translocating into the nucleus, upregulating profibrogenic genes and inhibitory Smad7, which serves as a negative regulator of fibrogenesis (Conidi et al., 2011; Xie et al., 2014). Both profibrogenic Smads 2 and 3 and inhibitory Smad7 are required to modulate fibrogenesis. Otherwise, excessive accumulation of scar tissue may drive fibrosis, cirrhosis and HCC (Tao & Sampath, 2010). Smad3 is the most profibrogenic protein in the TGF-β pathway that drives fibrogenesis in the liver (Zhangdi et al., 2019). Many components of the ECM, α-SMA and tissue inhibitors of metalloproteinase-1 (TIMP-1), which prevents ECM degradation, are upregulated by Smad3 (Latella et al., 2009; Masszi & Kapus, 2011). By contrast, Smad2 blocks the Smad3 pathway, providing another negative regulator of fibrogenesis (Massague, 2000; Tsuchida et al., 2003).

4.3 Increasing the Nrf2-ARE signaling pathway may attenuate oxidative stress and hepatic fibrosis

Prolonged elevated ROS formation may saturate antioxidant defenses, leading to hepatic necrosis, inflammation, and fibrosis. Therefore, researchers have focused on enhancing endogenous antioxidant systems or administering exogenous antioxidants against liver damage and progression to fibrosis (Casas-Grajales & Muriel, 2017; Girish & Pradhan, 2017; Ramos-Tovar & Muriel, 2020b). Additionally, abundant evidence exists that antioxidants from plants that increase the Nrf2-ARE signaling pathway provide beneficial effects in experimental models of liver injury and fibrosis (Hernández-Aquino et al., 2017, 2019; Ramos-Tovar, Casas-Grajales, et al., 2019; Ramos-Tovar, Flores-Beltrán, et al., 2018, 2019; Ramos-Tovar, Hernández-Aquino, et al., 2018). Furthermore, several phytodrugs and active components from plants, including quercetin, diallyl disulfide, sulforaphane, and curcumin, have been demonstrated to attenuate hepatic maladies (Shimozono et al., 2013; Shin et al., 2009; Yu et al., 2011). Sulforaphane, which is found in cruciferous vegetables, is a well-known Nrf2 activator that acts by disrupting cysteine residues in Keap-1 and attenuates ethanol-induced hepatic steatosis in in vivo and in vitro models (Wu, Liu, & Klaassen, 2012; Zhou, Zhang, & Qiao, 2014). Nrf2-knockout mice chronically treated with CCl_4 showed decreased antioxidant and protective enzymes and enhanced hepatic fatty accumulation, inflammation, and fibrosis compared with control littermates (Khor et al., 2006; Wang et al., 2008; Xu et al., 2019), evidencing the protective role of the Nrf2-ARE signaling pathway. A crucial Nrf2-ARE-regulated gene is NQO1, which exerts a cytoprotective effect by scavenging superoxide radicals, catalyzing reductive metabolism of chemicals, and maintaining the redox balance (Itoh et al., 1997; Walters, Cho, & Kleeberger, 2008). GSH, critical endogenous antioxidant in the liver, is in charge of preserving hepatic redox and protects against various hepatotoxicants and diverse liver pathologies (Fig. 1) (Pall & Levine, 2015). The synthesis of GSH depends on the pool of cysteine, and the GCL enzyme, comprising a catalytic (GCLC) and modifier (GCLM) subunit (Zhou et al., 2014), is upregulated by the Nrf2-ARE signaling pathway (Aleksunes & Manautou, 2007). Therefore, increasing knowledge indicates that the Nrf2-ARE signaling pathway

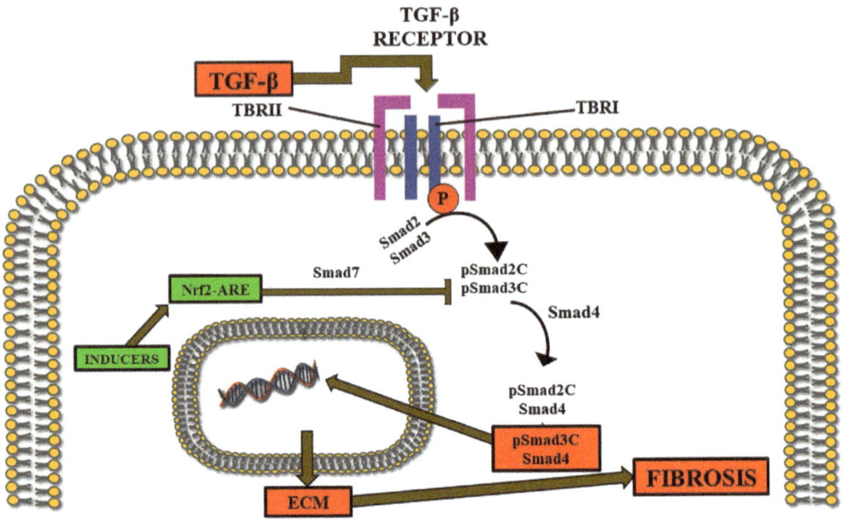

Fig. 4 The nuclear factor-E2-related factor-2 (Nrf2)-antioxidant response element (ARE) signaling pathway blocks the canonical transforming growth factor-β (TGF-β) fibrogenic pathway. TGF-β binds to its receptor in the plasma membrane of hepatic stellate cells, activating the TGF-β receptor I (TBRI) subunit, which phosphorylates Smad2 or Smad3 in the carboxyl terminus, forming pSmad2C or pSmad3C, respectively. pSmad3C binds to Smad4 and translocates to the nucleus to induce extracellular matrix (ECM) genes, leading to fibrosis. The Nrf2-ARE signaling pathway upregulates Smad7, which blocks the phosphorylation of Smad2/3, thereby inhibiting ECM production and fibrosis.

protects against liver damage and provides a novel strategy to attenuate hepatic injury-driven fibrosis (Fig. 4).

However, studies in liver cancer indicate that an increased Nrf2-ARE signaling pathway may protect cancer cells from the cytotoxic effects of anticancer drugs and increase malignant cell survival by upregulating antioxidant genes (Lee et al., 2005; Nguyen, Sherratt, & Pickett, 2003; Niture, Khatri, & Jaiswal, 2014; Yates & Kensler, 2007).

Lipid accumulation and adipogenesis play essential roles in the development of liver diseases. Lipid metabolism is regulated by peroxisome proliferator-activated receptor (PPAR)-γ, and Nrf2 decreases PPAR-γ in a leptin-deficient model (Xu et al., 2012). Activation of the Nrf2 pathway improves insulin resistance and prevents fatty liver and dyslipidemia in an experimental model (Chartoumpekis et al., 2018). Thus, activation of Nrf2 attenuates NASH (Sharma et al., 2018). This broad protective effect of the Nrf2 pathway also encompasses its improvement in carcinogenesis.

As previously mentioned, evidence indicates that activation of Nrf2 by natural compounds ameliorates liver fibrosis, underlying the antifibrotic properties of Nrf2 target genes (Liu et al., 2010; Surh, Kundu, & Na, 2008; Wang et al., 2017), and some studies aimed to investigate the molecular mechanisms involved in this effect. Several signaling pathways play a role in HSC activation, such as TGF-β, Smads and free radicals (Krithika, Jyothilakshmi, & Verma, 2016; Lei et al., 2016). TGF-β1/Smad3 is the most important signaling pathway of scar tissue production (Matsuzaki, 2012; Sekimoto et al., 2007; Yamaguchi et al., 2013); notably, Nrf2 crosstalks with TGF-β to attenuate fibrogenesis (Guan et al., 2018; Latella, 2018; Michaeloudes et al., 2011; Nakamura et al., 2011). Evidence from diverse researchers has shown the interaction of the Nrf2 and TGF-β pathways (Fig. 4). The protective effect of rosuvastatin is associated with crosstalk between the TGF-β/Smad and Nrf2-ARE signaling pathways during free radical-induced myocardial damage (Wang et al., 2018). Sulforaphane, which is an Nrf2 activator, attenuates mouse dystrophic muscle fibrosis by decreasing the phosphorylation of Smad2 and Smad3 and downregulating TGF-β in an Nrf2-dependent manner (Sun, Li, & Li, 2016). Testosterone propionate decreased kidney fibrogenesis by modulating oxidative stress by activating the Nrf2 signaling pathway and decreasing the TGF-β/Smad3 pathway (Rui et al., 2016). Additionally, increasing Nrf2 activity by idazoxan inhibited liver fibrogenesis in vivo and reduced fibrotic tissue HSC line (LX2 cells) production by attenuating the TGF-β/Smad signaling pathway (Xuanfei et al., 2017). Schisandrin B decreased CCl_4-induced fibrogenesis by attenuating ROS production and modulating the Nrf2 and TGF-β/Smad signaling pathways (Chen et al., 2017). We have investigated the effect of several plants and compounds derived from plants on liver fibrosis induced by CCl_4 or thioacetamide intoxication and found that the antifibrotic effect was associated with both the attenuation of oxidative stress and upregulation of the Nrf2 signaling pathway (Casas-Grajales et al., 2019; Hernández-Aquino & Muriel, 2018; Hernández-Aquino et al., 2020; Ramos-Tovar, Buendia-Montaño, et al., 2019; Ramos-Tovar, Casas-Grajales, et al., 2019; Ramos-Tovar, Flores-Beltrán, et al., 2018, 2019; Ramos-Tovar, Hernández-Aquino, et al., 2018; Ramos-Tovar & Muriel, 2019; Vargas-Pozada & Muriel, 2020).

5. Conclusions and future directions

Although ROS may play a signaling role under physiological concentrations, excessive production of these radicals alters cell redox homeostasis, surpassing the cell's antioxidant capacity, leading to structural and functional disruption of the tissue. ROS induce liver inflammation, fibrosis, cirrhosis, and HCC. Free radical overproduction causes activation of HSCs and the

TGF-β-Smad3 fibrogenic pathway, thereby exacerbating ECM production, which is conducive to fibrosis. Additionally, the most important proinflammatory pathways, the NF-κB and NLP3 inflammasome pathways in the liver, are induced by ROS with the consequent increase in proinflammatory cytokine production, leading to cell death and necroinflammation that in turn further promotes fibrogenesis. Naturally, endogenous and exogenous antioxidants play important protective roles. In this context, the master regulator of the endogenous antioxidant response, the Nrf2-ARE signaling pathway, appears to be an attractive target to fight the harmful effects of ROS. Indeed, several researchers have demonstrated that compounds that induce the Nrf2-ARE system afford important protection against experimentally induced liver damage. However, further studies are required to better understand the mechanisms by which ROS induce necrosis and fibrosis and, importantly, the usefulness of antioxidants and Nrf2-ARE inducers as a therapy before recommending these treatments in humans. More studies are needed to clarify the role of ROS in hepatic maladies, particularly in liver cancer, where there is controversy regarding the beneficial or harmful effect of free radicals. The development of new molecules that can induce the Nrf2-ARE signaling pathway seems critical to prevent/attenuate the progression of liver diseases where oxidative stress plays a causative role. Additionally, human trials are required to evaluate the putative benefits of antioxidants and upregulation of Nrf2 in liver diseases.

Acknowledgments

E.R.-T. thanks the postdoctoral scholarship from Consejo Nacional de Ciencia y Tecnología.

Funding

This work was financially supported by Consejo Nacional de Ciencia y Tecnología through the Ciencia de Frontera funding program, grant number 53358.

References

Aleksunes, L. M., & Manautou, J. E. (2007). Emerging role of Nrf2 in protecting against hepatic and gastrointestinal disease. *Toxicologic Pathology*, *35*(4), 459–473. https://doi.org/10.1080/01926230701311344.

Angulo, P. (2002). Nonalcoholic fatty liver disease. *The New England Journal of Medicine*, *346*(16), 1221–1231. https://doi.org/10.1056/NEJMra011775.

Arauz, J., Ramos-Tovar, E., & Muriel, P. (2016). Redox state and methods to evaluate oxidative stress in liver damage: From bench to bedside. *Annals of Hepatology*, *15*(2), 160–173. https://doi.org/10.5604/16652681.1193701.

Asrani, S. K., et al. (2019). Burden of liver diseases in the world. *Journal of Hepatology*, *70*(1), 151–171. https://doi.org/10.1016/j.jhep.2018.09.014.

Ayala, A., Muñoz, M. F., & Argüelles, S. (2014). Lipid peroxidation: Production, metabolism, and signaling mechanisms of malondialdehyde and 4-hydroxy-2-nonenal. *Oxidative Medicine and Cellular Longevity, 2014*, 31. https://doi.org/10.1155/2014/360438.

Bataille, A. M., & Manautou, J. E. (2012). Nrf2: A potential target for new therapeutics in liver disease. *Clinical Pharmacology & Therapeutics, 92*(3), 340–348. https://doi.org/10.1038/clpt.2012.110.

Begriche, K., et al. (2013). Mitochondrial adaptations and dysfunctions in nonalcoholic fatty liver disease. *Hepatology, 58*(4), 1497–1507. https://doi.org/10.1002/hep.26226.

Bellezza, I., et al. (2018). Nrf2-Keap1 signaling in oxidative and reductive stress. *Biochimica et Biophysica Acta (BBA)—Molecular Cell Research, 1865*(5), 721–733. https://doi.org/10.1016/j.bbamcr.2018.02.010.

Bertini, R., et al. (1999). Thioredoxin, a redox enzyme released in infection and inflammation, is a unique chemoattractant for neutrophils, monocytes, and T cells. *Journal of Experimental Medicine, 189*(11), 1783–1789. https://doi.org/10.1084/jem.189.11.1783.

Browning, J. D., et al. (2004). Prevalence of hepatic steatosis in an urban population in the United States: Impact of ethnicity. *Hepatology, 40*(6), 1387–1395. https://doi.org/10.1002/hep.20466.

Brunt, E. M. (2012). Histological assessment of nonalcoholic fatty liver disease in adults and children. *Clinical Liver Disease, 1*(4), 108–111. https://doi.org/10.1002/cld.31.

Casas-Grajales, S., & Muriel, P. (2017). The liver, oxidative stress and antioxidants. In P. Muriel (Ed.), *Liver pathophysiology: Therapies and antioxidants* (pp. 583–604). Waltham, MA: Elsevier. https://doi.org/10.1016/B978-0-12-804274-8.00043-6.

Casas-Grajales, S., et al. (2019). Stevioside inhibits experimental fibrosis by down-regulating profibrotic Smad pathways and blocking hepatic stellate cell activation. *Basic and Clinical Pharmacology and Toxicology, 124*(6), 670–680. https://doi.org/10.1111/bcpt.13194.

Castera, L., Friedrich-Rust, M., & Loomba, R. (2019). Noninvasive assessment of liver disease in patients with nonalcoholic fatty liver disease. *Gastroenterology, 156*(5), 1264–1281. https://doi.org/10.1053/j.gastro.2018.12.036.

Catanzaro, E., et al. (2017). Nrf2: A potential therapeutic target for naturally occurring anticancer drugs? *Expert Opinion on Therapeutic Targets, 21*(8), 781–793. https://doi.org/10.1080/14728222.2017.1351549.

Chalasani, N., et al. (2018). The diagnosis and management of nonalcoholic fatty liver disease: Practice guidance from the American Association for the Study of Liver Diseases. *Hepatology, 67*(1), 328–357. https://doi.org/10.1002/hep.29367.

Chartoumpekis, D. V., et al. (2018). Nrf2 prevents Notch-induced insulin resistance and tumorigenesis in mice. *JCI Insight, 3*(5). https://doi.org/10.1172/jci.insight.97735.

Chen, Q., et al. (2017). Schisandrin B attenuates CCl4-induced liver fibrosis in rats by regulation of Nrf2-ARE and TGF-β/Smad signaling pathways. *Drug Design, Development and Therapy, 11*, 2179–2191. https://doi.org/10.2147/DDDT.S137507.

Conidi, A., et al. (2011). Few Smad proteins and many Smad-interacting proteins yield multiple functions and action modes in TGFβ/BMP signaling in vivo. *Cytokine & Growth Factor Reviews, 22*(5–6), 287–300. https://doi.org/10.1016/j.cytogfr.2011.11.006.

Cooke, M. S., et al. (2003). Oxidative DNA damage: Mechanisms, mutation, and disease. *The FASEB Journal, 17*(10), 1195–1214. https://doi.org/10.1096/fj.02-0752rev.

Ekstedt, M., et al. (2006). Long-term follow-up of patients with NAFLD and elevated liver enzymes. *Hepatology, 44*(4), 865–873. https://doi.org/10.1002/hep.21327.

Evans, P., & Halliwell, B. (1999). Free radicals and hearing: Cause, consequence, and criteria. *Annals of the New York Academy of Sciences, 884*(1), 19–40. https://doi.org/10.1111/j.1749-6632.1999.tb08633.x.

Forlano, R., et al. (2020). High-throughput, machine learning–based quantification of steatosis, inflammation, ballooning, and fibrosis in biopsies from patients with nonalcoholic fatty liver disease. *Clinical Gastroenterology and Hepatology, 18*(9), 2081–2090. https://doi.org/10.1016/j.cgh.2019.12.025.

Gebhardt, R. (2002). Oxidative stress, plant-derived antioxidants and liver fibrosis. *Planta Medica, 68*(4), 289–296. https://doi.org/10.1055/s-2002-26761.

Girish, C., & Pradhan, S. C. (2017). Herbal drugs on the liver. In P. Muriel (Ed.), *Liver pathophysiology: Therapies and antioxidants* (pp. 605–620). Waltham, MA: Elsevier.

Gong, Y., & Yang, Y. (2020). Activation of Nrf2/AREs-mediated antioxidant signalling, and suppression of profibrotic TGF-β1/Smad3 pathway: A promising therapeutic strategy for hepatic fibrosis—A review. *Life Sciences, 256*, 117909. https://doi.org/10.1016/j.lfs.2020.117909.

Guan, Y., et al. (2018). NF-E2-related factor 2 suppresses intestinal fibrosis by inhibiting reactive oxygen species-dependent TGF-β1/SMADs pathway. *Digestive Diseases and Sciences, 63*(2), 366–380. https://doi.org/10.1007/s10620-017-4710-z.

Gupta, R. K., et al. (2014). Oxidative stress and antioxidants in disease and cancer: A review. *Asian Pacific Journal of Cancer Prevention, 15*(11), 4405–4409. https://doi.org/10.7314/APJCP.2014.15.11.4405.

Gutteridge, J. M. C., & Halliwell, B. (2018). Mini-review: Oxidative stress, redox stress or redox success? *Biochemical and Biophysical Research Communications, 502*(2), 183–186. https://doi.org/10.1016/j.bbrc.2018.05.045.

Halliwell, B. (2007). Biochemistry of oxidative stress. *Biochemical Society Transactions, 35*(5), 1147–1150. https://doi.org/10.1042/BST0351147.

Hernández-Aquino, E., & Muriel, P. (2018). Beneficial effects of naringenin in liver diseases: Molecular mechanisms. *World Journal of Gastroenterology, 24*(16), 1679–1707. https://doi.org/10.3748/wjg.v24.i16.1679.

Hernández-Aquino, E., et al. (2017). Naringenin prevents experimental liver fibrosis by blocking TGFβ-Smad3 and JNK-Smad3 pathways. *World Journal of Gastroenterology, 23*(24), 4354. https://doi.org/10.3748/wjg.v23.i24.4354.

Hernández-Aquino, E., et al. (2019). Naringenin attenuates the progression of liver fibrosis via inactivation of hepatic stellate cells and profibrogenic pathways. *European Journal of Pharmacology, 865*, 172730. https://doi.org/10.1016/j.ejphar.2019.172730.

Hernández-Aquino, E., et al. (2020). Curcumin downregulates Smad pathways and reduces hepatic stellate cells activation in experimental fibrosis. *Annals of Hepatology, 19*(5), 497–506. https://doi.org/10.1016/j.aohep.2020.05.006.

Higashi, T., Friedman, S. L., & Hoshida, Y. (2017). Hepatic stellate cells as key target in liver fibrosis. *Advanced Drug Delivery Reviews, 121*, 27–42. https://doi.org/10.1016/j.addr.2017.05.007.

Ignat, S.-R., et al. (2020). Cellular interplay as a consequence of inflammatory signals leading to liver fibrosis development. *Cells, 9*(2), 461. https://doi.org/10.3390/cells9020461.

Itoh, K., et al. (1997). An Nrf2/small Maf heterodimer mediates the induction of phase II detoxifying enzyme genes through antioxidant response elements. *Biochemical and Biophysical Research Communications, 236*(2), 313–322. https://doi.org/10.1006/bbrc.1997.6943.

Jaeschke, H. (2011). Reactive oxygen and mechanisms of inflammatory liver injury: Present concepts. *Journal of Gastroenterology and Hepatology, 26*, 173–179. https://doi.org/10.1111/j.1440-1746.2010.06592.x.

Jaganjac, M., et al. (2020). The NRF2, thioredoxin, and glutathione system in tumorigenesis and anticancer therapies. *Antioxidants, 9*(11), 1151. https://doi.org/10.3390/antiox9111151.

Kabe, Y., et al. (2005). Redox regulation of NF-κB activation: Distinct redox regulation between the cytoplasm and the nucleus. *Antioxidants & Redox Signaling,* 7(3–4), 395–403. https://doi.org/10.1089/ars.2005.7.395.

Karanjia, R. N., et al. (2016). Hepatic steatosis and fibrosis: Non-invasive assessment. *World Journal of Gastroenterology*, *22*(45), 9880. https://doi.org/10.3748/wjg.v22.i45.9880.

Keum, Y.-S., & Choi, B. (2014). Molecular and chemical regulation of the Keap1-Nrf2 signaling pathway. *Molecules*, *19*(7), 10074–10089. https://doi.org/10.3390/molecules190 710074.

Khor, T. O., et al. (2006). Nrf2-deficient mice have an increased susceptibility to dextran sulfate sodium–induced colitis. *Cancer Research*, *66*(24), 11580–11584. https://doi.org/10.1158/0008-5472.CAN-06-3562.

Krajka-Kuźniak, V., Paluszczak, J., & Baer-Dubowska, W. (2017). The Nrf2-ARE signaling pathway: An update on its regulation and possible role in cancer prevention and treatment. *Pharmacological Reports*, *69*(3), 393–402. https://doi.org/10.1016/j.pharep.2016.12.011.

Krithika, R., Jyothilakshmi, V., & Verma, R. J. (2016). Phyllanthin inhibits CCl4-mediated oxidative stress and hepatic fibrosis by down-regulating TNF-α/NF-κB, and pro-fibrotic factor TGF-β1 mediating inflammatory signaling. *Toxicology and Industrial Health*, *32*(5), 953–960. https://doi.org/10.1177/0748233714532996.

Latella, G. (2018). Redox imbalance in intestinal fibrosis: Beware of the TGFβ-1, ROS, and Nrf2 connection. *Digestive Diseases and Sciences*, *63*(2), 312–320. https://doi.org/10.1007/s10620-017-4887-1.

Latella, G., et al. (2009). Targeted disruption of Smad3 confers resistance to the development of dimethylnitrosamine-induced hepatic fibrosis in mice. *Liver International*, *29*(7), 997–1009. https://doi.org/10.1111/j.1478-3231.2009.02011.x.

Lee, J., et al. (2005). Nrf2, a multi-organ protector? *The FASEB Journal*, *19*(9), 1061–1066. https://doi.org/10.1096/fj.04-2591hyp.

Lei, X.-F., et al. (2016). Hic-5 deficiency attenuates the activation of hepatic stellate cells and liver fibrosis through upregulation of Smad7 in mice. *Journal of Hepatology*, *64*(1), 110–117. https://doi.org/10.1016/j.jhep.2015.08.026.

Li, S., et al. (2015). The role of oxidative stress and antioxidants in liver diseases. *International Journal of Molecular Sciences*, *16*(11), 26087–26124. https://doi.org/10.3390/ijms1611 25942.

Li, S., et al. (2016). Insights into the role and interdependence of oxidative stress and inflammation in liver diseases. *Oxidative Medicine and Cellular Longevity*, *2016*, 1–21. https://doi.org/10.1155/2016/4234061.

Liu, S., et al. (2010). Quercetin protects against ethanol-induced oxidative damage in rat primary hepatocytes. *Toxicology In Vitro*, *24*(2), 516–522. https://doi.org/10.1016/j.tiv.2009.03.006.

Liu, J., et al. (2019). Praziquantel ameliorates CCl4-induced liver fibrosis in mice by inhibiting TGF-β/Smad signalling via up-regulating Smad7 in hepatic stellate cells. *British Journal of Pharmacology*, *176*(24), 4666–4680. https://doi.org/10.1111/bph.14831.

Liu, J., et al. (2021). Sex-, age-, and race/ethnicity-dependent variations in drug-processing and NRF2-regulated genes in human livers. *Drug Metabolism and Disposition*, *49*(1), 111–119. https://doi.org/10.1124/dmd.120.000181.

Lu, J., & Holmgren, A. (2014). The thioredoxin antioxidant system. *Free Radical Biology and Medicine*, *66*, 75–87. https://doi.org/10.1016/j.freeradbiomed.2013.07.036.

Luangmonkong, T., et al. (2018). Targeting oxidative stress for the treatment of liver fibrosis. *Reviews of Physiology, Biochemistry and Pharmacology*, *175*, 71–102. https://doi.org/10.1007/112_2018_10.

Łukawska-Kuźma, K., Podgórska, A., & Rychter, A. M. (2012). Plasma membrane-generated ROS and their possible contribution to leaf cell growth of cucumber (Cucumis sativus) MSC16 mitochondrial mutant. *Acta Physiologiae Plantarum*, *34*(2), 721–730. https://doi.org/10.1007/s11738-011-0872-z.

Magesh, S., Chen, Y., & Hu, L. (2012). Small molecule modulators of Keap1-Nrf2-ARE pathway as potential preventive and therapeutic agents. *Medicinal Research Reviews*, *32*(4), 687–726. https://doi.org/10.1002/med.21257.

Márquez-Quiroga, L. V., et al. (2022). Models of nonalcoholic steatohepatitis potentiated by chemical inducers leading to hepatocellular carcinoma. *Biochemical Pharmacology*, *195*, 114845. https://doi.org/10.1016/j.bcp.2021.114845.

Massague, J. (2000). Transcriptional control by the TGF-beta/Smad signaling system. *The EMBO Journal*, *19*(8), 1745–1754. https://doi.org/10.1093/emboj/19.8.1745.

Masszi, A., & Kapus, A. (2011). Smaddening complexity: The role of Smad3 in epithelial-myofibroblast transition. *Cells, Tissues, Organs*, *193*(1–2), 41–52. https://doi.org/10.1159/000320180.

Matsuzaki, K. (2012). Smad phosphoisoform signals in acute and chronic liver injury: Similarities and differences between epithelial and mesenchymal cells. *Cell and Tissue Research*, *347*(1), 225–243. https://doi.org/10.1007/s00441-011-1178-6.

Michaeloudes, C., et al. (2011). Transforming growth factor-β and nuclear factor E2–related factor 2 regulate antioxidant responses in airway smooth muscle cells. *American Journal of Respiratory and Critical Care Medicine*, *184*(8), 894–903. https://doi.org/10.1164/rccm.201011-1780OC.

Muriel, P. (2009a). NF-kappaB in liver diseases: A target for drug therapy. *Journal of Applied Toxicology*, *29*(2), 91–100. https://doi.org/10.1002/jat.1393.

Muriel, P. (2009b). Role of free radicals in liver diseases. *Hepatology International*, *3*(4), 526–536. https://doi.org/10.1007/s12072-009-9158-6.

Muriel, P. (2017). The liver: General aspects and epidemiology. In P. Muriel (Ed.), *Liver pathophysiology: Therapies and antioxidants* (pp. 3–22). Waltham, MA: Elsevier. https://doi.org/10.1016/B978-0-12-804274-8.00001-1.

Muriel, P., López-Sánchez, P., & Ramos-Tovar, E. (2021). Fructose and the liver. *International Journal of Molecular Sciences*, *22*(13), 6969. https://doi.org/10.3390/ijms22136969.

Nakamura, T., et al. (2011). Attenuation of transforming growth factor–β–stimulated collagen production in fibroblasts by quercetin-induced heme oxygenase–1. *American Journal of Respiratory Cell and Molecular Biology*, *44*(5), 614–620. https://doi.org/10.1165/rcmb.2010-0338OC.

Nguyen, T., Sherratt, P. J., & Pickett, C. B. (2003). Regulatory mechanisms controlling gene expression mediated by the antioxidant response element. *Annual Review of Pharmacology and Toxicology*, *43*(1), 233–260. https://doi.org/10.1146/annurev.pharmtox.43.100901.140229.

Niture, S. K., Khatri, R., & Jaiswal, A. K. (2014). Regulation of Nrf2—An update. *Free Radical Biology and Medicine*, *66*, 36–44. https://doi.org/10.1016/j.freeradbiomed.2013.02.008.

Pall, M. L., & Levine, S. (2015). Nrf2, a master regulator of detoxification and also antioxidant, anti-inflammatory and other cytoprotective mechanisms, is raised by health promoting factors. *Acta Physiologica Sinica*, *67*(1), 1–18. Available at: http://www.ncbi.nlm.nih.gov/pubmed/25672622.

Parola, M., & Robino, G. (2001). Oxidative stress-related molecules and liver fibrosis. *Journal of Hepatology*, *35*(2), 297–306. https://doi.org/10.1016/S0168-8278(01)00142-8.

Peluso, M., et al. (2020). Oxidative stress and DNA damage in chronic disease and environmental studies. *International Journal of Molecular Sciences*, *21*(18), 6936. https://doi.org/10.3390/ijms21186936.

Pisoschi, A. M., & Pop, A. (2015). The role of antioxidants in the chemistry of oxidative stress: A review. *European Journal of Medicinal Chemistry*, *97*, 55–74. https://doi.org/10.1016/j.ejmech.2015.04.040.

Prestigiacomo, V., & Suter-Dick, L. (2018). Nrf2 protects stellate cells from Smad-dependent cell activation. *PLoS One*, *13*(7), e0201044. https://doi.org/10.1371/journal.pone.0201044. Edited by P. Mukhopadhyay.

Ramos-Tovar, E., Buendia-Montaño, L. D., et al. (2019). Stevia prevents experimental cirrhosis by reducing hepatic myofibroblasts and modulating molecular profibrotic pathways. *Hepatology Research*, *49*(2), 212–223. https://doi.org/10.1111/hepr.13275.

Ramos-Tovar, E., Casas-Grajales, S., et al. (2019). Cirrhosis induced by thioacetamide is prevented by stevia. Molecular mechanisms. *Journal of Functional Foods*, *52*, 552–564. https://doi.org/10.1016/j.jff.2018.11.039.

Ramos-Tovar, E., Flores-Beltrán, R. E., et al. (2018). Stevia rebaudiana tea prevents experimental cirrhosis via regulation of NF-κB, Nrf2, transforming growth factor beta, Smad7, and hepatic stellate cell activation. *Phytotherapy Research*, *32*(12), 2568–2576. https://doi.org/10.1002/ptr.6197.

Ramos-Tovar, E., Flores-Beltrán, R. E., et al. (2019). An aqueous extract of Stevia rebaudiana variety Morita II prevents liver damage in a rat model of cirrhosis that mimics the human disease. *Annals of Hepatology*, *18*(3), 472–479. https://doi.org/10.1016/j.aohep.2018.10.002.

Ramos-Tovar, E., Hernández-Aquino, E., et al. (2018). Stevia prevents acute and chronic liver injury induced by carbon tetrachloride by blocking oxidative stress through Nrf2 upregulation. *Oxidative Medicine and Cellular Longevity*, *2018*, 1–12. https://doi.org/10.1155/2018/3823426.

Ramos-Tovar, E., & Muriel, P. (2019). Phytotherapy for the liver. In R. Watson, & V. Preedy (Eds.), *Dietary interventions for liver disease and treatment* (pp. 101–121). London, UK: Elsevier.

Ramos-Tovar, E., & Muriel, P. (2020a). Free radicals, antioxidants, nuclear factor-E2-related factor-2 and liver damage. *Journal of Applied Toxicology*, *40*(1), 151–168. https://doi.org/10.1002/jat.3880.

Ramos-Tovar, E., & Muriel, P. (2020b). Molecular mechanisms that link oxidative stress, inflammation, and fibrosis in the liver. *Antioxidants*, *9*(12), 1279. https://doi.org/10.3390/antiox9121279.

Roehlen, N., Crouchet, E., & Baumert, T. F. (2020). Liver fibrosis: Mechanistic concepts and therapeutic perspectives. *Cells*, *9*(4), 875. https://doi.org/10.3390/cells9040875.

Rolo, A. P., Teodoro, J. S., & Palmeira, C. M. (2012). Role of oxidative stress in the pathogenesis of nonalcoholic steatohepatitis. *Free Radical Biology and Medicine*, *52*(1), 59–69. https://doi.org/10.1016/j.freeradbiomed.2011.10.003.

Rowe, I. A. (2017). Lessons from epidemiology: The burden of liver disease. *Digestive Diseases*, *35*(4), 304–309. https://doi.org/10.1159/000456580.

Rui, W., et al. (2016). Nuclear factor erythroid 2-related factor 2 deficiency results in amplification of the liver fat-lowering effect of estrogen. *Journal of Pharmacology and Experimental Therapeutics*, *358*(1), 14–21. https://doi.org/10.1124/jpet.115.231316.

Rushmore, T. H., Morton, M. R., & Pickett, C. B. (1991). The antioxidant responsive element. Activation by oxidative stress and identification of the DNA consensus sequence required for functional activity. *Journal of Biological Chemistry*, *266*(18), 11632–11639. https://doi.org/10.1016/S0021-9258(18)99004-6.

Sanchez-Valle, V., et al. (2012). Role of oxidative stress and molecular changes in liver fibrosis: A review. *Current Medicinal Chemistry*, *19*(28), 4850–4860. https://doi.org/10.2174/092986712803341520.

Sass, G., Barikbin, R., & Tiegs, G. (2012). The multiple functions of heme oxygenase-1 in the liver. *Zeitschrift für Gastroenterologie*, *50*(01), 34–40. https://doi.org/10.1055/s-0031-1282046.

Schuppan, D., Surabattula, R., & Wang, X. Y. (2018). Determinants of fibrosis progression and regression in NASH. *Journal of Hepatology, 68*(2), 238–250. https://doi.org/10.1016/j.jhep.2017.11.012.

Sekimoto, G., et al. (2007). Reversible Smad-dependent signaling between tumor suppression and oncogenesis. *Cancer Research, 67*(11), 5090–5096. https://doi.org/10.1158/0008-5472.CAN-06-4629.

Shah, N. D., et al. (2019). Alcohol-related liver disease is rarely detected at early stages compared with liver diseases of other etiologies worldwide. *Clinical Gastroenterology and Hepatology, 17*(11), 2320–2329.e12. https://doi.org/10.1016/j.cgh.2019.01.026.

Sharma, R. S., et al. (2018). Experimental nonalcoholic steatohepatitis and liver fibrosis are ameliorated by pharmacologic activation of Nrf2 (NF-E2 p45-related factor 2). *Cellular and Molecular Gastroenterology and Hepatology, 5*(3), 367–398. https://doi.org/10.1016/j.jcmgh.2017.11.016.

Shimozono, R., et al. (2013). Nrf2 activators attenuate the progression of nonalcoholic steatohepatitis–related fibrosis in a dietary rat model. *Molecular Pharmacology, 84*(1), 62–70. https://doi.org/10.1124/mol.112.084269.

Shin, S., et al. (2009). Role of Nrf2 in prevention of high-fat diet-induced obesity by synthetic triterpenoid CDDO-imidazolide. *European Journal of Pharmacology, 620*(1–3), 138–144. https://doi.org/10.1016/j.ejphar.2009.08.022.

Slavin, J. L., & Lloyd, B. (2012). Health benefits of fruits and vegetables. *Advances in Nutrition, 3*(4), 506–516. https://doi.org/10.3945/an.112.002154.

Sun, C., Li, S., & Li, D. (2016). Sulforaphane mitigates muscle fibrosis in mdx mice via Nrf2-mediated inhibition of TGF-β/Smad signaling. *Journal of Applied Physiology, 120*(4), 377–390. https://doi.org/10.1152/japplphysiol.00721.2015.

Surh, Y. J., Kundu, J. K., & Na, H. K. (2008). Nrf2 as a master redox switch in turning on the cellular signaling involved in the induction of cytoprotective genes by some chemopreventive phytochemicals. *Planta Medica, 74*(13), 1526–1539. https://doi.org/10.1055/s-0028-1088302.

Tang, W. (2014). Role of Nrf2 in chronic liver disease. *World Journal of Gastroenterology, 20*(36), 13079. https://doi.org/10.3748/wjg.v20.i36.13079.

Tao, S., & Sampath, K. (2010). Alternative splicing of SMADs in differentiation and tissue homeostasis. *Development, Growth & Differentiation, 52*(4), 335–342. https://doi.org/10.1111/j.1440-169X.2009.01163.x.

Tao, Y., et al. (2020). The role of autophagy and NLRP3 inflammasome in liver fibrosis. *BioMed Research International, 2020*, 1–8. https://doi.org/10.1155/2020/7269150.

Thanan, R., et al. (2014). Oxidative stress and its significant roles in neurodegenerative diseases and cancer. *International Journal of Molecular Sciences, 16*(1), 193–217. https://doi.org/10.3390/ijms16010193.

Tsai, M.-K., Lin, Y.-L., & Huang, Y.-T. (2010). Effects of salvianolic acids on oxidative stress and hepatic fibrosis in rats. *Toxicology and Applied Pharmacology, 242*(2), 155–164. https://doi.org/10.1016/j.taap.2009.10.002.

Tsuchida, K.-I., et al. (2003). Role of Smad4 on TGF-β–induced extracellular matrix stimulation in mesangial cells. *Kidney International, 63*(6), 2000–2009. https://doi.org/10.1046/j.1523-1755.2003.00009.x.

Vargas-Pozada, E. E., & Muriel, P. (2020). Herbal medicines for the liver. *European Journal of Gastroenterology & Hepatology, 32*(2), 148–158. https://doi.org/10.1097/MEG.0000000000001485.

Walters, D. M., Cho, H.-Y., & Kleeberger, S. R. (2008). Oxidative stress and antioxidants in the pathogenesis of pulmonary fibrosis: A potential role for Nrf2. *Antioxidants & Redox Signaling, 10*(2), 321–332. https://doi.org/10.1089/ars.2007.1901.

Wang, X.-J., et al. (2008). Nrf2 enhances resistance of cancer cells to chemotherapeutic drugs, the dark side of Nrf2. *Carcinogenesis, 29*(6), 1235–1243. https://doi.org/10.1093/carcin/bgn095.

Wang, P., et al. (2017). Hepatoprotective effect of 2′-O-galloylhyperin against oxidative stress-induced liver damage through induction of Nrf2/ARE-mediated antioxidant pathway. *Food and Chemical Toxicology*, *102*, 129–142. https://doi.org/10.1016/j.fct.2017.02.016.
Wang, P., et al. (2018). Rosuvastatin improves myocardial hypertrophy after hemodynamic pressure overload via regulating the crosstalk of Nrf2/ARE and TGF-β/ smads pathways in rat heart. *European Journal of Pharmacology*, *820*, 173–182. https://doi.org/10.1016/j.ejphar.2017.12.013.
Wong, R. J., & Singal, A. K. (2020). Trends in liver disease etiology among adults awaiting liver transplantation in the United States, 2014-2019. *JAMA Network Open*, *3*(2), e1920294. https://doi.org/10.1001/jamanetworkopen.2019.20294.
Wree, A., Eguchi, A., et al. (2014). NLRP3 inflammasome activation results in hepatocyte pyroptosis, liver inflammation, and fibrosis in mice. *Hepatology*, *59*(3), 898–910. https://doi.org/10.1002/hep.26592.
Wree, A., McGeough, M. D., et al. (2014). NLRP3 inflammasome activation is required for fibrosis development in NAFLD. *Journal of Molecular Medicine*, *92*(10), 1069–1082. https://doi.org/10.1007/s00109-014-1170-1.
Wu, D., & Cederbaum, A. (2009). Oxidative stress and alcoholic liver disease. *Seminars in Liver Disease*, *29*(02), 141–154. https://doi.org/10.1055/s-0029-1214370.
Wu, K. C., Liu, J., & Klaassen, C. D. (2012). Role of Nrf2 in preventing ethanol-induced oxidative stress and lipid accumulation. *Toxicology and Applied Pharmacology*, *262*(3), 321–329. https://doi.org/10.1016/j.taap.2012.05.010.
Wu, J., et al. (2019). Pyroptosis in liver disease: New insights into disease mechanisms. *Aging and Disease*, *10*(5), 1094–1108. https://doi.org/10.14336/AD.2019.0116.
Xie, F., et al. (2014). Regulation of TGF-β superfamily signaling by SMAD monoubiquitination. *Cells*, *3*(4), 981–993. https://doi.org/10.3390/cells3040981.
Xu, J., et al. (2012). Enhanced Nrf2 activity worsens insulin resistance, impairs lipid accumulation in adipose tissue, and increases hepatic steatosis in leptin-deficient mice. *Diabetes*, *61*(12), 3208–3218. https://doi.org/10.2337/db11-1716.
Xu, F., et al. (2016). TGF-β/SMAD pathway and its regulation in hepatic fibrosis. *Journal of Histochemistry & Cytochemistry*, *64*(3), 157–167. https://doi.org/10.1369/0022155415627681.
Xu, D., et al. (2019). The role of Nrf2 in liver disease: Novel molecular mechanisms and therapeutic approaches. *Frontiers in Pharmacology*, *9*, 1428. https://doi.org/10.3389/fphar.2018.01428.
Xuanfei, L., et al. (2017). Imidazoline I2 receptor inhibitor idazoxan regulates the progression of hepatic fibrosis via Akt-Nrf2-Smad2/3 signaling pathway. *Oncotarget*, *8*(13), 21015–21030. https://doi.org/10.18632/oncotarget.15472.
Yamaguchi, T., et al. (2013). Phosphorylated Smad2 and Smad3 signaling: Shifting between tumor suppression and fibro-carcinogenesis in chronic hepatitis C. *Hepatology Research*, *43*(12), 1327–1342. https://doi.org/10.1111/hepr.12082.
Yates, M. S., & Kensler, T. W. (2007). Keap1 eye on the target: Chemoprevention of liver cancer. *Acta Pharmacologica Sinica*, *28*(9), 1331–1342. https://doi.org/10.1111/j.1745-7254.2007.00688.x.
Younossi, Z. M., et al. (2016). Global epidemiology of nonalcoholic fatty liver disease—Meta-analytic assessment of prevalence, incidence, and outcomes. *Hepatology*, *64*(1), 73–84. https://doi.org/10.1002/hep.28431.
Younossi, Z. M., et al. (2019). Global perspectives on nonalcoholic fatty liver disease and nonalcoholic steatohepatitis. *Hepatology*, *69*(6), 2672–2682. https://doi.org/10.1002/hep.30251.
Yu, Z., et al. (2011). Oltipraz upregulates the nuclear respiratory factor 2 alpha subunit (NRF2) antioxidant system and prevents insulin resistance and obesity induced by a high-fat diet in C57BL/6J mice. *Diabetologia*, *54*(4), 922–934. https://doi.org/10.1007/s00125-010-2001-8.

Zhang, J., et al. (2016). ROS and ROS-mediated cellular signaling. *Oxidative Medicine and Cellular Longevity*, *2016*, 1–18. https://doi.org/10.1155/2016/4350965.
Zhangdi, H.-J., et al. (2019). Crosstalk network among multiple inflammatory mediators in liver fibrosis. *World Journal of Gastroenterology*, *25*(33), 4835–4849. https://doi.org/10.3748/wjg.v25.i33.4835.
Zhou, W. C., Zhang, Q. B., & Qiao, L. (2014). Pathogenesis of liver cirrhosis. *World Journal of Gastroenterology*, *20*(23), 7312–7324. https://doi.org/10.3748/wjg.v20.i23.7312.
Zhou, S.-F., et al. (2014). Schisandra chinensis regulates drug metabolizing enzymes and drug transporters via activation of Nrf2-mediated signaling pathway. *Drug Design, Development and Therapy*, *9*, 127. https://doi.org/10.2147/DDDT.S68501.

CHAPTER ELEVEN

Mechanisms of action of vitamin D in delaying aging and preventing disease by inhibiting oxidative stress

Dengshun Miao[a,*] **and David Goltzman**[b,*]
[a]Nanjing Medical University, Nanjing, China
[b]McGill University Health Centre and McGill University, Montreal, QC, Canada
*Corresponding authors: e-mail address: dsmiao@njmu.edu.cn; david.goltzman@mcgill.ca

Contents

Vitamins and Hormones, Volume 121
ISSN 0083-6729
https://doi.org/10.1016/bs.vh.2022.09.004

Abstract

Although several recent studies have shown that vitamin D supplementation beneficially decreases oxidative stress parameters, there is no consensus on this subject in humans. Thus the role of vitamin D supplementation has recently become a controversial topic because large intervention studies in humans have not shown significant benefits. These studies have indicated that supplementation with precursor forms of active vitamin D has no effect on all-cause mortality, cannot reduce the fracture risk of the elderly, cannot reduce the incidence of cancer or cardiovascular disease in the elderly, and cannot significantly reduce the incidence risk of diabetes in the elderly. However, a link between several age-related diseases and enhanced oxidative stress has been found in mice with insufficient or deficient 1,25-dihydroxyvitamin D (1,25$(OH)_2$D), the active form of vitamin D, which indicates that reduced active vitamin D accelerates aging and age-related diseases by increasing oxidative stress. Furthermore, supplementation of exogenous 1,25$(OH)_2D_3$, or antioxidants, could dramatically postpone aging, prevent osteoporosis and spontaneous tumor development induced by 1,25$(OH)_2$D insufficiency or deficiency, by inhibiting oxidative stress. Mechanistically, the anti-oxidative effects of 1,25$(OH)_2D_3$ are carried out via the vitamin D receptor (VDR) by activation of the Nrf2 oxidative stress response pathway though transcriptional or post-transcriptional activation of Nrf2 or transcriptional upregulation of Sirt1 and Bmi1 expression. Whether discrepancies between studies in humans and in mice reflect the different forms of vitamin D examined remains to be determined.

1. Introduction

After ultraviolet B (UVB) irradiation of skin by sunlight, vitamin D_3 (cholecalciferol) can be produced by photolysis of its skin precursor, 7-dehydrocholesterol (7DHC). Vitamin D_3 synthesized in skin or absorbed in the intestine from the diet or supplements, or vitamin D_2 (ergocalciferol) absorbed in the intestine from food sources or dietary supplements, can the transported in the circulation to the liver bound to vitamin D binding protein. In liver, it is hydroxylated at the C-25 position of the side chain, mainly, but not exclusively, via the cytochrome P450 (CYP)-containing microsomal enzyme, Cyp2r1. This produces 25-hydroxyvitamin D [25(OH)D], the most abundant circulating form of vitamin D. 25(OH)D can then be converted to the active form of vitamin D, 1,25$(OH)_2$D (calcitriol) by the action of the mitochondrial enzyme Cyp27b1, which encodes 25-hydroxyvitamin D 1α-hydroxylase [1α(OH)ase]. Renal Cyp27b1 is the major source of circulating 1,25$(OH)_2$D but extra-renal synthesis of 1,25 $(OH)_2$D can occur in other tissues. The active form of vitamin D, 1,25 $(OH)_2$D, then binds to its nuclear receptor, the vitamin D receptor (VDR) to carry out its functions.

Vitamin D is known to contribute to maintaining normal calcium metabolism, but may also be crucial for an extensive range of "non-classical" actions. Vitamin D deficiency may increase the incidence and severity of several age-related common diseases, including metabolic disorders that are linked to oxidative stress. These include obesity, insulin resistance and type 2 diabetes, hypertension, pregnancy complications, memory disorders, osteoporosis, autoimmune diseases, certain cancers, and systemic inflammatory diseases (Wimalawansa, 2019). The link between vitamin D status and risk of age-related diseases could be due, at least in part, to effects on oxidative stress and antioxidant status. Thus, vitamin D deficiency can impair mitochondrial function, and enhance oxidative stress and systemic inflammation. The antioxidant effect of vitamin D is among the newest suggested roles of this compound.

The antioxidant properties of vitamin D reportedly may protect cultured human endothelial cells and retinal cone cells from oxidative challenge (Tohari, Zhou, & Shu, 2016; Uberti et al., 2014). Vitamin D deficiency in early life has been reported to increase arterial blood pressure, promotes vascular oxidative stress, and induces changes in cardiac gene expression (Argacha et al., 2011). Vitamin D deficiency exacerbates hepatic oxidative stress and inflammation during acetaminophen-induced acute liver injury in mice (Wang et al., 2021), causes insulin resistance by provoking oxidative stress in hepatocytes (Tao et al., 2017) and aggravates hepatic oxidative stress and inflammation during chronic alcohol-induced liver injury in mice (Hu et al., 2020).

In humans, reduced vitamin D status was found to be associated with higher levels of oxidative stress biomarkers in obese children (Codoner-Franch et al., 2012). Depleted antioxidant status predisposes to oxidative stress, and logistic regression analysis of data from a study of forty preeclampsia patients revealed a positive association between 25(OH)D levels and plasma total antioxidant capacity (Pourghassem Gargari, Pourteymour Fard Tabrizi, Sadien, Asghari Jafarabadi, & Farzadi, 2016). Significantly lower plasma total antioxidant capacity was found in 58 vitamin D-deficient chronic hepatitis patients compared with nondeficient patients with the same disease, and a significant inverse correlation was seen between levels of plasma 25(OH)D and lipid hydroperoxides (Almeida et al., 2012). In 65 elderly subjects with type 2 diabetes or impaired fasting glucose, 25(OH)D correlated inversely with two oxidative stress biomarkers in plasma, namely, oxidized LDL and advanced protein peroxidation products (Gradinaru et al., 2012).

2. The effect of 1,25-dihydroxyvitamin D deficiency and 1,25-dihydroxyvitamin D supplementation on aging and age-related diseases and on oxidative stress parameters in studies in animals in vivo and in vitro

2.1 Deficiency of 1,25-dihydroxyvitamin D and lifespan

1α(OH)ase homozygous null mice [1α(OH)ase$^{-/-}$] in which the serum 1,25$(OH)_2$D levels were undetectable and serum 25(OH)D levels were increased, develop hypocalcemia, hypophosphatemia and elevated parathyroid hormone (PTH) levels, when fed a normal diet, and survive on average for only 3 months (Chen et al., 2019). Increased tissue oxidative stress and DNA damage, downregulated Bmi1 and upregulated p16, p53 and p21 expression levels, reduced cell proliferation, and induced cell senescence and a senescence-associated secretory phenotype (SASP) were observed. Homozygous ablation of p16 or heterozygous ablation of p53 prolonged the average lifespan of 1α(OH)ase$^{-/-}$ mice on the normal diet from 3 to 6 months by enhancing cell proliferative ability and reducing cell senescence or apoptosis (Chen et al., 2019). Supplementation of 1α(OH)ase$^{-/-}$ mice with dietary calcium and phosphate (a "rescue" diet) (Chen et al., 2019; Yang et al., 2020) normalized their serum calcium, phosphorus and PTH levels and prolonged their average lifespan to more than 8 months but the 1,25$(OH)_2$D deficiency could still be demonstrated to accelerate aging by increasing oxidative stress and SASP. When 1,25$(OH)_2$D deficient 1α (OH)ase$^{-/-}$ mice were injected with 1,25$(OH)_2D_3$ subcutaneously three times per week, their average lifespan was prolonged from 91 days to 487 days; Body size and body weight were near normal, the thickness of skin and collagen deposition in skin were increased dramatically; ROS levels and cells positive for the DNA damage marker, gamma histone H2AX (γ-H2AX$^+$) in the skin and kidney were markedly reduced; SOD2 and peroxiredoxin-1 (Prdx I) protein expression levels in skin and liver were increased significantly, while Bmi1, p16, p53 and p21 protein expression levels in the skin and liver were normalized; The percentage of Ki67 positive cells in the skin, liver and kidney was increased; senescence-associated β-galactosidase (SA-β-gal) positive areas in the skin, kidney and lung and SASP molecules including TNFα, IL-1α and β, IL-6, and Mmp 3 and 13 in the skin, kidney, liver and lung were reduced significantly (Chen et al., 2019). These results demonstrate that supplementation with exogenous $1,25(OH)_2D_3$ could dramatically postpone aging induced by $1,25(OH)_2D$-deficiency by inhibiting oxidative stress, DNA damage and p16 and p53 signaling pathways, subsequently stimulating cell proliferation and inhibiting cell senescence and SASP. These studies suggested that 1,25 $(OH)_2D$ deficiency accelerates aging by increasing oxidative stress and DNA damage, activating p53-p21 and p16-Rb signaling pathways, and inhibiting cell senescence and SASP.

2.2 Insufficiency of 1,25-dihydroxyvitamin D and bone aging and osteoporosis

We have reported that heterozygous 25-hydroxyvitamin D 1α-hydroxylase null [1α(OH)ase$^{+/-}$ or Cyp27b1$^{+/-}$] mice, in which the serum 1,25 $(OH)_2D$ levels were significantly reduced, while serum calcium, phosphorus, parathyroid hormone (PTH) and 25(OH)D levels were not altered (Qiao et al., 2020; Sun et al., 2020), could be used as a model for 1,25 $(OH)_2D$ insufficiency. Results from these $1,25(OH)_2D$ insufficient mice showed that ROS levels in bone marrow mesenchymal stem cells (BM-MSCs) and serum lipid peroxidation levels were significantly increased, whereas the protein expression levels of Nuclear factor-erythroid factor 2-related factor 2 (Nrf2), Sirt1 and Bmi1 and the protein and mRNA expression levels of antioxidant enzymes in bone tissue were significantly reduced. Furthermore, the parameters of DNA damage, senescence and SASP in bone marrow mesenchymal stem cells (BM-MSCs) and osteocytes were augmented, and osteoclastic bone resorption was significantly increased. In contrast, the parameters of osteogenesis of BM-MSCs, and osteoblastic bone formation were significantly reduced. These results indicate that $1,25(OH)_2D$ insufficiency accelerates age-related bone loss by downregulating Nrf2, Sirt1 and Bmi1, increasing oxidative stress and DNA damage, inducing bone cell senescence and SASP, and subsequently inhibiting osteoblastic bone formation while stimulating osteoclastic bone resorption (Qiao et al., 2020; Sun et al., 2020). The resulting osteoporosis and bone aging phenotypes induced by $1,25(OH)_2D$ insufficiency in mice were significantly rescued in vivo upon the supplementation of oltipraz, an inhibitor of Nrf2 degradation (Yang et al., 2022) or overexpression of Bmi1 in MSCs (Sun et al., 2020).

Homoxygous null 1α(OH)ase$^{-/-}$ mice on a rescue diet, which have normalized serum calcium, phosphorus and PTH levels, but a high serum 25(OH)D levels and an undetectable serum $1,25(OH)_2D$, still demonstrate accelerated age-related bone loss (Yang et al., 2020), similar to

1α(OH)ase$^{+/-}$ mice on a normal diet. Bone mineral density (BMD), trabecular bone volume, trabecular number and thickness were decreased progressively in 3-, 6- and 12-month-old 1α(OH)ase$^{-/-}$ mice on the rescue diet, and were dramatically decreased at each age compared with age-matched wild-type mice. The osteoblast number, mineral apposition rate (MAR), bone formation rate (BFR), and long bone strength were all significantly decreased, whereas tartrate resistant acid phosphatase (TRAP) positive osteoclast numbers and C-telopeptide of type 1 collagen (CTX) levels were increased significantly in 1α(OH)ase$^{-/-}$ mice on the rescue diet compared with wild-type mice. The percentages of β-gal^{+}, p16^{+} and Mmp3^{+} osteocytes, the protein expression levels of p16, p19, p21 and p53, and the mRNA expression levels of *TNFα, IL-6, IL-1α, IL-1β, Mmp3, Mmp13,* and *p16* were all increased significantly in 1α(OH)ase$^{-/-}$ mice on the rescue diet compared with wild-type mice. All bone phenotypes induced by 1,25 $(OH)_2D$ deficiency were largely rescued by deletion of p16. To determine whether the acceleration of bone loss induced by 1,25$(OH)_2D$ deficiency is linked to oxidative stress, the ROS levels, proliferation and senescence of BM-MSCs derived from 6-month-old wild-type and VDR$^{-/-}$ mice were compared. We found that the proliferation capacity, including population doubling and percentage of EdU^{+} cells were significantly decreased, whereas cell sizes and the percentages of cells positive for SA-β-gal, the fluorogenic dye DCFDA^{+} and γ-H2AX were significantly increased; additionally, ectopic bone formation was significantly reduced in the second-passaged VDR$^{-/-}$ BM-MSCs compared to the second-passaged wild-type BM-MSCs. These results demonstrated that 1,25$(OH)_2D$ deficiency accelerates age-related bone loss by increasing oxidative stress, DNA damage and senescence of BM-MSCs, thus reducing osteogenesis and osteoblastic bone formation, and increasing osteoclastic bone resorption.

When 1,25$(OH)_2D$ deficient 1α(OH)ase$^{-/-}$ mice were injected with 1,25$(OH)_2D_3$ subcutaneously three times per week, the BMD, trabecular bone volume, trabecular number and thickness, osteoblast number, mineral apposition rate and bone formation rate were all increased significantly, whereas TRAP positive osteoclast number, serum CTX levels, the percentages of β-gal^{+}, p16^{+} and IL-6^{+} osteocytes, the mRNA levels of *TNFα, IL-6, IL-1α, IL-1β, Mmp3, Mmp13, p16* and *p21* and the protein expression levels of p16 and p21 were all comparable to those of wild-type mice (Yang et al., 2020). These results demonstrated that supplementation with exogenous 1,25$(OH)_2D_3$ rescued bone loss induced by 1,25$(OH)_2D$ deficiency by increasing osteoblastic bone formation, inhibiting osteoclastic bone resorption, osteocyte senescence and SASP. When 12-month-old wild-type mice were injected with $1,25(OH)_2D_3$ subcutaneously thrice weekly or with vehicle for 6 months, BMD, trabecular bone volume, trabecular number and thickness, osteoblast number, mineral apposition rate and bone formation rate were increased significantly, whereas TRAP positive osteoclastic number, the percentages of β-gal$^+$, p16$^+$ and IL-6$^+$ osteocytes, and the mRNA levels of *TNFα, IL-1α, IL-1β, IL-6, Mmp3* and *p16* were all significantly decreased. These results demonstrated that supplementation of exogenous $1,25(OH)_2D_3$ could also prevent bone loss induced by natural aging (Yang et al., 2020).

2.3 Decreased 1,25-dihydroxyvitamin D and muscle

Decreased skeletal muscle mass and strength, and increased incidence rates of sarcopenia and risk of falls have been related to a decrease of serum 25(OH)D levels in humans (Annweiler, Schott, Berrut, Fantino, & Beauchet, 2009; Garcia, Seelaender, Sotiropoulos, Coletti, & Lancha, 2019). VDR deficient mice showed decreased cross-sectional area of muscle fibers, accompanied by dysregulated expression of myogenic transcription factors and myosin heavy chain (MyHC) isomers (Endo et al., 2003). A recent study reported that muscle mass, grip strength and muscle fiber size were significantly decreased in 1α(OH)ase$^{-/-}$ mice (Yu et al., 2021). The level of gene expression of SOD1, catalase, NAD(*P*)H dehydrogenase, quinone 1 (Nqo1) and glutamylcysteine synthetase were significantly downregulated in skeletal muscle tissue of 1α(OH)ase$^{-/-}$ mice, demonstrating that the activity of genes related to antioxidant activity and mitochondrial metabolic activity were reduced. In contrast, the level of factors associated with skeletal muscle cell senescence and SASP, including protein expression levels of p16, p19, p21 and p53, and mRNA levels of p16, p21, TNFα, IL6 and Mmp3, were significantly increased. Furthermore, the percentage of p16$^+$ and p21$^+$ myofibers was significantly augmented. The numbers of newly formed regenerating central nucleated fibers, the percentage of BrdU$^+$ cells and the expression of MyoD, MyHC and myogenic factor 5 at mRNA levels were significantly downregulated in the injured skeletal muscle tissue of 1α(OH)ase$^{+/-}$ mice. This study in mice indicates that $1,25(OH)_2D$ deficiency can result in the development of age-related sarcopenia by inducing oxidative stress, skeletal muscular cell senescence and SASP, and by inhibiting skeletal muscle regeneration.

A study on muscle cells with oxidative stress induced by *tert*-butyl hydroperoxide, showed potent antioxidant effects of $1,25(OH)_2D_3$ that are considered beneficial for muscle tissue homeostasis (Chang, 2019). Increase in total protein degradation, 20S proteasomal enzyme activity, muscle atrophy gene markers and expression of proteasome subunit genes induced by oxidative stress were corrected both by pre- and post-treatment of C2C12 muscle cells with $1,25(OH)_2D_3$ (Bhat & Ismail, 2015).

2.4 Insufficiency of 1,25-dihydroxyvitamin D and tumors

Human epidemiological studies suggest that $1,25(OH)_2D$ deficiency might increase cancer incidence, but no spontaneous tumors had been reported in mice lacking $1,25(OH)_2D$, or deficient in its receptor, until a recent report of Chen et al. (Chen et al., 2018). $1\alpha(OH)ase^{+/-}$ mice and their wild-type littermates on a normal diet, $1\alpha(OH)ase^{-/-}$ mice and their wild-type littermates on a rescue diet, or on the rescue diet supplemented with the antioxidant *N*-Acetyl Cysteine (NAC) in drinking water were studied. Spontaneous tumors were found in 6 of 52 $1\alpha(OH)ase^{+/-}$ mice on a normal diet between 13 and 20 months of age at which time serum $1,25(OH)_2D$ levels were significantly reduced; this demonstrated a spontaneous tumor formation rate of 11.5%; no spontaneous tumors were found in 52 wild-type littermates of comparable age. When $1\alpha(OH)ase^{-/-}$ mice were fed a rescue diet, survival was increased up to 465 days. Among 21 $1\alpha(OH)ase^{-/-}$ mice that survived over 12 months, spontaneous tumors were found in 5, demonstrating a spontaneous tumor formation rate of 23.8% (Chen et al., 2018). The rescue diet combined with NAC not only prolonged the lifespan of $1\alpha(OH)ase^{-/-}$ mice to 320–540 days, but also prevented spontaneous tumor formation. The pathological diagnosis of the 11 spontaneous tumors included breast cancer, squamous-cell carcinoma of skin, soft tissue sarcoma, hepatocellular carcinoma, granulocytic sarcoma, endometrial stromal tumor, gastrointestinal stromal tumor and pulmonary adenocarcinoma. These results demonstrated that $1,25(OH)_2D$ deficiency could increase diverse types of spontaneous tumor formation, and these could be prevented with antioxidant NAC. Cells immunopositive for the antioxidant enzyme SOD2 were reduced in spontaneous tumor tissue compared with the paracancerous tissue, whereas the percentage of cells positive for the DNA damage marker, phosphorylation of histone protein H2AX (γ-H2AX) and for the cell proliferation marker, Ki67, were increased significantly in spontaneous tumor tissue compared with the paracancerous tissue.

Furthermore, protein expression levels of the C-fos, Ras and Bmi1 were upregulated, whereas protein expression levels of p16, p21 and p27 were downregulated in spontaneous tumor tissue compared with the para-cancerous tissue, and even compared with corresponding normal tissues of 1,25$(OH)_2D$ sufficient mice. 1,25$(OH)_2D$ deficiency accelerated allograft tumor initiation and growth by increasing oxidative stress and DNA damage, activating oncogenes, inactivating tumor suppressor genes, stimulating malignant cell proliferation and inhibiting their senescence. These results suggested that 1,25$(OH)_2D$ deficiency increased spontaneous tumors by stimulating cell proliferation through increased oxidative stress and DNA damage, activating oncogenes and inhibiting tumor suppressor genes (Chen et al., 2018).

This study also demonstrated that supplementation with exogenous 1,25 $(OH)_2D_3$ prevented spontaneous tumor development in 1,25$(OH)_2D$ deficient 1α(OH)ase$^{-/-}$ mice (Chen et al., 2018). An allograft tumor experiment was performed using TM40D mouse breast cancer cells, which expressed VDR, but not 1α(OH)ase at both gene and protein levels. These cells were transplanted into 1α(OH)ase$^{-/-}$ mice and their WT littermates on a normal diet with without subcutaneous injection of 1,25 $(OH)_2D_3$ or on a rescue diet supplemented with the antioxidant NAC in drinking water. In the tumors transplanted into 1α(OH)ase$^{-/-}$ mice with either exogenous 1,25$(OH)_2D_3$ supplementation or combined supplementation with the rescue diet and NAC (Chen et al., 2018), the augmentation of tumor formation rate and tumor volume seen in 1α(OH)ase$^{-/-}$ mice on a normal diet were prevented. In addition, the increased levels of ROS and DNA damage, the protein expression levels of Prdx I/IV and SOD2, and SA-β-gal positive areas, the increased percentage of Ki67 and c-Met positive cells, the upregulated protein expression levels of oncogenes including Bmi1 and Ras and the downregulated protein expression levels of tumor suppressor genes including p16, p21 and p27 were all reversed. These results indicate that supplementation with exogenous 1,25$(OH)_2D_3$ inhibited allograft tumor initiation and growth by inhibiting oxidative stress and DNA damage, inactivating oncogenes, activating tumor suppressor genes, inhibiting malignant cell proliferation and stimulating their senescence.

2.5 Decreased 1,25-dihydroxyvitamin D and hypertension

An association between low serum levels of 25(OH)D and hypertension has been suggested in humans (McGreevy & Williams, 2011; Sorriento, De

Luca, Trimarco, & Iaccarino, 2018; Tamez, Kalim, & Thadhani, 2013), and both, 1α(OH)ase knockout mice (Zhou et al., 2008) and VDR knockout mice (Li et al., 2002) develop arterial hypertension and myocardial hypertrophy, which persists even after normalization of calcium homeostasis. Blocking of the renin–angiotensin–aldosterone system (RAAS) with angiotensin converting enzyme inhibitors normalized blood pressure and cardiac abnormalities, however, vitamin D may affect blood pressure through several mechanisms. Thus, endothelial function has a pivotal role in cardiovascular health and oxidative stress alters many functions of the endothelium, including modulation of vasomotor tone and inactivation of nitric oxide by ROS (Cai & Harrison, 2000). Vitamin D deficiency-induced hypertension has been reported to be associated with increased vascular oxidative stress (Argacha et al., 2011) and may exert an effect on endothelial function by protection against oxidative stress. Thus, active vitamin D may be an antihypertensive effector through the activation of antioxidant mechanisms.

One study demonstrated that $1,25(OH)_2D_3$ reversed the oxidative cardiac injury induced by isoproterenol in a rat model. These effects were mediated by the reduction in H_2O_2 levels accumulated in cardiac tissue; the increase in superoxide dismutase and catalase activities as antioxidant mechanisms (Wang, Zhu, Wang, Yang, & Cheng, 2013). $1,25(OH)_2D_3$ considerably restored vascular function, attenuated oxidative stress, and reduced monoamine oxidase expression in vascular preparations from diabetic rats (Sturza et al., 2019).

2.6 Decreased 1,25-dihydroxyvitamin D and reproductive function

Significant vitamin D deficiency was noted in males coming to infertility centers, however, how vitamin D impacted reproduction was not known until the VDR was found to be expressed in many organs of the male reproductive tract. It is now thought that vitamin D deficiency decreases male fertility by contributing to oxidative stress and gonadal insufficiency, disrupting spermatogenesis, and affecting sperm morphology and normal testicular calcium homeostasis (Shahid, Hashmi, & Rehman, 2021). Thus, when young male $1\alpha(OH)ase^{-/-}$ mice on a normal diet developed infertility, this was characterized by hypergonadotropic hypogonadism, with downregulation of testicular calcium channels, lower intracellular calcium levels, decreased sperm count and motility, and histological abnormalities of the testes (Sun et al., 2015). The proliferation of spermatogenic cells was decreased with downregulation of cyclin E and CDK2 and upregulation of p53 and p21 expression, whereas apoptosis of spermatogenic cells was increased with upregulation of Bax and p-caspase 3 expression and downregulation of Bcl-xl expression. When serum calcium and phosphorus were normalized by a rescue diet, the defective reproductive phenotype in the male 1α(OH)ase$^{-/-}$ mice, including the hypergonadotropic hypogonadism, the decreased sperm count and motility, histological abnormalities of the testis, and defective spermatogenesis, were reversed (Sun et al., 2015). However, when 1α(OH)ase$^{-/-}$ male mice were fed a rescue diet since weaning, and the reproductive phenotypes were evaluated at 12–18 months of age, it was found that 1,25(OH)$_2$D deficiency accelerated male reproductive aging, characterized by lower fertility efficiency and gonadal hormone levels, reduced cell proliferation, and increased cell apoptosis, cellular senescence, and SASP; supplementation with the antioxidant pyrroloquinoline quinone (PQQ) could largely rescue the reproductive aging phenotype (He, Wang, Shi, Shi, & Sun, 2021). These results suggest that the regulation by 1,25(OH)$_2$D of the male reproductive system is mediated through calcium and phosphate homeostasis in young animals, but is mediated through redox homeostasis in middle age.

Young female 1α(OH)ase$^{-/-}$ mice on a normal diet also developed infertility accompanied by decreased estrogen and progestogen levels, elevated follicle-stimulating hormone and luteinizing hormone levels, defects in follicular development and corpus luteum formation, uterine hypoplasia, and decreased ovarian expression of angiogenic factors including vascular endothelial growth factor (VEGF), angiopoietin-1 and -2, and Tie-2. When serum calcium and phosphorus were normalized by the rescue diet, the defective reproductive phenotype in the female 1α(OH)ase$^{-/-}$ mice, including the dysfunction in the hypothalamic-pituitary-ovarian axis, and ovarian angiogenesis were reversed (Sun et al., 2010). Whether 1,25(OH)$_2$D$_3$ deficiency accelerated female reproductive aging remains to be investigated.

2.7 Other abnormalities of 1,25-dihydroxyvitamin D deficiency

1,25(OH)$_2$D$_3$ deficient mice also display osteoarthritis (Shen et al., 2013) and rheumatoid arthritis (Gu et al., 2016), colon inflammation (Liu et al., 2016), abnormal adult neurogenesis (Zhu et al., 2012), and disordered glucose metabolism and insulin resistance (Yuan et al., 2022), however, whether these disorders induced by 1,25(OH)$_2$D deficiency are linked to oxidative stress remains to be investigated. 1,25$(OH)_2D_3$ has, however, been reported to cause a reduction of oxidative stress induced by high glucose levels in a human renal tubular cell line through upregulation of the AKT/uncoupling protein 2 signaling pathway (Zhu, Wu, & Guo, 2019). These antioxidant effects were achieved by increasing the activity of superoxide dismutase and matrix metalloproteinase and decreasing MAD levels. Additionally, the antioxidant potential of vitamin D was also observed in Alzheimer's disease induced by streptozotocin in animals. Active vitamin D reduced neuronal oxidative stress as well as mitochondrial aberrations induced in these animals. Therefore, vitamin D could also be useful in the prevention or treatment of Alzheimer's disease and other neurobehavioral disorders associated with oxidative stress (Yamini, Ray, & Chopra, 2018).

3. Mechanisms of action of 1,25-dihydroxyvitamin D in reducing oxidative stress

3.1 1,25$(OH)_2D$/VDR and UCP-2/NF-κB

More than two decades ago, Wiseman reported that vitamin D reduced iron-dependent liposomal lipid peroxidation. Structural similarities between cholesterol, vitamin D_3 and vitamin D_2 may underlie the membrane antioxidant properties of vitamin D (Wiseman, 1993). Vitamin D may act as an antioxidant by mitochondrial function stabilization. For example, it is known that cyanide causes neurotoxicity and neuronal cell death through mitochondrial dysfunction, which at low doses of cyanide is potentiated by the induced upregulation of uncoupling protein-2 (UCP-2). In rat primary cortical cell culture, active vitamin D was able to attenuate the mitochondrial dysfunction provoked by cyanide. This effect was mediated by suppressing the redox-mediated transcriptional upregulation of UCP-2 through the inhibition of NF-κB, resulting in reduced mitochondrial proton leak and restoration of mitochondrial membrane potential and cellular ATP (Li et al., 2008).

3.2 1,25$(OH)_2D$/VDR and Nrf2/Keap1

Nrf2 is a critical transcription factor and a crucial regulator of the expression of antioxidant enzymes. It was postulated that Nrf2 activation can be a possible mechanism by which 1,25$(OH)_2D_3$, can exert an antioxidant effect (Berridge, 2015). Nrf2, has been reported to be regulated by active vitamin D through its overexpression, nuclear translocation, or through reduction of Keap1 (Kelch-like ECH-associated protein 1), an adaptor subunit of Cullin 3-based E3 ubiquitin ligase (Nakai et al., 2014). Under physiological conditions, Nrf2 is sequestered in the cytoplasm by Keap1, an actin-binding repressor protein. Through this negative regulation of Nrf2 and ARE/EpRE activity, Keap1 contributes to increase oxidative stress. Reduced expression of Nrf2 and increased oxidative stress have been observed in diabetes, although the detailed mechanism of Nrf2 down-regulation has not been studied (Bitar & Al-Mulla, 2011; Tan et al., 2011).

A recent study reported that the antioxidative effects of $1,25(OH)_2D_3$ are mediated though transcriptional activation of Nrf2 via the VDR (Chen et al., 2019). Mouse embryonic fibroblasts (MEFs) from wild-type and VDR knockout mice were treated with 10^{-9}–10^{-7} M $1,25(OH)_2D_3$ for 24h, and the mRNA expression levels of Nrf2, determined by real-time RT-PCR, were significantly upregulated in $1,25(OH)_2D_3$-treated MEFs from wild-type mice in a dose-dependent manner, but not in those from VDR knockout mice. A VDRE like sequence in the 5′-flanking regions of the Nrf2 promoter, sequence NT039207 (retrieved from the NCBI mouse genome data base), was identified by computer-aided analysis, at position −460. Using mouse genomic DNA as a template, PCR was used to amplify the whole promoter segment −613 to −361. The PCR products without and with mutated VDRE were then cloned into pGL3-basic vectors (Nrf2-PGL3, Nrf2-PGL3 mutant), which were transiently transfected into MEFs. Luciferase activities were increased significantly in MEFs transfected with Nrf2-PGL3 plasmid compared with the empty plasmid, and were increased more dramatically in $1,25(OH)_2D_3$ treated MEFs transfected with Nrf2-PGL3 plasmid, however, luciferase activities were not increased in MEFs transfected with Nrf2-PGL3 mutant plasmid compared with the empty plasmid, and $1,25(OH)_2D_3$ failed to activate the mutant reporter. These results confirmed that the promoter region containing the predicted VDR-binding sites of the Nrf2 gene is sufficient to promote transcription and supported the hypothesis that Nrf2 transcriptional activation could be mediated by the VDR. Furthermore, using a chromatin immunoprecipitation (ChIP) approach, the VDR was shown to have the ability to physically bind the Nrf2 promoter. These findings provide a molecular mechanism underlying VDR-dependent Nrf2 transcriptional activation (Yang et al., 2022).

Unexpectedly, $1,25(OH)_2D_3$ did not significantly alter Nrf2 mRNA levels in BM-MSCs, but significantly increased Nrf2 and VDR protein levels. Additional, VDR-knockdown or VDR-knockout in human or mouse BM-MSCs resulted in reduced Nrf2 protein levels, but did not significantly change Nrf2 mRNA levels. To determine whether the 1,25 $(OH)_2D_3$-induced upregulation of Nrf2 expression at the transcriptional level is cell-type dependent, mouse embryonic fibroblasts (MEF), articular chondrocytes, BM-MSCs and MC3T3-E1 osteoblasts were treated with 1,25$(OH)_2D_3$ and the mRNA expression levels of the VDR target genes were examined, including vitamin D 24-hydroxylase (CYP24A1), Nrf2 and its target Nqo1. Results showed that CYP24A1, Nrf2 and Nqo1 mRNA levels were all significantly upregulated in 1,25$(OH)_2D_3$-treated MEF and articular chondrocytes; in contrast, CYP24A1, but not Nrf2 and Nqo1 mRNA levels were significantly upregulated in 1,25 $(OH)_2D_3$-treated BM-MSCs and MC3T3-E1 osteoblasts. These findings indicate that 1,25$(OH)_2D_3$ can upregulate Nrf2 mRNA expression at a transcriptional level in MEF and articular chondrocytes, but not in BM-MSCs and MC3T3-E1 osteoblasts. In view of the fact that intracellular levels of Nrf2 are mainly controlled by ubiquitination and proteasome-mediated protein degradation, whether 1,25$(OH)_2D_3$ can elevate Nrf2 by regulating its protein degradation was assessed. Human BM-MSCs were pre-treated with 1,25$(OH)_2D_3$ or control vehicle for 72 h and then treated with the protein translation inhibitor, cycloheximide (CHX). The results showed that 1,25$(OH)_2D_3$ delayed the degradation of Nrf2 in hBM-MSCs. In addition, the proteasome inhibitor MG132 significantly rescued the decline in Nrf2 induced by VDR deficiency. Moreover, 1,25$(OH)_2D_3$ significantly reduced the ubiquitination of the endogenous Nrf2 in BM-MSCs. Conversely, VDR knockdown increased the level of ubiqutin-Nrf2. These results indicate that 1,25$(OH)_2D_3$ increased Nrf2 accumulation in BM-MSCs and osteoblasts by inhibiting its ubiquitin-proteasome degradation (Yang et al., 2022).

Under basal conditions, Nrf2 is sequestered in the cytosol and marked for proteasomal degradation by Keap1 which acts as an adapter for E3-ubiquitin ligase (Cullinan, Gordan, Jin, Harper, & Diehl, 2004). Thus, previous studies have shown that Nrf2 degradation by the ubiquitin-proteasome pathway depends on its binding to Keap1, and that interference with Keap1-Nrf2 complex formation can reduce Nrf2 degradation. In the absence of electrophilic/oxidative stress, Nrf2 is ubiquitinated and targeted for degradation by the 26S-proteasome, keeping cellular levels of Nrf2 low and preventing activation of its response element. The expression of Keap1 protein was significantly reduced following 1,25$(OH)_2D_3$ treatment, with a significant decline of Keap1 mRNA levels. In contrast, protein and mRNA levels of

Keap1 were not altered in $VDR^{-/-}$ BM-MSCs treated with 1,25$(OH)_2D_3$, suggesting the inhibitory effect of 1,25$(OH)_2D_3$ on Keap1 is VDR-dependent. The Keap1 promoter was then cloned into a PGL-4.17 luciferase reporter vector to validate that 1,25$(OH)_2D_3$ represses Keap1 transcription. The luciferase activity driven by the Keap1 promoter containing a putative nVDRE was significantly inhibited by VDR/RXR overexpression plus 1,25$(OH)_2D_3$ treatment, suggestive of transcriptional repression of Keap1. ChIP-qPCR analysis showed that the VDR binds to this putative nVDRE of the Keap1 promoter and this binding was strengthened in the presence of 1,25$(OH)_2D_3$. Given that intracellular Keap1 is regulated by DNA methylation and histone modification, 1,25$(OH)_2D_3$ may repress Keap1 transcription by VDR-mediated hypermethylation of DNA or histones in the Keap1 promoter complex. To test this hypothesis, human BM-MSCs treated with 1,25$(OH)_2D_3$ or vehicle were subjected to DNA methylation analysis of the Keap1 promoter; the results showed that 1,25 $(OH)_2D_3$ did not alter the DNA methylation of the Keap1 promoter. Then H3K27me3 levels in the promoter region of Keap1 were examined in human BM-MSCs treated with 1,25$(OH)_2D_3$ using ChIP-qPCR assay; the H3K27me3 level in the Keap1 promoter region was significantly elevated in human BM-MSCs treated with 1,25$(OH)_2D_3$ compared to vehicle-treated control. Next, whether VDR might recruit histone methyltransferase (i.e., Ezh2) to the Keap1 promoter were examined using co-immunoprecipitation; we found that HA-VDR was immunoprecipitated with Flag-Ezh2. The interaction between endogenous VDR and Ezh2 was also evidenced in human BM-MSCs treated with vehicle or 1,25$(OH)_2D_3$. Moreover, VDR was able to mediate the recruitment of Ezh2 to the Keap1 promoter as demonstrated by ChIP-qPCR. Furthermore, the decline of Keap1 induced by 1,25$(OH)_2D_3$ was significantly blocked following GSK503 (an Ezh2 inhibitor) treatment. These data suggest that 1,25$(OH)_2D_3$ can repress Keap1 transcription by VDR-mediated histone hyper-methylation in the Keap1 promoter region by recruiting Ezh2 (Yang et al., 2022).

Osteoporosis and the bone aging phenotype induced by 1,25$(OH)_2D$ deficiency in male mice were significantly rescued in vivo upon the supplementation of oltipraz, an inhibitor of Nrf2 degradation. Increased oxidative stress, cellular senescence and reduced osteogenesis of BM-MSCs from VDR knockout mice were also significantly rescued when the cells were pre-treated with oltipraz. To further validate that the Nrf2-Keap1 pathway serves as the key mediator in the anabolic effect of 1,25$(OH)_2D_3$ on bone,

$Nrf2^{-/-}$ mice, or hBM-MSCs with shRNA mediated Nrf2-knockdown, were treated with $1,25(OH)_2D_3$. The results demonstrated that Nrf2 knockout largely blocked the bone anabolic effect of $1,25(OH)_2D_3$ in vivo and ex vivo, and Nrf2 knockdown in hBM-MSCs markedly blocked the role of $1,25(OH)_2D_3$ in inhibiting oxidative stress and promoting osteogenic differentiation and bone formation. This study therefore disclosed a mechanism whereby $1,25(OH)_2D_3$ postpones age-related osteoporosis via VDR-mediated activation of Nrf2-antioxidant signaling and inhibition of oxidative stress (Yang et al., 2022).

3.3 $1,25(OH)_2D$/VDR and Sirt1

Sirt1 is the most extensively studied member of the highly conserved sirtuin family and acts as a nicotinamide adenine dinucleotide (NAD)-dependent histone deacetylase, which regulates various biological processes including oxidative stress, metabolism, and aging (Zhang et al., 2017). Growing evidence has indicated that Sirt1 acts as a sensing regulator in response to oxidative stress and attenuates vascular dysfunction via cooperating with adenosine-monophosphate-activated protein kinase (AMPK) to activate antioxidant signals through various downstream effectors, including peroxisome proliferator-activated receptor-gamma co-activator 1 (PGC-1α), forkhead transcription factors (FOXOs), and peroxisome proliferative-activated receptor α (PPARα) (Meng, Qin, & Liu, 2020). The beneficial effects of Sirt1 on mitochondrial function are regulated by vitamin D, which acts by increasing the formation of Sirt1. Vitamin D insufficiency significantly decreases Sirt1 mRNA levels and activity in obese adipose tissue (Chang & Kim, 2017), whereas $1,25(OH)_2D_3$ supplementation inhibits oxidative stress and upregulates a Sirt1/AMPK/GLUT4 cascade in high glucose-treated 3T3L1 adipocytes and in adipose tissue of high fat diet-fed diabetic mice (Manna, Achari, & Jain, 2017). However, the detailed mechanism of $1,25(OH)_2D_3$ in regulating Sirt1 expression was not investigated in these studies.

A recent study has reported that $1,25(OH)_2D_3$ upregulates Sirt1 expression via VDR-mediated gene transcription (Chen et al., 2020). Results from this study demonstrated that Sirt1 mRNA and protein expression levels in mandibular tissue were downregulated dramatically in $1\alpha(OH)ase^{-/-}$ mice compared with their wild-type littermates. $1,25(OH)_2D_3$ significantly upregulated mRNA expression levels of Sirt1 in a dose-dependent manner in BM-MSCs derived from mandibular bone. A VDRE like sequence at position −884 in the 5′-flanking regions of the Sirt1 promoter, retrieved from the NCBI human genome data base, was identified by computer-aided analysis, and a ChIP-PCR assay demonstrated that the VDR could directly bind to the Sirt1 promoter on the predicted binding site. Using human genomic DNA as a template, PCR was used to amplify the whole promoter segment −863 to −1055. The PCR products without and with mutated VDRE were then cloned into pGL3-basic vectors (pGL3-Sirt1, pGL3-Sirt1-mut), which were transiently transfected into mandible derived BM-MSCs. Luciferase reporter assays demonstrated that luciferase expression levels were increased significantly in mandible derived BM-MSCs transfected with a Sirt1 binding sequencing plasmid compared with the vehicle, and the luciferase expression levels were augmented by 1,25 $(OH)_2D_3$; in contrast, luciferase activity was not increased in mandible derived BM-MSCs transfected with a pGL3-Sirt1 mutant plasmid compared with the vehicle. Furthermore, the Sirt1 agonist resveratrol could increase the interaction between Sirt1 and FOXO3a, reduce the level of FOXO3a acetylation, and promote its nuclear translocation, thereby promoting the proliferation and osteogenic differentiation of BM-MSCs and inhibiting their senescence. The protein expression levels of Sirt1 and FoxO3a were significantly downregulated in mandibular tissue from 1α $(OH)ase^{-/-}$ mice with an increase in oxidative stress as reflected by a reduction of the antioxidant enzymes Prdx I, SOD1 and SOD2 expression, and an increase of markers for osteocyte senescence and SASP, including β-gal, p16, p53 and p21. The targeted overexpression of Sirt1 in the $1\alpha(OH)ase^{-/-}$ MSCs restored the expression levels of these molecules. Overall, results from this study support the concept that $1,25(OH)_2D$-mediated increases in Sirt1/FOXO3a signaling levels can inhibit oxidative stress and greatly improve the bone loss caused by $1,25(OH)_2D$ deficiency.

3.4 $1,25(OH)_2D$/VDR and Bmi1

It has been previously reported that Bmi1 plays a vital role in maintaining stem cell self-renewal in many tissues through transcriptional repression of the p16/Rb and p19/p53/p21 signaling pathways (Anna et al., 2003; In-Kyung et al., 2003). It has also been reported that abnormal ROS elevation occurs in Bmi1 knockout mice, and that *N*-acetylcysteine (NAC) supplementation may correct the premature aging phenotype of Bmi1 null mice (Liu et al., 2009). Bmi1 is, therfore, another emerging target of the 1,25 $(OH)_2D$/VDR complex in maintaining redox balance (Sun et al., 2020).

$1,25(OH)_2D$ insufficient mice display increased oxidative stress and DNA damage, with activated p16/Rb and p19/p53/p21 signaling pathways and downregulated Bmi1. Human BM-MSCs from wild-type mice treated with $1,25(OH)_2D_3$ at a concentration of 10^{-9}–10^{-8} M for 24h revealed significant dose dependent increases in Bmi1 mRNA levels. A VDRE-like sequence at 2000kb upstream of the Bmi1 gene (http://jaspar.genereg.net/) was detected, and ChIP-PCR and EMSA assays demonstrated that the VDR could directly bind to the Bmi1 promoter on the predicted binding site. Luciferase reporter assays demonstrated that luciferase expression levels were increased significantly in BM-MSCs transfected with a Bmi1 binding sequencing plasmid compared with the vehicle, and the luciferase expression levels were augmented by $1,25(OH)_2D_3$; in contrast, luciferase activity was not increased in BM-MSCs transfected with a pGL3-Bmi1 mutant plasmid compared with the vehicle, and $1,25(OH)_2D_3$ could not activate the mutant reporter. Therefore, $1,25(OH)_2D_3$ upregulated Bmi1 expression via VDR at the transcriptional level (Sun et al., 2020). Bmi1 overexpression in MSCs also corrected $1,25(OH)_2D$ deficiency-induced oxidative stress and DNA damage, and cellular senescence of 1,25 $(OH)_2D$ insufficient mice by reducing levels of reactive oxygen species, elevating serum total superoxide dismutase levels, reducing the percentage of γH2A.X, p16, IL-1β and TNF-α positive cells and decreasing γH2A.X, p16, p19, p53, p21, IL-1β and IL-6 expression levels. Furthermore, 1,25 $(OH)_2D$ stimulated the osteogenic differentiation of MSCs, both ex vivo and in vitro, from wild-type mice but not from $Bmi1^{-/-}$ mice and 1,25 $(OH)_2D$ administration in vivo increased osteoblastic bone formation in wild-type mice, but not in $Bmi1^{-/-}$ mice. These results indicate that Bmi1 is a key downstream target of $1,25(OH)_2D$, that plays a crucial role in preventing bone loss induced by $1,25(OH)_2D$ deficiency by reducing oxidative stress.

In summary, $1,25(OH)_2D$, mediated by VDR, can directly upregulate gene transcription of Nrf2 or downregulate Keap1 by epigenetic regulation of its promoter, in order to activate Nrf2 antioxidant signaling, and can upregulate gene expression of Sirt1 and Bmi1 to inhibit oxidative stress and DNA damage. In normal cells, DNA damage activates p16/p53/p21, inactivates oncogenes, induces cell senescence and SASP, inhibits cell proliferation, and accelerates the occurrence of aging and age-related diseases, simultanously, decreasing $1\alpha,25(OH)_2D$ production by reducing 1α-hydroxylase expression levels. This sequence can be reversed by the effects of $1,25(OH)_2D$ on reducing oxidative stress and DNA damage. In

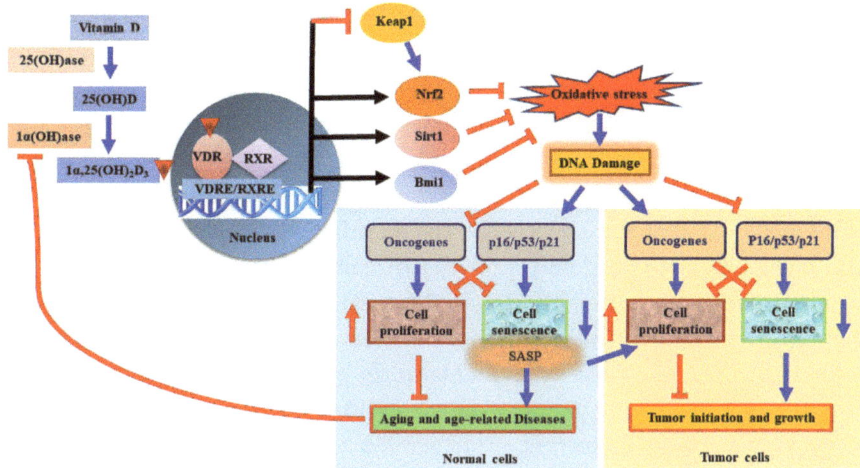

Fig. 1 Model of mechanisms leading from $1,25(OH)_2D_3$ to prevent aging and age-related diseases as well as tumorigenesis.

contrast, in tumor cells, DNA damage inactivates p16/p53/p21, activates oncogenes, inhibits tumor cell senescence, stimulates tumor cell proliferation, and leads to tumor occurrence and growth (Fig. 1). This too can be reversed by the effects of $1,25(OH)_2D$ on reducing oxidative stress and DNA damage.

4. The effect of vitamin D supplementation on aging and age-related diseases and on oxidative stress parameters in humans

Vitamin D deficiency in humans has been associated with an increased risk of all-cause mortality and a variety of chronic diseases, including a variety of cancers, cardiovascular diseases, diabetes, osteoporosis, autoimmune diseases and infectious diseases (Bouillon et al., 2019; Holick, 2007). Recent studies in humans have reported that vitamin D supplementation beneficially decreases oxidative stress parameters, however, there is no consensus on this subject. A systematic review and meta-analysis of 13 clinical trials found that vitamin D supplementation increased serum levels of total antioxidant capacity (TAC) and glutathione (GSH) and significantly decreased malondialdehyde (MDA) concentration compared to placebo, however, its effects on nitric oxide (NO) were not significant (Sepidarkish et al., 2019). A systematic review and meta-analysis of randomized controlled trials in diabetic patients, found that vitamin D supplementation significantly reduced serum high-sensitivity C-reactive protein and MDA levels and increased markers of NO release, serum levels of TAC and GSH in diabetic patients (Mansournia et al., 2018). Another meta-analysis demonstrated that vitamin D supplementation in women with polycystic ovary syndrome resulted in an improvement in high-sensitivity C-reactive protein, MDA and TAC, but did not affect NO and GSH levels (Akbari et al., 2018). Another meta-analysis demonstrated that taking vitamin D supplements among patients with psychiatric disorders had beneficial effects on Beck Depression Inventory (BDI), Pittsburgh Sleep Quality Index (PSQI), GSH, TAC and C-reactive protein levels (Jamilian et al., 2019).

As a result of a clinical trial performed with pregnant women, it has also been suggested that vitamin D could act as a competitive inhibitor of cytochrome P450scc in placental cells, avoiding the lipid peroxidation and oxidative stress phenomenon that usually contributes to the development and pathogenesis of preeclampsia (Zabul et al., 2015).

5. Controversies of the use of vitamin D supplementation in humans to reduce its anti-oxidant effect and ameliorate age-related disease

Despite the evidence above, the role of vitamin D supplementation has become a controversial topic because large intervention studies have not demonstrated significant benefits. Vitamin D_3 supplementation has been reported to show no effect on all-cause mortality (Neale et al., 2022), cannot reduce the fracture risk of the elderly (Zhao, Zeng, Wang, & Liu, 2017; LeBoff et al., 2022), cannot reduce the incidence of cancer or cardiovascular disease in the elderly (Manson et al., 2019), and cannot significantly reduce the incidence risk of diabetes in the elderly (Pittas et al., 2019). However, these studies all used inactive vitamin D and were performed in the elderly. Both skin synthesized vitamin D_3 (cholecalciferol) and oral supplemented vitamin D (which can be either vitamin D_3 or vitamin D_2, i.e., ergocalciferol) need to be converted to 25(OH)D by liver 25-hydroxylase and then to 1,25 $(OH)_2D$ by renal or extrarenal 1α-hydroxylase. 1,25 $(OH)_2D$, which can bind to VDR and exert its biological functions (Deeb, Trump, & Johnson, 2007). In human studies, as a result of declining renal function with age, a decrease in 1,25$(OH)_2D$ production by approximately 50% has been reported (Gallagher, 2013), with resultant decreased circulating levels. In a recent study, the protein expression levels of 1α-hydroxylase were examined in kidney, intestine and bone of 3-, 9- and 18-month-old wild-type mice using Western blots and revealed that they were progressively downregulated with age (Qiao et al., 2020). These results might help to explain why so many human clinical trials of vitamin D supplementation have failed (Boucher, 2020), i.e., vitamin D supplementation in the elderly may not have prevented age-related diseases, at least in part, because the expression levels of 1α (OH)ase may be too low in the elderly to synthesize sufficient active vitamin D, i.e., $1,25(OH)_2D$. Active vitamin D might be required for the prevention and treatment of aging and age-related diseases, but due to its narrow dose window, the effect is not obvious when the dose is small, and there are side effects such as hypercalcemia when the dose is increased (Tebben, Singh, & Kumar, 2016). In the past 30 years, many structurally modified $1,25(OH)_2D_3$ analogs have been synthesized in order to find new vitamin D analogs with selective biological activity and no adverse side effects, but this goal has not to date been optimally achieved (Jones & Kaufmann, 2019). Future developments may correct this need.

Acknowledgments

This work was supported by grants from the National Key R&D Program of China (2018YFA0800800 to D.M.), the National Natural Science Foundation of China (81730066 to D.M.) and the Canadian Institutes of Health Research (PJT-152963 to D.G.).

References

Akbari, M., Ostadmohammadi, V., Lankarani, K. B., Tabrizi, R., Kolahdooz, F., Heydari, S. T., et al. (2018). The effects of vitamin D supplementation on biomarkers of inflammation and oxidative stress among women with polycystic ovary syndrome: A systematic review and meta-analysis of randomized controlled trials. *Hormone and Metabolic Research*, *50*(4), 271–279. https://doi.org/10.1055/s-0044-101355.

Almeida, D. D., Scortecci, K. C., Kobashi, L. S., Agnez-Lima, L. F., Medeiros, S. R., Silva-Junior, A. A., et al. (2012). Profiling the resting venom gland of the scorpion *Tityus stigmurus* through a transcriptomic survey. *BMC Genomics*, *13*, 362. https://doi.org/10.1186/1471-2164-13-362.

Anna, V. M., Ricardo, P., Toshihide, I., In-Kyung, P., Michael, F. C., & Sean, J. M. (2003). Bmi-1 dependence distinguishes neural stem cell self-renewal from progenitor proliferation. *Nature*, *425*(6961), 962–967.

Annweiler, C., Schott, A. M., Berrut, G., Fantino, B., & Beauchet, O. (2009). Vitamin D-related changes in physical performance: A systematic review. *The Journal of Nutrition, Health & Aging*, *13*(10), 893–898.

Argacha, J. F., Egrise, D., Pochet, S., Fontaine, D., Lefort, A., Libert, F., et al. (2011). Vitamin D deficiency-induced hypertension is associated with vascular oxidative stress and altered heart gene expression. *Journal of Cardiovascular Pharmacology*, *58*(1), 65–71. https://doi.org/10.1097/FJC.0b013e31821c832f.

Berridge, M. J. (2015). Vitamin D: A custodian of cell signalling stability in health and disease. *Biochemical Society Transactions*, *43*(3), 349–358. https://doi.org/10.1042/BST20140279.

Bhat, M., & Ismail, A. (2015). Vitamin D treatment protects against and reverses oxidative stress induced muscle proteolysis. *The Journal of Steroid Biochemistry and Molecular Biology*, *152*, 171–179. https://doi.org/10.1016/j.jsbmb.2015.05.012.

Bitar, M. S., & Al-Mulla, F. (2011). A defect in Nrf2 signaling constitutes a mechanism for cellular stress hypersensitivity in a genetic rat model of type 2 diabetes. *American Journal of Physiology. Endocrinology and Metabolism*, *301*(6), E1119–E1129. https://doi.org/10.1152/ajpendo.00047.2011.

Boucher, B. J. (2020). Why do so many trials of vitamin D supplementation fail? *Endocrine Connections*, *9*(9), R195–R206. https://doi.org/10.1530/EC-20-0274.

Bouillon, R., Marcocci, C., Carmeliet, G., Bikle, D., White, J. H., Dawson-Hughes, B., et al. (2019). Skeletal and extraskeletal actions of vitamin D: Current evidence and outstanding questions. *Endocrine Reviews*, *40*(4), 1109–1151. https://doi.org/10.1210/er.2018-00126.

Cai, H., & Harrison, D. G. (2000). Endothelial dysfunction in cardiovascular diseases: The role of oxidant stress. *Circulation Research*, *87*(10), 840–844. https://doi.org/10.1161/01.res.87.10.840.

Chang, E. (2019). 1,25-Dihydroxyvitamin D decreases tertiary butyl-hydrogen peroxide-induced oxidative stress and increases AMPK/SIRT1 activation in C2C12 muscle cells. *Molecules*, *24*(21). https://doi.org/10.3390/molecules24213903.

Chang, E., & Kim, Y. (2017). Vitamin D insufficiency exacerbates adipose tissue macrophage infiltration and decreases AMPK/SIRT1 activity in obese rats. *Nutrients*, *9*(4). https://doi.org/10.3390/nu9040338.

Chen, H., Hu, X., Yang, R., Wu, G., Tan, Q., Goltzman, D., et al. (2020). SIRT1/FOXO3a axis plays an important role in the prevention of mandibular bone loss induced by 1,25(OH)2D deficiency. *International Journal of Biological Sciences*, *16*(14), 2712–2726. https://doi.org/10.7150/ijbs.48169.

Chen, L., Yang, R., Qiao, W., Yuan, X., Wang, S., Goltzman, D., et al. (2018). 1,25-Dihydroxy vitamin D prevents tumorigenesis by inhibiting oxidative stress and inducing tumor cellular senescence in mice. *International Journal of Cancer*, *143*(2), 368–382. https://doi.org/10.1002/ijc.31317.

Chen, L., Yang, R., Qiao, W., Zhang, W., Chen, J., Mao, L., et al. (2019). 1,25-Dihydroxyvitamin D exerts an antiaging role by activation of Nrf2-antioxidant signaling and inactivation of p16/p53-senescence signaling. *Aging Cell*, *18*(3), e12951. https://doi.org/10.1111/acel.12951.

Codoner-Franch, P., Tavarez-Alonso, S., Simo-Jorda, R., Laporta-Martin, P., Carratala-Calvo, A., & Alonso-Iglesias, E. (2012). Vitamin D status is linked to biomarkers of oxidative stress, inflammation, and endothelial activation in obese children. *The Journal of Pediatrics*, *161*(5), 848–854. https://doi.org/10.1016/j.jpeds.2012.04.046.

Cullinan, S. B., Gordan, J. D., Jin, J., Harper, J. W., & Diehl, J. A. (2004). The Keap1-BTB protein is an adaptor that bridges Nrf2 to a Cul3-based E3 ligase: Oxidative stress sensing by a Cul3-Keap1 ligase. *Molecular and Cellular Biology*, *24*(19), 8477–8486. https://doi.org/10.1128/MCB.24.19.8477-8486.2004.

Deeb, K. K., Trump, D. L., & Johnson, C. S. (2007). Vitamin D signalling pathways in cancer: Potential for anticancer therapeutics. *Nature Reviews. Cancer*, 7(9), 684–700. https://doi.org/10.1038/nrc2196.

Endo, I., Inoue, D., Mitsui, T., Umaki, Y., Akaike, M., Yoshizawa, T., et al. (2003). Deletion of vitamin D receptor gene in mice results in abnormal skeletal muscle development with deregulated expression of myoregulatory transcription factors. *Endocrinology*, *144*(12), 5138–5144. https://doi.org/10.1210/en.2003-0502.

Gallagher, J. C. (2013). Vitamin D and aging. *Endocrinology and Metabolism Clinics of North America*, *42*(2), 319–332. https://doi.org/10.1016/j.ecl.2013.02.004.

Garcia, M., Seelaender, M., Sotiropoulos, A., Coletti, D., & Lancha, A. H., Jr. (2019). Vitamin D, muscle recovery, sarcopenia, cachexia, and muscle atrophy. *Nutrition*, *60*, 66–69. https://doi.org/10.1016/j.nut.2018.09.031.

Gradinaru, D., Borsa, C., Ionescu, C., Margina, D., Prada, G. I., & Jansen, E. (2012). Vitamin D status and oxidative stress markers in the elderly with impaired fasting glucose and type 2 diabetes mellitus. *Aging Clinical and Experimental Research*, *24*(6), 595–602. https://doi.org/10.3275/8591.

Gu, X., Gu, B., Lv, X., Yu, Z., Wang, R., Zhou, X., et al. (2016). 1,25-Dihydroxy-vitamin D3 with tumor necrosis factor-alpha protects against rheumatoid arthritis by promoting p53 acetylation-mediated apoptosis via Sirt1 in synoviocytes. *Cell Death & Disease*, 7(10), e2423. https://doi.org/10.1038/cddis.2016.300.

He, J., Wang, H., Shi, J., Shi, M., & Sun, W. (2021). 1,25-Dihydroxyvitamin D deficiency accelerates male reproductive senescence in aging mice and 1,25(OH)2D3 alleviates oxidative stress via NF-kappaB/SOD. *American Journal of Physiology. Endocrinology and Metabolism*, *320*(4), E732–E746. https://doi.org/10.1152/ajpendo.00531.2020.

Holick, M. F. (2007). Vitamin D deficiency. *The New England Journal of Medicine*, *357*(3), 266–281. https://doi.org/10.1056/NEJMra070553.

Hu, C. Q., Bo, Q. L., Chu, L. L., Hu, Y. D., Fu, L., Wang, G. X., et al. (2020). Vitamin D deficiency aggravates hepatic oxidative stress and inflammation during chronic alcohol-induced liver injury in mice. *Oxidative Medicine and Cellular Longevity*, *2020*. https://doi.org/10.1155/2020/5715893.

In-Kyung, P., Dalong, Q., Mark, K., Becker, M. W., Michael, P., Weissman, I. L., et al. (2003). Bmi-1 is required for maintenance of adult self-renewing haematopoietic stem cells. *Nature*, *423*(6937), 302–305.

Jamilian, H., Amirani, E., Milajerdi, A., Kolahdooz, F., Mirzaei, H., Zaroudi, M., et al. (2019). The effects of vitamin D supplementation on mental health, and biomarkers of inflammation and oxidative stress in patients with psychiatric disorders: A systematic review and meta-analysis of randomized controlled trials. *Progress in Neuro-Psychopharmacology & Biological Psychiatry*, *94*, 109651. https://doi.org/10.1016/j.pnpbp.2019.109651.

Jones, G., & Kaufmann, M. (2019). Update on pharmacologically-relevant vitamin D analogues. *British Journal of Clinical Pharmacology*, *85*(6), 1095–1102. https://doi.org/10.1111/bcp.13781.

LeBoff, M. S., Chou, S. H., Ratliff, K. A., Cook, N. R., Khurana, B., Kim, E., et al. (2022). Supplemental vitamin D and incident fractures in midlife and older adults. *N Engl J Med*, *387*(4), 299–309. https://doi.org/10.1056/NEJMoa2202106.

Li, Y. C., Kong, J., Wei, M., Chen, Z. F., Liu, S. Q., & Cao, L. P. (2002). 1,25-Dihydroxyvitamin D(3) is a negative endocrine regulator of the renin-angiotensin system. *The Journal of Clinical Investigation*, *110*(2), 229–238. https://doi.org/10.1172/JCI15219.

Li, L., Prabhakaran, K., Zhang, X., Zhang, L., Liu, H., Borowitz, J. L., et al. (2008). 1Alpha,25-dihydroxyvitamin D3 attenuates cyanide-induced neurotoxicity by inhibiting uncoupling protein-2 up-regulation. *Journal of Neuroscience Research*, *86*(6), 1397–1408. https://doi.org/10.1002/jnr.21596.

Liu, J., Cao, L., Chen, J., Song, S., Lee, I. H., Quijano, C., et al. (2009). Bmi1 regulates mitochondrial function and the DNA damage response pathway. *Nature*, *459*(7245), 387.

Liu, Y., Chen, L., Zhi, C., Shen, M., Sun, W., Miao, D., et al. (2016). 1,25(OH)2D3 deficiency induces Colon inflammation via secretion of senescence-associated inflammatory cytokines. *PLoS One*, *11*(1), e0146426. https://doi.org/10.1371/journal.pone.0146426.

Manna, P., Achari, A. E., & Jain, S. K. (2017). Vitamin D supplementation inhibits oxidative stress and upregulate SIRT1/AMPK/GLUT4 cascade in high glucose-treated 3T3L1 adipocytes and in adipose tissue of high fat diet-fed diabetic mice. *Archives of Biochemistry and Biophysics*, *615*, 22–34. https://doi.org/10.1016/j.abb.2017.01.002.

Manson, J. E., Cook, N. R., Lee, I. M., Christen, W., Bassuk, S. S., Mora, S., et al. (2019). Vitamin D supplements and prevention of cancer and cardiovascular disease. *The New England Journal of Medicine*, *380*(1), 33–44. https://doi.org/10.1056/NEJMoa1809944.

Mansournia, M. A., Ostadmohammadi, V., Doosti-Irani, A., Ghayour-Mobarhan, M., Ferns, G., Akbari, H., et al. (2018). The effects of vitamin D supplementation on biomarkers of inflammation and oxidative stress in diabetic patients: A systematic review and Meta-analysis of randomized controlled trials. *Hormone and Metabolic Research*, *50*(6), 429–440. https://doi.org/10.1055/a-0630-1303.

McGreevy, C., & Williams, D. (2011). New insights about vitamin D and cardiovascular disease: A narrative review. *Annals of Internal Medicine*, *155*(12), 820–826. https://doi.org/10.7326/0003-4819-155-12-201112200-00004.

Meng, T., Qin, W., & Liu, B. (2020). SIRT1 antagonizes oxidative stress in diabetic vascular complication. *Frontiers in Endocrinology*, *11*, 568861. https://doi.org/10.3389/fendo.2020.568861.

Nakai, K., Fujii, H., Kono, K., Goto, S., Kitazawa, R., Kitazawa, S., et al. (2014). Vitamin D activates the Nrf2-Keap1 antioxidant pathway and ameliorates nephropathy in diabetic rats. *American Journal of Hypertension*, *27*(4), 586–595. https://doi.org/10.1093/ajh/hpt160.

Neale, R. E., Baxter, C., Romero, B. D., McLeod, D. S. A., English, D. R., Armstrong, B. K., et al. (2022). The D-health trial: A randomised controlled trial of the effect of vitamin D on mortality. *The Lancet Diabetes and Endocrinology*, *10*(2), 120–128. https://doi.org/10.1016/S2213-8587(21)00345-4.

Pittas, A. G., Dawson-Hughes, B., Sheehan, P., Ware, J. H., Knowler, W. C., Aroda, V. R., et al. (2019). Vitamin D supplementation and prevention of type 2 diabetes. *The New England Journal of Medicine*, *381*(6), 520–530. https://doi.org/10.1056/NEJMoa1900906.

Pourghassem Gargari, B., Pourteymour Fard Tabrizi, F., Sadien, B., Asghari Jafarabadi, M., & Farzadi, L. (2016). Vitamin D status is related to oxidative stress but not high-sensitive C-reactive protein in women with pre-eclampsia. *Gynecologic and Obstetric Investigation*, *81*(4), 308–314. https://doi.org/10.1159/000441781.

Qiao, W., Yu, S., Sun, H., Chen, L., Wang, R., Wu, X., et al. (2020). 1,25-Dihydroxyvitamin D insufficiency accelerates age-related bone loss by increasing oxidative stress and cell senescence. *American Journal of Translational Research*, *12*(2), 507–518.

Sepidarkish, M., Farsi, F., Akbari-Fakhrabadi, M., Namazi, N., Almasi-Hashiani, A., Maleki Hagiagha, A., et al. (2019). The effect of vitamin D supplementation on oxidative stress parameters: A systematic review and meta-analysis of clinical trials. *Pharmacological Research*, *139*, 141–152. https://doi.org/10.1016/j.phrs.2018.11.011.

Shahid, M., Hashmi, S. A., & Rehman, R. (2021). Oxidative stress, vitamin D deficiency and male infertility: An under-looked aspect. *The Journal of the Pakistan Medical Association*, *71*(1(B)), 786–787. https://doi.org/10.47391/JPMA.736.

Shen, M., Luo, Y., Niu, Y., Chen, L., Yuan, X., Goltzman, D., et al. (2013). 1,25(OH)2D deficiency induces temporomandibular joint osteoarthritis via secretion of senescence-associated inflammatory cytokines. *Bone*, *55*(2), 400–409. https://doi.org/10.1016/j.bone.2013.04.015S8756-3282(13)00158-0 (pii).

Sorriento, D., De Luca, N., Trimarco, B., & Iaccarino, G. (2018). The antioxidant therapy: New insights in the treatment of hypertension. *Frontiers in Physiology*, *9*, 258. https://doi.org/10.3389/fphys.2018.00258.

Sturza, A., Vaduva, A., Utu, D., Ratiu, C., Pop, N., Duicu, O., et al. (2019). Vitamin D improves vascular function and decreases monoamine oxidase A expression in experimental diabetes. *Molecular and Cellular Biochemistry*, *453*(1–2), 33–40. https://doi.org/10.1007/s11010-018-3429-2.

Sun, W., Chen, L., Zhang, W., Wang, R., Goltzman, D., & Miao, D. (2015). Active vitamin D deficiency mediated by extracellular calcium and phosphorus results in male infertility in young mice. *American Journal of Physiology. Endocrinology and Metabolism*, *308*(1), E51–E62. https://doi.org/10.1152/ajpendo.00076.2014. ajpendo.00076.2014 [pii].

Sun, H., Qiao, W., Cui, M., Yang, C., Wang, R., Goltzman, D., et al. (2020). The polycomb protein Bmi1 plays a crucial role in the prevention of 1,25(OH)2 D deficiency-induced bone loss. *Journal of Bone and Mineral Research*, *35*(3), 583–595. https://doi.org/10.1002/jbmr.3921.

Sun, W., Xie, H., Ji, J., Zhou, X., Goltzman, D., & Miao, D. (2010). Defective female reproductive function in 1,25(OH)2D-deficient mice results from indirect effect mediated by extracellular calcium and/or phosphorus. *American Journal of Physiology. Endocrinology and Metabolism*, *299*(6), E928–E935. https://doi.org/10.1152/ajpendo.00378.2010.

Tamez, H., Kalim, S., & Thadhani, R. I. (2013). Does vitamin D modulate blood pressure? *Current Opinion in Nephrology and Hypertension*, *22*(2), 204–209. https://doi.org/10.1097/MNH.0b013e32835d919b.

Tan, Y., Ichikawa, T., Li, J., Si, Q., Yang, H., Chen, X., et al. (2011). Diabetic downregulation of Nrf2 activity via ERK contributes to oxidative stress-induced insulin resistance in cardiac cells in vitro and in vivo. *Diabetes*, *60*(2), 625–633. https://doi.org/10.2337/db10-1164.

Tao, S., Yuan, Q., Mao, L., Chen, F. L., Ji, F., & Cui, Z. H. (2017). Vitamin D deficiency causes insulin resistance by provoking oxidative stress in hepatocytes. *Oncotarget*, *8*(40), 67605–67613. https://doi.org/10.18632/oncotarget.18754.

Tebben, P. J., Singh, R. J., & Kumar, R. (2016). Vitamin D-mediated hypercalcemia: Mechanisms, diagnosis, and treatment. *Endocrine Reviews*, *37*(5), 521–547. https://doi.org/10.1210/er.2016-1070.

Tohari, A. M., Zhou, X., & Shu, X. (2016). Protection against oxidative stress by vitamin D in cone cells. *Cell Biochemistry and Function*, *34*(2), 82–94. https://doi.org/10.1002/cbf.3167.

Uberti, F., Lattuada, D., Morsanuto, V., Nava, U., Bolis, G., Vacca, G., et al. (2014). Vitamin D protects human endothelial cells from oxidative stress through the autophagic and survival pathways. *The Journal of Clinical Endocrinology and Metabolism*, *99*(4), 1367–1374. https://doi.org/10.1210/jc.2013-2103.

Wang, Y. Q., Geng, X. P., Wang, M. W., Wang, H. Q., Zhang, C., He, X., et al. (2021). Vitamin D deficiency exacerbates hepatic oxidative stress and inflammation during acetaminophen-induced acute liver injury in mice. *International Immunopharmacology*, *97*, 107716. https://doi.org/10.1016/j.intimp.2021.107716.

Wang, X., Zhu, Y., Wang, X., Yang, Y., & Cheng, S. (2013). Cardioprotective effect of calcitriol on myocardial injury induced by isoproterenol in rats. *Journal of Cardiovascular Pharmacology and Therapeutics*, *18*(4), 386–391. https://doi.org/10.1177/1074248413482754.

Wimalawansa, S. J. (2019). Vitamin D deficiency: Effects on oxidative stress, epigenetics. *Gene Regulation, and Aging. Biology (Basel)*, *8*(2). https://doi.org/10.3390/biology8020030.

Wiseman, H. (1993). Vitamin D is a membrane antioxidant. Ability to inhibit iron-dependent lipid peroxidation in liposomes compared to cholesterol, ergosterol and tamoxifen and relevance to anticancer action. *FEBS Letters*, *326*(1-3), 285–288. https://doi.org/10.1016/0014-5793(93)81809-e.

Yamini, P., Ray, R. S., & Chopra, K. (2018). Vitamin D3 attenuates cognitive deficits and neuroinflammatory responses in ICV-STZ induced sporadic Alzheimer's disease. *Inflammopharmacology*, *26*(1), 39–55. https://doi.org/10.1007/s10787-017-0372-x.

Yang, R., Chen, J., Zhang, J., Qin, R., Wang, R., Qiu, Y., et al. (2020). 1,25-Dihydroxyvitamin D protects against age-related osteoporosis by a novel VDR-Ezh2-p16 signal axis. *Aging Cell*, *19*(2), e13095. https://doi.org/10.1111/acel.13095.

Yang, R., Zhang, J., Li, J., Qin, R., Chen, J., Wang, R., et al. (2022). Inhibition of Nrf2 degradation alleviates age-related osteoporosis induced by 1,25-Dihydroxyvitamin D deficiency. *Free Radical Biology & Medicine*, *178*, 246–261. https://doi.org/10.1016/j.freeradbiomed.2021.12.010.

Yu, S., Ren, B., Chen, H., Goltzman, D., Yan, J., & Miao, D. (2021). 1,25-Dihydroxyvitamin D deficiency induces sarcopenia by inducing skeletal muscle cell senescence. *American Journal of Translational Research*, *13*(11), 12638–12649.

Yuan, Q., Zhang, R., Sun, M., Guo, X., Yang, J., Bian, W., et al. (2022). Sirt1 mediates vitamin D deficiency-driven gluconeogenesis in the liver via mTorc2/Akt signaling. *Journal Diabetes Research*, *2022*. https://doi.org/10.1155/2022/1755563.

Zabul, P., Wozniak, M., Slominski, A. T., Preis, K., Gorska, M., Korozan, M., et al. (2015). A proposed molecular mechanism of high-dose vitamin D3 supplementation in prevention and treatment of preeclampsia. *International Journal of Molecular Sciences*, *16*(6), 13043–13064. https://doi.org/10.3390/ijms160613043.

Zhang, W., Huang, Q., Zeng, Z., Wu, J., Zhang, Y., & Chen, Z. (2017). Sirt1 inhibits oxidative stress in vascular endothelial cells. *Oxidative Medicine and Cellular Longevity*, *2017*, 7543973. https://doi.org/10.1155/2017/7543973.

Zhao, J. G., Zeng, X. T., Wang, J., & Liu, L. (2017). Association between calcium or vitamin D supplementation and fracture incidence in community-dwelling older adults: A systematic review and Meta-analysis. *JAMA*, *318*(24), 2466–2482. https://doi.org/10.1001/jama.2017.19344.

Zhou, C., Lu, F., Cao, K., Xu, D., Goltzman, D., & Miao, D. (2008). Calcium-independent and 1,25(OH)2D3-dependent regulation of the renin-angiotensin system in 1alpha-hydroxylase knockout mice. *Kidney International*, *74*(2), 170–179. https://doi.org/10.1038/ki.2008.101.

Zhu, X., Wu, S., & Guo, H. (2019). Active vitamin D and vitamin D receptor help prevent high glucose induced oxidative stress of renal tubular cells via AKT/UCP2 signaling pathway. *BioMed Research International*, *2019*, 9013904. https://doi.org/10.1155/2019/9013904.

Zhu, Y., Zhou, R., Yang, R., Zhang, Z., Bai, Y., Chang, F., et al. (2012). Abnormal neurogenesis in the dentate gyrus of adult mice lacking 1,25-dihydroxy vitamin D3 (1,25-(OH)2 D3). *Hippocampus*, *22*(3), 421–433. https://doi.org/10.1002/hipo.20908.

CHAPTER TWELVE

Antioxidant conjugated metal complexes and their medicinal applications

Anindya Roy and Jugun Prakash Chinta*

Department of Chemistry, National Institute of Technology-Warangal, Telangana, India
*Corresponding author: e-mail address: jugun@nitw.ac.in

Contents

Abstract

Antioxidants are naturally available and man-made substances have the ability to protect cells from damage due to a number of intracellular redox activities. Moreover, Antioxidants such as α-lipoic acid, curcumin and catechin are good anticancer agents. In recent years, the usage of metal complexes as therapeutic agents is gaining importance due to their useful biological properties. Most of the metal ions act as the essential components in building drug molecules that serve as medicines for cancer and neurodegenerative diseases. In particular, metals like copper, gold, ruthenium, and platinum have adequate anticancer properties at both micro- and nano-levels. Hence, conjugation of antioxidants with metals and metal-based compounds results in hybrid bioactive materials with improved anticancer properties. In this chapter, medicinal applications of antioxidant conjugated metal complexes are reviewed and discussed.

Abbreviations

BODIPY boron-dipyrromethene
BSA bovine albumin serum
CT-DNA calf thymus-DNA
Cur curcumin
EB ethidium bromide
FACS fluorescence-activated cell sorting

Vitamins and Hormones, Volume 121
ISSN 0083-6729
https://doi.org/10.1016/bs.vh.2022.10.003

Hcur	diglucosylcurcumin
HIV	human immunodeficiency virus infection
ISC	intersystem crossing
JNK	c-Jun N-terminal kinase
MB-DNA	megabase DNA
Mitocur	mitocurcumin
PARP	polyadenosine diphosphate-ribose polymerase
PDT	photodynamic therapy
PTA	1,3,5-triaza-7-phosphaadamantane
RAPTA	ruthenium arene PTA
ROS	reactive oxygen species
TBARS assay	thiobarbituric acid reactive substance assay

1. Introduction

Cisplatin, a platinum-containing anti-cancer drug, is useful for cancer patients around the world. The compound showed tremendous anticancer activity towards various types of cancers (Alderden, Hall, & Hambley, 2006). Consequently, various transition metal complexes of organic ligands have been well studied in recent years for medicinal applications (Mjos & Orvig, 2014). Transition metals have variable valence states and are capable of binding with various types of ligands. Hence, they can form metal complexes with distinctive geometries based on their oxidation states, co-ordination sites available, bond lengths, bond angles, and the strength of metal-ligand interactions (Monro et al., 2019). Complexes of transition metals display diverse magnetic and electronic characteristics that depend upon their electronic configurations of d- or f- shell (Chipman & Berry, 2020). Due to their unique properties mentioned above, they are important in designing and developing drugs for pharmacological applications such as anticancer, anti-inflammatory, antimalarial, antifungal, antipyretic, and antidiabetic drugs, cellular imaging, and sensors (Banerjee & Chakravarty, 2015; Englinger et al., 2019; Wenger, 2013). Metals such as copper, silver, gold, ruthenium, and also platinum possess potential anticancer properties as well as antimicrobial activities in both micro- and nano-levels (Gill & Vallis, 2019). Due to the immeasurable medicinal applicability of transition metal complexes, increased focus has been on the development of metal-based complexes as potential multitargeted drugs (Kenny & Marmion, 2019). Antioxidants are naturally available as well as man-made substances that are capable of protecting cells from damage by various redox active components (Helberg & Pratt, 2021). They are abundantly found in many food materials, including vegetables, fruits and also in dietary supplements (Nimse & Pal, 2015). The principal characteristic of an antioxidant is the prevention of intracellular oxidation, a chemical reaction that produces free radicals and reactive oxygen species and hence healthy cells are protected from oxidative damage in living organisms (Liu & Mabury, 2020).

Antioxidants, such as α-lipoic acid, curcumin and catechin are potential anticancer agents (Sen & Chakraborty, 2011). When conjugated with metal complexes, an antioxidant is a good anticancer agent with activity close to cis-platin (Suntharalingam, Song, & Lippard, 2014). Transition metals as well as the antioxidants have both anticancer activity and scavenging properties (Konkankit, Marker, Knopf, & Wilson, 2018). Hence, the conjugation of antioxidants with metal and metal-based compounds may result in hybrid compounds with improved anti-cancer activity. In This chapter we discuss the enhanced anticancer effects of antioxidant conjugated metal complexes.

2. Antioxidant conjugated metal complexes

2.1 Flavonoid metal complexes

Flavonoids, are phenol containing moieties which are isolated from a wide variety of vascular plants. Besides their antioxidant activities in plants, they act as antimicrobials and photo-receptors (Kasprzak, Erxleben, & Ochocki, 2015). In addition, they exhibit numerous biological actions, such as antiviral, anti-allergic, and anti-inflammatory activities. However, flavonoids are antioxidants that scavenge and diminish free radical formation. In vivo antioxidant action of flavonoids has been less studied because of their minimal uptake in humans, whereas the antioxidant activity of them has thoroughly been studied in the recent years, and predominant structure-activity relationships have been established (Pietta, 2000). Flavonoids also have been shown to possess a wide variety of anticancer effects. They attenuate reactive oxygen species and reduce enzyme activities that take part in arresting the cell cycle. Therefore, they induce autophagy, apoptosis, and suppress the uncontrolled growth of cancer cells (García & Castillo, 2008). Most of the flavonoids degrade to phenolic compounds in vivo but still possess the ability to scavenge radicals (Zou, Lu, & Wei, 2004). Examples of flavonoids are quecertin, kaempferol, morin, Chrysin and Genistein. Metal complexes of Flavonoids has been well studied in recent years and also have been recognized as potential anticancer agents due to their improved activity when coordinated with transition metal ions (Santini et al., 2014).

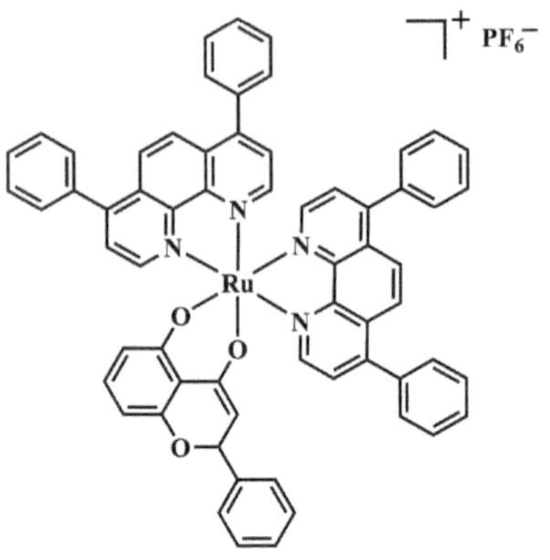

1a

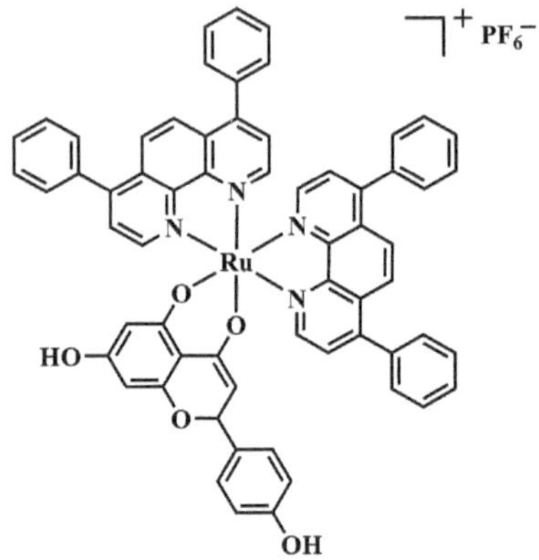

1b

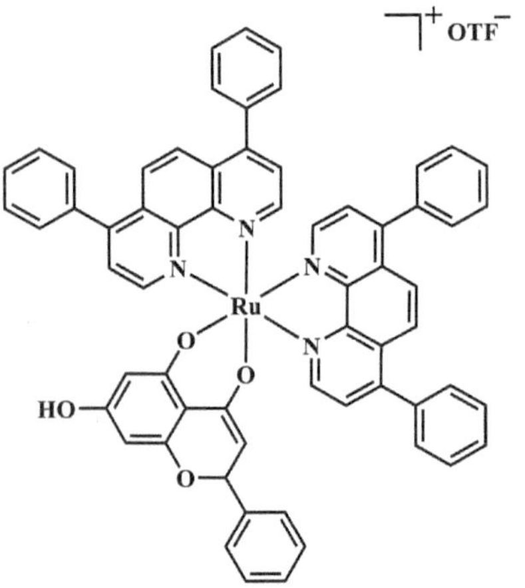

1c

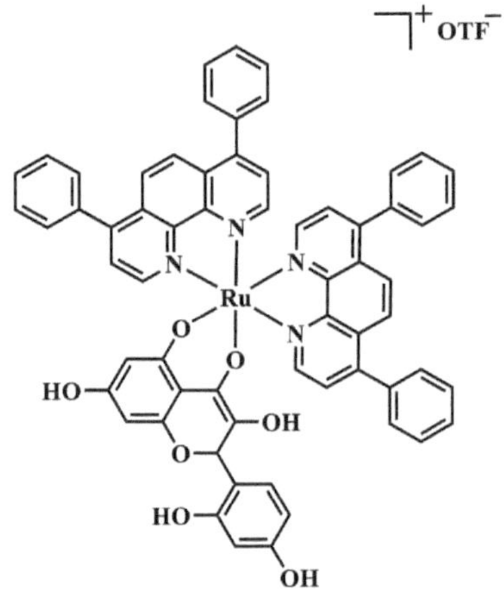

1d

Fig. 1 Antioxidant conjugated ruthenium complexes [Ru-(DIP)$_2$flv]X, where DIP is 4,7-diphenyl-1,10-phenanthroline, flv is a flavonoid ligand (5-hydroxyflavone in 1(a)-[Ru(DIP)$_2$(5-OHF)](PF$_6$), genistein in 1(b)-[Ru(DIP)$_2$(gen)](PF$_6$), chrysin in 1(c)-[Ru(DIP)$_2$(chr)](OTf), and morin in 1(d)-[Ru(DIP)$_2$(mor)](OTf)).

A set of four flavonoid containing mono-cationic Ru(II) polypyridyl complexes with general formulae [Ru-(DIP)$_2$flv]X (Fig. 1), were developed by Gasser and co-workers (Munteanu et al., 2020). Their cytotoxic effect was investigated with various cancer cell lines and promising outcomes were obtained for the complex 1b, which is comparable to doxorubicin and cisplatin (Table 1). The complex exhibits superior activity towards the MDA-MB-435S cell line (IC_{50} ~2.64 μM), a melanoma cell line derived from the pleural effusion of a female with metastatic breast adenocarcinoma.

Table 1 Summary of the antioxidant metal complexes and their activity.

S. no.	Complex	Activity (IC_{50})	Cell lines	Reference
1.	Cisplatin	3–19 μM	MDA-MB-435S	Munteanu et al. (2020)
	Doxorubicin	0.25–10 μM		
	$Ru(DIP)_2(5\text{-}OHF)](PF_6)$(1a)	30–50 μM		
	$Ru(DIP)_2(5\text{-}OHF)](PF_6)$(1b)	1–16 μM		
	$[Ru(DIP)_2(chr)](OTf)$(1c)	35–50 μM		
	$[Ru(DIP)_2(mor)](OTf)$(1d)	>50 μM		
2.	Chrysin-organogermanium(IV) complex (2)	N.A.	MCF-7	Fonseca et al. (2015)
3.	Ge(IV)-polyphenol complexes (3a and 3b)	N.A.	HepG2 cells	Pi, Zeng, Luo, Yang, and Cai (2013)
4	$[VO(chrysin)_2EtOH]_2$ (4)	16 μM	MG-63	León, Díez, Etcheverry, and Fuentes (2016)
5.	Kaempferol-zinc(II) complex (5)	N.A.	Oesophageal cancer cell	Tu, Pi, Jin, Cai, and Deng (2016)
			EC9706	
6.	$Ni(Que)_2(H_2O)_2$. (6)	N.A.	Plasmid DNA cleavage	Tan, Zhua, and Wang (2009)
7.	$Ni(Que)_2(H_2O)_2$. (6)	15–35 μM	A549, SMMC7721 and HepG2	Tan, Zhu, and Wang (2010)
	Quercetin	42–72 μM		
8.	Que-Cu^{II}-Sn_2^{IV} (7a)	<8.5 μM	pBR322 DNA	Tabassum, Zaki, Afzal, and Arjmand (2013)
	Que-Zn^{II}-Sn_2^{IV} (7b)	(GI_{50})		
9.	[Pd{(C,N)-C6H4CH2NH(Et) (Qu)] (8a)	1.5–6 μM	MCF7& TCC	Karami, Lighvan, Farrokhpour, Jahromi, and Momtazi-borojeni (2017)
	[Pd{(C,N)-C6H4CH2NH(Et) (Nar)] (8b)			

Continued

Table 1 Summary of the antioxidant metal complexes and their activity.—cont'd

S. no.	Complex	Activity (IC_{50})	Cell lines	Reference
10.	Rutin-copper complex	N.A.	Time-dependent oxidative DNA cleavage	Cai et al. (2019)
11.	$Na_2[VO(rut)(OH)_2]5H_2O$ (9)	95 μM	A549 cells	Goitia et al. (2019)
12.	4′′-Methoxy-5,7-dihydroxy-isoflavone metal complexes (10a–10e)	N.A.	MDA-MB-435, HepG2, Hela, SW620 and A549	Chen, Tang, Sun, Qiu, and Liang (2010)
13.	La (III) complex of L (Fig. 12)	17–23 μM	HepG-2 and HL-60 cells	Wang, Huang, Zhang, and Yang (2014)
14	Lipoic acid conjugated metal complexes (11a, 11b)	3–5 μM	Cervical and breast cancer cell lines and also T-cells	Velugula, Kumar, and Chinta (2020)
15.	[VO(Cur)(L)Cl] (12a, 12b)	<5 μM	MCF-7, HaCaT, and HeLa	Banerjee et al. (2015)
	[VO(scur)(L)Cl] (12c, 12d)			
16.	$[(bipy)Pd(Cur)][CF_3SO_3]$ (13)	20–24 μM	DU145, LnCaP, and PC3	Valentini et al. (2009)
17.	[(bpy-9)Zn-(Cur)Cl] (14)	11.79 μM	SHSY5Y cells	Pucci et al. (2013)
	$[(bpy\text{-}9)Zn\text{-}(Cur)Cl]^+BF_4^-$ (14a)	7.55 μM		
	$[(bpy\text{-}9)Zn\text{-}(Cur)Cl]_2SO_2$ (14b)	11.28 μM		
	$[(L^1)Zn(Cur)Cl]$ (14c)	19.31 μM		
	$[(L^2)Zn(Cur)Cl]$ (14d)	6.84 μM		
	$[(L^3)Zn(Cur)Cl]$ (14e)	20.33 μM		

18.	(p-cymene)Ru(demethoxycurcuminato) metal complex (15)	20–25 μM	MCF7 & A2780	Caruso et al. (2011)
19.	Oxovanadium(IV) complex of curcumin		HeLa cells	Banerjee et al. (2012)
	16a	8.1 ± 0.3 μM		
	16b	3.3 ± 0.4 μM		
20.	[Ln(R-tpy) (Cur)$(NO_3)_2$] (17a–17h)	7–60 μM	HeLa cells	Hussain et al. (2013)
21.	Copper(II) complexes of *N*-ferrocenyl methyl-L-amino acids with curcumin (18a–18c)	3 μM	MCF-7 and HeLa cancer cells	Goswami, Gadadhar, Gole, Karande, and Chakravarty (2013)
22.	Vanadyl complexes ferrocenyl conjugated terpyridine and curcumin (19a–19d)	2–10 μM	HepG2 and HeLa cancer cells	Balaji, Balakrishnan, Perumalla, Karande, and Chakravarty (2014)
23.	[Cu(L_1)Cur)]Cl (20a)	7.9 ± 0.3 μM	HeLa cells	Bhattacharyya et al. (2015)
	[Cu(L_2)(Cur)]Cl (20b)			
24.	Ternary oxidovanadium(IV) complexes of dipicolylamine derivatives (22a–22c, 22′a–22′c)	2–6 μM	MCF-7, HeLa cells	Bhattacharyya et al. (2017)
25.	RAPTA complexes containing curcuminoid ligands (23a and 23b)	≤1 μM	A2780cisR and A2780	Pettinari et al. (2014)
26.	Cobalt complexes of Cur and Mitocur (24a and 24b)	3.9 μM	MCF-7 cells	Garai et al. (2016)
27.	[Pt(Cur)$(NH_3)_2$](NO_3) (25a)	15 μM	HaCaT&BT474 cells	Mitra, Gautam, Kondaiah, and Chakravarty (2015)
	[Pt(acac)$(NH_3)_2$](NO_3) (25b)			
28.	[(bpy-9)Zn(Cur)] (26a)	12 μM	SD SHSY-5Y, SD Pc3, SD LNCaP	Pucci et al. (2011)
	[(bpy-OH)Zn(Cur)] (26b)	36 μM		

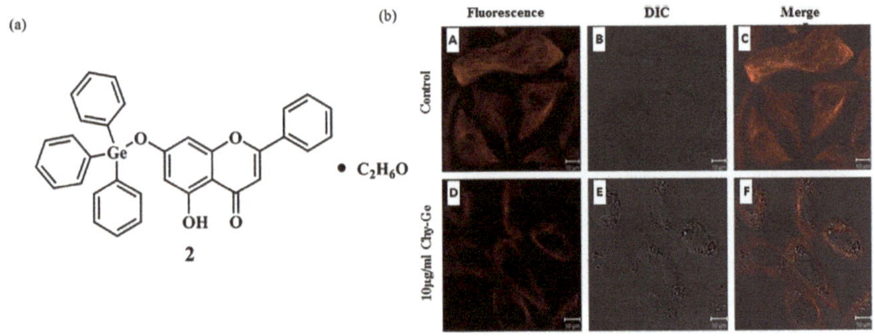

Fig. 2 (A) Chrysin-organogermanium(IV) complex (Chry-Ge). (B) Immunofluorescent data of MCF-7 cells treated Chry-Ge. *Reprinted with permission of Bioorganic & Medicinal Chemistry Letters, Elsevier from Yang, F., Jin, H., Pi, J., Jiang, J.-H., Liu, L., Bai, H.-H., Yang, P.-H., & Cai, J.-Y. (2013).Anti-tumor activity evaluation of novel chrysin-organogermanium(IV) complex in MCF-7 cells.* Bioorganic & Medicinal Chemistry Letters, 23, *5544–5551. https://doi.org/10.1016/j.bmcl.2013.08.055.*

In addition, it showed preferential accumulation in a human melanoma cell line (MDA-MB-435S) when compared to MCF-7 (human, ductal carcinoma) via a passive transport mechanism. Addition of the flavonoid moiety appears to control the performance of the complex in living cells and allows a more complex mode of action that ultimately leads to cell death.

Germanium (Ge) is an ingredient of many medicinal plants, such as ginseng, which is understood to play vital roles in the pharmaceutical effects of these plants. After appreciating the clinical application of carboxyethyl germanium sesquioxide (Ge-132) as a potential anti-cancer agent, Ge-containing compounds attracted much attention in the field of metal based anti-cancer drugs (Maier & Koepf, 1987). Yang et.al. have synthesized a novel chrysin-organogermanium(IV) complex (Chry-Ge) (Fig. 2A) and studied its anticancer activity (Yang et al., 2013). Chrysin is a natural product which is extracted from plants, propolis, and honey, and possesses anti-inflammatory, anticancer and antioxidant properties (Fonseca et al., 2015). It is clear from MTT assay that Chry-Ge inhibited the propagation of HepG2, Colo205, and also MCF-7 human cancer cell lines whereas a little cytotoxic effect occurred on human normal cells like MCF-10A under similar conditions. These results suggest that Chry-Ge exhibits improved anticancer effect and adequate selectivity towards cancer cells. Immuno-staining showed that the nuclei of MCF-7 cells consisted of fragmented morphology with a cluttered cytoskeletal system after treating with Chry-Ge (Fig. 2B). The microscopy data suggested that the cell rigidity

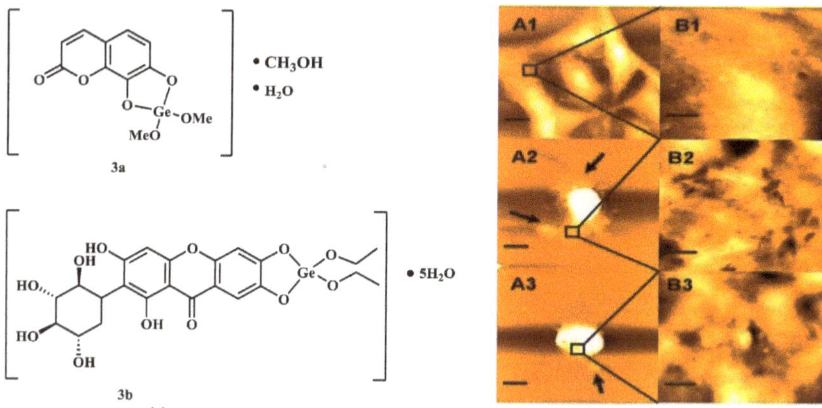

(a) (b)

Fig. 3 (A) Ge(IV)-polyphenol complexes. (B) Ge(IV)-polyphenol complexes induced HepG2 cell morphology changes imaged by AFM. Control (A1), Da-Ge(IV) complex (A2) and M-Ge(IV) complex (A3) treated HepG2 cell in 24h. *Reprinted with permission of Bioorganic & Medicinal Chemistry Letters, Elsevier from Pi, J., Zeng, J., Luo, J.-J., Yang, P.-H. & Cai, J.-Y. (2013). Synthesis and biological evaluation of germanium(IV)-polyphenol complexes as potential anti-cancer agents.* Bioorganic & Medicinal Chemistry Letters, *23, 2902–2908. doi:10.1016/j.bmcl.2013.03.061.*

and adhesion force of MCF-7 cells had been drastically decreased on treatment with Chry-Ge, which infers that membrane toxicity in MCF-7 cells might be a major target of Chry-Ge. Moreover, induced apoptosis in MCF-7 cells via ROS-dependent mitochondrial pathway in presence of Chry-Ge has been shown by fluorescence-based flow cytometric analysis. These results suggest that this Chry-Ge complex may play a leading role in next generation anticancer drug development.

Polyphenols are a type of micronutrient that contain antioxidants and can act as an anticancer agents through wide-range of mechanisms including elimination of cancer cells by modifying the signaling pathways, apoptosis induction and silencing of cell cycle events. Polyphenols also modulate the enzymatic activities concerned with proliferation of cancerous cells (Bhosale et al., 2020). In this context, Ge(IV)-polyphenol complexes (Fig. 3A) were prepared and tested their anti-oxidative activity. These complexes have excellent anti-oxidative activity with stronger hydroxyl scavenging effects and exhibit strong intercalation into calf thymus-DNA, in turn, show good effects on HepG2 cells (Pi et al., 2013). The flow cytometric studies suggest that Ge(IV)-polyphenol complexes instigate G0/G1 phase arrest in HepG2 cells and, by AFM analysis, Ge(IV)-polyphenol complexes induce morphological modification of HepG2 cells by significantly expanding their cell membrane particles(Fig. 3B). These results provide a basis for designing Ge(IV)-polyphenol complexes with good bioactivities and highlighting their potential applications as a promising anti-cancer drugs.

Vanadium complexes are considered to be good substitutes for platinum for the preparation of non-platinum antitumor agents. As discussed, flavonoid chrysin is a naturally available bioactive compound that possesses good anticancer activity. Moreover, the previous knowledge acquired from in vitro analysis of oxovanadium and vanadium containing drugs may allow these anticancer active compounds to be tested preclinically. On the basis of these facts, Fuentes and his co-workers prepared an oxovanadium(IV) complex VO-Chrys(4) (Fig. 4) and achieved the modification of the intracellular signaling pathways inside a human osteosarcoma cell line (MG-63) (León et al., 2016). The results evidenced that the VO-Chrys complex causes up-regulation of some proteins and the down-regulation of proteins such us FADD, AP2, NAK, and JNK, PAK, DAPK, CdK4, 6 and 7, PKB/AKT. Moreover, some proteins of PKC family, FAK, and PTK2B

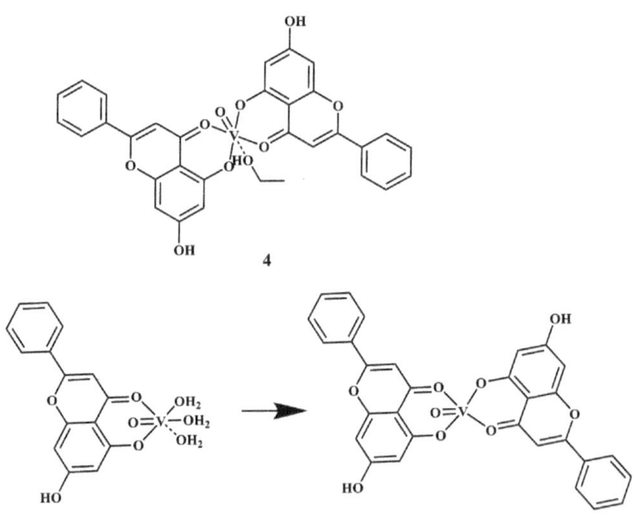

4

Acidic PH

Physiological PH

4a

4b

Fig. 4 Vanadyl-chrysin complexes$[VO(chrysin)_2EtOH]_2$ (VO-Chrys). (4a) Monomeric oxovanadium(IV)-chrysin complex under acidic conditions, (4b) dimeric oxovanadium(IV)-chrysin complex at physiological pH.

were found to be commonly expressed in all the treatments suggesting the important role of these proteins in the anticancer activity of the complex. The newly synthesized complex exhibited good activity in MG-63 cells with an IC_{50} of 16 μM while the chrysin and $VO^{(+2)}$ showed IC_{50}values of >100 μM. Taken together, these results indicate that vanadium-chrysin has superior antitumor activity that can pave a way towards the development of vanadium complexes as potential anticancer drugs.

Zinc Complex of Kaempferol (Fig. 5a′) has been prepared and the anticancer activity of the complex was tested on human oesophageal cancer cell (EC9706). MTT assay results show that the cytotoxic effect of the newly synthesized complex was nearly twice that of free Kaempferol (Fig. 5b′). Morphological- and ultrastructural-changes of the cellular membrane under the influence of kaempferol-Zn at subcellular- or nanometer-level were examined by AFM analysis data (Tu et al., 2016). Also, the complex induced apoptosis in oesophageal cells by regulating intracellular calcium ions as depicted by flow cytometric analysis. These results confirm the activity of the Kaempferol-Zn complex as a potential drug candidate for cancer therapy.

Plasmid DNA cleavage at physiological pH and temperature to a single or double stranded DNA via an hydrolytic pathway has been attained by quercetin-nickel(II) complex 6 (Fig. 6) (Tan et al., 2009). Thiobarbituric acid reactive substance assay (TBARS assay) along with agarose gel electrophoresis techniques have been applied to study the DNA cleavage activity

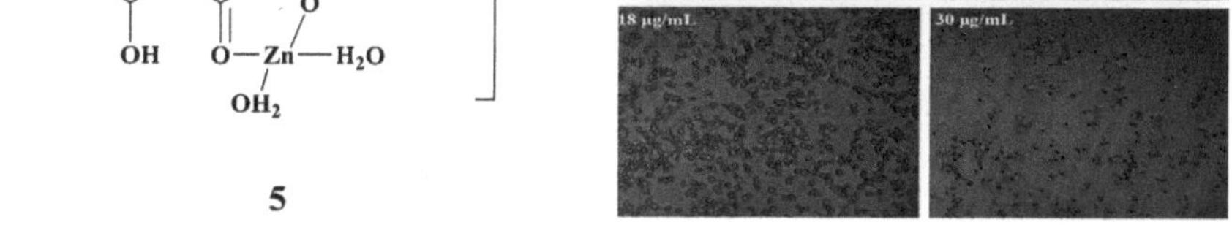

Fig. 5 (A) Kaempferol-zinc(II) complex. (B) Optical microscopy images of EC9706 cells treated with different concentrations of kaempferol-Zn for 24 h. *Reprinted with permission of Bioorganic & Medicinal Chemistry Letters, Elsevier from Tu. L.-Y., Pi, J., Jin, H., Cai, J.-Y. & Deng, S.-P. (2016). Synthesis, characterization and anticancer activity of kaempferol-zinc(II) complex.* Bioorganic & Medicinal Chemistry Letters, 26, *2730–2734. doi:10.1016/j.bmcl.2016.03.091.*

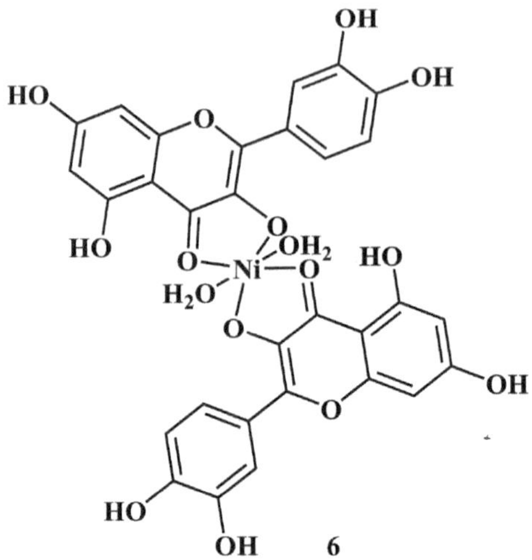

6

Fig. 6 Quercetin(Que) nickel(II) complex[$Ni(Que)_2(H_2O)_2$].

of the complex 6. The transformation of the supercoiled form of plasmid DNA to a nicked circular form depends on the concentration of the complex with the maximum rate of conversion at 100 μM of the complex. The rate of conversion (i.e. $0.76 \times 10^{-4}\ s^{-1}$) is 10^7-fold higher than that of uncatalyzed transformation of supercoiled DNA. The observed Stern-Volmer quenching constant (Ksq) for this complex is 1.0, which suggests that the $Ni(Que)_2(H_2O)_2$ complex is a strong DNA binder with high intercalation activity as compared to ethidium bromide (EB). DNA cleavage activity of the complex may be explained by the hydrolysis mechanism where a pentacoordinate intermediate is formed as a result of nucleophilic attack of a phosphorus atom in the phosphodiester bond of the DNA backbone.

Quercetin complexes are known to cleave DNA through breaking of both single- and double-strands. Therefore, the DNA cleavage and antitumor activity properties of nickel (II) quercetin metal complex 6 were studied (Tan et al., 2010). The IC_{50} values of the complex are depicted in Table 1. The IC_{50} values of $Ni(Que)_2(H_2O)_2$ complex were found to be higher than that of quercetin when tested on A549, SMMC7721 and HepG2 cells. The treatment of HepG2 cells with complex 6 resulted in unique morphological transformations as identified by nuclear contraction, fragmentation or chromatin condensation. Immunocytochemistry analysis reveals that the levels of p53 protein increased considerably whereas survivin and bcl-2 protein expression in HepG2 cells are reduced concurrently on treatment with $Ni(Que)_2(H_2O)_2$. The conclusion is that quercetin nickel

(II) complex selectively binds to the GC-rich promoter region, surviving through intercalation, and in turn, inhibits survivin gene expression and stimulating tumor cell apoptosis.

Subsequently, mixed metal complexes of quercetin such as complex 7a and 7b were designed, synthesized and the DNA cleavage and antitumor activities were reported. The complexes 7(a) and 7(b) bind CT DNA through an intercalation mode and cleaved pBR322 DNA via an oxidative mechanism and a hydrolytic pathway respectively (Tabassum et al., 2013). The hydroxy radical ($OH^{\bullet}$), hydrogen peroxide (H_2O_2) and superoxide radical (O_2^{-}) are reactive oxygen species (ROS) and these are the major metabolites causing the chronic diseases of respiratory disorders, diabetes, HIV, and cancer. Hence, removing ROS by molecular scaffolds involving SOD enzymatic activity has emerged as a key pathway to develop an innovative class of drugs. So, in vitro superoxide dismutase activity of redox active complex 7a was assessed by means of a xanthine/xanthine oxidase-NBT assay with an IC_{50} in the range of 2.26 μM. Furthermore, the cytotoxicity of both of the complexes were screened on a panel of human carcinoma cell lines (GI_{50} values <8.7 μM); and the results indicated that 7a has a better prospect as an anticancer agent (Fig. 7).

Karami et al. have systematically synthesized two mononuclear palladium(II) complexes of bioactive O,O-chelated ligands Quercetin and Naringin 8a an 8b (Fig. 8) and studied their binding interactions towards CT DNA (calf thymus DNA) and BSA (bovine serum albumin) (Karami et al., 2017). Complexes 8a and 8b showed binding constant (Kb) values 2.5×10^6 and 3.2×10^6 respectively which indicates the intercalative mode of binding between the DNA and the complexes. Moreover, the compounds are capable to deform MB-DNA complex by displacing MB from it. Hence, it can be concluded that the most likely binding mode of the complexes with DNA is through intercalation. In vitro cytotoxicity of the complexes on two cancer cells lines *i.e.*, breast cancer (MCF7) and bladder carcinoma (TCC) exhibits good anti-cancer activity with lower IC_{50} values (1.35–5.8 μM).

It is known that the water decoction process causes organic ingredients of plants to form complexes with trace metal elements, and this promotes them to show strong medicinal activities and disease treatment effects. Rutin (Fig. 9) is an extensively used naturally available antioxidant which is present in cherries, grapes, plums, and oranges, and owns metal ion chelation properties. The metal complexation properties of rutin was studied by HPLC and the results demonstrated the stable complex formation with iron followed by

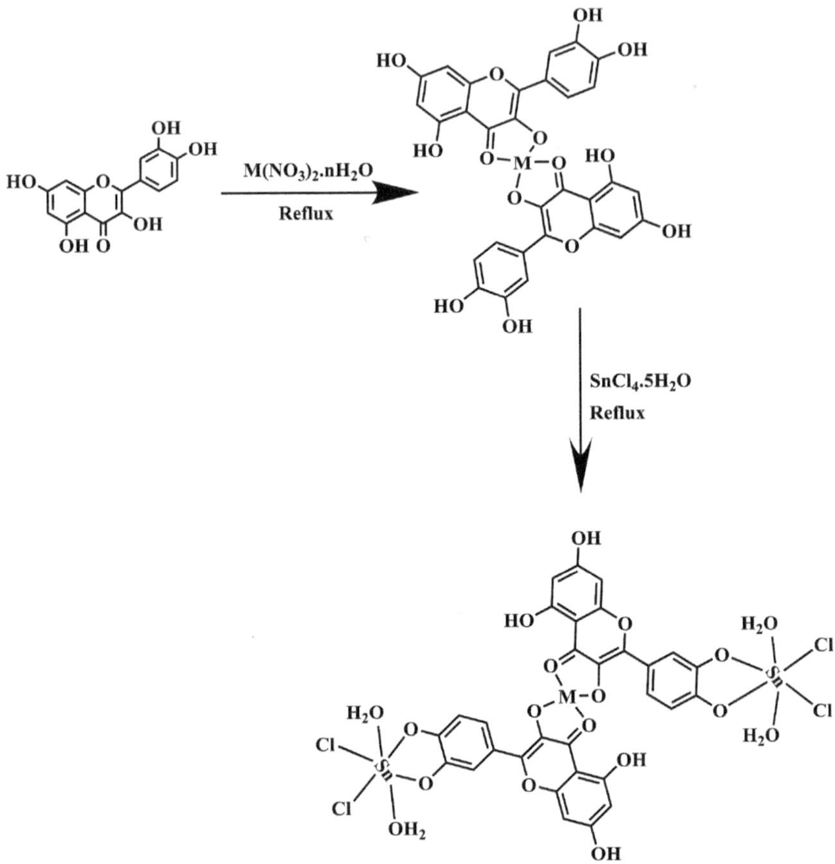

7

M - Cu(II) 7a M - Zn(II) 7b

Fig. 7 Synthetic route for the formation of monometallic and heterobimetallic complexes Que-Cu^{II}-Sn_2^{IV} (7a) and Que-Zn^{II}-Sn_2^{IV} (7b).

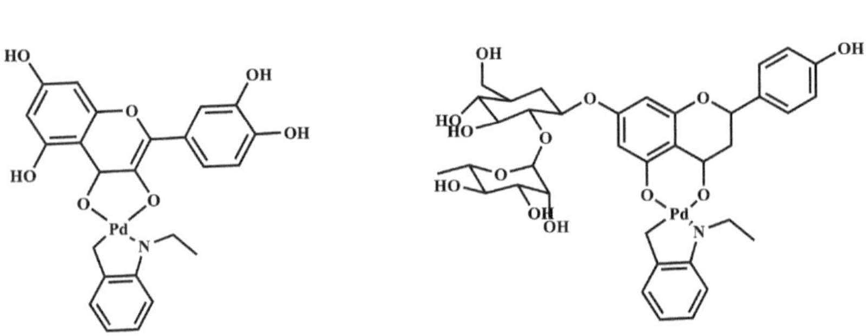

8a 8b

Fig. 8 Palladium (II) complexes of Quercetin and Naringin [Pd{(C,N)-$C_6H_4CH_2NH$(Et) (Qu)] (8a) and [Pd{(C,N)-$C_6H_4CH_2NH$(Et) (Nar)] (8b) (Qu, Quercetin; Nar, Naringin).

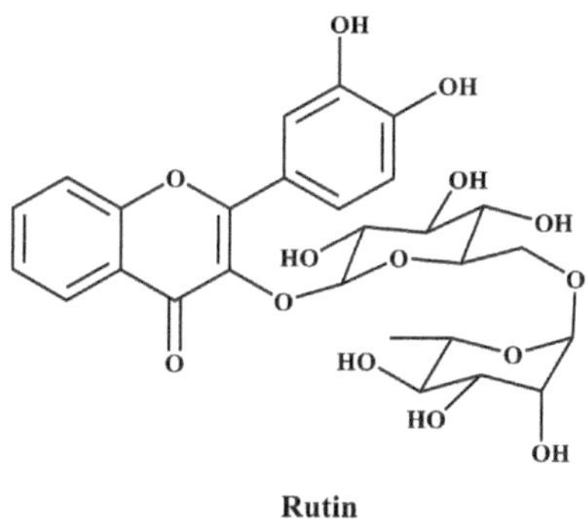

Fig. 9 Chemical structure of antioxidant rutin.

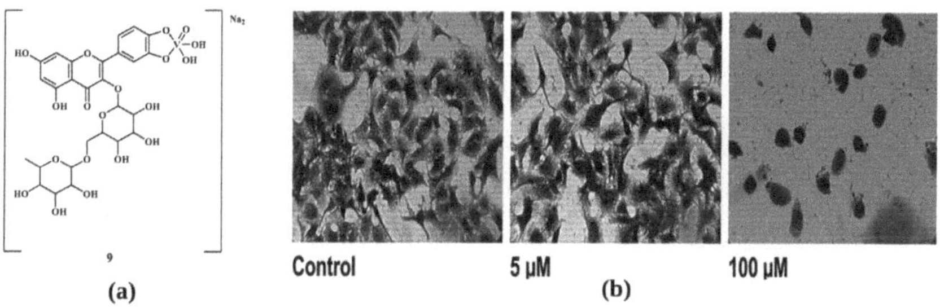

Fig. 10 (A)Vanadyl complex of rutin $Na_2[VO(rut)(OH)_2]5H_2O$ (rut, rutin, a glycosylated flavonoid). (B) Effect of VOrut on the cell morphology of the A549 cell line incubated for 24h. *Reprinted with permission of New Journal of Chemistry, Royal Society of Chemistry from Goitia, H., Quispe, P., Naso, G., Valeria R., Martínez Rey, M., Rizzi, A. C., Ferrer, E. G., & Williams, P. A. M. (2019). Interactions of rutin with the oxidovanadium(IV) cation. Anticancer improvement effects of glycosylated flavonoids.* New Journal of Chemistry, 43, *17636–17646. https://doi.org/10.1039/C9NJ01039D.*

strontium and copper (Cai et al., 2019). The rutin-copper complex showed time-dependent oxidative DNA cleavage which is unique among all other complexes studies in this report.

Improved antioxidant activity was achieved by vanadyl complex of rutin due to its unique catechol-like coordination geometry (Fig. 10A) (Goitia et al., 2019). The complex 9 improved the anticancer activity (IC_{50} ~95 μM) and reactive oxygen generation without effecting the normal lung cells while 100 μM of $V^{(IV)}O$ and rutin did not show any cytotoxicity on A549 cells (Fig. 10B). Rutin and its vanadyl complex bind with BSA and thus transported as well as stored by the protein. The antioxidant effect of the complex which involves C=O and deprotonated 5—OH groups in metal coordination was found to be superior to the ligand. This is due to the stabilization of the phenoxy radical generated from the free radical interaction.

Isoflavones are also known as phytoestrogens, as they are structurally similar to the estrogen. They are also considered to have adequate antioxidant as well as anticancer activities as flavonoids. Therefore, complexes of Mn^{2+}, Co^{2+}, Ni^{2+}, Cu^{2+} and Zn^{2+} with 4′-methoxy-5,7-dihydroxy-isoflavone (Fig. 11) were synthesized and studied their biological activities. In vitro anticancer activities of these complexes were assessed in five human cancer cell lines, *i.e.* MDA-MB-435, HepG2, Hela, SW620 and A549 using the flow cytometry and also MTT assay. Among all these metal complexes, Mn^{2+} (10a) and Cu^{2+} (10d) complexes exhibit better antitumor effect and selectivity than other analogues, and found to be more effective than the cisplatin under similar conditions (Chen et al., 2010). Furthermore, it is evidenced that the complexes display G2/M phase arrest, followed by early apoptosis as revealed by flow cytometry. The findings indicated that the

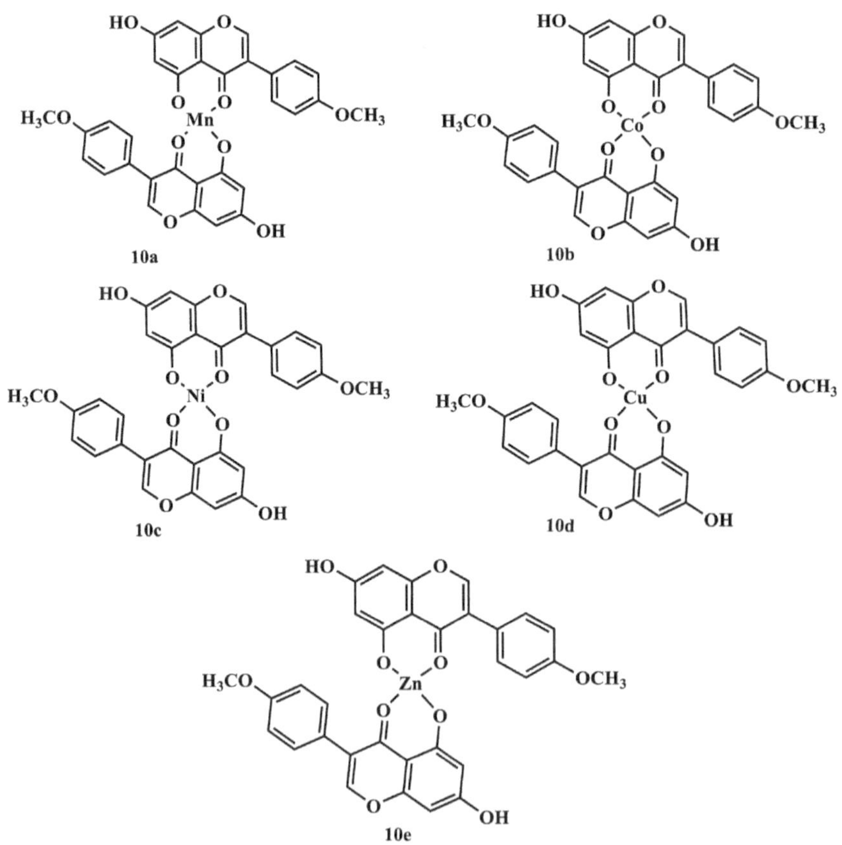

Fig. 11 Transition metal complexes of 4′-methoxy-5,7-dihydroxy-isoflavone (Mn^{2+}, Co^{2+}, Ni^{2+}, Cu^{2+} and Zn^{2+}).

superior activity of the Mn^{2+}complex is linked with cellular apoptosis and cell cycle arrest and helps in the development of distinctive anti-cancer remedies.

An antioxidant containing La(III)-Schiff base complex has been synthesized, and binding between the complex and the DNA has been examined. The results show that the complex binds through intercalation mode of binding and has a higher binding ability than the ligand (Wang et al., 2014). The binding constants (K_b) obtained from dye displacement studies resulted in the values ranging from 1.71 to 17.3×10^5 and 0.454 to $0.659 \times 10^5 \, L\,mol^{-1}$ for the La (III) complex and ligand L (Fig. 12) respectively. La (III) complex exhibited enhanced cytotoxicity in comparison to free ligand, against the HepG-2 and HL-60 cells tumor cell lines, that can be related with the enhanced DNA binding and cleaving abilities of the complex.

2.2 α-Lipoic acid metal complexes

α-Lipoic acid is known to play an indispensable role in mitochondrial dehydrogenase reactions and has acquired considerable curiosity as an antioxidant in recent years. It reacts as Lipoate or in its reduced form dihydrolipoate, with the ROS species such as peroxyl radicals, hydroxyl radicals, singlet oxygen or hypochlorous acid. It also prevents cell membrane damage by interacting with glutathione and vitamin C that may in turn regenerate vitamin E (Packer, WittE, & Tritschler, 1995). A distinctive hybrid compound consisting of antioxidant α-lipoic acid along with transition metal complex has been designed and prepared and also found to be effective anti-cancer agent as cisplatin (Velugula et al., 2020). Amongst the transition metal complexes invented in the present years, copper complexes depicted

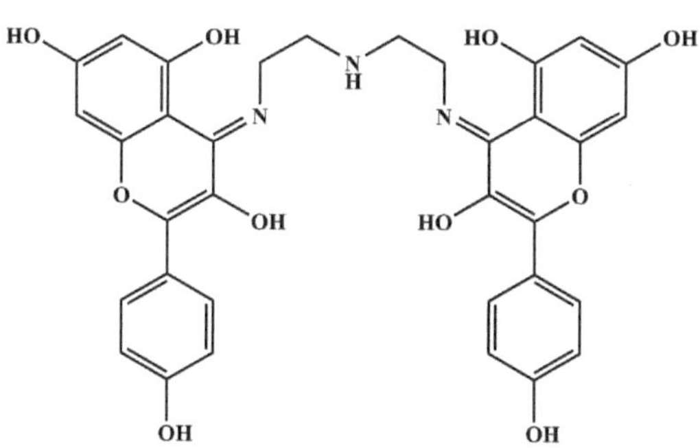

L

Fig. 12 Ligand derived from kaempferol and diethylenetriamine.

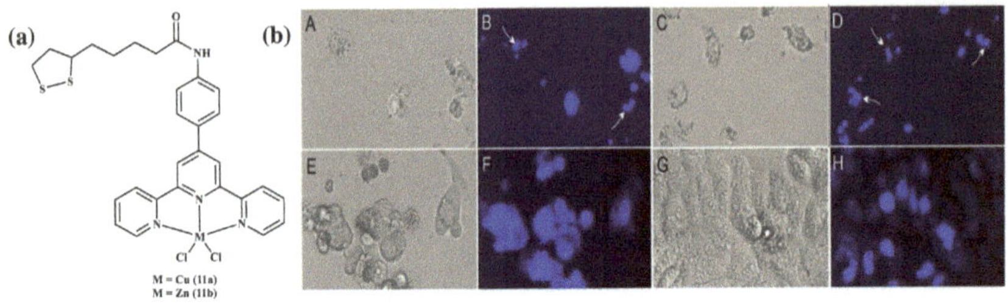

Fig. 13 (A) α-Lipoic acid conjugated terpyridine metal complexes. (B) Nuclear fragmentation exhibited by MCF-7 (A and B) and HeLa cells (C and D). Arrow indicated the fragmented nucleus. (E and F) untreated MCF 7 cells; and (G and H) untreated HeLa cells. *Reprinted with permission of Inorganica Chimica Acta, Elsevier from Velugula, K., Kumar, A., & Chinta, J. P. (2020). Nuclease and anticancer activity of antioxidant conjugated terpyridine metal complexes.* Inorganica Chimica Acta, 507, *119596. https://doi.org/10.1016/j.ica.2020.119596.*

cheering perceptions due to their high toxicity. Henceforth, α-lipoic acid conjugated terpyridine Cu(II) and Zn(II) complexes were developed (Fig. 13A). Both the complexes show good binding ability towards DNA through multiple binding modes. The Cu-complex (11a) displayed momentous chemical nuclease property as checked by plasmid DNA without using any stimulant. The cytotoxicity of these complexes in the cervical and breast cancer cell lines and also T-cells were investigated (Fig. 13B). The copper complex is proven to be as effective as cisplatin and also found to be more superior than some of the terpyridine-copper complexes ($IC_{50} \sim 4.95 \pm 0.5$ and $3.25 \pm 0.5\,\mu M$ for MCF-7 and HeLa cells). Mechanistic experiments revealed that the complexes impede the cell growth through apoptotic cell death.

2.3 Curcumin metal complexes

Curcumin is a natural product that has substantially been utilized as a food constituent in Asian region and also as a traditional medicine. It is a brilliant yellow colored compound, called Curcumin, is found in the plants of the Curcuma longa class. It is also the principal curcuminoid present in turmeric (Curcuma longa), a member of the ginger family, Zingiberaceae. It is popular for its spell-bound antioxidant as well as anti-cancer activities. Its known health benefits make this natural product an attractive ligand for metal coordination. Transition metal complexes of curcumin gathered the

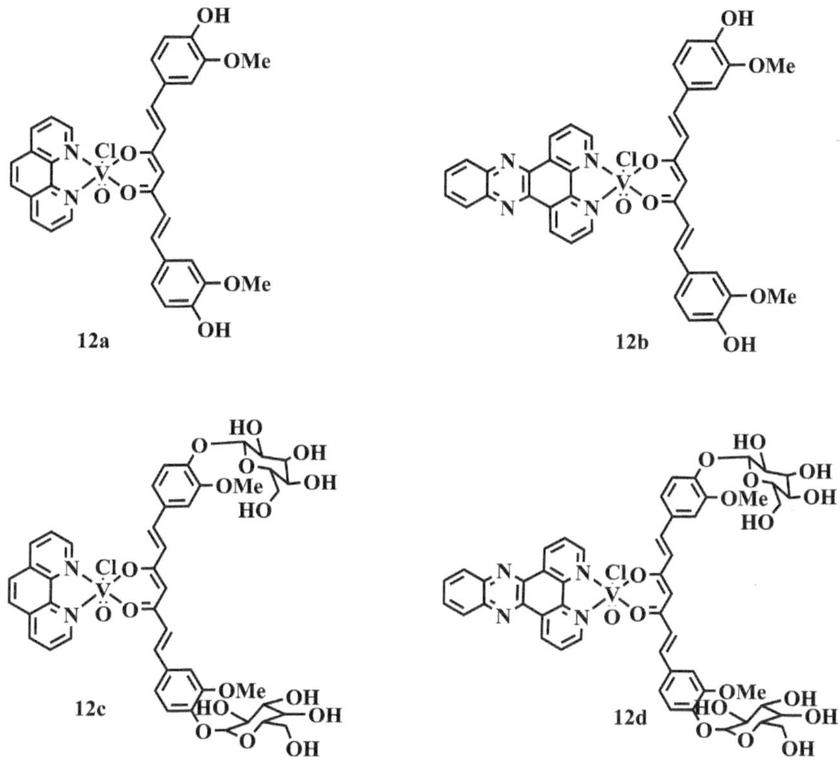

Fig. 14 Oxovanadium(IV) complexes containing curcumin-based ligands, viz. [VO(Cur)(L)Cl] (12a, 12b) and [VO(Hcur)(L)Cl] (12c, 12d), where L is 1,10-phenanthroline (phen in 12aand 12c), dipyrido[3,2-a:2′,3′-c]phenazine (dppz in 12b and 12d), Cur is curcumin then Hcur is diglucosylcurcumin.

attention of the researchers in present years for their lofty anti-cancerous effects. Thus, ternary oxovanadium(IV) complexes containing curcumin or diglucosylcurcumin and phenanthroline bases were studied for their anticancer activities (Fig. 14) (Banerjee et al., 2015). The sugar appended complexes of curcumin rendered extraordinarily high cytotoxicity in presence of visible light (400–700 nm) towards cancer cells like MCF-7, HaCaT, and HeLa when compared to their normal body cell lines ($IC_{50}<5\,\mu M$), and found to be non-toxic under dark conditions. The green emission of the complexes comes out to be useful for imaging cellular components. The complexes has strong binding capacity to CT-DNA, consequently it exhibited cleavage of supercoiled pUC19 DNA in red light by generating

ROS (•OH radical). The cell death takes place via an apoptotic pathway with the involvement of the ROS. Moreover, due to the presence of the vanadyl moiety, curcumin becomes more hydrolytically stable accompanied by an improvement in its photocytotoxicity.

Pucci and his co-workers have prescribed a new synthetic procedure for ionic Pd(II) leading to next generation anticancer agent with unique oxidation properties (Valentini et al., 2009). The newly synthesized complex 13, consists of curcumin as primary ligand and bipyridine as an auxiliary ligand (Fig. 15). The metal center is coordinated to two dissimilar chelating ligands, i.e.,4,4′-dinonyl-2,2′-bipyridine (bipy) and the curcumin. The complex exhibited cytotoxicity and induces apoptosis of human prostate cancer cells (DU145, LnCaP, and PC3) through the production of ROS and JNK (c-Jun N-terminal Kinase) phosphorylation linked with GSTp1 down regulation. This study indicates that curcumin itself has an ability to obstruct cell magnification of prostate cancer cells due to the G2 phase arrest whereas the complex 13 show cytotoxicity along with apoptosis in hormone-independent prostate cancer cells. It is observed from the IC_{50} values ranging from 19.7 to 24 μM that the complex is not as much toxic as curcumin (IC_{50} ~2.2–7.8 μM) in all cell lines. The complex does not elicit any activation of PARP enzyme whereas curcumin induces DNA damage and PARP cleavage. The data suggest that the complexation of curcumin with heavy metal center and its conjugation with another bioactive ligand resulted in apoptosis signal transduction and increment of anticancer activity in prostate cancer cell lines. Hence, the current study suggests a potential design strategy in the development of the metal-based drugs cytotoxicity.

Furthermore, heteroleptic metal complexes consisting of Zn(II) and curcumin have been synthesized by using the *N,N* chelating moieties as

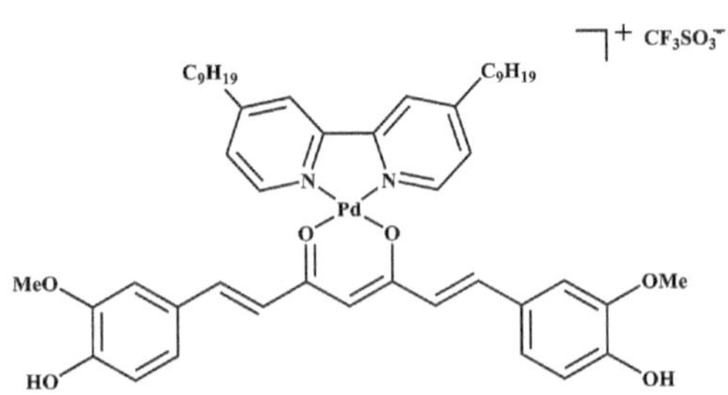

13

Fig. 15 Pd(II) complex of Curcumin and 4,4′-dinonyl-2,2′-bipyridine (bipy) [(bipy)Pd(Cur)][CF_3SO_3].

an auxiliary ligand (Pucci et al., 2013). Two dissimilar Zn(II) complexes were obtained, i.e., neutral complex where the sulfate anion acts as a bridging group through two Zn(II) centers and tetracoordinated Zn(II) cation with tetrafluoroborate as counter ion (Fig. 16). In addition, different N, N chelating units resulted in clusters, (14c–14e), in which the Zn(II)

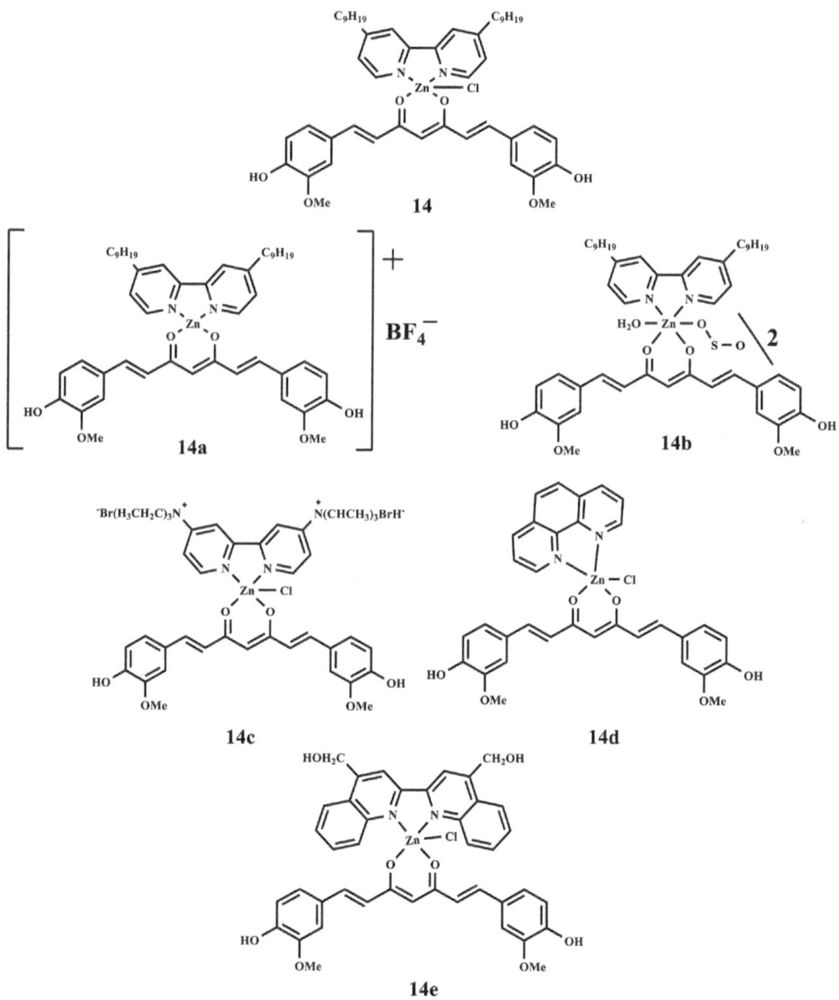

Fig. 16 Zinc(II) complexes of curcumin and N,N-chelating ligands: 14-[(bpy-9)Zn-(Cur)Cl], 14a-[(bpy-9)Zn-(Cur)Cl]$^+BF_4^-$, 14b-[(bpy-9)Zn-(Cur)Cl]$_2SO_2$, 14c-[(L^1)Zn(Cur)Cl], 14d-[(L^2)Zn(Cur)Cl] & 14e-[(L^3)Zn(Cur)Cl], where, L^1, L^2, L^3 are Modified bipyridine and phenanthroline based N, N chelating ligands.

ion maintains the same penta-coordination as seen in the previous complex (14). The complexes 14a and 14d induces cell apoptosis by following a molecular mechanism that involves the activation of some enzymes such as JNK, caspase-3 and change in mitochondrial membrane potential. The detailed studies revealed that the complexes exhibit good cytotoxic activity, IC_{50} values in micromolar level, and found to be superior than the reference compound 14.

A Ru-Curcumin complex, (*p*-cymene)Ru(demethoxycurcuminato) (Complex 15), has been isolated by using commercial curcumin reagent as a β-diketonato ligand (Fig. 17) (Caruso et al., 2011). The anti-proliferative action of the complex on the five Gastrointestinal tumors cell lines depicts the propensity of the complex toward the colon-rectal tumor cell line HCT116 with IC_{50} ~13.98 ± 1.503 μM followed by breast cancer cell MCF7 (IC_{50} ~19.58 ± 2.367 μM) and ovarian cancer cell A2780 (IC_{50} ~23.38 ± 3.334 μM) cell lines. The effect on lung carcinoma A549 and human glioblastoma U-87 cell lines are less significant.

Oxovanadium(IV) complex of curcumin (Fig. 18A) has been proven for it's significant photocytotoxicity on human cervical HeLa cells in visible light of 400–700 nm (Banerjee et al., 2012). Curcumin manifests remarkable photocytotoxicity in HeLa cells undervisible light providing IC_{50} ~ 8.2 ± 0.2 μM (Fig. 18B). The photocytotoxicity of complex 16b is quite comparable to that of reported for Photofrins (IC_{50} ~ 4.3 ± 0.2 μM,) and better than its analogue [VOCl(dppz)$_2$]Cl (IC_{50} ~12.0 μM). The MTT and confocal imaging data delineates that curcumin as well as its complexes innocuously persist in the interior of the cells and demolish the cancer cells

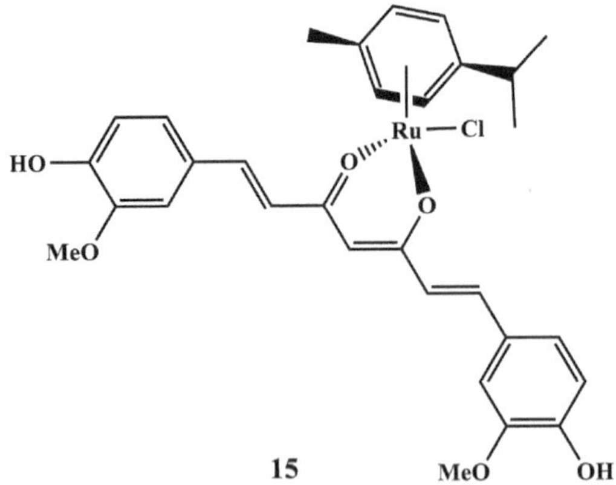

15

Fig. 17 (p-cymene)Ru(demethoxycurcuminato) metal complex.

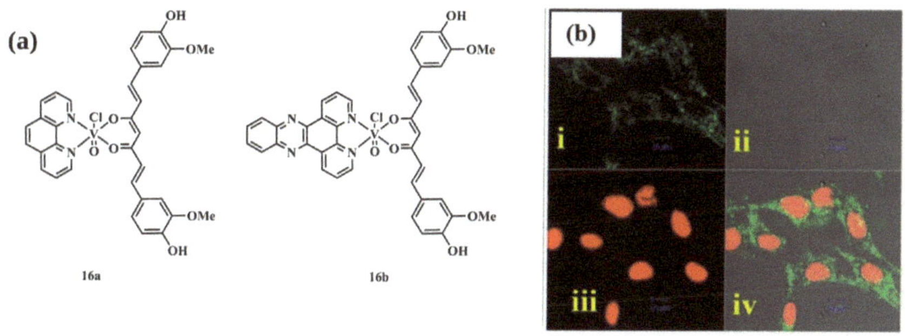

Fig. 18 (A) VO^{2+} complexes of curcumin and the phenanthroline ligands. Ligand = phen in 16a; dppz in 16b. (B) Confocal images of HeLa cells treated with curcumin (10 μM), 16b (10 μM) and propidium iodide (PI, 10 mg mL^{-1}). Panels (i) show the green emission of curcumin. Panels (ii) show the bright field images. Panels (iii) correspond to the red emission of PI. Panels (iv) show the merged images. *Reprinted with permission of Chemical Communications, Royal Society of Chemistry from Banerjee, S., Prasad, P., Hussain, A., Khan, I., Kondaiah, P., & Chakravarty, A. R. (2012). Remarkable photocytotoxicity of curcumin in HeLa cells in visible light and arresting its degradation on oxovanadium(IV) complex formation.* Chemical Communications, 48, *7702–7704. doi:10.1039/C2CC33576J.*

only upon photo-irradiation. The DNA melting data indicate the complexes binds to Calf thymus-DNA in a partial intercalative mode and show photo-activated DNA cleavage via apoptotic pathway.

Hussain et. al. has incorporated Lanthanide(III) metal ions into curcumin(Cur) and diglucosylcurcumin(Hcur) moieties and studied their DNA photocleavage activity and photocytotoxicity (Fig. 19) (Hussain et al., 2013). The hydrolytic decomposition of curcumin was found to be decreased significantly upon binding to the lanthanides while preserving its photocytotoxic potential. The complexes exhibit good binding ability towards ct-DNA and cleave the supercoiled plasmid DNA to its nicked circular form in presence of visible (454 nm) and UV-A (365 nm) light via hydroxy radical and singlet oxygen pathways. Moreover, the complexes show remarkable photocytotoxicity to HeLa cells in visible light (IC_{50} ~7–60 μM) and persist its non-toxicity in the dark. Fluorescence-activated cell sorting (FACS) analysis of the complex-treated HeLa cells illustrated the apoptotic pathway of cell death.

Complexes of bio-essential copper metal ion chelated with *N*-ferrocenylmethyl-L-amino acids where *N*-ferrocenylmethyl-L-amino acids are ferrocenyl methyl-L-tyrosine (18a, 18d), ferrocenylmethyl-L-tryptophan (18b, 18e) and ferrocenylmethyl-L-methionine (18c, 18f) and

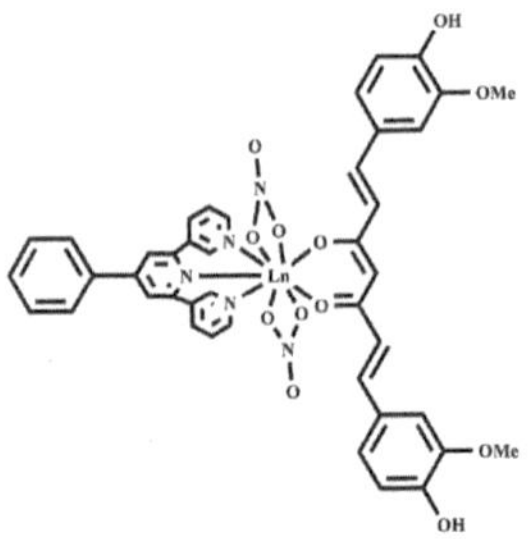

17a

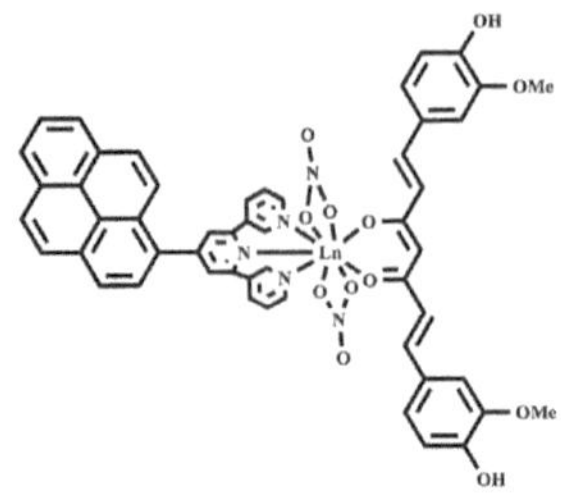

17b

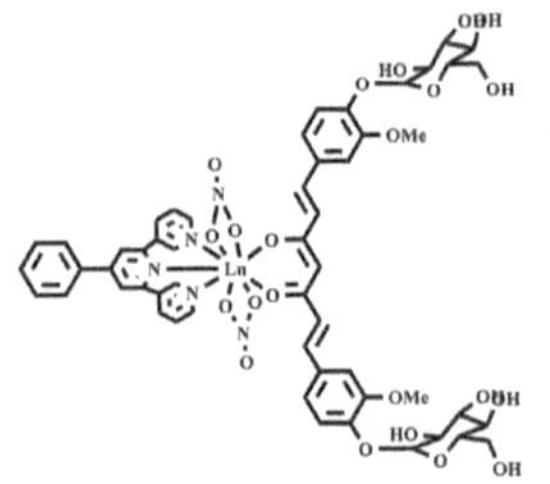

17c

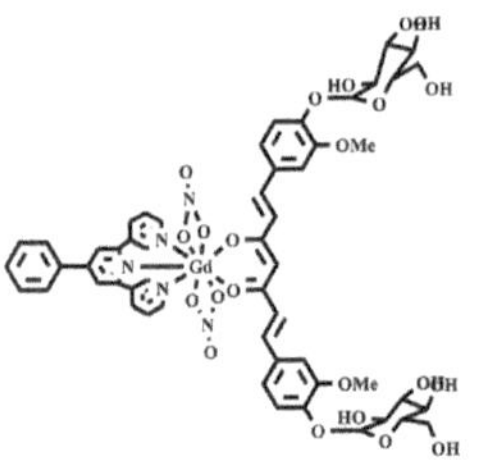

17d

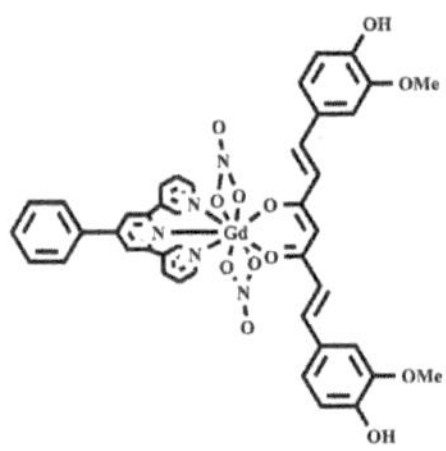

17e

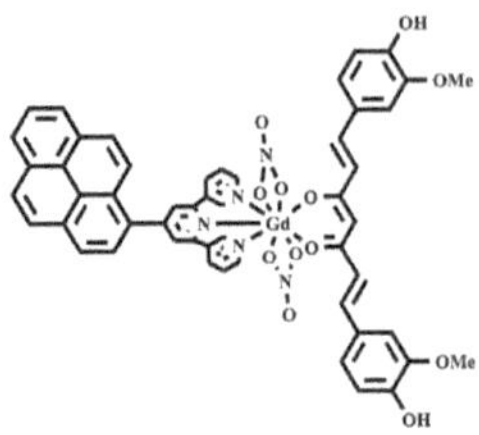

17f

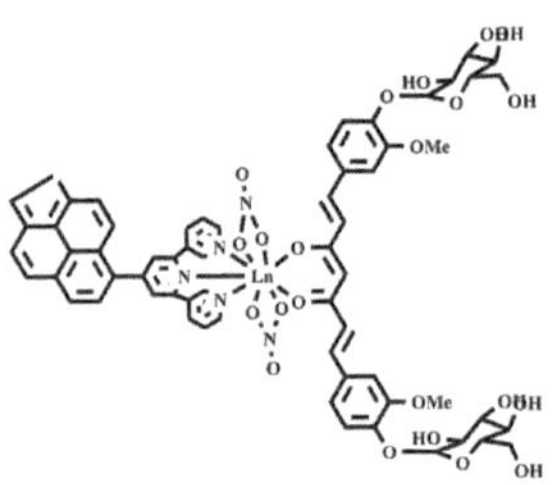

17g

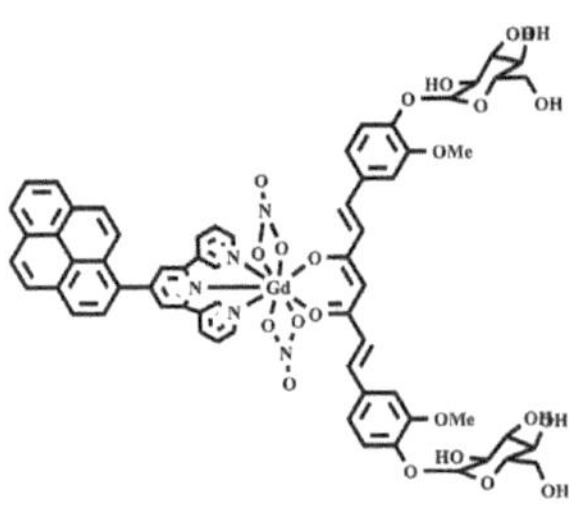

17h

Fig. 19 Lanthanide complexes of curcumin[Ln(R-tpy) (Cur)$(NO_3)_2$]. (Ln=La(III) in 17a, 17b; Gd(III) in 17e, 17f) and [Ln(R-tpy)(Cur)$(NO_3)_2$] (Ln=La(III) in 17c, 17d; Gd(III) in 17g, 17h), where R-tpy is 4′-phenyl-2,2′:6′,2″-terpyridine (ph-tpy in 17a, 17c, 17e, 17g), is 4′-(1-pyrenyl-2,2′:6′,2″-terpyridine (py-tpy in 17b, 17d, 17f, 17h).

curcumin are synthesized (Fig. 20). The DNA cleavage and cytotoxicity under visible light, and cellular localization in cancer cells of these complexes were studied by Chakravarty and his co-workers (Goswami et al., 2013). The curcumin-Cu complexes showed good photocytotoxicity (IC_{50} ~3 μM) with low dark toxicity (IC_{50} ~48 μM). The DNA photocleavage activity and photocytotoxicity of the above-mentioned complexes were examined in MCF-7 and HeLa cancer cells, and found to kill cells through an apoptotic pathway.

The photocytotoxicity and photo-induced DNA cleavage of Oxovanadium(IV) containing curcumin complexes (19a to 19d) in visible

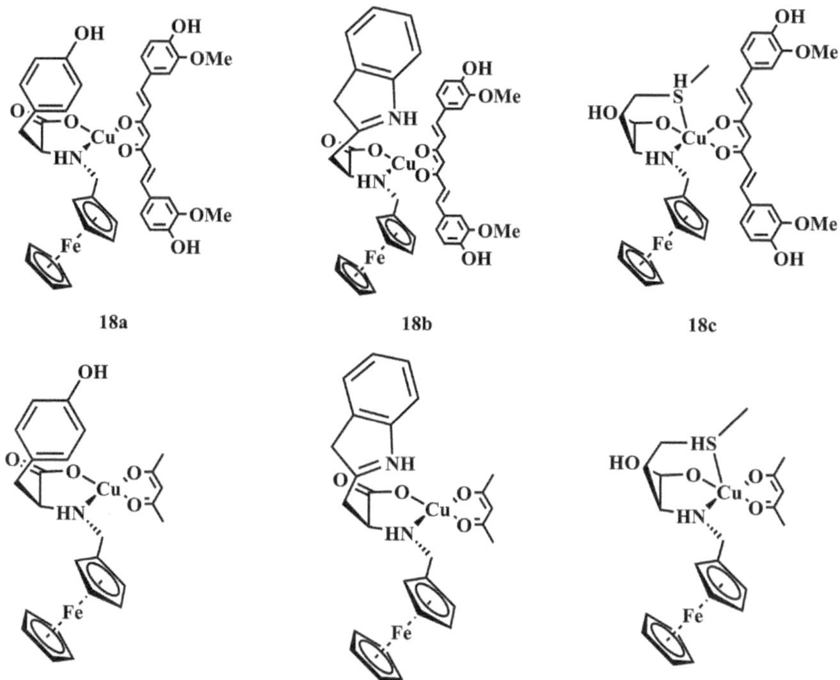

Fig. 20 Copper(II) complexes of *N*-ferrocenyl methyl-L-amino acids with curcumin (18a to 18c) and with acetylacetonate (18d to 18f).

light were studied (Fig. 21) (Balaji et al., 2014). The complexes displayed PDT effect in HepG2 and HeLa cancer cells in visible light with IC_{50} values from 2 to 10 μM with low dark toxicity. The complexes generate hydroxy radicals to damage plasmid DNA when irradiated with 647 nm red light. The photocytotoxicity (IC_{50} value) of the complex 19a was found to be equivalent to that of Photofrin. The complex 19a showed as much as seven-fold higher activity than its phenyl analogue due to the presence of the lipophilic ferrocenyl moiety as evidenced by ICP-MS and fluorescence microscopic studies. The complex generates reactive oxygen species inside the cells and induces apoptotic cell death. These studies suggest that metal-based photocytotoxic agents of curcumin and curcuminoids play an important role in medicinal applications of bioorganometallic chemistry.

BODIPY dyes are known for its photosensitizing effect and has been widely used in PDT activity (Pederzoli, Baig, Kývala, Pittner, & Cwiklik, 2019). It is also known that hydrolytic stability of curcumin in buffer medium can be increased by binding with copper(II) (Xu et al., 2014). By combining these two effects, Bhattacharyya et. al. has synthesized

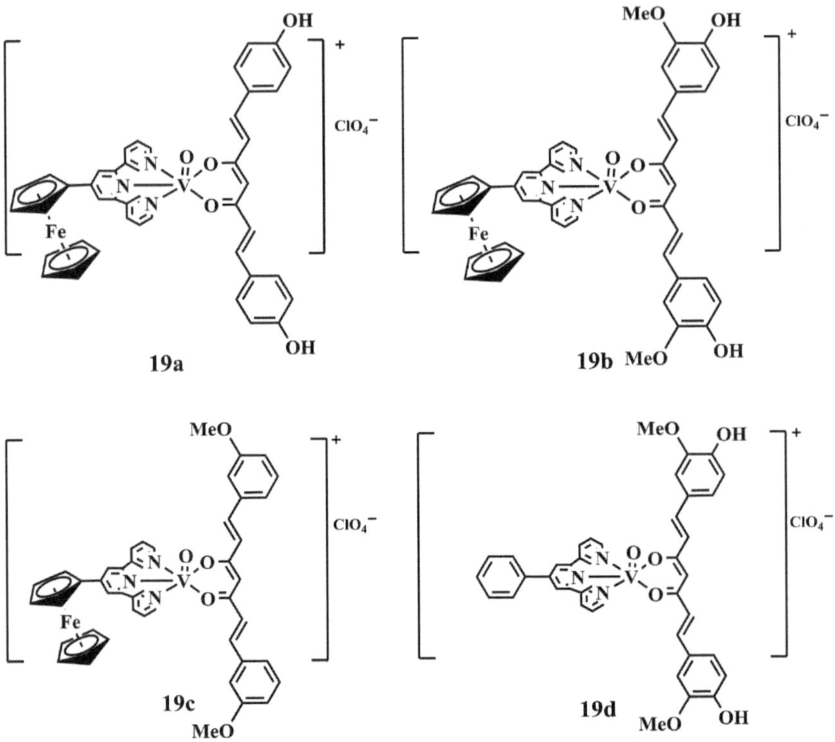

Fig. 21 Vanadyl complexes of ferrocenyl conjugated terpyridine and curcumin. [VO(Fc-tpy)(Cur)](ClO_4) (19a), [VO(Fc-tpy)(bDHC)](ClO_4) (19b), [VO(Fctpy)(bDMC)](ClO_4) (19c) and [VO(Ph-tpy)(Cur)](ClO_4) (19d) where Fc-tpy is 4′-ferrocenyl-2,2′:6′,2″-terpyridine, Ph-tpy is4′-phenyl-2,2′:6′,2″-terpyridine, Cur is monoanionic curcumin, bDHC is bis-dehydroxycurcmin and bDMC is bis-demethoxy curcumin.

photocytotoxic copper(II) complexes of BODIPY derivatives and curcumin (20a, 20b and Fig. 22A) (Bhattacharyya et al., 2015). The complexes damage the cancerous cells through mitochondrial targeting (Fig. 22B). Complex 20a show IC_{50} values of $7.9 \pm 0.3\,\mu M$ in visible light and $29.1 \pm 0.5\,\mu M$ in the dark. Complex 20b exhibit better photocytoxicity with IC_{50} values of $3.8 \pm 0.2\,\mu M$ in visible light and $32.1 \pm 0.4\,\mu M$ in the dark. Further, the PDT effect of complex 20b is quite similar to FDA approved PDT drug Photofrin. The remarkable PDT effect of 20b is originated from diiodinated BODIPY ligand that facilitates the inter system Crossing (ISC) giving rise to triplet state of the photosensitizer which results in dual mode photosensitizing effect from both curcumin and BODIPY derivative.

The water soluble and highly stable ternary oxidovanadium(IV) complexes of dipicolylamine (dpa) and curcumin (Cur) were synthesized

(a)

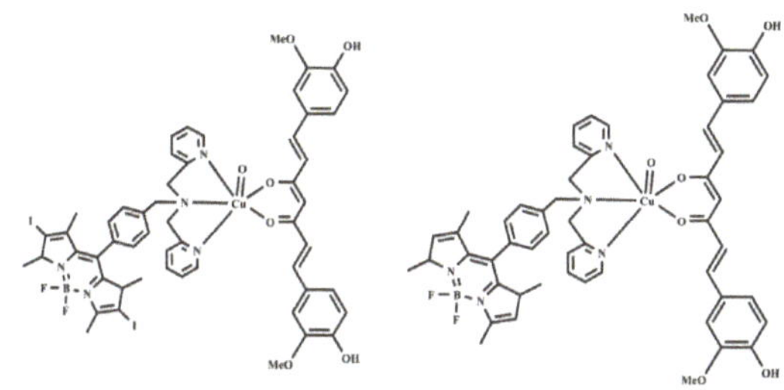

(b)

20a 20b

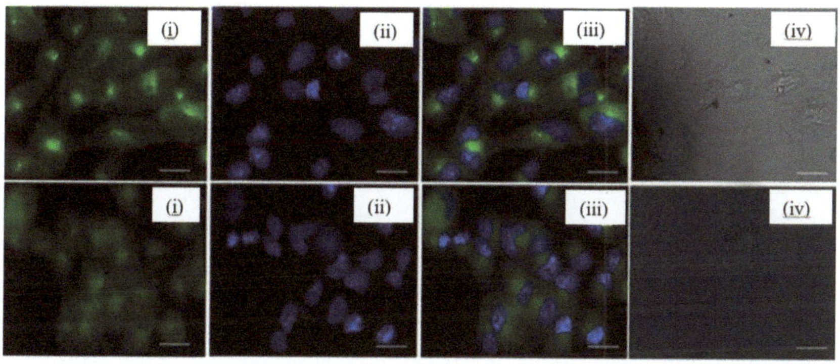

Fig. 22 (A) Copper(II) complexes of BODIPY (boron dipyrromethene) derivatives (L_1, L_2) and curcumin (Cur), viz.[Cu(L_1)Cur)]Cl (20a) and [Cu(L_2)(Cur)]Cl (20b), where L_1 and L_2 are the non-iodinated and diiodinated BODIPY appended dipicolylamine ligands respectively. (B) Fluorescence microscopic images of the HeLa cells treated with complexes 20a (top panels) and 20b (10 μM) (bottom panels) followed by nuclear staining with Hoechst 33342 dye (5 μg mL^{-1}) for 4 h incubation: panels (i) show the green emission of the complexes; panels (ii) show the blue emission of the Hoechst dye; panels (iii) are the merged images and panels (iv) are the bright field images. The merged image panels show cytosolic localization of the complexes. *Reprinted with permission of Medicinal Chemistry Communications, Royal Society of Chemistry from Bhattacharyya, A., Dixit, A., Mitra, K., Banerjee, S., Karande, A. A., & Chakravarty, A. R. (2015). BODIPY appended copper(II) complexes of curcumin showing mitochondria targeted remarkable photocytotoxicity in visible light.* Medicinal Chemistry Communications, 6, *846–851. doi:10.1039/C4MD00425F.*

(Fig. 23) (Bhattacharyya et al., 2017). Some complexes contain non-iodinated (21b, 21′b) and di-iodinated (21c, 21′c) BODIPY conjugated dipicolylamine as ligands. The fluorescent BODIPY containing complex 21′b displayed mitochondrial localization in MCF-7 cancer cells. The BODIPY appended complex 21′c presented good PDT effect with IC_{50}values of 2–6 μM in cancer cells like MCF-7, HeLa cells with low dark toxicity (~100 μM). The complexes display cell via apoptotic pathway

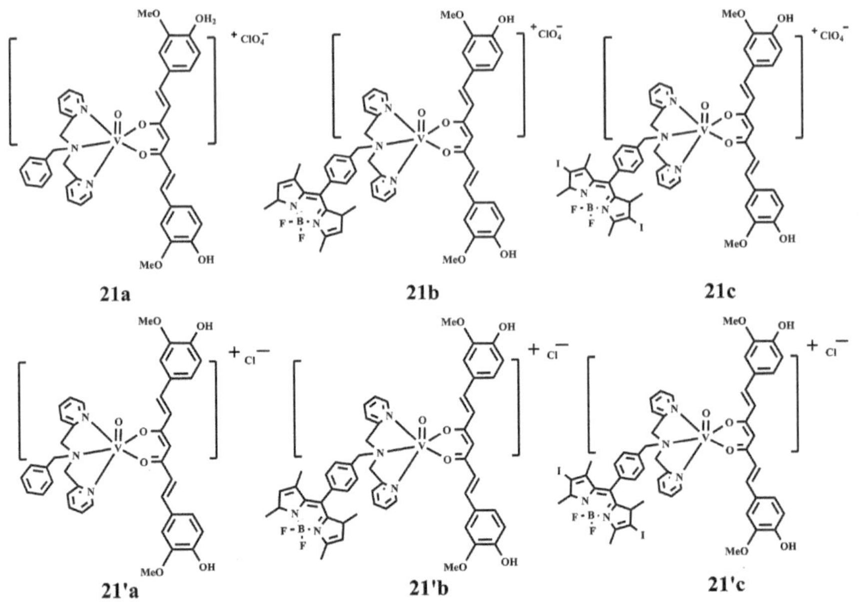

Fig. 23 Ternary oxidovanadium(IV) complexes of dipicolylamine derivatives (dpa) and curcumin (Cur), [VO(dpa)(Cur)]ClO_4 (21a), [VO(L_1)(Cur)]ClO_4 (21b), [VO(L_2)(Cur)] ClO_4 (21c), [VO(dpa)(Cur)]Cl (21′a), [VO(L1)(Cur)]ClO_4 (21′b), [VO(L2)(Cur)]ClO_4 (21′c), where L_1 and L_2 are the non-iodinated and diiodinated BODIPY appended dipicolylamine ligands, respectively.

with the involvement of ROS. These findings are of prime importance, as several anticancer agents like cisplatin and its analogues suffer from resistance of cells.

RAPTA complexes containing curcuminoid ligands (22a and 22b) as an inherent leaving group are endowed with good cytotoxicity and also better solubility properties (Fig. 24) (Pettinari et al., 2014). In contrast to other sets of analogous compounds, these advanced complexes illustrate 100-fold higher cytotoxicity on the cancer cell lines with quite satisfactory IC_{50} values ($\leq 1\ \mu M$). Apart from this, they exhibit an exceptional selectivity, as they show less cytotoxicity towards nontumorous human embryonic kidney cells. The selectivity and cytotoxicity of these complexes are enhanced due to the presence PTA ligand in comparison to earlier reported compounds. The complexes 22a and 22b displayed good in vitro antitumor action in human ovarian carcinoma cells (A2780cisR and A2780). The good cytotoxicity and selectivity in contrast to clinically used cisplatin allows the further synthetic improvement of the novel complexes.

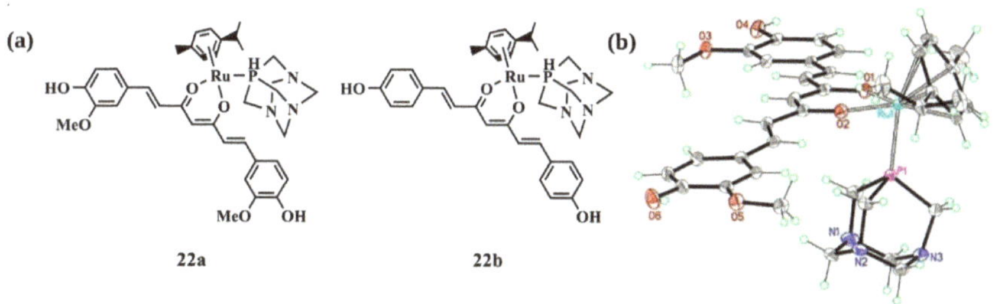

Fig. 24 (A) Ruthenium(II) arene(arene = cymene, hexamethylbenzene) complexes containing PTA (1,3,5-triaza-7-phosphaadamantane) derivatives and curcumin-based ligands (Cur, curcumin; bdcur, bisdemethoxycurcumin). (B) Molecular structure of 22a. *Reprinted with permission of Organometallics, American Chemical Society from Pettinari, R., Marchetti, F., Condello, F., Pettinari, C., Lupidi, G.,Scopelliti, R., Mukhopadhyay, S., Riedel, T., & Dyson, P. J. (2014). Ruthenium(II)-arene RAPTA type complexes containing curcuminand bisdemethoxycurcumin display potent and selective anticancer activity.* Organometallics, 33, *3709–3715. doi:10.1021/om500317b.*

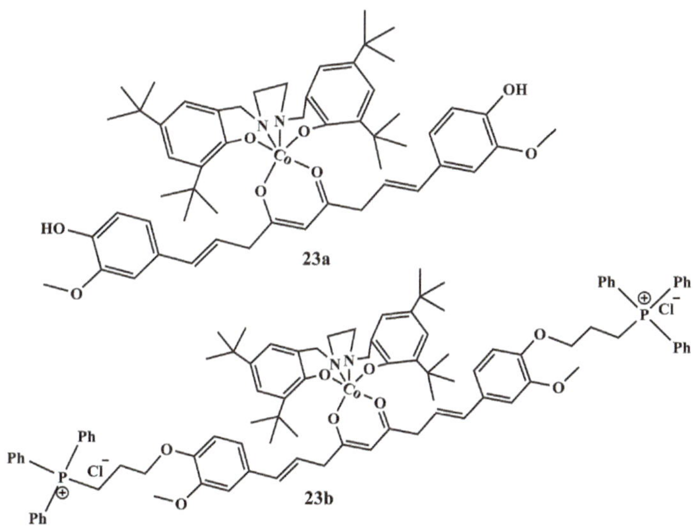

Fig. 25 Cobalt complexes of Cur and Mitocur [dicationic bis-(triphenylphosphonium) derivative of curcumin] with tetra dentate phenolate-based ligand (H_2L), [Co(Cur)(L)] (23a) and [Co(Mitocur)(L)]Cl_2 (23b).

Light-assisted transport of mitocurcumin, a curcuminoid, into the tumor cells has been achieved by ternary cobalt(III) complexes of curcumin (Cur) and mitocurcumin (Mitocur) (Fig. 25) (Garai et al., 2016). Two lipophilic cationic triphenyl-phosphonium moieties has been appended to the curcumin dye to increase cellular uptake with targeted delivery to the mitochondria of the cells. A dicationic curcuminoid containing complex 23b manifests notable cellular uptake and photoinduced cytotoxicity in contrast to its neutral curcumin counter part. The complex 23b liberates the mitocurcumin dye upon irradiation of 400–700 nm visible light in MCF-7 cells by means of photo-reduction of Co(III) to Co(II). The delivered mitocurcumin acts as a phototoxin by producing intracellular ROS. Photocytotoxicity with lower dark toxicity is an inseparable character of curcuminoids and in this case also it is true. Thus, the complex 23b exhibits a notable PDT effect with IC_{50} values ~3.9 μM under visible light in MCF-7 cells and remains nontoxic in dark (IC_{50} >50 μM).

The application of curcumin and cisplatin in medicinal chemistry is limited due to their poor bioavailability and hydrolytic instability that leads to non-specific interactions with biomolecules. To overcome these demerits, Chakravarty and his co-workers have synthesized a novel photoactivated chemotherapeutic agent, named as Platicur, Complex 24a (Fig. 26A), a curcumin-ligated cisdiamine-platinum(II) complex, that releases an anti-cancer active platinum(II) along with photoactive curcumin under-visible light (Mitra et al., 2015). The complex forms DNA adduct on irradiation with visible light and induces apoptotic photocytotoxicity (IC_{50} ~15 μM) with low toxicity (IC_{50}>200 μM) in dark. Cytosolic

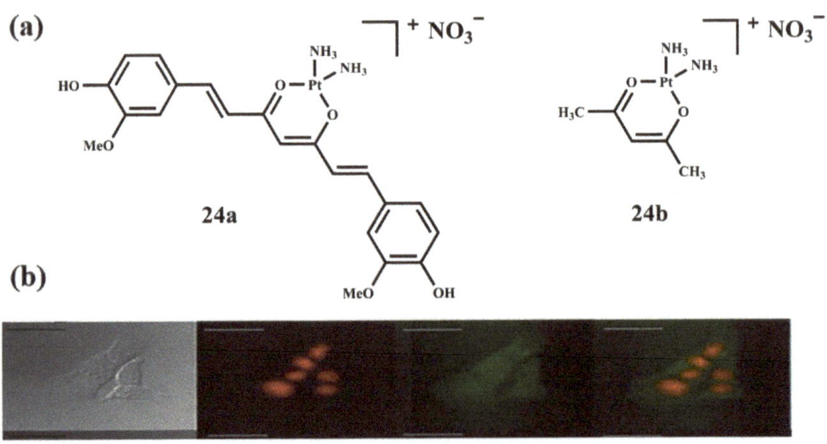

Fig. 26 (A) Platinum complexes of Cur (Platicur), $[Pt(Cur)(NH_3)_2](NO_3)$ and $[Pt(acac)(NH_3)_2](NO_3)$. (B) Fluorescence images of HaCaT cells upon 4h incubation with 24a (20 μM). *Reprinted with permission of Angewandte Chemie, Wiley Online Library from Mitra, K., Gautam, S., Kondaiah, P., & Chakravarty, A. R. (2015).The cis diammineplatinum(II) complex of curcumin: A dual action DNA crosslinking and photochemotherapeutic agent.* AngewandteChemie, 54, *13989–13993. https://doi.org/10.1002/anie.201507281.*

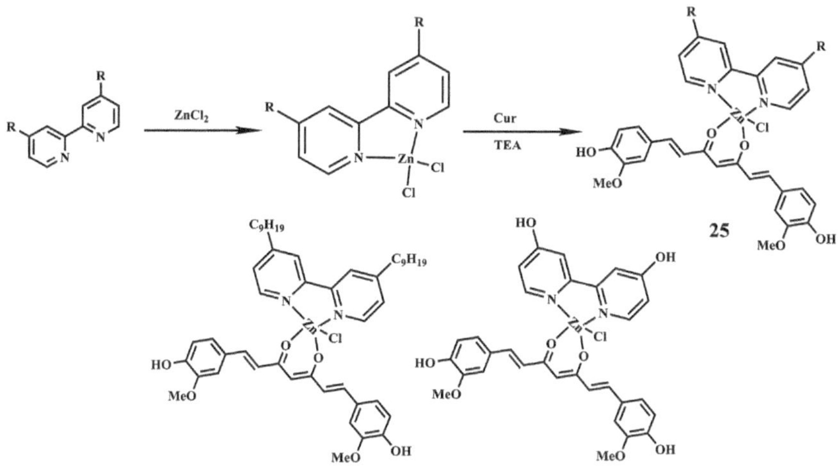

Fig. 27 Curcumin-Zn(II) complexes, [(bpy-9)Zn(Cur)] (25a), [(bpy-OH)Zn(Cur)] (25b).

localization of curcumin has been asserted through fluorescence microscopy images, thus suggesting the dual mode action of the complex as a photo and chemotherapeutic agent (Fig. 26B).

Curcumino-Zn(II) complexes containing penta-coordinated Zn(II) along with 4,4′-disubstituted 2,2′-bipyridines (Fig. 27) as the primary ligand and pure curcumin as an auxiliary ligand have been produced. The curcumin-Zn(II) complex 25a and 25b depicts the good growth inhibition in various cancer cell lines and more effective than the pure curcumin. The presence of curcumin moiety not only makes the complex as fluorescent but also as potent anticancer drug with IC_{50} ~12 and 36 μM; thus, paves the way for the prepared complex to be a unique one (Pucci et al., 2011). The DNA binding studies revealed that the complexes align perpendicular to DNA axis and bind through partial intercalation mode, in contrast with preceding reports of curcumin as a groove-binder.

3. Conclusions

In summary, this chapter provides a general overview of the emerging and rapidly expanding area of antioxidant-metal complexes and their medicinal applications. Besides their toxic side effects, metal complexes are extensively used as life-saving drugs in medicinal research. Here, the usage of antioxidant coupled metal complexes as anticancer agents are concisely described. Most of these metal complexes conjugated with antioxidants showed good activity in various types of cancers by reducing their toxicity towards healthy cells. Therefore, these hybrid metal complexes have the ability to replace a number of older drugs that have deadly side effects. Antioxidants metal complexes of α-lipoic acid, along with some of the flavonoids evolve as prospective anticancer drugs due to their low IC_{50} values. Metal complexes of curcumin are widely acceptable due its high selective photocytotoxicity with low dark toxicity. To conclude, the improved activity ensued by conjugation of antioxidants to the metals and metal complexes can pave the path towards development of the next generation anticancer drugs.

Acknowledgments

The authors would like to thank National Institute of Technology-Warangal for providing infrastructural facilities to carry out this research work.

References

Alderden, R. A., Hall, M. D., & Hambley, T. W. (2006). The discovery and development of cisplatin. *Journal of Chemical Education, 83*, 728. https://doi.org/10.1021/ed083p728.

Balaji, B., Balakrishnan, B., Perumalla, S. K., Karande, A. A., & Chakravarty, A. R. (2014). Photoactivated cytotoxicity of ferrocenyl-terpyridineoxovanadium(IV) complexes of curcuminoids. *European Journal of Medicinal Chemistry, 85*, 458–467. https://doi.org/10.1016/j.ejmech.2014.07.098.

Banerjee, S., & Chakravarty, A. R. (2015). Metal complexes of curcumin for cellular imaging, targeting, and photoinduced anticancer activity. *Accounts of Chemical Research, 48*, 2075–2083. https://doi.org/10.1021/acs.accounts.5b00127.

Banerjee, S., Pant, I., Khan, I., Prasad, P., Hussain, A., Kondaiah, P., et al. (2015). Remarkable enhancement in photocytotoxicity and hydrolytic stability of curcumin on binding to an oxovanadium(IV) moiety. *Dalton Transactions, 44*, 4108–4122. https://doi.org/10.1039/C4DT02165G.

Banerjee, S., Prasad, P., Hussain, A., Khan, I., Kondaiah, P., & Chakravarty, A. R. (2012). Remarkable photocytotoxicity of curcumin in HeLa cells in visible lightand arresting its degradation on oxovanadium(IV) complex formation. *Chemical Communications, 48*, 7702–7704. https://doi.org/10.1039/C2CC33576J.

Bhattacharyya, A., Dixit, A., Mitra, K., Banerjee, S., Karande, A. A., & Chakravarty, A. R. (2015). BODIPY appended copper(II) complexes of curcumin showing mitochondria targeted remarkable photocytotoxicity in visible light. *Medicinal Chemistry Communications, 6*, 846–851. https://doi.org/10.1039/C4MD00425F.

Bhattacharyya, U., Kumar, B., Garai, A., Bhattacharyya, A., Kumar, A., Banerjee, S., et al. (2017). Curcumin "Drug" stabilized in oxidovanadium(IV)-BODIPY conjugates for mitochondria-targeted photocytotoxicity. *Inorganic Chemistry, 56*, 12457–12468. https://doi.org/10.1021/acs.inorgchem.7b01924.

Bhosale, P. B., Ha, S. E., Vetrivel, P., Kim, H. H., Kim, S. M., & Kim, G. S. (2020). Functions of polyphenols and its anticancer properties in biomedical research: A narrative review. *Translational Cancer Research, 9*, 7619–7631. https://doi.org/10.21037/tcr-20-2359.

Cai, W., Zheng, K., Li, Z., Peng, L., Yin, Q., & Zeng, H. (2019). ESI-TOF MS analysis and DNA cleavage activity of rutin-metal complexes in aqueous extracts of medicinal plants. *Inorganic Chemistry Frontiers*, *6*, 3184–3195. https://doi.org/10.1039/C9QI00878K.

Caruso, F., Rossi, M., Benson, A., Opazo, C., Freedman, D., Monti, E., et al. (2011). Ruthenium-Arene complexes of curcumin: X-ray and density functional theory structure, synthesis, and spectroscopic characterization, in vitro antitumor activity, and DNA docking studies of (p Cymene)Ru(curcuminato)chloro. *Journal of Medicinal Chemistry*, *55*, 1072–1081. https://doi.org/10.1021/jm200912j.

Chen, X., Tang, L.-J., Sun, Y.-N., Qiu, P.-H., & Liang, G. (2010). Syntheses, characterization and antitumor activities of transition metal complexes with isoflavone. *Journal of Inorganic Biochemistry*, *104*, 379–384. https://doi.org/10.1016/j.jinorgbio.2009.11.008.

Chipman, J. A., & Berry, J. F. (2020). Paramagnetic metal–metal bonded heterometallic complexes. *Chemical Reviews*, *5*, 2409–2447. https://doi.org/10.1021/acs.chemrev.9b00540.

Englinger, B., Pirker, C., Heffeter, P., Terenzi, A., Kowol, C. R., Keppler, B. K., et al. (2019). Metal drugs and the anticancer immune response. *Chemical Reviews*, *119*, 1519–1624. https://doi.org/10.1021/acs.chemrev.8b00396.

Fonseca, S. F., Lima, D. B., Alves, D., Jacob, R. G., Perin, G., Lenardão, E. J., et al. (2015). Synthesis, characterization and antioxidant activity of organoselenium and organotellurium compound derivatives of chrysin. *New Journal of Chemistry*, *39*, 3043–3050. https://doi.org/10.1039/C4NJ02329C.

Garai, A., Pant, I., Banerjee, S., Banik, B., Kondaiah, P., & Chakravarty, A. R. (2016). Photorelease and cellular delivery of mitocurcumin from its cytotoxic cobalt(III) complex in visible light. *Inorganic Chemistry*, *55*, 6027–6035. https://doi.org/10.1021/acs.inorgchem.6b00554.

García, O. B., & Castillo, J. (2008). Update on uses and properties of citrus flavonoids: New findings in anticancer, cardiovascular, and anti-inflammatory activity. *Journal of Agricultural and Food Chemistry*, *56*, 6185–6205. https://doi.org/10.1021/jf8006568.

Gill, M. R., & Vallis, K. A. (2019). Transition metal compounds as cancer radiosensitizers. *Chemical Society Reviews*, *48*, 540–557. https://doi.org/10.1039/C8CS00641E.

Goitia, H., Quispe, P., Naso, G., Martínez, V. R., Rey, M., Rizzi, A. C., et al. (2019). Interactions of rutin with the oxidovanadium(IV)cation. Anticancer improvement effects of glycosylated flavonoids. *New Journal of Chemisrty*, *43*, 17636–17646. https://doi.org/10.1039/C9NJ01039D.

Goswami, T. K., Gadadhar, S., Gole, B., Karande, A. A., & Chakravarty, A. R. (2013). Photocytotoxicity of copper(II) complexes of curcumin and N-ferrocenylmethyl-L-amino acids. *European Journal of Medicinal Chemistry*, *63*, 800–810.

Helberg, J., & Pratt, D. A. (2021). Autoxidation vs. antioxidants-the fight for forever. *Chemical Society Reviews*, *50*, 7343–7358. https://doi.org/10.1039/D1CS00265A.

Hussain, A., Somyajit, K., Banik, B., Banerjee, S., Ganesh, N., & Chakravarty, A. R. (2013). Enhancing the photocytotoxic potential of curcumin on terpyridyllanthanide(III) complex formation. *Dalton Transactions*, *42*, 182–195. https://doi.org/10.1039/C2DT32042H.

Karami, K., Lighvan, Z. M., Farrokhpour, H., Jahromi, M. D., & Momtazi-borojeni, A. A. (2017). Synthesis and spectroscopic characterization study of new palladium complexes containing bioactive O,O-chelated ligands: Evaluation of the DNA/protein BSA interaction, in vitro antitumoral activity and molecular docking. *Journal of Biomolecular Structure and Dynamics*, *36*. https://doi.org/10.1080/07391102.2017.1391125.

Kasprzak, M. M., Erxleben, A., & Ochocki, J. (2015). Properties and applications of flavonoid metal complexes. *RSC Advances*, *5*, 45853–45877. https://doi.org/10.1039/C5RA05069C.

Kenny, R. G., & Marmion, C. J. (2019). Toward multi-targeted platinum and ruthenium drugs—A new paradigm in cancer drug treatment regimens. *Chemical Reviews, 119*, 1058–1137. https://doi.org/10.1021/acs.chemrev.8b00271.

Konkankit, C. C., Marker, S. C., Knopf, K. M., & Wilson, J. J. (2018). Anticancer activity of complexes of the third row transition metals, rhenium, osmium, and iridium. *Dalton Transactions, 47*, 9934–9974. https://doi.org/10.1039/C8DT01858H.

León, I. E., Díez, P., Etcheverry, S. B., & Fuentes, M. (2016). Deciphering the effect of an oxovanadium(IV) complex with the flavonoid chrysin (VOChrys) on intracellular cell signalling pathways in an osteosarcoma cell line. *Metallomics, 8*, 739–749. https://doi.org/10.1039/C6MT00045B.

Liu, R., & Mabury, S. A. (2020). Synthetic phenolic antioxidants: A review of environmental occurrence, fate, human exposure, and toxicity. *Environmental Science & Technology, 54*, 11706–11719. https://doi.org/10.1021/acs.est.0c05077.

Maier, P. K., & Koepf, H. (1987). Non-platinum-group metal antitumor agents: History, current status, and perspectives. *Chemical Reviews, 87*, 1137–1152. https://doi.org/10.1021/cr00081a012.

Mitra, K., Gautam, S., Kondaiah, P., & Chakravarty, A. R. (2015). The cis diammineplatinum(II) complex of curcumin: A dual action DNA crosslinking and photochemotherapeutic agent. *Angewandte Chemie, 54*, 13989–13993. https://doi.org/10.1002/anie.201507281.

Mjos, K. D., & Orvig, C. (2014). Metallodrugs in medicinal inorganic chemistry. *Chemical Reviews, 114*, 4540–4563. https://doi.org/10.1021/cr400460s.

Monro, S., Colón, K. L., Yin, H., Roque, J., III, Konda, P., Gujar, S., et al. (2019). Transition metal complexes and photodynamic therapy from a tumor-centered approach: Challenges, opportunities, and highlights from the development of TLD1433. *Chemical Reviews, 119*, 797–828. https://doi.org/10.1021/acs.chemrev.8b00211.

Munteanu, A.-C., Notaro, A., Jakubaszek, M., Cowell, J., Tharaud, M., Goud, B., et al. (2020). Synthesis, characterization, cytotoxic activity, and metabolic studies of ruthenium(II) polypyridyl complexes containing flavonoid ligands. *Inorganic Chemistry, 59*, 4424–4434. https://doi.org/10.1021/acs.inorgchem.9b03562.

Nimse, S. B., & Pal, D. (2015). Free radicals, natural antioxidants, and their reaction mechanisms. *RSC Advances, 5*, 27986–28006. https://doi.org/10.1039/C4RA13315C.

Packer, L., WittE, H., & Tritschler, H. J. (1995). Alpha-lipoic acid as a biological antioxidant. *Free Radical Biology & Medicine, 19*, 227–250. https://doi.org/10.1016/0891-5849(95)00017-R.

Pederzoli, M., Baig, M. W., Kývala, M., Pittner, J., & Cwiklik, L. (2019). Photophysics of BODIPY-based photosensitizer for photodynamic therapy: Surface hopping and classical molecular dynamics. *Journal of Chemical Theory and Computation, 15*, 5046–5057. https://doi.org/10.1021/acs.jctc.9b00533.

Pettinari, R., Marchetti, F., Condello, F., Pettinari, C., Lupidi, G., Scopelliti, R., et al. (2014). Ruthenium(II)-arene RAPTA type complexes containing curcumin and bisdemethoxycurcumin display potent and selective anticancer activity. *Organometallics, 33*, 3709–3715. https://doi.org/10.1021/om500317b.

Pi, J., Zeng, J., Luo, J.-J., Yang, P.-H., & Cai, J.-Y. (2013). Synthesis and biological evaluation of germanium(IV)-polyphenol complexes as potential anti-cancer agents. *Bioorganic & Medicinal Chemistry Letters, 23*, 2902–2908. https://doi.org/10.1016/j.bmcl.2013.03.061.

Pietta, P. G. (2000). Flavonoids as antioxidants. *Journal of Natural Products, 63*, 1035–1042. https://doi.org/10.1021/np9904509.

Pucci, D., Bellini, T., Crispini, A., D'Agnano, I., Liguori, P. F., Garcia-Orduña, P., et al. (2011). DNA binding and cytotoxicity of fluorescent curcumin based Zn (II) complexes. *Medicinal Chemistry Communications, 3*, 462–468. https://doi.org/10.1039/C2MD00261B.

Pucci, D., Crispini, A., Mendiguchía, B. S., Pirillo, S., Ghedini, M., Morelli, S., et al. (2013). Improving the bioactivity of Zn(II)-curcumin based complexes. *Dalton Transactions*, *42*, 9679–9687. https://doi.org/10.1039/C3DT50513H.

Santini, C., Pellei, M., Gandin, V., Porchia, M., Tisato, F., & Marzano, C. (2014). Advances in copper complexes as anticancer agents. *Chemical Reviews*, *114*, 815–862. https://doi.org/10.1021/cr400135x.

Sen, S., & Chakraborty, R. (2011). The role of antioxidants in human health. *ACS Symposium Series*, *1*(1083), 1–37. https://doi.org/10.1021/bk-2011-1083.ch001.

Suntharalingam, K., Song, Y., & Lippard, S. J. (2014). Conjugation of vitamin E analog a-TOS to Pt(IV) complexes for dual-targeting anticancer therapy. *Chemical Communications*, *50*, 2465–2468. https://doi.org/10.1039/C3CC48740G.

Tabassum, S., Zaki, M., Afzal, M., & Arjmand, F. (2013). New modulated design and synthesis of quercetin-$Cu^{II}/Zn^{II}-Sn_2^{IV}$scaffold as anticancer agents: In vitro DNA binding profile, DNA cleavage pathway and Topo-I activity. *Dalton Transactions*, *2013*(42), 10029–10041. https://doi.org/10.1039/C3DT50646K.

Tan, J., Zhu, L., & Wang, B. (2010). From GC-rich DNA binding to the repression of surviving gene for quercetin nickel (II) complex: Implications for cancer therapy. *Biometals*, *23*, 1075–1084. https://doi.org/10.1007/s10534-010-9353-x.

Tan, J., Zhua, L., & Wang, B. (2009). DNA binding and cleavage activity of quercetin-nickel(II) complex. *Dalton Transactions*, *2009*, 4722–4728. https://doi.org/10.1039/B901353A.

Tu, L.-Y., Pi, J., Jin, H., Cai, J.-Y., & Deng, S.-P. (2016). Synthesis, characterization and anticancer activity of kaempferol-zinc(II) complex. *Bioorganic & Medicinal Chemistry Letters*, *26*, 2730–2734. https://doi.org/10.1016/j.bmcl.2016.03.091.

Valentini, A., Conforti, F., Crispini, A., Martino, A. D., Condello, R., Stellitano, C., et al. (2009). Synthesis, oxidant properties, and antitumoral effects of a heteroleptic palladium(II) complex of curcumin on human prostate cancer cells. *Journal of Medicinal Chemistry*, *52*, 484–491. https://doi.org/10.1021/jm801276a.

Velugula, K., Kumar, A., & Chinta, J. P. (2020). Nuclease and anticancer activity of antioxidant conjugated terpyridine metal complexes. *Inorganica Chimica Acta*, *507*, 119596. https://doi.org/10.1016/j.ica.2020.119596.

Wang, Q., Huang, Y., Zhang, J.-S., & Yang, X.-B. (2014). Synthesis, characterization, DNA interaction, and antitumor activities of La (III) complex with Schiff base ligand derived from kaempferol and diethylenetriamine. *Bioinorganic Chemistry and Applications*, *2014*. Article ID 354138 https://doi.org/10.1155/2014/354138.

Wenger, O. S. (2013). Vapochromism in organometallic and coordination complexes: Chemical sensors for volatile organic compounds. *Chemical Reviews*, *113*, 3686–3733. https://doi.org/10.1021/cr300396p.

Xu, G., Wang, J., Liu, T., Wang, M., Zhou, S., Wu, B., et al. (2014). Synthesis and crystal structure of a novel copper(II) complex of curcumin-type and its application in *in vitro* and *in vivo* imaging. *Journal of Materials Chemistry B*, *2*, 3659–3666. https://doi.org/10.1039/c4tb00133h.

Yang, F., Jin, H., Pi, J., Jiang, J.-H., Liu, L., Bai, H.-H., et al. (2013). Anti-tumor activity evaluation of novel chrysin-organogermanium(IV) complex in MCF-7 cells. *Bioorganic & Medicinal Chemistry Letters*, *23*, 5544–5551. https://doi.org/10.1016/j.bmcl.2013.08.055.

Zou, Y., Lu, Y., & Wei, D. (2004). Antioxidant activity of a flavonoid-rich extract of Hypericumperforatum L. in vitro. *Journal of Agricultural and Food Chemistry*, *52*, 5032–5039. https://doi.org/10.1021/jf049571r.

CHAPTER THIRTEEN

Natural–product–inspired bioactive alkaloids agglomerated with potential antioxidant activity: Recent advancements on structure-activity relationship studies and future perspectives

Pooja Prakash Atpadkar[a,†], **Sumanth Gopavaram**[a,†], **and Sandeep Chaudhary**[a,b,*]

[a]Laboratory of Bioactive heterocycles and Catalysis (BHC Lab), Department of Medicinal Chemistry, National Institute of Pharmaceutical Education and Research, Lucknow, UP, India

[b]Laboratory of Organic and Medicinal Chemistry (OMC Lab), Department of Chemistry, Malaviya National Institute of Technology, Jaipur, India

*Corresponding author: e-mail addresses: sandeep.chaudhary2311@gmail.com; schaudhary.chy@niperraebareli.edu.in

Contents

† These authors contributed equally to this work.

Vitamins and Hormones, Volume 121
ISSN 0083-6729
https://doi.org/10.1016/bs.vh.2022.10.002

Abstract

Alkaloids derived from natural sources have been shown to have substantial antioxidant activity, suggesting that these natural-product-inspired bioactive entities may have major beneficial influence on human health and food processing sector. The primary process intricates in the etiology of several disorders such as neurodegenerative, inflammatory cardiovascular, and other chronic diseases appear to be either oxidative injury or a cellular damage caused by reactive oxygen species (ROS) or free-radicals. The alkaloid class of bio-heterocycles have been divided into numerous groups based on their biosynthetic precursor and heterocyclic ring systems i.e., piperidine, imidazole, purine, pyrrolizidine, indole, quinolozidine, isoquinoline, tropane, and pyrrolidine alkaloids. Distinct biological properties have been attributed to various compounds belonging to this chemical groups, including antirheumatic, cardiovascular, antispasmodic, anti-ulcer, anti-inflammatory, antibacterial, antinociceptive etc. For many years, natural products and their analogs have been recognized as a possible source of medicinal agents. Recently, research has been concentrated on the synthesis, separation/purification, and identification of new alkaloids derived from a variety of natural sources. This book chapter aims to summarize on the latest developments on the current knowledge on the relationship between the structural features of promising class of bioactive alkaloids with their antioxidant activities.

List of abbreviations

AAPH 2,2′-azobis(2-amidinopropane)dihydrochloride
ABTS (2,2′-azino-bis(3-ethylbenzothiazoline-6-sulfonic acid))
AF acidic fraction
BF basic fraction

BHA	butylated hydroxyanisole
BHT	butylated hydroxytoluene
CAA	cellular antioxidant activity assay
CS	coffee silverskin
CUPRAC	cupric ion (Cu^{2+}) reducing antioxidant power
DA	diterpenoid alkaloid
DCM	dichloromethane
DMPD	*N,N*-dimethyl-*p*-phenylenediamine
DNA	deoxyribose nucleic acid
DPPH	2,2-diphenyl-1-picrylhydrazyl
EDG	electron donating group
EWG	electron withdrawing group
FRAP	ferric cyanide (Fe^{3+}) reducing antioxidant power assay
HAT	hydrogen-atom-transfer
LDL	low density lipoprotein
NAC	*N*-acetyl cysteine
NF	neutral fraction
ORAC	oxygen radical absorbance capacity
QCM	quinoline-4-methylamine
ROS	reactive oxygen species
SET	single-electron-transfer
SGAF	steroidal glycoalkaloids fractions
SM	solamargine
SN	solasonine
SOD	super-oxide dismutase
TBARS	thiobarbituric acid reactive substances
TBHQ	*tert*-butyl hydroquinone
TEMPO	2,2,6,6-tetramethylpiperidin-1-yl-oxyl
TEMPOL	2,2,6,6-tetramethyl-4-piperidinol-*N*-oxyl
THBA	(−)-tetrahydroberberrubine•acetate
TPTZ	ferric-tripyridyltriazine

1. Introduction

Alkaloids belongs to the category of naturally occurring bioactive compounds containing nitrogen atom in its ring structure, which is usually found inside the heterocyclic ring structure, is responsible for a number of associated compounds with neutral and mildly acidic characteristics. Alkaloids may also include carbon, oxygen, hydrogen, sulfur, and, in rare cases, other elements in addition to the nitrogen (Dey et al., 2020). The main secondary metabolites recognized to have medicinal properties belongs to alkaloid class of bio-heterocycles. The substances are divided into numerous clusters based upon their biosynthetic antecedent and heterocyclic ring system, including indole, imidazole, isoquinoline, piperidine, pyrrolizidine, purine, tropane, pyrrolidine and quinolozidine alkaloids (Roy, 2017). At physiological amounts, ROS may be essential for appropriate cell activity. They can also harm important macromolecules such carbohydrates, proteins, nucleic acids, lipids, and polyunsaturated fatty acids, as well as induce DNA damage that can result in mutations. When the cellular components adequately scavenges ROS, then, illness symptoms develop (Halliwell & Gutteridge, 1990). Antioxidants shields the human body from the detrimental effects of reactive oxygen species (ROS) and free-radicals. Several chronic illnesses and lipid peroxidation processes are reported to be slowed by these alkaloids (Gülçin, Berashvili, & Gepdiremen, 2005). As a result, there is a growing need to uncover alternative safe and natural sources of dietary antioxidants. During the last decades, the hunt for plant-derived natural antioxidants has been accelerated tremendously (Škerget et al., 2005). Antioxidants are commonly added to meals to stop oxidative free-radical chain reactions by suppressing the initiation stage followed by the propagation stage which, then, leads to stop the reaction and delays oxidation process (Kahl & Kappus, 1993; Lobo, Patil, Phatak, & Chandra, 2010). The *tert*-butyl hydroquinone (TBHQ), butylated hydroxyanisole (BHA), butylated hydroxytoluene (BHT) and propylgallate, etc. have been recognized as the most widely utilized antioxidants till date. BHT and BHA, on the other hand, are subject to legal restrictions because of concerns about their toxicity and carcinogenic potential (Kahl & Kappus, 1993). As a result, interest has been grown tremendously in safe and natural antioxidants in food applications which encourage researchers to search for better natural antioxidant sources (Anwar, Hussain, & Mustafa, 2018; Lourenço, Moldão-Martins, & Alves, 2019; Xu et al., 2017).

2. Mechanism of antioxidant potential and its evaluation methods

In literature, it has been stated that the antioxidant activity has been divided into two types: primary and secondary. The primary antioxidants are known to scavenge free-radicals to stop the chain initiation step which further breaks the chain propagation step; the secondary antioxidants are known to suppress free-radical formation, thus provides protection against oxidative damage (Lim, Lim, & Tee, 2007). The stable free radical

2,2-diphenyl-1-picrylhydrazyl (DPPH) has been frequently used to assess the free radical-scavenging activity of natural antioxidants. The compound's capacity to donate a proton is connected to its radical-scavenging action (Yin et al., 2016). Because DPPH is a stable free radical, it quickly converts to a stable molecular entity when an electron or hydrogen radical is accepted. The two molecular mechanisms hypothesized for DPPH free-radical scavenging antioxidant activity assays are single-electron transfer (SET) and hydrogen-atom transfer (HAT), which have been thoroughly described (Kolak, Öztürk, Özgökçe, & Ulubelen, 2006). The structural design of the quenching molecule has been linked to the DPPH free-radical quenching phenomenon. Several critical parameters, such as steric hindrance, stiffness, and electron density, have been shown to promote this molecule's entry to the radical site of DPPH, resulting in increased antioxidant activity (Mathew, Abraham, & Zakaria, 2015).

Metal chelating activity, which lowers the catalyzing transition metal concentration in lipid peroxidation, has been proposed as one of the antioxidant processes. The Fe^{2+} ion, the most potent pro-oxidant amongst the different metal ions, produces extremely reactive hydroxyl radicals, contributing to "oxidative stress" events. The oxyradicals that follow have been shown to harm carbohydrates, cellular lipids, proteins, nucleic acids, and resulting in cellular dysfunction. Chemically, compounds having the following functional groups with two or more in a constructive structure-function configuration, such as —OH, —COOH, —SH, —C=O, —O— and —NR_2 have a tendency to chelate with metal ions. For example, diterpenoid alkaloids may chelate Fe^{2+} with the —O— or —NR_2— groups (Faria, Calhau, De Freitas, & Mateus, 2006).

The schematic representation of various evaluation methods that are being used to find the antioxidant activity has been illustrated in Fig. 1. The details of various evaluation methods are given below:

2.1 By ferric thiocyanate method (Kikuzaki & Nakatani, 1993)

The employed conventional procedure involves a vial with a screw lid filled with a combination of the desired compound or plant extract (4.0 mg) dissolved in 100% ethanol (4 mL), 2.5% linolenic acid (4.1 mL), 8.0 mL of 0.05 M phosphate buffer (pH 7.0), and water (3.9 mL) in a dark oven at 40 °C. To 0.1 mL of the desired chemical solution, ammonium thiocyanate (30%, 0.1 mL) and ethanol (75%, 9.7 mL) were added. At 500 nm the red color of absorbance was measured every 24 h until the day after the

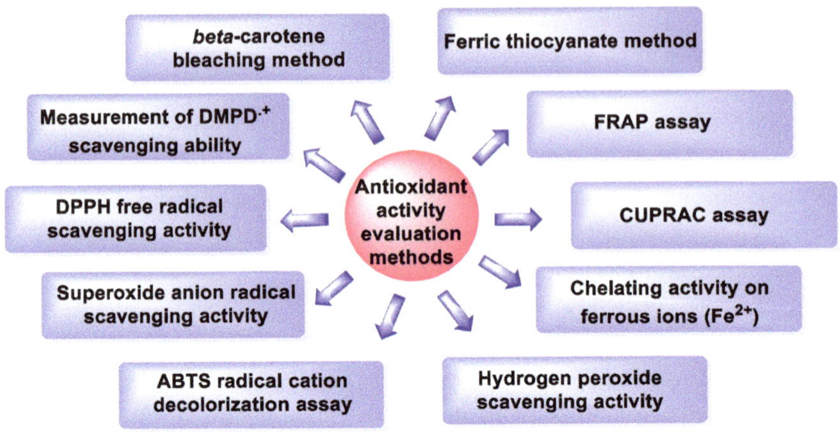

Fig. 1 Schematic representation showing various evaluation methods for antioxidant activity.

maximum absorbance of the control reached. To the reaction mixture 0.02 M ferrous chloride (0.1 mL) in 3.5 percent HCl solution was added. BHT and α-tocopherol served as positive controls, whereas a combination devoid of plant material served as a negative control (Kikuzaki & Nakatani, 1993; Sharma & Vig, 2013).

2.2 Ferric cyanide (Fe^{3+}) reducing antioxidant power assay (FRAP)

Freshly prepared ferrictripyridyltriazine (TPTZ) solution (10.0 mM), $FeCl_3.6H_2O$ solution (20.0 mM), sodium acetate buffer (300 mM, pH 3.6), and were combined in a 1:1:10 (v/v/v) ratio to prepare the FRAP reagent. Following that, 3 mL of newly made FRAP reagent was added to the chosen sample, and the reaction mixture was incubated at 37 °C for 30 min, at 593 nm the absorbance was measured. It has been notified that newly generated $FeSO_4$ solution was used to calibrated the standard curve. $C_{0.5}$FRAP

(the antioxidant capability according to their samples concentration, which is corresponding to $FeSO_4$ at 0.5 mmol/L) was used to assess FRAP antioxidant capability (Ben Ahmed et al., 2016).

2.3 Cupric ion (Cu^{2+}) reducing power: CUPRAC assay

The Cu^{2+} based CUPRAC assay was utilized with minor modifications to study the decreasing capability of alkaloids (such as cepharanthine and fangchinoline). In a test tube, 0.25 mL ethanolic neocuproine solution (7.5103 M), 0.25 mL NH_4Ac buffer solution (1 M), and 0.25 mL $CuCl_2$ solution (0.01 M), were added, then mixed with various concentrations of alkaloids (such as cepharanthine and fangchinoline) ranging from 10 to 30 g/mL. By using distilled water, the whole quantity was well mixed and reduced to 2 mL. After 30 min, at 450 nm the absorbance maxima were measured against the reagent blank. A rise in the absorbance of the reaction mixture has demonstrated an increase in the reduction capacity (Gülçin, Elias, Gepdiremen, Chea, & Topal, 2010).

2.4 Chelating activity on ferrous ions (Fe^{2+})

Dinis and his colleagues used a modified procedure to test the Fe^{2+} chelation of the sample (compounds or plant extracts). In a 1 mL ferrous sulphate solution, 50–150 μL extracts were poured (2 mM). Reaction was start by adding 1 mL of ferrozine (0.25 mM), shaking briskly, and leaving the liquid for 10 min at room temperature. Ferrozine forms stable magenta complex when it interacts with divalent iron. To measure the reaction mixture in control, 1 mL of deionized distilled water was utilized. At 517 nm the absorbance mixtures were measured in the presence (A_i) as well as absence (A_o) of the compounds or plant extracts. The chelating activity was determined using the following formula: Chelating rate (percent) = $A_o - Ai/A_o*100$ (Adusei, Otchere, Oteng, Mensah, & Tei-Mensah, 2019; Ghous, Aziz, Mehmood, & Andleeb, 2015).

2.5 Hydrogen peroxide (H_2O_2) scavenging activity

H_2O_2 is rapidly decomposed by catalase enzyme into water. When heated in the presence of H_2O_2, dichromate in acetic acid was transformed to chromic acetate and subsequently to perchloric acid. At 610 nm, the chromic acetate generated is spectrophotometrically measured. The said sample (0.1 mL) and H_2O_2 (0.4 mL) were added to phosphate buffer (0.9 mL). The dichromate-acetic acid mixture (2.0 mL) was poured at 0 and 60 s. The color generated by the synthesis of chromic acetate was measured at 620 nm after the tubes were immersed in a boiling water bath for 10 min. Standards ranging from 1.2 to 6.0 mol were collected and processed into reagent-only blank and test samples. For erythrocytes, catalase activity was measured in mol of H_2O_2 decomposed/min/g of Hb (Sasikumar, Subramaniam, Aneesh, & Saravanan, 2015; Sundaram et al., 2015).

2.6 DPPH free-radical scavenging activity

For performing the free-radical scavenging procedure using DPPH assay, the desired sample (0.75 mL) was combined with 1.5 mL of a DPPH methanolic solution (20 mg/L) at various concentrations ranging from 10 to 100 g/mL. Vitamin A, C and E were employed as reference substances with pure methanol serving as a reference control. After 20 min of treatment, at 517 nm the absorbance was recorded. The formula was used to calculate the percent of DPPH decoloration in the sample. % of decoloration of sample = [1 (Abs SAMPLE/Abs CONTROL)]*100. To get the IC_{50} the decoloration was strategized vs the sample concentration and a logarithmic regression line was generated. The outcomes are given as anti-radical efficiency (AE), which would be 1000 times the IC_{50} value ($AE = 1000/IC_{50}$). (Sharma & Bhat, 2009).

2.7 ABTS radical cation decolorization assay

The spectrophotometric measurement of radical scavenging activity of $ABTS^{\bullet +}$ was identified by Re et al. (1999). By reacting ABTS with potassium persulfate (2.45 mM) in H_2O (7 mM) in the dark at room temperature for 12 h, the $ABTS^{\bullet +}$ cation radical was formed. Before using, the $ABTS^{\bullet +}$ combination was diluted with phosphate buffer to get an absorbance of 0.700, 0.025 at 734 nm (0.1 M, pH 7.4). The required $ABTS^{\bullet +}$ was created by marinating ABTS with potassium persulfate, as previously stated. The generation of $ABTS^{\bullet +}$ may be reduced by chemical substances that block potassium persulfate action. The overall $ABTS^{\bullet +}$ in the system is reduced as a result of this reduction, which adds to the overall $ABTS^{\bullet +}$ scavenging capacity. The alkaloid sample (10 mg) were diluted in 10 mL distilled water for stock solutions. Further, at varying concentrations (10–20 g/mL), the solution of $ABTS^{\bullet +}$ (1 mL) was added to alkaloid sample in ethanol (3 mL). Half an hour later, the amount of inhibitory activity in the absorbance at 734 nm compared to a blank absorbance was determined for each concentration (ethanol). All the tests were done in duplicate at least three times. In the reaction mixture, the subsequent calibration curve, derived by linear regression analysis (R^2: 0.9922), was used to compute the $ABTS^{\bullet +}$ concentration (mM).

$$\text{Absorbance} = 0 : 0116\left[\text{ABTS}^{\cdot +}\right] + 0 : 0479 :$$

To calculate the ability to scavenge the $ABTS^{\bullet +}$ radical the following equation was used:

$$\text{ABTS}^{\bullet+} \text{ scavenging effect } (\%) = (A_C - A_S)/A_C \times 100$$

A_C = initial ABTS$^{\bullet+}$ concentration

A_S = absorbance of residual ABTS$^{\bullet+}$ concentration in the presence of scavengers (Gülçin, Elias, Gepdiremen, Taoubi, & Köksal, 2009; Re et al., 1999) [2L-2].

2.8 Superoxide anion radical scavenging activity

The technique used to measure the superoxide anion radical scavenging activity of sample cuttings was slightly modified from Liu, Ooi, and Chang (1997). In PMS–NADH systems, the NADH oxidation produces superoxide radicals, which are detected by the nitroblue tetrazolium (NBT) reduction. The superoxide radicals were created in a Tris–HCl buffer (16 mM; pH 8.0; 3 mL) having NBT solution (50 mM; 1 mL), NADH solution (78 mM; 1 mL), and a solution of sample extracts (varying from 12.5 to 62.5 mg/mL) in water. To begin the reaction, in the mixture phenazine methosulphate (PMS) solution (10 mM; 1 mL) was added. For 5 min at 25 °C, the reaction mixture was incubated, and at 560 nm the absorbance was compared to the blank samples. The reference control was L-ascorbic acid. The enhanced superoxide anion scavenging property is shown by the decreased absorbance of the reaction mixtures. The following formula was used to compute the % suppression of superoxide anion generation:

$$\% \text{ Inhibition} = [(A_o - A_1)/A_o] \times 100$$

where A_1 seems to be the absorbance of sample extract or standards, and A_o would be the absorbance of the control (Gülçin, Oktay, Kireçci, & Küfrevioğlu, 2003; Liu et al., 1997; Zhishen, Mengcheng, & Jianming, 1999).

2.9 Measurement of DMPD$^{\bullet+}$ scavenging ability

According to Fogliano and colleagues, the capacity of sample (alkaloid) to scavenge DMPD radicals was discovered. The colorful radical cation (DMPD$^{\bullet+}$) was discovered by diluting DMPD (209 mg) in deionized water and adding a 0.05 M ferric chloride solution (0.2 mL) to a DMPD solution (100 mM) (10 mL). After that, 1 mL of this solution was added to both 100 mL of 0.1 M acetate buffer (pH 5.25) and 0.1 M $FeCl_3$ solution (100 mL). The absorbance of all this freshly produced solution remained constant for up to 12 h at room temperature. The test tubes were filled with various doses of conventional antioxidants, sample (10–30 g/mL), and the whole volume was made to 0.5 mL with distilled water. At 505 nm the absorbance was recorded after 10 min. The absorbance of one milliliter of $DMPD^{\bullet+}$ solution was measured at 505 nm after it was introduced immediately to the reaction mixture. As a blank sample, the buffer solution was employed. The concentration of $DMPD^{\bullet+}$ in the reaction media (mM) was obtained using the calibration curve below, which was established using linear regression. ($r^2 = 0.9993$):

$$\text{Absorbance}\ (\lambda_{505}) = 0.0088 \times [\text{DMPD}^{\bullet+}]$$

The $DMPD^{\bullet+}$ radical's scavenging ability was determined using the equation:

$$\text{Scavenged DMPD}^{\bullet+}\ (\%) = (1 - \lambda_{505-S}/\lambda_{505-c}) * 100$$

$\lambda_{505\text{-}S}$ = absorbance of remaining concentration of $DMPD^{\bullet+}$ in the presence of sample

$\lambda_{505\text{-}C}$ = absorbance of initial concentration of $DMPD^{\bullet+}$

(Fogliano, Verde, Randazzo, & Ritieni, 1999; Gülçin et al., 2010).

2.10 *β*-carotene bleaching method

By dissolving carotene (0.1 mg) in chloroform (5 mL), a version of Pratt's technique was used. To *β*-carotene solution, tween 40 (200 mg) and linoleic acid (25 L) were added (1 mL). After the chloroform had evaporated, added oxygenated distilled water (50 mL), and the liquid was emulsified in a sonicator for 1 min to produce an emulsion. Before each experiment, the emulsion was freshly made. In ethanol, several extract solutions (200 L) were made, and that each extract concentration was combined with 5 mL of the emulsion. All samples were read immediately (t = 0 min) and then every 15 min for 120 min (Kherkhache et al., 2020; Pratt, 2022).

$$\%\ \text{inhibition} = [(\text{Abs sample}(120\ \text{min})/\text{Abs control}(120\ \text{min}))/(\text{Abs sample}(0\ \text{min})/\text{Abs control}(0\ \text{min}))] \times 100$$

3. Structure-activity relationship studies of natural bioactive alkaloids agglomerated with potential antioxidant property

This section summarizes the recent advancements and developments in the antioxidant activity of natural bioactive alkaloids, their detailed synthetic studies accompanied, in particular, with their structure-activity relationship analysis.

3.1 Diterpenoid alkaloid

The naturally occurring diterpenoid alkaloids (DAs) having complex architectural framework with several stereocenters have been derived from the amination of counterpart tetracyclic diterpenes and separation from some plants belongs to the *Ranunculaceae* family, particularly, of the genera *Delphinium, Consolida,* and *Aconitum* etc. DAs have been differentiated based on the number of carbon atoms constituting their fundamental framework as molecules comprising moreover 18, 19 and 20 cycled carbon atoms (C18, C19, C20 diterpenoid) (Thawabteh, Thawabteh, Lelario, Bufo, & Scrano, 2021).

Yin et al. extracted C_{20}-diterpenoid alkaloid **1-4** from the roots of *Aconitum handelianum* (Fig. 2) along with 27 known alkaloids. The DPPH radical-scavenging activity of compound **4** possess approximately equal antioxidant potency as compared to the positive control rutin ($6.64 \pm 0.19\,\mu g/mL$). The results shown that the decreasing order of free-radical scavenging activity follows the order: *Rutin* > **3** > **4** > **2** > **1**. The presence of aryl hydroxyl/phenolic groups in compounds **3** and **4** was identified as a plausible reason for their anti-radical activities. Furthermore, Compounds **1** and **2** demonstrated modest anti-radical activity, which was explained by the presence of the vicinal triol

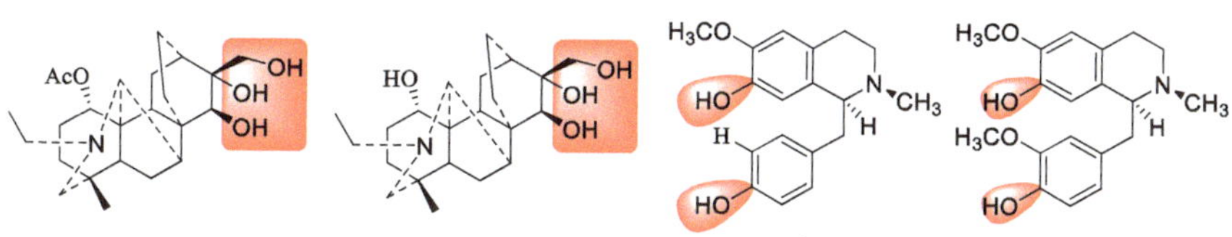

1 2 3 4

Fig. 2 The structures of C_{20}-diterpenoid alkaloids **1-4** from the roots of *Aconitum handelianum.*

system. Surprisingly, both the compounds **1** and **2** have almost identical structures, with the exception of an OH group at the C-1 position in **2**; nonetheless, the anti-radical activity of compound **2** is substantially greater than that of compound **1**. This suggests that the OH-1 in **2** may have played a major role in free-radical scavenging phenomena.

The outcomes specified that the vicinal triol system containing denudatine-type C_{20}-diterpenoid alkaloids and benzyltetrahydroisoquinoline alkaloids had been reported to serve as natural antioxidant as they revealed substantial antioxidant potencies measured by ABTS, DPPH and Fe^{2+} chelating assays. Despite the fact that aconitine-type C_{19}-diterpenoid alkaloids do not showed anti-radical potency, powerful secondary antioxidant potential had been demonstrated due to their strong binding affinities to the metal ions (Yin et al., 2016).

From the aerial portions of the *Aconitum* laeve Royle, S. Begum isolated two novel lycoctonine type C19-diterpenoid alkaloids, Swatinine-A **5** and Swatinine-B **6**, as well as four known C19-diterpenoid alkaloids (foresticine, neoline, delvestine, and chasmanine) (Fig. 3). Swatinine-A **5** (63.4%) and Neoline **7** (65.3%) displayed significant DPPH radical scavenging activities at concentrations of 1 mM, in comparison with reference anti-oxidant BHA which inhibited 92.1% at the same concentration. Over the last 10 years, there had been substantial advancement in understanding the phytochemistry of diterpenoid alkaloids (DAs), identification of new natural DAs and their biological activities. The diterpenoid alkaloids exhibited favorable 2,2-diphenyl-1-picrylhydrazyl (DPPH)-like scavenging activity (Begum et al., 2014).

Because of their capacity to bind to metal ions, aconitine-type C19-DAs might have been proven to be useful antioxidants. At 1 M, Swatinine compounds (**8** and **9**) had a 65.3% and 63.4% effective DPPH radical scavenging ratio, respectively, whereas butylated hydroxytoluene (standard antioxidant) inhibited to 92.1% (Fig. 4). These findings demonstrated that C19-DAs might be identified as new antioxidant agents that were chosen for their

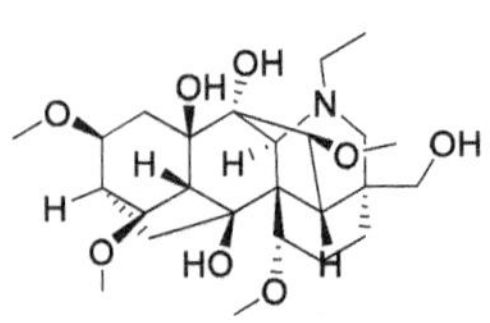

Swatinine-A 5

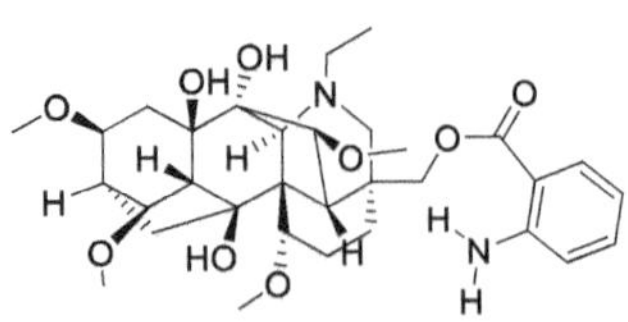

Swatinine-B 6

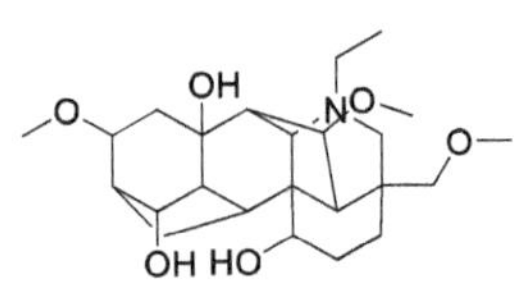

Neoline 7

Fig. 3 Structures of C_{19}-diterpenoid alkaloids Swatinine-A **5**, Swatinine-B **6** & neoline **7**.

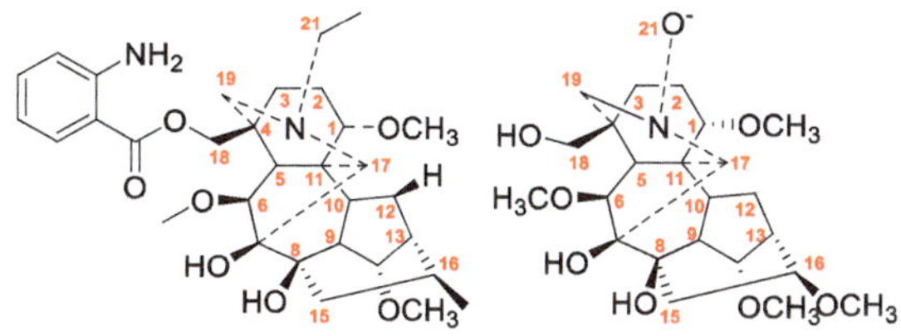

8 9

Fig. 4 Structures of Swatinine compounds **8** and **9**.

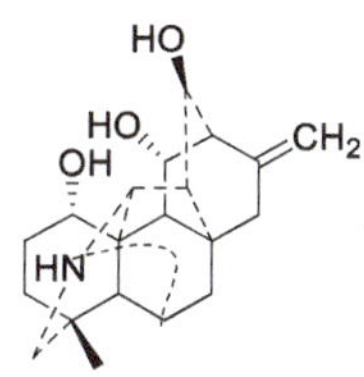

10

Fig. 5 The chemical structure of 1α,11,13β-Trihydroxylhetisine **10** isolated from an *Aconitum heterophyllum*.

low toxicity in this group. Compound **8** has an *O*-acetamidobenzoate moiety, whereas compound **9** has a nitrone (*N*-oxide) functional group due to the N=C19 moiety (Thawabteh et al., 2021).

Ahmad et al. demonstrated significant antioxidant potential of new diterpenoid alkaloid i.e. 1α,11,13β-trihydroxylhetisine **10** isolated from an *Aconitum heterophyllum* (Fig. 5) (Ahmad et al., 2017).

3.2 Aporphine alkaloids

On the basis of results obtained by Sharma et al., it had been observed that in the series of aporphine analog i.e. isoquinoline alkaloids, *N*-methylsulfonyl substituent demonstrated a vital character in anti-oxidant potency, and these aporphines were recorded as the most active compounds of the series (Fig. 6). The antioxidant activity was reduced in aporphine analogs with a *N*-arylsulfonyl, *N*-acetyl, or *N*-benzoyl group. As a result of the higher electronegative character of the S-atom in *N*-methylsulfonyl-containing aporphine analogs, it was speculated that accumulated electron density restricts delocalization of the bonds; as a result, free electrons are not available to quench the DPPH radical, which accounted for the good antioxidant potency.

A philosophical difference in the antioxidant potential of the aporphines was observed when the *N*-acetyl group at the N6 position was changed to an

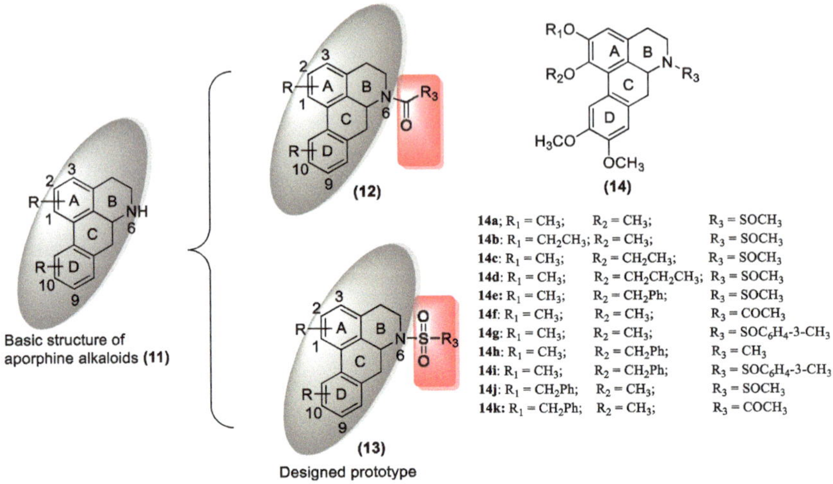

Fig. 6 Basic structure of aporphine alkaloids and their designed prototype.

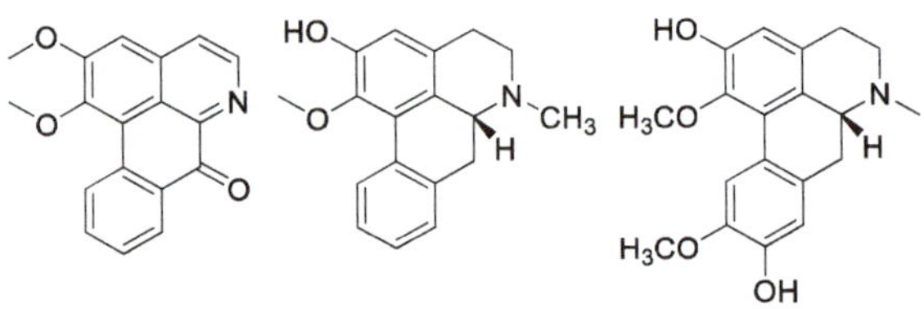

Boldine

15 **16** **17**

Fig. 7 Structures of compounds lysicamine **15**, (−)-*N*-methylasimilobine **16**, lysicamine **15** and boldine **17**.

N-methylsulfonyl group, as seen in the case of **14e** (IC_{50} = 4.360.09 μg mL^{-1}), which indicated higher activity than the standard ascorbic acid (IC_{50} = 4.57 μg mL^{-1}). As a result, **14e** was selected as the series' most active molecule (Sharma et al., 2018).

In the two ABTS and DPPH antioxidant experiments, Chi-Ming Liu et al. found that (-)-*N*-methylasimilobine **16** and lysicamine **15** show considerable antioxidant activity when compared to vitamin C at the same dosage (100 M) (Fig. 7) (Liu et al., 2014).

Aporphines usually underwent oxidation afforded dehydroaporphines and oxoaporphines. Regarding the antioxidant potential of non-phenolic aporphines (absence of —OH groups); the stabilization of the benzylic C-6a radical with the nitrogen lone pair upheld its antioxidant property. Several aporphines acts as an effective antioxidant such as Boldine **17**.

Owing to the phenolic structure, such phytochemicals employ biphasic effect. In particular to the cancerous cell lines, these phenolic aporphines scavenge significantly high amounts of H_2O_2 that encourage proliferation. Conversely, under particular experimental conditions, the similar type of phytochemicals could exhibit pro-oxidant activity and generate beyond threshold ROS for its tolerance. Furthermore, lipoperoxidation induced by ischemia could be potentiated by Boldine, thereby, showing a pro-oxidant consequence in the hippocampal slices of rat. Boldine restores NO bioavailability in diabetic conditions and, overall, decreases the dysfunction of the endothelial cells in addition to the oxidative stress. Boldine suppressed the quick pain response and employed a central anti-nociceptive property (Muthna, Cmielova, Tomsik, & Rezacova, 2013).

3.3 Indole alkaloid

Kherkhache et al., isolated a novel alkaloid having acylated indole glucoside i.e., (6′-*O*-caffeoyl) indole-3-carboxylic acid-b-D-glucoside **18** from Saccocalyx satureioides Coss. & Dur. (Lamiaceae) EtOAc extract, together with eight recognized secondary metabolites, including five methylated flavone aglycones, two indoles, and one monoterpene glucoside. The two sequestered indoles **19** and **20** from the Lamiaceae family were identified for the first time (Fig. 8).

The ethyl acetate and chloroform extracts displayed remarkable free-radical scavenging activity with the IC_{50} value of 20.81 ± 3.52 mg/mL and 23.93 ± 4.15 mg/mL, respectively. Quercetin was taken as a positive control having IC_{50} value of 48.2 ± 0.91 mg/mL. Alternatively, the chloroform extract's lesser effectiveness in contrast to the ethyl acetate extract has been attributed to the chloroform extract's lower availability of active components capable of donating H-atoms or electrons and arresting free radicals. Furthermore, the results of the -Carotene bleaching procedure suggest that the chloroform extract's antioxidant effectiveness is attributable to the

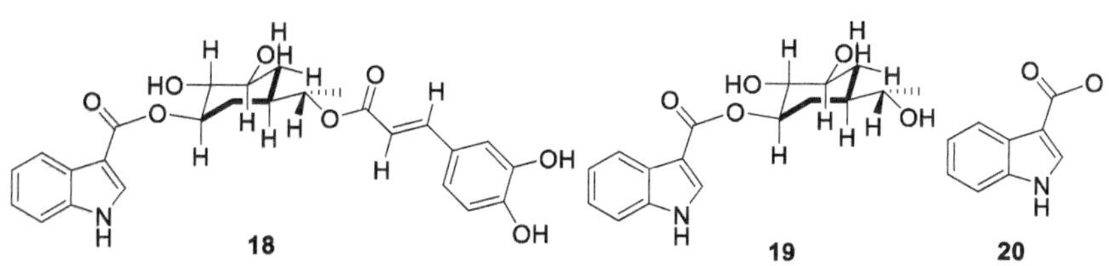

Fig. 8 The structures of acylated indole glucoside **18** along with two indoles **19, 20** from the family *Lamiaceae*.

presence of lipophilic active molecules (Kherkhache et al., 2020; Kulisic, Radonic, Katalinic, & Milos, 2004).

Fragoso et al. investigated the antioxidant capacity of a monoterpene Psychollatine **21**, which is an indole-based alkaloid and are produced by *Psychotria umbellata Vell.* (*Rubiaceae*) leaves (Fig. 9). The occurrence of double bonds, two secondary amines and the glucose residue in its structure can be the possible reason for the defensive role of Psychollatine against ROS (Fragoso et al., 2008).

Cardoso et al., isolated nine indole glycoalkaloids from the plant *Chimarrhis turbinate* and evaluated their antioxidant potency. Alkaloids Cordifoline **22** displayed strong scavenging activity (Fig. 10). However other alkaloids exhibited reasonable to feeble activity. There result specified that the free-radical scavenging potency of **22** has been identified due to its higher H-donating ability (to DPPH), which is given by the hydroxy group at C-10 position, resulting in a more efficient free-radical scavenging activity (Cardoso et al., 2004).

3.4 Oxazine alkaloids

Sharma and colleagues found C-3-tied 2-oxo-benzo[1,4]oxazines as potent antioxidants that showed promising *in vitro* antioxidant activity when tested

Psychollatine. Glc – glucose residue. 21

Fig. 9 The chemical structure of Indole alkaloid Psychollatine **21** from *Psychotria umbellata Vell.*

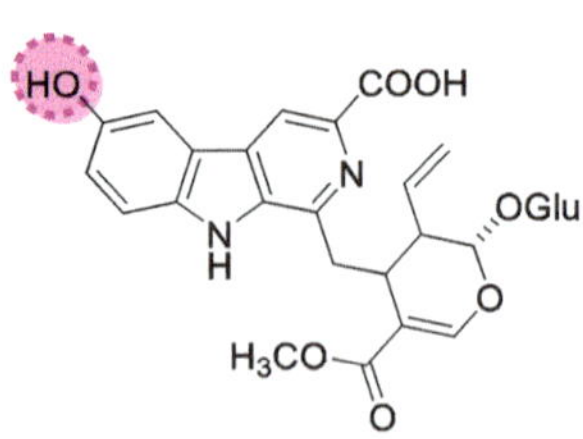

Cordifoline 22

Fig. 10 The chemical structure of Indole glycoalkaloid Cordifoline **22** from the plant *Chimarrhis turbinate.*

with the DPPH free-radical scavenging assay in addition to the FRAP assay. The produced C-3 attached 2-oxo-benzo[1,4]oxazines have two fused cyclic rings I and II connected to ring III by a,-unsaturated ketone integrating electron-donating group (EDG) and/or electron-withdrawing group (EWG) at ring I or III. The —CO—C=C—NH— group allows resonance between rings II and III, resulting in numerous resonance structures that can be further triggered by ring I and III's connected substituents.

It was discovered that whether the compounds contain EWG/EDG at ring I and III or do not have substitution at ring I and III plays a significant influence in determining their DPPH radical scavenging capabilities. The results indicated that the antioxidant activities were favorable, with IC_{50} values ranging from 4.74 ± 0.08 to 92.20 ± 1.54 μg/mL, using ascorbic acid (IC50 = 4.57 μg/mL) as a standard reference. In general, no substitution or EWG at ring I or ring III increases the antioxidant activity of all produced 2-oxo-benzo[1,4]oxazines, it was assumed. EDG, on the other hand, decreases antioxidant activity at either ring I or III. When compared to Ascorbic acid as a standard reference, SAR data show that **23a** and **23b**, the best compounds in the series (Fig. 11), have antioxidant activity (Sharma, Jaiswal, Kumar, et al., 2018; Sharma et al., 2018).

3.5 Isoquinoline alkaloids

Because of its ethnomedicinal effects, (−)-tetrahydroberberrubineacetate (THBA) **24**, a new form of protoberberine alkaloid, was isolated from *Nandina domestica* (Grycová, Dostál, & Marek, 2007; Taha, Khalil,

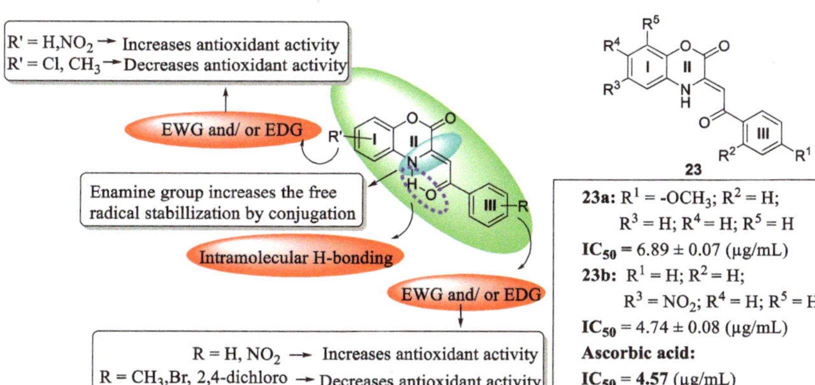

Fig. 11 SAR studies on C-3 tethered 2-oxo-benzo[1,4]oxazines and effect of EWG/EDG at ring I and III on antioxidant activity.

Abubakr, & Shawky, 2020) The protoberberine quaternary alkaloids, which belong to the isoquinoline class of alkaloids, are derivatives of the 5,6-dihydrodibenzo[a,g]quinolizinium system. Antioxidant proteins such as superoxide dismutase (SOD1) and superoxide dismutase (SOD2) were shown to be reduced in H_2O_2 treated samples, but elevated in a dose-dependent manner in THBA treated samples (P*** 0.001). As a result, it shows that THBA can control the oxidative stress generated by H_2O_2 in NIH 3T3 fibroblasts. The NIH 3T3 fibroblast is protected against cell death by THBA, which also decreases the oxidative stress caused by H_2O_2. These findings imply that THBA might be employed in the food manufacturing industry to control and prevent oxidative stress-induced food oxidation and illnesses linked to the production of reactive oxygen species (ROS), such as food contamination caused by food-borne bacteria (Wang, Zhou, Shi, Liu, & Yu, 2020) (Fig. 12).

From the DCM crude extract of Alphonsea cylindrica bark, Obaid et al. isolated two novel isoquinoline alkaloids, iraqiine **25a** and kareemine **26**, as well as five known alkaloids *O*-methylmoschatoline **25b**, atherospermidine **25c**, muniranine **27a**, kinabaline **27b**, and *N*-methylouregidione **28**, and tested them for their antioxidant activities (Fig. 13). The presence of hydroxyl group (donating hydrogen atoms) in the skeleton of compounds **25a**, **27a**, and **27b** may explain their higher activity in comparison to the other compounds. Alkaloids without hydroxyl group showed weak activity against the DPPH radicals scavenging, as evidenced by the results obtained for compounds **26**, **25b**, and **28** (Kareem et al., 2018; Nurul, Wan, Sivasothy, & Yee, 2017).

Khamtache et al., isolated neutral fraction (NF), basic fraction (BF) and acidic fraction (AF) with alkaloid levels of 300, 180 and 240 mg/100 g of dry weight, correspondingly from *Fumaria officinalis*. In total, 11 isoquinoline alkaloids were recognized (Fig. 14). The neutral fraction extract had an advanced scavenging activity in comparison to the extracts of other. The free-radical scavenging activity of DPPH may be ascribed to the presence

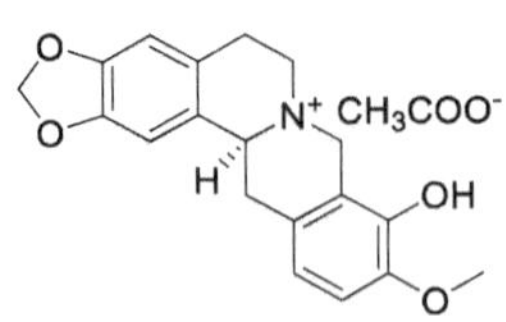

(THBA) (-)-tetra hydroberberrubine.acetate 24

Fig. 12 Structure of protoberberine alkaloid (−)-tetrahydroberberrubine•acetate (THBA) **24**.

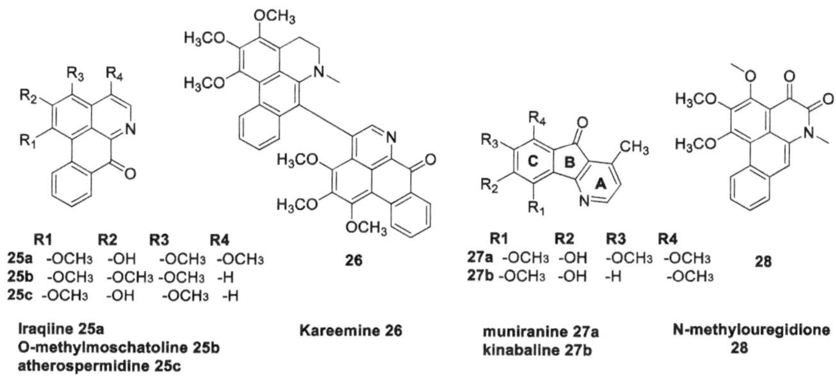

Fig. 13 Structure of isoquinoline alkaloids, iraqiine **25a**, *O*-methylmoschatoline **25b**, atherospermidine **25c**, kareemine **26**, muniranine **27a**, kinabaline **27b** and *N*-methylouregidione **28** from the Alphonsea cylindrica bark.

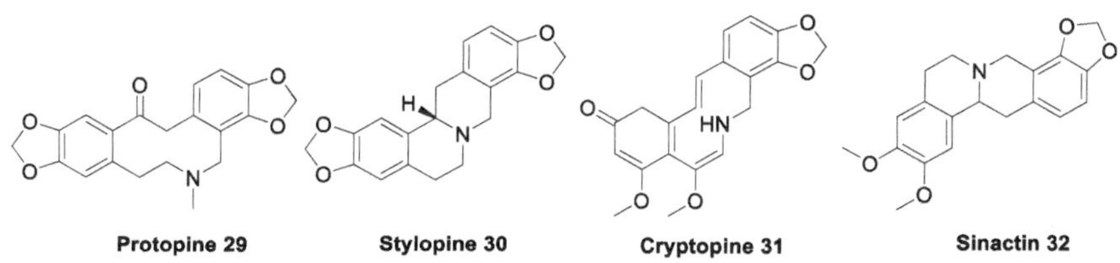

Fig. 14 Structure of isoquinoline alkaloids protopine **29**, stylopine **30**, cryptopine **31**, and sinactin **32**.

of —OH groups in the compound (presence of the *N*-containing group) and the availability of H-atom. It had been also vital to highpoint that the groups position and their degree of methylation could affect these alkaloids antiradical activity (Khamtache-abderrahim et al., 2016).

3.6 Purine-based alkaloid

Shyamlal et al., investigated the advancement of oxidant-promoted, metal-free cross-dehydrogenative Csp^2–Csp^3 coupling on purine-based alkaloid C_8-Caffeine through C—H bond activation. The diverse varieties of alcohols (1°/2°/alicyclic) on reaction with under optimal reaction circumstances, caffeine, theobromine, and theophylline provided the intended results. 8-(1-hydroxyethyl)-caffeine, 8-(1-hydroxyethyl)-theobromine, and 8-(1-hydroxyethyl)-theophylline derivatives. **33a–f**, respectively (Fig. 15).

Compound **33f** (IC_{50}=8.330.08 μg/mL) showed promising antioxidant capability when compared to ascorbic acid (IC_{50}=4.56 μg/mL).

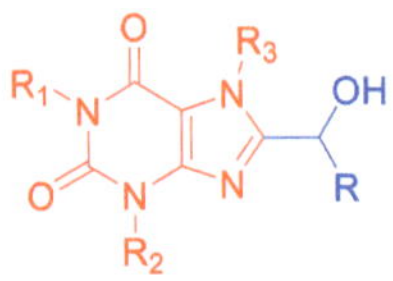

33a: $R_1 = CH_3$; $R_2 = CH_3$; $R_3 = CH_3$; R = H
33b: $R_1 = CH_3$; $R_2 = CH_3$; $R_3 = CH_3$; R = OH
33c: $R_1 = CH_3$; $R_2 = CH_3$; $R_3 = CH_3$; $R = CH_2CH_3$
33d: $R_1 = H$; $R_2 = CH_3$; $R_3 = CH_3$; R = H
33e: $R_1 = CH_3$; $R_2 = CH_3$; $R_3 = H$; $R = CH_3$
33f: $R_1 = CH_3$; $R_2 = CH_3$; $R_3 = CH_3$; $R = C\text{-}(CH_3)_2$

Fig. 15 Synthesis of substituted caffeine, theobromine and theophylline derivatives **33a-f**.

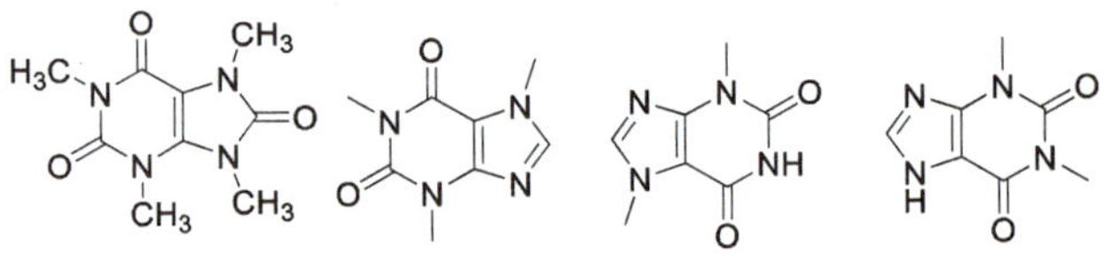

Theacrine 34 **Caffeine 35** **Theobromine 36** **Theophylline 37**

Fig. 16 Structure of purine alkaloid theacrine, caffeine, theobromine and theophylline.

Furthermore, when compared to ascorbic acid, compounds **33a** and **33b** exhibit IC_{50} values of 110.06 μg/mL and 10.620.05 μg/mL, respectively. Compounds **33c, 33d**, and **33e** (the series' three most active compounds) showed arachidonic acid (AA)-induced antiplatelet activities with IC_{50} values of 6.67, 0.05 μg/mL, 6.99, 0.34 μg/mL, and 6.32, 0.27 μg/mL, respectively, and were three-times more active than aspirin (IC_{50}=21.39 μg/mL) as the standard reference drug (Shyamlal, Mathur, Yadav, & Chaudhary, 2020).

The antioxidant potential of purine alkaloids was assessed using the cellular antioxidant activity assay (CAA), oxygen radical absorbance capacity (ORAC), and the chick embryo model as an efficient trio method. The four purine alkaloids (caffeine, theobromine, theophylline, and theacrine) were found to have enormous antioxidant potential (Fig. 16). Among these, theobromine **36** has a higher ORAC value. Previous research had demonstrated that the quantity of OH groups in a molecule might affect peroxyl scavenging activity, therefore the ORAC and CAA results were expected. Infact, this screening also highlighted that theacrine **34** and caffeine **35** possess potent antioxidant capacity *in vivo*. In a chick embryo model, theacrine and caffeine can greatly enhance vascular density on myocardial apoptosis and the chorioallantoic membrane (Tsoi et al., 2015).

3.7 Imidazole alkaloids

It is essential to discuss that the acid-base equilibrium shows a key role in their free-radical scavenging activity. The Marcus theory and density functional theory (DFT) were used to investigate the single-electron transfer

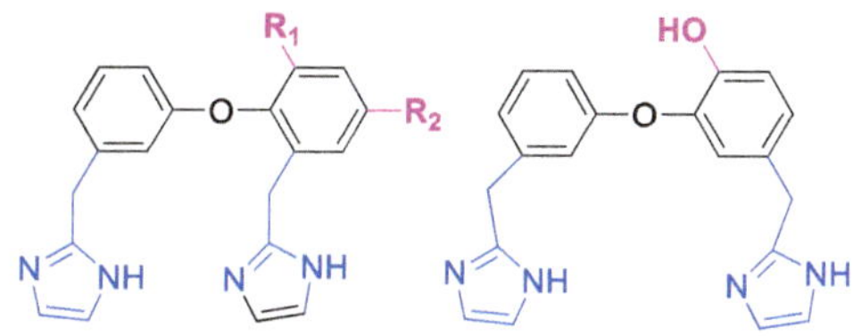

38a: R_1 = -OCH_3 R_2 = H
38b: R_1 = -OH; R2 = H
38c: R_1 = H R_2 = -OCH_3
38d: R_1 = H; R_2 = H

39

Fig. 17 The chemical structures of imidazole-based alkaloids lepidines **38a-d** and lepidine **39**.

(SET) reactions from the mono-anion and neutral species of five imidazole alkaloids (Fig. 17) (lepidines **38a**, **38b**, **38c**, **38d** and **39**) against hydroperoxyl radicals. The deprotonated species of three alkaloids (**38b**, **38d**, **39**) were demonstrated to exhibit free radical scavenging activity. It had been mentioned that the utmost active species was the mono-anion, however, it relies on their molar fraction. Lepidines which containing phenolic group showed significant scavenging activity. The imidazole moiety in the neutral species, according to the reactivity parameters, acts an important role once an electron transfer process is carried out; nevertheless, it is not thermodynamically favorable, while the phenolic oxygen group was identified as the mediator in the mono-anion species.

As the pK_a values of lepidines, **38b** and **39** had been calculated to be close to the physiological pH, for such conditions, it was predicted that the significant amounts of both their neutral species and mono-anion are present. This implies that the most reactive would give one electron to the free-radical further deactivating it, and its population would be re-established through the acid-base equilibrium (Pérez-González, García-Hernández, & Chigo-Anota, 2020).

3.8 Steroidal alkaloid

An intact or modified steroid skeleton present with nitrogen are the characterization of steroidal alkaloids. Because *N*-atom is introduced into a non-amino acid residues, these compounds fit into the subgroup of pseudoalkaloids (or isoprenoid alkaloids). The widely held plant species use solasodine as the main aglycone of glycoalkaloids in the form of water soluble triglycosides solamargine (SM) and solasonine (SN), which follows in almost 200 species of *Solanum* plants (Fig. 18). Many *Solanaceae* species have

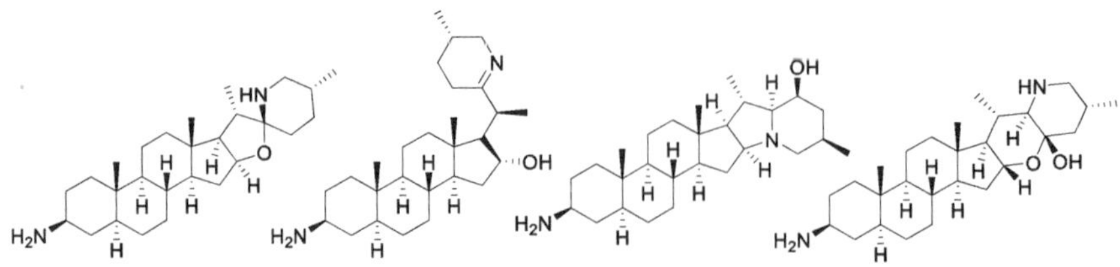

Soladunalinidine 40 **Soladunalinidine 41** **Solanopubamine 42** **Solanocapsine 43**

Fig. 18 The structures of 3-amino steroidal alkaloids **40–43** found in Solanaceae.

been identified as oxidative substances. Such as, plant *S. fastigiatum* demonstrated enormous antioxidant activity in the DPPH assay (Sabir & Rocha, 2008). Muthuvel et al. studied the effect of aqueous extract of *S. nigrum* as an antioxidant agent.

Steroidal glycoalkaloids fractions (SGAFs) of *Solanum incanum L.*, *S. schimperianum Hochst*, *S. nigrum L.*, *Physalis lagascae Roem. & Schult.* and *Withania somnifera (L) Dunal.* investigated promising scavenging activity than methanolic extracts of their respective counterparts. The SGAF of *S. schimperianum* exhibited the strongest antioxidant potential in both DPPH and ABTS assays. Many such studies have identified differences in the biological activities of extracts prepared by employing different extraction techniques, due to many factors such as the solubility of the extracts in different testing systems or stereoselectivity of the radicals hamper the capacity of extracts to react and quench different radicals (Fadl Almoulah et al., 2017).

3.9 Pyridine and piperidine

Karthik et al. revealed that piperidone derivatives have strong antioxidant activity *via* electron-donating mechanism i.e. it gives an unshared pair of electrons on carbonyl oxygen atom. Piperidone exhibit strong antioxidant potential by reduction of C=O into —OH which resulted in a decreased scavenging activity of isomeric alcohols as comparing to piperidone according to *in vitro* antioxidant evaluation using DPPH assay. Amongst the isomeric alcohols, compound **46** having equatorial alcohol displayed high antiradical activity than that of compound **45** having axial alcohol (Fig. 19). Because of 1,3-diaxial interaction, **45** having axial alcohol appeared lesser antiradical scavenging effect to DPPH radical (Karthik, Nithiya, & Jayabharathi, 2011).

Piperine **47**, a piperidine alkaloid (Fig. 20), enhances the bioavailability of numerous therapeutical drugs (e.g., theophylline, propranolol, β-carotene,

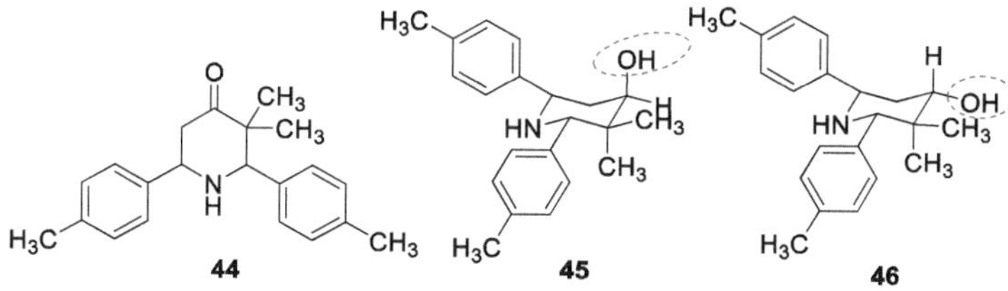

Fig. 19 The chemical structures of Piperidone derivatives **44-46**.

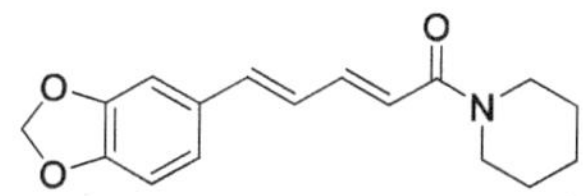

Fig. 20 Structure of Piperidine alkaloid Piperine **47**.

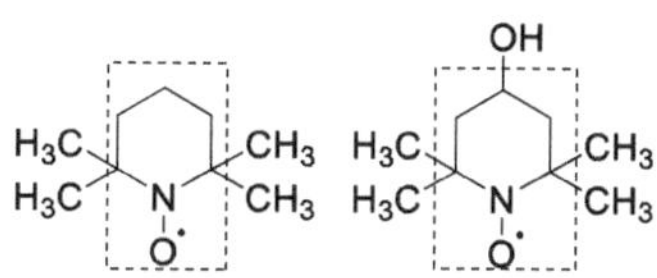

Fig. 21 Structure of piperidine nitroxides TEMPO **48** and TEMPOL **49**.

curcumine,) possibly by boosting intestinal absorption otherwise by employing an antioxidant effect in the first pass through the liver. Piperine has been proven in *in vitro* studies to safeguard against oxidative damage by quenching reactive oxygen species and free-radicals. Piperine or black pepper treatment has also been demonstrated to lower lipid peroxidation *in vivo* and constructively affect antioxidant molecules, antioxidant enzymes and cellular thiol status in many experimental conditions of oxidative stress (Ramawat & Mérillon, 2013; Srinivasan, 2007).

3.9.1 Piperidine nitroxides

Piperidine nitroxides 2,2,6,6-tetramethyl-4-piperidinol-*N*-oxyl **49** (TEMPOL) and 2,2,6,6-tetramethylpiperidin-1-yl-oxyl **48** (TEMPO), are potent antioxidant agents owing to their capacity to scavenge reactive free-radicals (Fig. 21). The antioxidant potential of TEMPOL may be because of hydroxylamine and together they shield mitochondria from oxidative damage (Trnka, Blaikie, Logan, Smith, & Murphy, 2009).

3.9.1.1 Substituted piperidines

Alexidis et al. prepared various piperidine derivatives **50** by substituting several substituent groups in conjunction with chelating properties. Piperidine derivatives revealed strong antioxidant activity in the assays such as hydroxyl radical scavenging assay, lipid peroxidation inhibitory assay and DPPH assay because of their oxidizable SH group (Fig. 22). Inspite derivatives which contain SH group was substituted by amine and hydroxy functionality, displayed reduced antioxidant activity (Alexidis, Rekka, Demopoulos, & Kourounakis, 1995).

Kumar et al. synthesized distinct derivatives of ethyl *N*-aryl-2,6-dioxo-piperid-3-ene-4-carboxylates **51** and evaluated for their *in vitro* antioxidant and antimicrobial potential. Amongst the synthesized compounds, those derivatives which have 4-nitro substituted phenyl ring and also unsubstituted phenyl ring exhibited highest DPPH free-radical scavenging activity (Fig. 23). The majority of the compounds have adequate antimicrobial and antioxidant activities (Kumar, Lokanatha Rai, Vasanth Kumar, & Mylarappa, 2012).

3.9.1.2 *N*-acyl substituted piperidines

Among the synthesized piperamides, which having —OH group at C-4 position of piperidine ring demonstrated best antioxidant potency, whereas 4-phenyl substituted piperidine derivatives **52** displayed reduced antioxidant activity specifying that this substitution is inauspicious for the activity (Fig. 24) (Prashanth, Revanasiddappa, Lokanatha Rai, & Veeresh, 2012).

Lee et al. studied that tricyclic pyridine alkaloid 4,6′-anhydrooxysporidinone **53** derived from *F. lateritium* (Fig. 25) had been more efficient than NAC as an antioxidant during the study on glutamate-induced cell death due to its capacity to decrease oxidative stress. Although, this had been

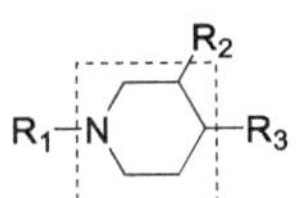

50

R_1 = $Et_2NCH_2CH_2$-, $PhCH_2CH_2$-, $PrOOCCH_2$-
R_2 = H, $COOCH_3$
R_3 = =O, -$NH(CH_2)_2SH$, $NH(CH_2)_2OH$, -$NH(CH_2)_2NH_2$

Fig. 22 Structure of substituted piperidines of type **50**.

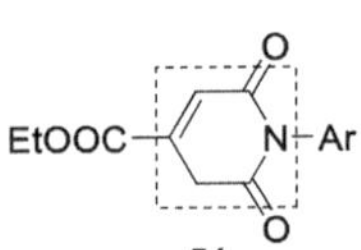

51

Ar = C_6H_5, 4-$CH_3C_6H_4$, 4-$NO_2C_6H_4$, 4-$(OCH_3)C_6H_4$, 2-$(OCH_3)C_6H_4$, 4-BrC_6H_4

Fig. 23 Structure of ethyl *N*-aryl-2,6-dioxo-piperid-3-ene-4-carboxylate derivative **51**.

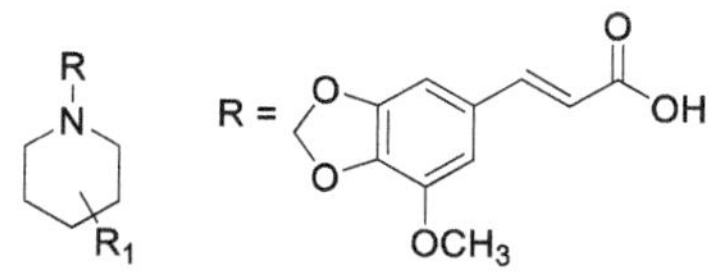

52

R_1 = 2-OH, 4-OH, 4-CH_3, 2,6-$(CH_3)_2$, 4-CH_3, 4-C_6H_5, 4-$C_6H_5OCH_3$, 4-C_6H_4N, 4-$C_4H_8N_2$

Fig. 24 *N*-acyl substituted piperidine derivative **52**.

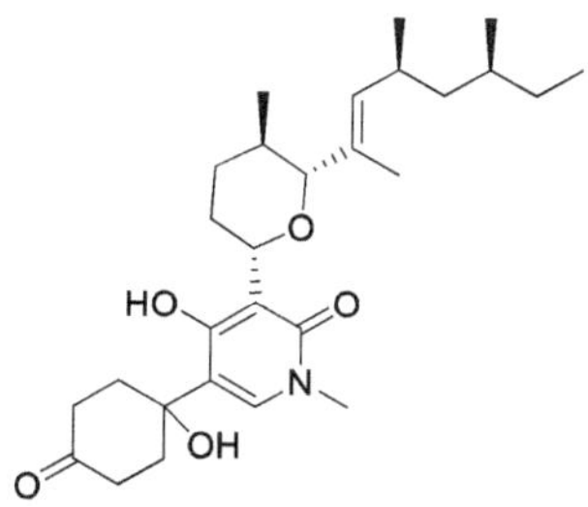

(-)-4,6'-anhydrooxysporidinone 53

Fig. 25 Structure of Tricyclic pyridine alkaloid 4,6′-anhydrooxysporidinone **53**.

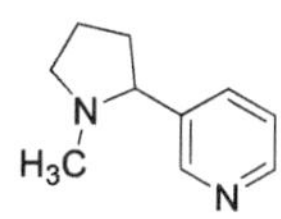

Nicotine 54

Fig. 26 Structure of Pyrrolidine alkaloid nicotine.

validated by assessing the increases in superoxide anion production, reduced intracellular ROS, apoptotic cell death and Ca^{2+} depolarization of mitochondrial membrane potential (Lee, Choi, Hwang, Shim, & Kang, 2020).

3.10 Pyrrolidine alkaloids

Malczewska-Jaskóła et al. stipulated that the nicotine **54** at the concentration of 0.01–1.0 mg/mL displayed significant DPPH-based activity and ferric reducing capacity (*in vitro*) (Fig. 26). However, nicotine, at the same concentrations, efficiently decreased hemolysis by subsiding oxidative stress in the human erythrocytes (Islam & Mubarak, 2020).

3.11 Pyrrolizidine alkaloids

Heliotropium indicum roots yielded helindicine **55** and lycopsamine **56** pyrrolizidine alkaloids (Boraginaceae) (Fig. 27). The antioxidant activity was measured and compared to that of conventional radical scavengers such as trolox and BHT. The alkaloid lycopsamine **56** has a higher DPPH free-radical scavenging activity than helindicine **55** owing to the free —OH group (Kelley & Seiber, 1992).

3.12 Quinoline alkaloid

Mi-Ae Yoon and colleagues investigated the antioxidant activities of quinoline alkaloids extracted from *Scolopendra subspinipes*. Compound **57** has been shown to have dual antioxidant properties, acting as both a radical scavenger and a Cu^{2+} chelator. According to Pinchuk and Lichtenberg's findings on the antioxidant properties of phenol and quinolines, the antioxidant capacity of quinoline **57** with two aromatic OH groups at positions 3 and 8 was extremely high against LDL-oxidation, while the antioxidant activity of *O*-substituted quinoline **58** at positions 3 and 4 was negligibly weak. In the TBARS assay, compounds **57-59** showed antioxidant activity against copper-mediated AAPH-mediated oxidation and SIN-1-mediated oxidation. Furthermore, compounds **57–59** displayed 1,1-diphenyl-2-picrylhydrasyl (DPPH) radical scavenging activity, as well as metal chelating activity (Fig. 28) (Yoon et al., 2006).

3.13 Tropane alkaloids

Al-Ashaal et al. investigated tropane alkaloids produced by *Atropa belladonna* L. *in vitro* cultures. The antioxidant, motor incoordination, antinociceptive, and anticonvulsant properties of original plant extracts and *in vitro* evaluation

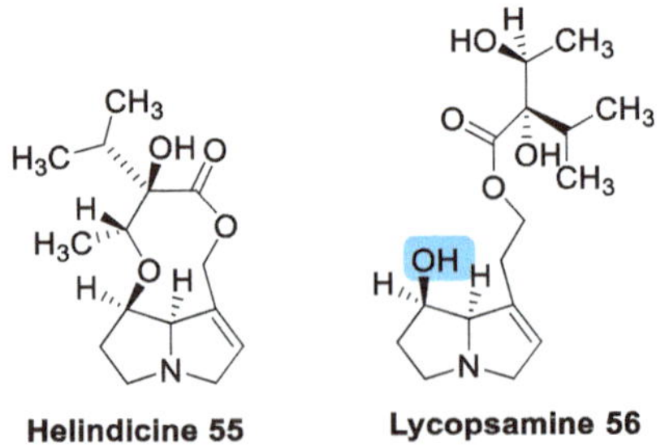

Fig. 27 The chemical structures of Pyrrolizidine alkaloids helindicine **55** and lycopsamine **56**.

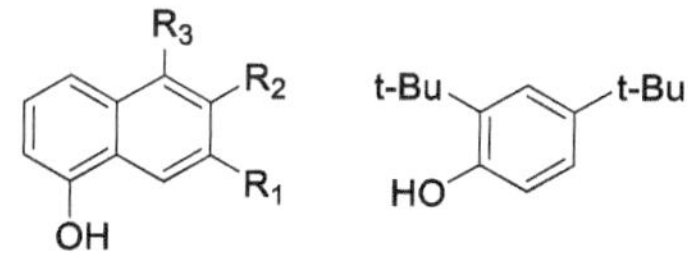

Fig. 28 The chemical structures of Quinoline alkaloids **57-59** isolated from *Scolopendra subspinipes.*

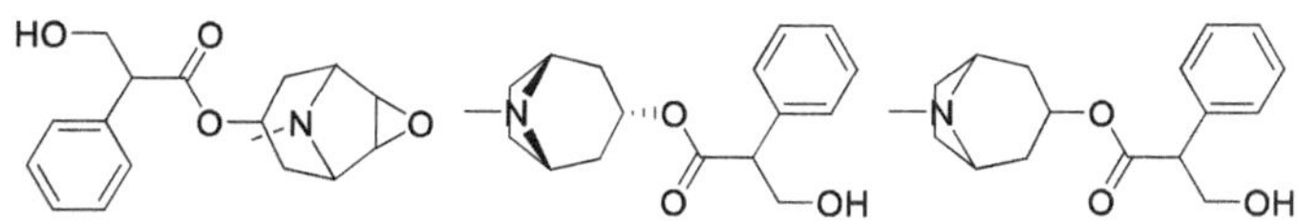

Fig. 29 Structure of tropane alkaloids **60–62** obtained by *Atropa belladonna* L.

had been performed. Hyoscine (scopolamine) **60**, hyoscyamine **61**, and its enantiomer atropine **62** are tropane alkaloids (Fig. 29). Plant tissue culture might be an effective and another source of tropane alkaloids with significant antioxidant, antinociceptive, anticonvulsant, and anti-motor incoordination properties (Al-Ashaal, Aboutabl, Maklad, & El-Beih, 2013).

4. Application of antioxidant potential of alkaloids (Table 1)

An inequity of prooxidants and antioxidants in the body causes oxidative stress. Inflammation, heart disease, hypertension, numerous neurological illnesses, and cancer are all linked to oxidative stress (Li, Wu, Chen, Liu, & Chen, 2013). A range of clinical problems, including ageing, atherosclerosis, diabetes, and neuro-degenerative illnesses, have been interconnected to oxidative stress (Bajpai et al., 2021). The various types of alkaloids isolated from different natural phytoconstituents, their sources, chemical structures and their antioxidant activities along with reference are given in Table 1, Figs. 30 and 31. Antioxidants are substances that can scavenge additional free-radicals, reduce oxidative stress, and therefore protect a biological system from harm. Antioxidants protect the body through a variety of methods (Zeynali, Hosseini, & Rezaei, 2016). The creation of novel molecules with the potential to prevent the organism from oxidative stress by neutralizing free radicals has become a crucial task in organic synthesis. Various *in vitro* approaches analyze the ability of substances for radical capture or prevention of radical generation

Table 1 Different types of alkaloids, their source, structure and antioxidant activity along with reference.

Sr. no.	Class	Source	Structure	Antioxidant activity	Ref.
1	Aporphine alkaloids	Plant species *Hernandiaceae*, *Lauraceae*, *Annonaceae*	Boldine, Glaucine, Apomorphine, Anonaine, bulbocapnine	Inhibition of microsomal lipid peroxidation	Li et al. (2013); Liu et al. (2014); Sharma, Jaiswal, Kumar, et al. (2018)
				Quenching of (DPPH) free-radical	
2	Pyrrolidine	*Nicotiana tabacum*	Nicotine derivatives	Several *in vivo* experiments suggest that the beneficial properties of nicotine in both Alzheimer's disease and Parkinson's disease may be due to antioxidant mechanisms	Linert et al. (1999); Malczewska-Jaskóła, Jasiewicz, and Mrówczyńska (2016)
				Iron (II) chelating activity of nicotine derivative is significant and can be used in different chelation therapy, e.g. treatment of Thalassemia (cases of an iron overload in the body)	
3	Bisbenzyl-isoquinoline	Bark of *Dehaasia longipedicellata*	(−)-*O*-*O*-dimethylgrisabine	Possess a high chelating activity as part of its antioxidant effect, it could decrease the presence of iron and provide a more effective treatment for malaria. As a good reductant with the ability to chelate metal and prevent pro-oxidant activity, while DPPH scavenging effect and metal chelating activities had IC_{50} values of 18.38 and 64.30 μg/mL, respectively	Zahari et al. (2014)
4	Indole alkaloids	*Tabernaemontana divaricata roots*	Coronaridine-hydroxyindolenine	DPPH scavenging activity	Nicola, Salvador, Escalona Gower, Moura, and Echeverrigaray (2013)
5	Isoquinoline	Extracts of *Nandina domestica*	Protoberberine alkaloid	Dose-dependent cellular antioxidant potential against hydrogen peroxide-induced oxidative stress in NIH 3T3 fibroblast cells and diminished the ROS levels and boosted the expression levels of SOD1 and SOD2 enzymes	Bajpai et al. (2021)
			(−)-tetrahydroberberrubine•acetate (THBA)		

		Roots of various *Berberis* species	Berberine	Act as neuroprotective by Attenuating oxidative stress in AD pathogenesis	Singh et al. (2019)
		Stephania rotunda	Cepharanthine and fangchinoline	Cepharanthine and fangchinoline were possessed effective antioxidants in different *in vitro* assays, so these compounds can be retarding the formation of toxic oxidation products, utilized for curtailing or stopping lipid oxidation in pharmaceutical products, prolonging the shelf life of pharmaceuticals and maintaining nutritional quality	Gülçin et al. (2010)
			Pyrrolino-tetrahydroberberines	Protecting DNA against Fe^{2+} -induced strand breaks	Mari et al. (2018)
				Protect plasmid DNA against oxidative damage	
6	Diterpenoid alkaloids	Roots of *Aconitum vilmorinianum Ranunculaceae*	Karakoline, 14-acetylsachaconitine	Chelate Fe^{2+} effectively	Yin et al. (2015)
7	Indolizidine	*Margaritaria indica* Dalz	Allomargaritarine	Metal chelating properties effectively reduces Fe^{3+}-induced LPO and blockes the process at its initial stage	Klochkov and Neganova (2021)
8	Steroidal	Berries of *solanum aculeastrum*	Solasodine, tomatidine	DPPH & ABTS radical scavenging activity, Solasodine and	Srinivas Koduru, Jimoh, and Afolayan (2007)
				Tomatidine have synergistic effect	
9	Purine based	Coffee silverskin (CS)	Caffeine,	Antioxidant potential of this alkaloids, implying that these extracts could signify a appropriate raw material for bioactive compounds to be used in functional foods, fortified foods, nutraceutical formulations, dietary supplements	Castaldo, Narváez, Izzo, Graziani, and Ritieni (2020)
			Theobromine,		
			Theophylline		

Continued

Table 1 Different types of alkaloids, their source, structure and antioxidant activity along with reference.—cont'd

Sr. no.	Class	Source	Structure	Antioxidant activity	Ref.
10.	Pyridine piperidine	–	2,6-Diphenylpiperidine-4-one compounds	Potential to protect the body from oxidative damage by neutralizing the free radicals	Siddiqui et al. (2018)
11	Pyrrolizidine	Extracts from *Senecio delphinifolius* root	Senecionine, seneciphylline, integerrimine, senkirkine	DPPH radical scavenging activity	Tidjani et al. (2013)
12	Tropane	*Hyoscyamus reticulatus*	Hyoscyamine and scopolamine	Antioxidant activity in elicited hairy roots with colchicine was increased to 27% (0.05 colchicine), in comparison to the antioxidant activity level in non-transgenic root (12%) and transgenic hairy roots (18%)	Zeynali et al. (2016)
13	Quinoline	*Quisqualis indica*	Quinoline-4-methylamine (QCM)	DPPH radical scavenging activity	Rout et al. (2021)
14	Imidazole or glyoxalin	*Lepidium genus*	Lepidines D, B, and E	Able to deactivate hydroperoxyl free radicals in aqueous solution under physiological condition	Pérez-González et al. (2020)
		Salvadora persica roots	Persicaline	DPPH, superoxide anion and nitric oxide radicals scavenging activity	Farag, Abdel-Mageed, Basudan, and El-Gamal (2018)
				Recommended for future screening for different biological activities, especially as antitumor due to its reasonable antioxidant activity	

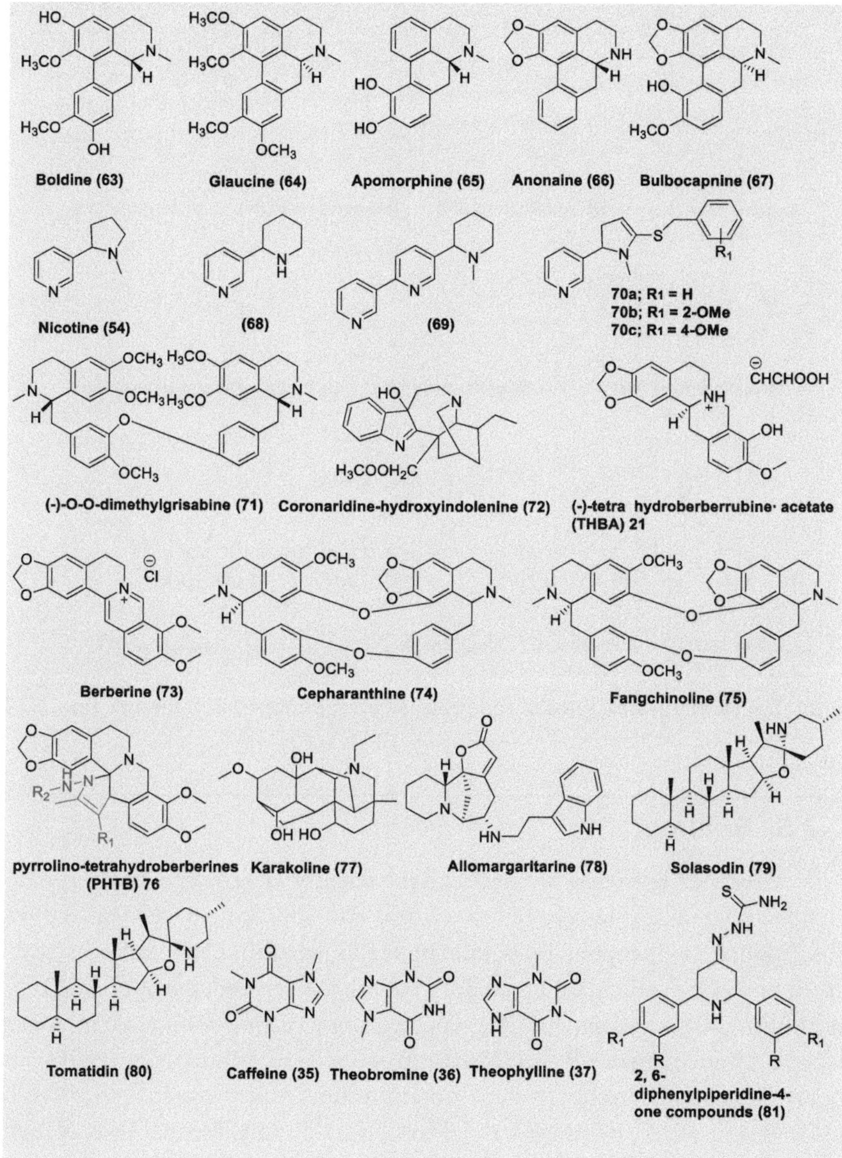

Fig. 30 The chemical structures of various alkaloids mentioned in Table 1 (S. No. 1-10).

to offer a meaningful indicator of antioxidant capacities. Because it is easy, reliable, and rapid, the *in vitro* DPPH technique is recommended over alternative methods. By acquiring hydrogen or an electron from an antioxidant, the DPPH free radical is scavenged and reduced to a colorless state (Siddiqui et al., 2018).

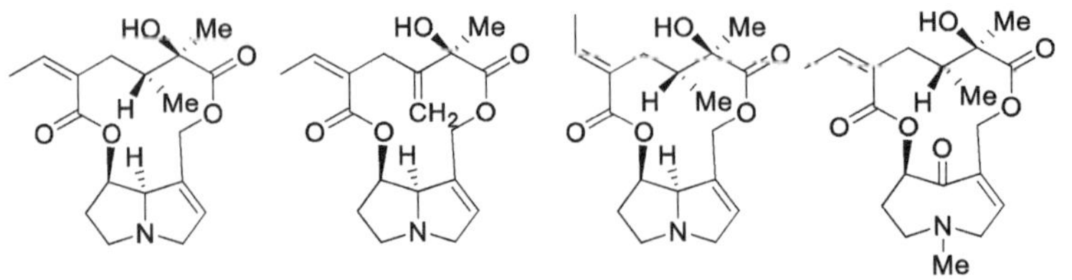

Senecionine (82) **Seneciphylline (83)** **Integerrimine (84)** **Senkirkine (85)**

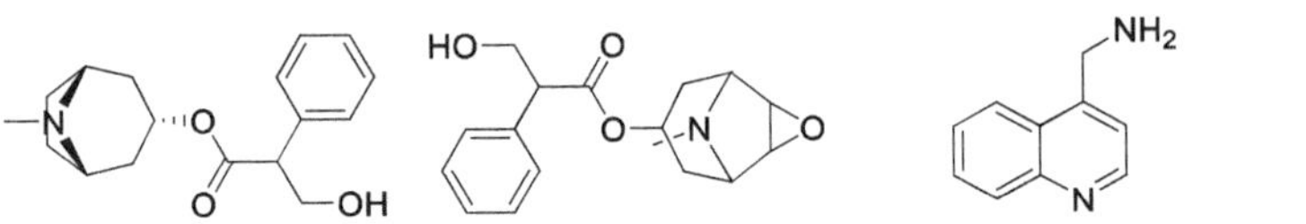

Hyoscyamine (86) **Scopolamine (87)** **Quinoline-4-methylamine (88)**

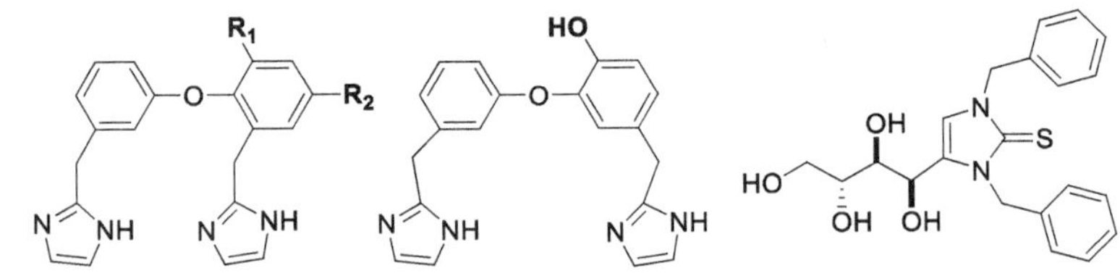

Lepidine 38b: R_1 = -OH; R2 = H **Lepidine 39** **Persicaline (89)**
Lepidine 38d: R_1 = H; R_2 = H

Fig. 31 The chemical structures of various alkaloids mentioned in Table 1 (S. No. 11-14).

5. Summary

A balance between the antioxidant defense system and the generated reactive oxygen species (ROS) is critical for appropriate cellular activity. This balance is disrupted by a rise in ROS, resulting in oxidative stress. Although ROS, which are created by partial oxygen reduction during rapid metabolism, are necessary for life, repeated stress causes an increase in energy usage and hence greater ROS formation, which can injure cells, tissues, and organs. Several plant extracts have potent antioxidant properties that can be used to treat a variety of illnesses and disorders, including liver damage, diabetic hepatotoxicity, and other problems. Natural products and their derivatives have been recognized as a potential source of therapeutic medicines for many years.

6. Conclusion

Free radicals are extremely reactive molecules that cause cell damage. Free-radicals cause cell damage in high quantities, including damage to

DNA, proteins, and cell membranes. Free radical damage to cells might have a role in the development of cancer and other disorders. Free radical scavengers are what antioxidants are renowned for. Antioxidants prevent free radicals from causing harm by neutralizing them. This book chapter focuses on the recent developments in the SAR studies of natural-product-inspired alkaloid-based antioxidants which are known to guards the human body from the damaging effects of ROS and free-radicals. As a result, there is a need to uncover alternative safe and natural alkaloidal sources of dietary antioxidants. The acceleration in the antioxidant research take a leap during the last decades.

7. Future perspectives

Antioxidants help to avoid oxidative stress and the cellular damage caused by the reactive oxygen species (ROS). As a result, natural substances and their derivatives have been utilized to treat oxidative stress-related disorders for a long time. The search for novel efficient antioxidants is an important area of investigation; numerous plant extracts or secondary metabolites have showed promise in protecting against oxidative damage. In particular, alkaloids, known for rich source of biological activities, had been identified to fulfil the requirements and could serve as better antioxidant agents.

Acknowledgments

P.P.A. and G.S. acknowledges Department of Pharmaceuticals, Ministry of Chemicals and Fertilizers for the financial assistance in terms of M.S. Pharm fellowships. The NIPER-Raebareli manuscript communication number is NIPER-R/Communication/303.

Declarations

Authors' contributions

SC conceived and designed the concept. PPA and GS carried out all the literature search. PPA, GS and SC carried out the writing of the manuscript. All the authors have read and approved the final version of the manuscript.

Ethics approval

Not applicable.

Competing interests

The authors(s) confirm that this chapter content has no conflicts of interest.

References

Adusei, S., Otchere, J. K., Oteng, P., Mensah, R. Q., & Tei-Mensah, E. (2019). Phytochemical analysis, antioxidant and metal chelating capacity of Tetrapleura tetraptera. *Heliyon, 5*, e02762. https://doi.org/10.1016/j.heliyon.2019.e02762.

Ahmad, H., Ahmad, S., Adnan, S., Shah, A., Ali, M., Khan, F. A., et al. (2017). *Faculty of Pharmacy, Universiti Teknologi MARA Puncak Alam Campus, 42300 Bandar Department of Biochemistry, Abdul Wali Khan University Mardan, Khyber Pakhtunkhwa, Bioorganic & Medicinal Chemistry*. https://doi.org/10.1016/j.bmc.2017.04.022.

Al-Ashaal, H., Aboutabl, M., Maklad, Y., & El-Beih, A. (2013). Tropane alkaloids of Atropa belladonna L.: in vitro production and pharmacological profile. *Egyptian Pharmaceutical Journal, 12*, 130. https://doi.org/10.4103/1687-4315.124012.

Alexidis, A. N., Rekka, E. A., Demopoulos, V. J., & Kourounakis, P. N. (1995). Novel 1,4 substituted piperidine derivatives. Synthesis and correlation of antioxidant activity with structure and lipophilicity. *Journal of Pharmacy and Pharmacology, 47*, 131–137. https://doi.org/10.1111/j.2042-7158.1995.tb05765.x.

Anwar, H., Hussain, G., & Mustafa, I. (2018). Antioxidants from natural sources. *Antioxidants in Foods and Its Applications, 3–28*. https://doi.org/10.5772/intechopen.75961.

Bajpai, V. K., Park, I. W., Khan, I., Alshammari, F. H., Kumar, P., Chen, L., et al. (2021). (−)-Tetrahydroberberrubine•acetate accelerates antioxidant potential and inhibits food associated Bacillus cereus in rice. *Food Chemistry, 339*, 127902. https://doi.org/10.1016/j.foodchem.2020.127902.

Begum, S., Ali, M., Latif, A., Ahmad, W., Alam, S., Nisar, M., et al. (2014). Pharmacologically active C-19 diterpenoid alkaloids from the aerial parts of Aconitum laeve royle. *Records of Natural Products, 8*, 83–92.

Ben Ahmed, Z., Yousfi, M., Viaene, J., Dejaegher, B., Demeyer, K., Mangelings, D., et al. (2016). Determination of optimal extraction conditions for phenolic compounds from: Pistacia atlantica leaves using the response surface methodology. *Analytical Methods, 8*, 6107–6114. https://doi.org/10.1039/c6ay01739h.

Cardoso, C. L., Castro-Gamboa, I., Siqueira Silva, D. H., Furlan, M., Epifanio, R. D. A., Da Cunha Pinto, Â., et al. (2004). Indole glucoalkaloids from Chimarrhis turbinata and their evaluation as antioxidant agents and acetylcholinesterase inhibitors. *Journal of Natural Products, 67*, 1882–1885. https://doi.org/10.1021/np049863m.

Castaldo, L., Narváez, A., Izzo, L., Graziani, G., & Ritieni, A. (2020). In vitro bioaccessibility and antioxidant activity of coffee silverskin polyphenolic extract and characterization of bioactive compounds using UHPLC-Q-Orbitrap HRMS. *Molecules, 25*. https://doi.org/10.3390/molecules25092132.

Dey, P., Kundu, A., Kumar, A., Gupta, M., Lee, B. M., Bhakta, T., et al. (2020). Analysis of alkaloids (indole alkaloids, isoquinoline alkaloids, tropane alkaloids). *Recent Advances in Natural Products Analysis*. https://doi.org/10.1016/B978-0-12-816455-6.00015-9.

Fadl Almoulah, N., Voynikov, Y., Gevrenova, R., Schohn, H., Tzanova, T., Yagi, S., et al. (2017). Antibacterial, antiproliferative and antioxidant activity of leaf extracts of selected Solanaceae species. *South African Journal of Botany, 112*, 368–374. https://doi.org/10.1016/j.sajb.2017.06.016.

Farag, M., Abdel-Mageed, W. M., Basudan, O., & El-Gamal, A. (2018). Persicaline, a new antioxidant sulphur-containing imidazoline alkaloid from salvadora persica roots. *Molecules, 23*, 1–13. https://doi.org/10.3390/molecules23020483.

Faria, A., Calhau, C., De Freitas, V., & Mateus, N. (2006). Procyanidins as antioxidants and tumor cell growth modulators. *Journal of Agricultural and Food Chemistry, 54*, 2392–2397. https://doi.org/10.1021/jf0526487.

Fogliano, V., Verde, V., Randazzo, G., & Ritieni, A. (1999). Method for measuring antioxidant activity and its application to monitoring the antioxidant capacity of wines. *Journal of Agricultural and Food Chemistry, 47*, 1035–1040. https://doi.org/10.1021/jf980496s.

Fragoso, V., Nascimento, N. C., do Moura, D. J., Silva, A. C. R., Richter, M. F., Saffi, J., et al. (2008). Antioxidant and antimutagenic properties of the monoterpene indole alkaloid psychollatine and the crude foliar extract of Psychotria umbellata Vell. *Toxicology in Vitro, 22*, 559–566. https://doi.org/10.1016/j.tiv.2007.11.010.

Ghous, T., Aziz, N., Mehmood, Z., & Andleeb, S. (2015). Comparative study of antioxidant, metal chelating and antiglycation activities of Momordica charantia flesh and pulp fractions. *Pakistan Journal of Pharmaceutical Sciences*, *28*, 1217–1223.

Grycová, L., Dostál, J., & Marek, R. (2007). Quaternary protoberberine alkaloids. *Phytochemistry*, *68*, 150–175. https://doi.org/10.1016/j.phytochem.2006.10.004.

Gülçin, I., Berashvili, D., & Gepdiremen, A. (2005). Antiradical and antioxidant activity of total anthocyanins from Perilla pankinensis decne. *Journal of Ethnopharmacology*, *101*, 287–293. https://doi.org/10.1016/j.jep.2005.05.006.

Gülçin, I., Elias, R., Gepdiremen, A., Chea, A., & Topal, F. (2010). Antioxidant activity of bisbenzylisoquinoline alkaloids from Stephania rotunda: Cepharanthine and fangchinoline. *Journal of Enzyme Inhibition and Medicinal Chemistry*, *25*, 44–53. https://doi.org/10.3109/14756360902932792.

Gülçin, I., Elias, R., Gepdiremen, A., Taoubi, K., & Köksal, E. (2009). Antioxidant secoiridoids from fringe tree (Chionanthus virginicus L.). *Wood Science and Technology*, *43*, 195–212. https://doi.org/10.1007/s00226-008-0234-1.

Gülçin, I., Oktay, M., Kireçci, E., & Küfrevioğlu, Ö. I. (2003). Screening of antioxidant and antimicrobial activities of anise (Pimpinella anisum L.) seed extracts. *Food Chemistry*, *83*, 371–382. https://doi.org/10.1016/S0308-8146(03)00098-0.

Halliwell, B., & Gutteridge, J. M. C. (1990). Role of free radicals and catalytic metal ions in human disease: An overview. *Methods in Enzymology*, *186*, 1–85. https://doi.org/10.1016/0076-6879(90)86093-B.

Islam, M. T., & Mubarak, M. S. (2020). Pyrrolidine alkaloids and their promises in pharmacotherapy. *Advances in Traditional Medicine*, *20*, 13–22. https://doi.org/10.1007/s13596-019-00419-4.

Kahl, R., & Kappus, H. (1993). Toxikologie der synthetischen antioxidantien BHA und BHT im Vergleich mit dem natürlichen antioxidans vitamin E. *Zeitschrift für Lebensmittel-Untersuchung und -Forschung*, *196*, 329–338. https://doi.org/10.1007/BF01197931.

Kareem, A., Aldulaimi, O., Salbiah, S., Abdul, S., Mhd, Y., Azlan, M., et al. (2018). Two new isoquinoline alkaloids from the bark of Alphonsea cylindrica King and their antioxidant activity phytochemistry letters two new isoquinoline alkaloids from the bark of Alphonsea cylindrica King and their antioxidant activity. *Phytochemistry Letters*, *29*, 110–114. https://doi.org/10.1016/j.phytol.2018.11.022.

Karthik, N., Nithiya, S., & Jayabharathi, J. (2011). Novel piperidone derivatives: Synthesis, spectral and evaluation of antioxidant activity. *International Journal of Drug Development and Research*, *3*, 122–127.

Kelley, R. B., & Seiber, J. N. (1992). Pyrrolizidine alkaloids from Amsinckia. *Phytochemistry*, *31*, 2513–2518. https://doi.org/10.1016/0031-9422(92)83312-M.

Khamtache-abderrahim, S., Lequart-pillon, M., Gontier, E., Gaillard, I., Pilard, S., Mathiron, D., et al. (2016). Isoquinoline alkaloid fractions of Fumaria officinalis: Characterization and evaluation of their antioxidant and antibacterial activities. *Industrial Crops & Products*, *94*, 1001–1008. https://doi.org/10.1016/j.indcrop.2016.09.016.

Kherkhache, H., Benabdelaziz, I., Silva, A. M. S., Lahrech, M. B., Benalia, M., & Haba, H. (2020). A new indole alkaloid, antioxidant and antibacterial activities of crude extracts from Saccocalyx satureioides. *Natural Product Research*, *34*, 1528–1534. https://doi.org/10.1080/14786419.2018.1519817.

Kikuzaki, H., & Nakatani, N. (1993). *Antioxidant Effects of Some Ginger Constituents*, *58*, 1407–1410.

Klochkov, S., & Neganova, M. (2021). Unique indolizidine alkaloid securinine is a promising scaffold for the development of neuroprotective and antitumor drugs. *RSC Advances*, *11*, 19185–19195. https://doi.org/10.1039/d1ra02558a.

Kolak, U., Öztürk, M., Özgökçe, F., & Ulubelen, A. (2006). Norditerpene alkaloids from Delphinium linearilobum and antioxidant activity. *Phytochemistry, 67*, 2170–2175. https://doi.org/10.1016/j.phytochem.2006.06.006.

Kulisic, T., Radonic, A., Katalinic, V., & Milos, M. (2004). Use of different methods for testing antioxidative activity of oregano essential oil. *Food Chemistry, 85*, 633–640. https://doi.org/10.1016/j.foodchem.2003.07.024.

Kumar, K. A., Lokanatha Rai, K. M., Vasanth Kumar, G., & Mylarappa, B. N. (2012). A facile route for the synthesis of ethyl N-aryl-2,6-dioxo-piperid-3-ene-4-carboxylates and their biological activity. *International Journal of Pharmacy and Pharmaceutical Sciences, 4*(Suppl. 4).

Lee, D., Choi, H. G., Hwang, J. H., Shim, S. H., & Kang, K. S. (2020). Neuroprotective effect of tricyclic pyridine alkaloids from fusarium lateritium ssf2, against glutamate-induced oxidative stress and apoptosis in the ht22 hippocampal neuronal cell line. *Antioxidants, 9*, 1–15. https://doi.org/10.3390/antiox9111115.

Li, H. T., Wu, H. M., Chen, H. L., Liu, C. M., & Chen, C. Y. (2013). The pharmacological activities of (-)-anonaine. *Molecules, 18*, 8257–8263. https://doi.org/10.3390/molecules 18078257.

Lim, Y. Y., Lim, T. T., & Tee, J. J. (2007). Antioxidant properties of several tropical fruits: A comparative study. *Food Chemistry, 103*, 1003–1008. https://doi.org/10.1016/j.foodchem. 2006.08.038.

Linert, W., Bridge, M. H., Huber, M., Bjugstad, K. B., Grossman, S., & Arendash, G. W. (1999). In vitro and in vivo studies investigating possible antioxidant actions of nicotine: Relevance to Parkinson's and Alzheimer's diseases. *Biochimica et Biophysica Acta - Molecular Basis of Disease, 1454*, 143–152. https://doi.org/10.1016/S0925-4439(99)00029-0.

Liu, C. M., Kao, C. L., Wu, H. M., Li, W. J., Huang, C. T., Li, H. T., et al. (2014). Antioxidant and anticancer aporphine alkaloids from the leaves of Nelumbo nucifera Gaertn. cv. Rosa-plena. *Molecules, 19*, 17829–17838. https://doi.org/10.3390/ molecules191117829.

Liu, F., Ooi, V. E. C., & Chang, S. T. (1997). Free radical scavanging activities of mashroom polysaccharide extract. *Elsevier, 60*, 763–771.

Lobo, V., Patil, A., Phatak, A., & Chandra, N. (2010). Free radicals, antioxidants and functional foods: Impact on human health. *Pharmacognosy Reviews, 4*, 118–126. https://doi. org/10.4103/0973-7847.70902.

Lourenço, S. C., Moldão-Martins, M., & Alves, V. D. (2019). Antioxidants of natural plant origins: From sources to food industry applications. *Molecules, 24*, 14–16. https://doi.org/ 10.3390/molecules24224132.

Malczewska-Jaskóła, K., Jasiewicz, B., & Mrówczyńska, L. (2016). Nicotine alkaloids as antioxidant and potential protective agents against in vitro oxidative haemolysis. *Chemico-Biological Interactions, 243*, 62–71. https://doi.org/10.1016/j.cbi.2015.11.030.

Mari, G., Catalani, S., Antonini, E., De Crescentini, L., Mantellini, F., Santeusanio, S., et al. (2018). Synthesis and biological evaluation of novel heteroring-annulated pyrrolinotetrahydroberberine analogues as antioxidant agents. *Bioorganic and Medicinal Chemistry, 26*, 5037–5044. https://doi.org/10.1016/j.bmc.2018.08.038.

Mathew, S., Abraham, T. E., & Zakaria, Z. A. (2015). Reactivity of phenolic compounds towards free radicals under in vitro conditions. *Journal of Food Science and Technology, 52*, 5790–5798. https://doi.org/10.1007/s13197-014-1704-0.

Muthna, D., Cmielova, J., Tomsik, P., & Rezacova, M. (2013). Boldine and related aporphines: From antioxidant to antiproliferative properties. *Natural Product Communications, 8*, 1797–1800. https://doi.org/10.1177/1934578x1300801235.

Nicola, C., Salvador, M., Escalona Gower, A., Moura, S., & Echeverrigaray, S. (2013). Chemical constituents antioxidant and anticholinesterasic activity of tabernaemontana catharinensis. *The Scientific World Journal, 2013*. https://doi.org/10.1155/2013/519858.

Nurul, W., Wan, N., Sivasothy, Y., & Yee, S. (2017). Alkaloids from Cryptocarya densi fl ora Blume (Lauraceae) and their cholinesterase inhibitory activity. *Phytochemistry Letters, 21*, 230–236. https://doi.org/10.1016/j.phytol.2017.07.002.

Pérez-González, A., García-Hernández, E., & Chigo-Anota, E. (2020). The antioxidant capacity of an imidazole alkaloids family through single-electron transfer reactions. *Journal of Molecular Modeling, 26*. https://doi.org/10.1007/s00894-020-04583-2.

Prashanth, M. K., Revanasiddappa, H. D., Lokanatha Rai, K. M., & Veeresh, B. (2012). Synthesis, characterization, antidepressant and antioxidant activity of novel piperamides bearing piperidine and piperazine analogues. *Bioorganic and Medicinal Chemistry Letters, 22*, 7065–7070. https://doi.org/10.1016/j.bmcl.2012.09.089.

Pratt, D. E. (2022). *Natural Antioxidants of Soybeans and Other Oil-Seeds*. (www document).

Ramawat, K. G., & Mérillon, J. M. (2013). *Natural products: Phytochemistry, botany and metabolism of alkaloids, phenolics and terpenes, Natural Products: Phytochemistry, Botany and Metabolism of Alkaloids, Phenolics and Terpenes*. https://doi.org/10.1007/978-3-642-22144-6.

Re, R., Pellegrini, N., Proteggenete, A., Pannala, A., Yang, M., & Rice-evans, C. (1999). Antioxidant activity applying an improved ABTS radical cation decolorization assay. *Elsevier, 26*, 1231–1237.

Rout, P. K., Kumar, P., Rao, Y. R., Kumar, A., Bawankule, D. U., Singh, R., et al. (2021). A quinoline alkaloid rich Quisqualis indica floral extract enhances the bioactivity. *Natural Product Research, 35*, 1632–1638. https://doi.org/10.1080/14786419.2019.1634709.

Roy, A. (2017). A review on the alkaloids an important therapeutic compound from plants antibacterial studies of medicinal plant view project. *International Journal of Plant Biotechnology, 3*, 1–9.

Sabir, S. M., & Rocha, J. B. T. (2008). Antioxidant and hepatoprotective activity of aqueous extract of Solanum fastigiatum (false "Jurubeba") against paracetamol-induced liver damage in mice. *Journal of Ethnopharmacology, 120*, 226–232. https://doi.org/10.1016/j.jep.2008.08.017.

Sasikumar, V., Subramaniam, A., Aneesh, A., & Saravanan, G. (2015). Protective effect of alkaloids from Amaranthus viridis Linn. against hydrogen peroxide induced oxidative damage in human erythrocytes (RBC). *International Journal of Clinical Endocrinology and Metabolism, 1*, 049–053.

Sharma, O. P., & Bhat, T. K. (2009). DPPH antioxidant assay revisited. *Food Chemistry, 113*, 1202–1205. https://doi.org/10.1016/j.foodchem.2008.08.008.

Sharma, V., Jaiswal, P. K., Kumar, S., Mathur, M., Swami, A. K., Yadav, D. K., et al. (2018). Discovery of aporphine analogues as potential antiplatelet and antioxidant agents: Design, synthesis, structure–activity relationships, biological evaluations, and in silico molecular docking studies. *ChemMedChem, 13*, 1817–1832. https://doi.org/10.1002/cmdc.201800318.

Sharma, V., Jaiswal, P. K., Saran, M., Yadav, D. K., Saloni Mathur, M., Swami, A. K., et al. (2018). Discovery of C-3 tethered 2-oxo-benzo[1,4]oxazines as potent antioxidants: Bio-inspired based design, synthesis, biological evaluation, cytotoxic, and in silico molecular docking studies. *Frontiers in Chemistry, 6*, 1–17. https://doi.org/10.3389/fchem.2018.00056.

Sharma, S., & Vig, A. P. (2013). Evaluation of in vitro antioxidant properties of methanol and aqueous extracts of Parkinsonia aculeata L. leaves. *The Scientific World Journal, 2013*. https://doi.org/10.1155/2013/604865.

Shyamlal, B. R. K., Mathur, M., Yadav, D. K., & Chaudhary, S. (2020). Microwave-assisted modified synthesis of C8-analogues of naturally occurring methylxanthines: Synthesis, biological evaluation and their practical applications. *Fitoterapia, 143*, 104533. https://doi.org/10.1016/j.fitote.2020.104533.

Siddiqui, R., Saify, Z. S., Akhter, S., Saeed, S. M. G., Haider, S., & Leghari, Q. A. (2018). Synthesis, characterization and evaluation of antioxidant potential of 2,6-diphenylpiperidine-4-one compounds and their novel imine derivatives. *Pakistan Journal of Pharmaceutical Sciences*, *31*, 2361–2365.

Singh, A. K., Singh, S. K., Nandi, M. K., Mishra, G., Maurya, A., Rai, A., et al. (2019). Berberine: A plant-derived alkaloid with therapeutic potential to combat Alzheimer's disease. *Central Nervous System Agents in Medicinal Chemistry*, *19*, 154–170. https://doi.org/10.2174/1871524919666190820160053.

Škerget, M., Kotnik, P., Hadolin, M., Hraš, A. R., Simonič, M., & Knez, Ž. (2005). Phenols, proanthocyanidins, flavones and flavonols in some plant materials and their antioxidant activities. *Food Chemistry*, *89*, 191–198. https://doi.org/10.1016/j.foodchem.2004.02.025.

Srinivas Koduru, F. O., Jimoh, D. S. G., & Afolayan, J. A. (2007). Antioxidant activity of two steroid alkaloid extracted from Solanum aculeastrum. *Journal of Pharmacology and Toxicology*, *2*, 160–167.

Srinivasan, K. (2007). Black pepper and its pungent principle-piperine: A review of diverse physiological effects. *Critical Reviews in Food Science and Nutrition*, *47*, 735–748. https://doi.org/10.1080/10408390601062054.

Sundaram, S., Radhakrishnan, A., Kanniappan, G. V., Bhaskaran, S. K., Palanisamy, C. P., & Kannappan, P. (2015). Comparative study on antioxidant activity of crude and alkaloid extracts of Hybanthus enneaspermus Linn F. Mull. *Analytical Chemistry Letters*, *5*, 291–299. https://doi.org/10.1080/22297928.2015.1135076.

Taha, K. F., Khalil, M., Abubakr, M. S., & Shawky, E. (2020). Identifying cancer-related molecular targets of Nandina domestica Thunb. by network pharmacology-based analysis in combination with chemical profiling and molecular docking studies. *Journal of Ethnopharmacology*, *249*. https://doi.org/10.1016/j.jep.2019.112413.

Thawabteh, A. M., Thawabteh, A., Lelario, F., Bufo, S. A., & Scrano, L. (2021). Classification, toxicity and bioactivity of natural diterpenoid alkaloids. *Molecules*, *26*. https://doi.org/10.3390/molecules26134103.

Tidjani, S., Okusa, P. N., Zellagui, A., Banuls, L. M. Y., Stévigny, C., Duez, P., et al. (2013). Analysis of pyrrolizidine alkaloids and evaluation of some biological activities of algerian senecio delphinifolius (Asteraceae). *Natural Product Communications*, *8*, 439–440. https://doi.org/10.1177/1934578x1300800406.

Trnka, J., Blaikie, F. H., Logan, A., Smith, R. A. J., & Murphy, M. P. (2009). Antioxidant properties of MitoTEMPOL and its hydroxylamine. *Free Radical Research*, *43*, 4–12. https://doi.org/10.1080/10715760802582183.

Tsoi, B., Yi, R., Cao, L., Li, S., Tan, R., Chen, M., et al. (2015). Comparing antioxidant capacity of purine alkaloids: A new, efficient trio for screening and discovering potential antioxidants in vitro and in vivo. *Food Chemistry*, *176*, 411–419. https://doi.org/10.1016/j.foodchem.2014.12.087.

Wang, R., Zhou, J., Shi, G., Liu, Y., & Yu, D. (2020). Aporphine and phenanthrene alkaloids with antioxidant activity from the roots of Stephania tetrandra. *Fitoterapia*, *143*, 104551. https://doi.org/10.1016/j.fitote.2020.104551.

Xu, D. P., Li, Y., Meng, X., Zhou, T., Zhou, Y., Zheng, J., et al. (2017). Natural antioxidants in foods and medicinal plants: Extraction, assessment and resources. *International Journal of Molecular Sciences*, *18*, 20–31. https://doi.org/10.3390/ijms18010096.

Yin, T. P., Cai, L., Fang, H. X., Fang, Y. S., Li, Z. J., & Ding, Z. T. (2015). Diterpenoid alkaloids from Aconitum vilmorinianum. *Phytochemistry*, *116*, 314–319. https://doi.org/10.1016/j.phytochem.2015.05.002.

Yin, T. P., Cai, L., Xing, Y., Yu, J., Li, X. J., Mei, R. F., et al. (2016). Alkaloids with antioxidant activities from Aconitum handelianum. *Journal of Asian Natural Products Research*, *18*, 603–610. https://doi.org/10.1080/10286020.2015.1114473.

Yoon, M. A., Jeong, T. S., Park, D. S., Xu, M. Z., Oh, H. W., Song, K. B., et al. (2006). Antioxidant effects of quinoline alkaloids and 2,4-di-tert-butylphenol isolated from Scolopendra subspinipes. *Biological and Pharmaceutical Bulletin*, *29*, 735–739. https://doi.org/10.1248/bpb.29.735.

Zahari, A., Cheah, F. K., Mohamad, J., Sulaiman, S. N., Litaudon, M., Leong, K. H., et al. (2014). Antiplasmodial and antioxidant isoquinoline alkaloids from Dehaasia longipedicellata. *Planta Medica*, *80*, 599–603. https://doi.org/10.1055/s-0034-1368349.

Zeynali, Z., Hosseini, B., & Rezaei, E. (2016). Effect of elicitation on antioxidant activity and production of tropane alkaloids in Hyoscyamus reticulatus hairy root cultures. *Research Journal of Pharmacognosy*, *3*, 43–53.

Zhishen, J., Mengcheng, T., & Jianming, W. (1999). The determination of flavonoid contents in mulberry and their scavenging effects on superoxide radicals. *Food Chemistry*. https://doi.org/10.1016/S0308-8146(98)00102-2.

Further reading

Neganova, M. E., Afanas Eva, S. V., Klochkov, S. G., & Shevtsova, E. F. (2012). Mechanisms of antioxidant effect of natural sesquiterpene lactone and alkaloid derivatives. *Bulletin of Experimental Biology and Medicine*, *152*, 720–722. https://doi.org/10.1007/s10517-012-1615-x.

CHAPTER FOURTEEN

Antioxidants: Structure–activity of plant polyphenolics

Aluru Rammohan[a,*], Grigory V. Zyryanov[a,b], Yerramathi Babu Bhagath[c], and Kola Manjula[c]

[a]Ural Federal University, Yekaterinburg, Russian Federation
[b]I. Ya. Postovsky Institute of Organic Synthesis, Ural Division of the RAS, Yekaterinburg, Russian Federation
[c]Food Technology Division, College of Science, Sri Venkateswara University, Tirupati, Andhra Pradesh, India
*Corresponding author: e-mail address: rammohan4ever@gmail.com

Contents

Abstract

The excessive accumulation of reactive oxygen species (ROS)/free radicals can lead to abnormal oxidation of biomolecules such as proteins, lipids, fats, carbohydrates and nucleic acids in human organisms. Accordingly, endogenous oxidative stress induces the progressive development of various chronic diseases like rheumatoid arthritis, cancers, cardiovascular risks, diabetes, digestive ulcers, hypertension, obesity, neurological disorders, and age-related complications. Therefore, anti-oxidant defense mechanisms are needed to control/prevent the unbalanced molecular oxidative damage. Indeed, the oxidative stress arises from both endogenous and exogenous factors such as smoking, alcohol, medications, air pollution, sunlight, lifestyle disorders, and metabolic processes. Therefore, consumption of fruits, vegetables, grains, beverages, and leafy vegetables rich in antioxidants may inhibit or treat oxidative damage accompanying diseases. From this aspect, dietary foods are rich in various antioxidant metabolites such as flavonoids, vitamin A, C, E, phenolic acids, curcumin, stilbenes, anthocyanins, etc., which promote healthy life and nutritional benefits. Additionally, various studies have

Vitamins and Hormones, Volume 121
ISSN 0083-6729
https://doi.org/10.1016/bs.vh.2022.10.001

also proven that foods rich in antioxidants interact with reactive species to prevent cell damage(s) or therapeutic pathways for diseases. Although, there are various myths about the antioxidant mechanism(s), the optimal dosage of antioxidants can show beneficial pharmacological activities against various molecular oxidation paths.

1. Introduction to polyphenolics

Plant phenolics are abundant in dietary agents, such as fruits, vegetables of *Leguminosae*, and *Citrus* members. Foods encompassing polyphenolics are crucial in the prevention of several oxidative stresses associated with chronic diseases *viz.* cancers, diabetes, cardiovascular and neurodegenerative diseases (Han, Shen, & Lou, 2007; Mrduljas, Kresic, & Bilusic, 2017). Consequently, researchers are focusing on foods and food beverages containing polyphenolics attributable to their extended human health benefits (Abbas et al., 2017). Consumption of foods that are rich in phenolic agents, i.e., fruits, tea, cocoa, nuts can reduce the risk of non-communicable diseases and promote healthy life (Kennedy, 2014; World Health Organization, 2003). Polyphenolics are "secondary metabolites" that are metabolized through both collective acetate and shikimate pathways of malonyl-CoA monomer units (Dewick, 2002; Kennedy, 2014). Thus, all phenolic compounds have a partial cinnamic acid monomeric unit with at least one aromatic ring or with a greater number of hydroxyl groups. Phenolic compounds commonly found in plants can occur as simple phenolic acids, complex organic structures and their derivative glycosides. Plant polyphenols can be classified mainly on the basis of their structure or pharmacological action. The basic unit in the polyphenolic compounds is a simple benzene ring with an hydroxyl group and thus is categorized (Fig.1) as simple phenolic acids, stilbenes, flavonoids, coumarins, lignins, tannins (Manach, Scalbert, Morand, Rémésy, & Jiménez, 2004). Nevertheless, flavonoids are the prime group of plant phenolic compounds with diverse structural substitution patterns, which are distributed in all parts of the plants including leaves, flowers, fruits, and roots (Crozier, Del Rio, & Clifford, 2010; Hollman & Katan, 1999). Furthermore, flavonoids can be subdivided into several groups as flavones, flavanones, flavanols, dihydro-flavonols, isoflavones and anthocyanins based on the substitution pattern of basic 15-carbon skeleton. Considering the structural diversity and potent pharmacological activities, flavonoids are well-known for their beneficial health promoting impacts.

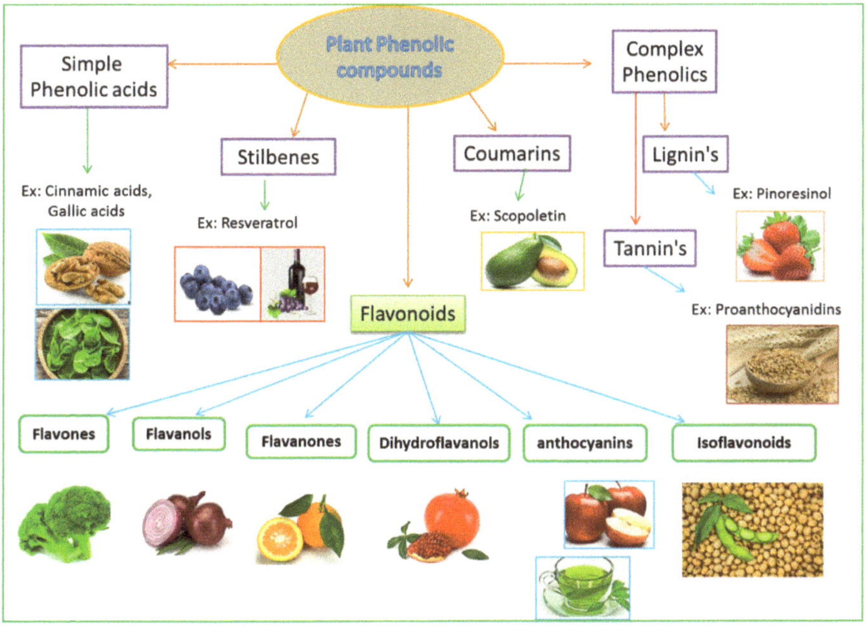

Fig. 1 An essential regular diet encompasses of natural polyphenolic compounds and their classification with illustrations.

2. Polyphenolics as antioxidants

Reactive oxygen species (ROS) or free radicals are formed in human tissue cells either through uncontrolled metabolic processes or by lifestyle disorders, such as smoking, pollution, radiation and persistent medications. These factors can cause extensive damage to cellular biomolecules, such as lipids, proteins and nucleic acids (Snezhkina et al., 2019; Tan, Norhaizan, & Liew, 2018; Tan, Norhaizan, Liew, & Sulaiman Rahman, 2018). Oxidative stress-related impairments result in major diseases like digestive ulcers, hypertension, certain types of cancers (including non-inflammatory tumors), diabetes, cardiovascular diseases, rheumatoid arthritis, neurodegenerative disorders, and aging (Aruoma, 1998; Bjørklund & Chirumbolo, 2017). Several decades of nutritional research suggest that vegetables, fruits and foods high in certain natural metabolites such as polyphenols, carotenoids, nitrogenous compounds, vitamin E, and ascorbic acid may reduce the risk of certain cancers and metabolic diseases through anti-oxidant principles (Hertog, Hollman, & Van de Putte, 1993; Key et al., 2004). Thus, plant

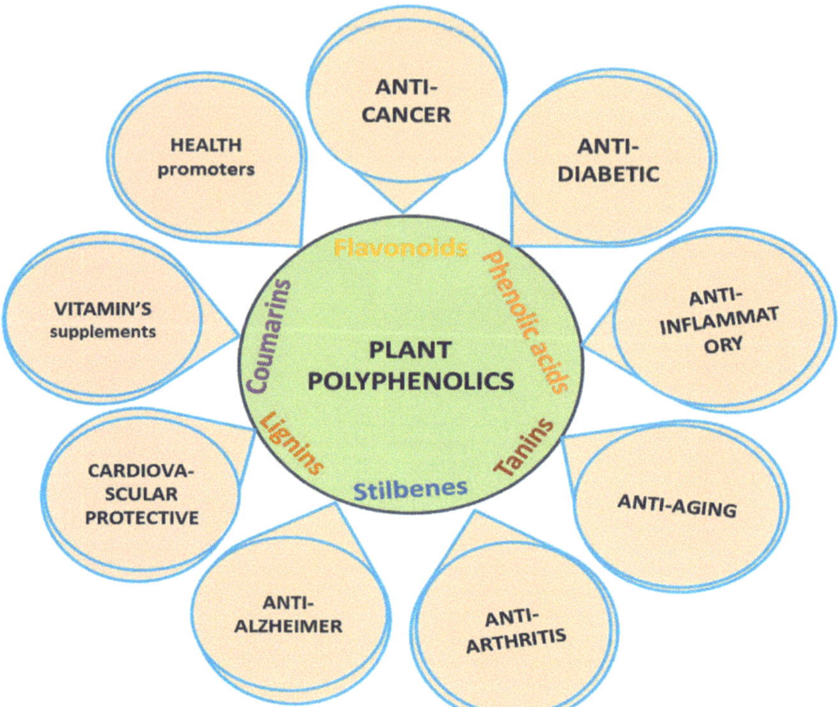

Fig. 2 Therapeutic potential of plant polyphenols against various chronic diseases through antioxidant principles.

metabolites, particularly polyphenols, are termed as "antioxidants" that have a unique potency to protect against non-communicable oxidative stress-related diseases (Fig. 2). Currently, plant phenolics are attracting special attention of researchers and the general public because of their health-promoting benefits (Aruoma, 1999; Tan, Norhaizan, & Liew, 2018; Tan, Norhaizan, Liew, & Sulaiman Rahman, 2018).

Over the past few decades, several natural polyphenolic constituents from edible plants, fruits and food beverages have been cited as potent "antioxidants" that endorse health benefits (Battino et al., 2019; Rammohan et al., 2020; Tresserra-Rimbau et al., 2014). Recently, the public has become progressively aware of the benefits of daily diets containing leafy vegetables, fruits, nuts, and whole grains that contain polyphenols, vitamins and minerals (Fig. 3). Leafy vegetables contain simple phenolic and ascorbic acids which are essential for reducing the ROS produced by metabolic processes in the human body (Tresserra-Rimbau et al., 2014).

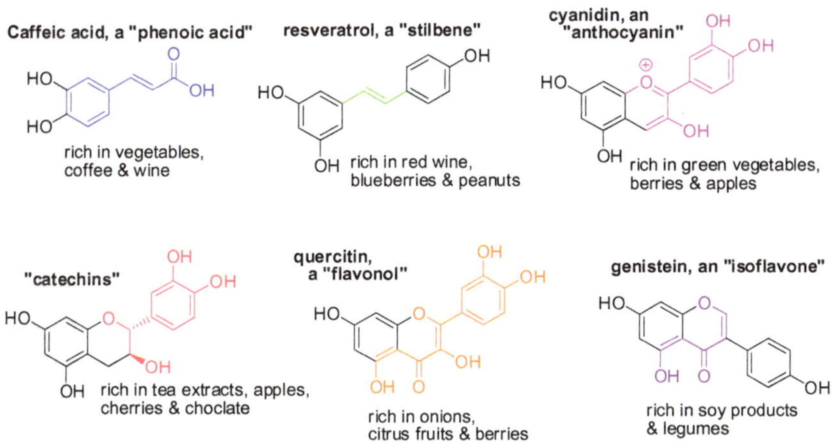

Fig. 3 Structures of some potent antioxidant "polyphenols" found in edible vegetables, fruits and food products.

For instance, spinach and walnuts are good sources of trace quantities of cinnamic and gallic acids, respectively (Nair & Vining, 1965; Pycia, Kapusta, Jaworska, & Jankowska, 2019). Likewise, dietary turmeric (curry powder) is a rich source of a key plant-phenol, i.e., curcumin, and it plays an important role in the pumping of blood vessels through lowering LDL cholesterol levels (Vasanthi & Parameswari, 2010). Red-wine and berries contain the important polyphenol "resveratrol," which plays a strategic role in controlling heart disease by dilating the arteries to speed-up blood flow (Dolinsky & Dyck, 2011; Vidavalur, Otani, Singal, & Maulik, 2006). Thus, curcumin and resveratrol are two potent naturally-occurring dietary phenolic compounds that play a crucial role in reducing/preventing cardiovascular diseases. Although phenolic acids of foods are beneficial, too much intake can cause problems, consequently, it is necessary to maintain ideal quantities of it (Działo et al., 2016).

Flavonoids are semi-essential food agents, also known as natural "oxygen heterocycles" with most promising antioxidant activities due to their structural substitution patterns (Kuhnau, 1976; Pietta, 2000). In addition, flavonoids exhibit pharmacological properties: antibacterial, anticancer, antidiabetic, anti-inflammatory, antiviral and heptaprotective assets by inhibition of oxidative enzymes or by chelation (Kumar & Pandey, 2013; Terao, 2009). Tea is a part of the daily diet in many countries; it promotes health benefits through its antioxidant activity (Yan, Zhong, Duan, Chen, & Li, 2020). Green tea, in particular, is a rich source of polyphenols including catechin, epicatechin, epigallocatechin, anthocyanins, and polymeric catechins (Zuo et al., 2014). Moreover, tea-polyphenols can prevent several diseases by regulating the various oxidase systems in the human body through hydrogen ion/radical donor or electron-donor or electron-conjugation mechanisms (Imran et al., 2018; Yan et al., 2020). Another natural flavonoid, quercetin and its glycoside (hesperidin) are found plentifully in vegetables and fruits, i.e., onions, green leafy vegetables, grapes, cherries, red grapes, apples (Ay et al., 2016; Shankar, Antony, & Anto, 2015). Chalconaringenin and naringenin are the two most important naturally occurring edible flavonoids found widely in *Citrus* fruits: oranges, grapefruits, tomatoes and pomelo (Salehi et al., 2019). Thus, these edible citrus fruits, containing flavonoids, have the potential for scavenging radicals, inhibiting lipid peroxidation, and reducing cytokine production as well as nitric oxide synthase activity (Ay et al., 2016; David, Arulmoli, & Parasuraman, 2016). Apples are another important perennial source of natural antioxidants including gallic acid, quercetin, catechin, phloretin (dihydrochalcone) and chlorogenic acid (Boyer & Liu, 2004; He & Liu, 2008). Owing to the various polyphenolics in apples, they play key roles in reducing the risk of cancer cell proliferation, lowering cholesterol levels, reducing lipid peroxidation and thereby preventing various chronic diseases through antioxidant activity (Boyer & Liu, 2004; Tu, Chen, & Ho, 2017). Further, the *Leguminosae* members of soybeans, mung beans, peanuts, chickpeas and other nuts are also the richest source of various concentrations of isoflavones like daidzein, genistein, formononetin and biochanin A (Patel et al., 2001; Song et al., 2007). Thus, these *Leguminosae* isoflavones have numerous health benefits including prevention of osteoporosis and cancers, expansion of immune systems, and reduction of the risks of cardiovascular diseases through lowering the LDL cholesterols (Patel et al., 2001; Wrigley, Corke, Seetharaman, & Faubion, 2015). Therefore, the bioactive principles of vegetables, fruits and foods rich in polyphenols improve the quality of life by reducing the occurrence of various types of chronic diseases.

3. Potent key targets behind antioxidant therapies

Free radicals/highly reactive species trigger "oxidative stress" deriving from endogenous and exogenous sources (Snezhkina et al., 2019). Further, these oxidative stresses and associated tissue damage paths play key roles in the development of various diseases (Fig. 4), such as atherosclerosis, cancers,

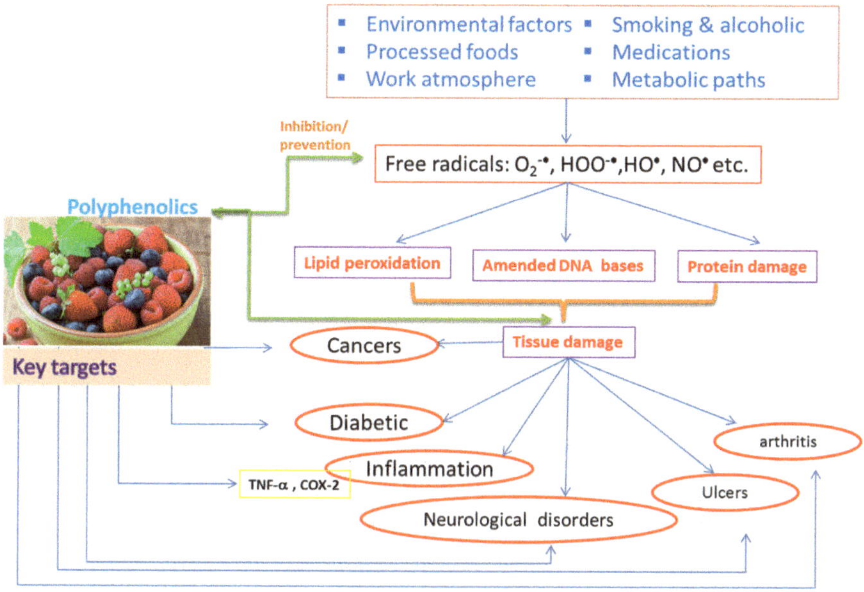

Fig. 4 Oxidative stress and associated tissue damage paths in the expansion of various diseases and therapeutic mechanisms by polyphenols.

cardiovascular diseases, cataracts, diabetes, hypertension, neurological disorders, and aging-linked muscular degenerative conditions (Lobo, Patil, Phatak, & Chandra, 2010; Nita & Grzybowski, 2016). Antioxidants are potent defensive mechanisms by suppressing free radical chain reactions, reducing ROS production, stalling free-radical imitators, and chelating metal ions (Salehi et al., 2020). Therefore, the following sections summarize the correlation of oxidative stress in the pathogenesis of various diseases and the essential role of anti-oxidants.

3.1 Arthritis

Oxidative stress is considered to be one of the most serious causes of the upsurge in non-communicable autoimmune disorders. Since, the reactive oxygen species (ROS) have been studied as key regulators of cell-damaging or as signaling molecules in the immune system (Hitchon & El-Gabalawy, 2004). Oxidative stress promotes various abnormal molecular paths/ transcription factors such as synovial inflammatory-proliferative response, hypoxia-inducible factor-1α (HIF-1α), nuclear factor-κB (NF-κB), tumor necrosis factor-α (TNF-α), and a functional mutation of oxidative burst, i.e., NADPH oxidase complex (neutrophil cytosolic factor 1-Ncf1).

(Hitchon & El-Gabalawy, 2004; Lepetsos & Papavassiliou, 2016). These transcriptional mutations and inflammatory cytokine molecular paths function for the pathogenesis of polygenic diseases such as rheumatoid arthritis (RA), osteoarthritis (OA), and inflammatory arthritis. In addition, various studies have established potential markers of oxidative damage in RA, relating synovitis, destruction of bone, cartilage and ligaments (Yoo, Go, Kim, Lee, & Kwon, 2016). Therefore, prevention/treatment of oxidative mechanisms is a potent target in the management of arthritis.

3.2 Cancers

The World Health Organization (WHO) estimates that 9.6 million people died of various cancers in 2018 (Ferlay et al., 2019; Prasad, Gupta, Pandey, Tyagi, & Deb, 2016). Excess reactive oxygen species (ROS) leads to the uncontrolled oxidation of nucleic acids (DNA, RNA), phospholipids, proteins, carbohydrates on the cell membranes (Hayes, Dinkova-Kostova, & Tew, 2020). These unbalanced oxidative stresses are involved in carcinogenesis through genetic mutations, signal transduction, and transcription factor-associated pathways (Noda & Wakasugi, 2001). Various studies have implicated the role of oxidative stress in tumor development through cell proliferation, metastasis, angiogenesis, and chronic inflammations (Arfin et al., 2021). Moreover, the budding role of oxidative stress in cell metabolism in various cancers such as breast, lung, prostate and ovarian cancers and their pathogenesis mechanisms are well established (Azad, Rojanasakul, & Vallyathan, 2008; Gupta-Elera, Garrett, Robison, & O'Neill, 2012; Nourazarian, Kangari, & Salmaninejad, 2014; Steelman et al., 2008). Consequently, the relative management of oxidative stress is also beneficial in the treatment/prevention of certain cancers.

3.3 Diabetes

Oxidative stress plays a key role in the development of diabetes mellitus through various biochemical mechanisms, such as advanced glycosylation end-products (AGEs), impaired glutathione metabolism, and superoxide dismutase (SOD) activity (Giacco & Brownlee, 2010; Rammohan & Bhaskar, 2021). Further, the AGEs show a major role in the pathogenesis of diabetic complications by initiating inflammatory and pro-inflammatory cytokines (Asmat, Abad, & Ismail, 2016; Rammohan & Bhaskar, 2021). Thus, the ROS induced micro- and macrovascular-barriers produce various diabetic complications, such as insulin resistance, hypertension, obesity, renal problems, and neuronal disorders (Rammohan & Bhaskar, 2021).

Thus, the oxidative stress leads to oxidation damage of biomolecules like proteins, lipids, nucleic acids that progress to hyperglycemia, insulin resistance and associated health complications.

3.4 Inflammatory diseases

Equally, the oxidative damage of cellular biomolecules is associated with the development of various carcinogenic and non-carcinogenic inflammatory syndromes. Since the inflammation is a critical/chronic component that activates cellular cytokines like interleukin (IL-1β), IL-6, IL-8, TNF-α, NF-κB, and NADPH oxidase-4 (NOX4), etc. under oxidative stress, it contributes to pathophysiological processes for various diseases (Biswas, 2016; Chatterjee, 2016). Thus, the oxidative stress associated inflammatory condition is a complex disorder that contributes to the incidence of chronic heart-attacks, arthritis, diabetes, cancers, and neurodegenerative disorders (Lugrin, Rosenblatt-Velin, Parapanov, & Liaudet, 2014). Considering the perpetual complex pathologies of inflammation associated with oxidative stress, a strong effective antioxidant defense mechanism is required.

3.5 Ulcers

In general, the gastric ulcers are caused either endogenous factors such as pepsin, hydrochloric acid, leukotrienes, refluxed bile, and reactive oxygen species (ROS) or exogenous elements like alcohol, non-steroidal anti-inflammatory drugs (NSAIDS), drugs which excite pepsin secretion and gastric acid stimulation, tension, stress and *Helicobacter pylori* (Bandyopadhyay, Biswas, Bhattacharyya, Reiter, & Banerjee, 2002). Oxidative stress/ROS has been found to initiate and enhance gastric mucosal lipid peroxidation (LPO), superoxide dismutase (SOD), and a decline in catalase (CAT) levels (Tandon, Khanna, Dorababu, & Goel, 2004). Moreover, gastric mucosa triggered by free radicals activates various oxidant-enzymes and inflammatory cytokines such as NADPH oxidase, inducible nitic oxidase synthase (iNOS) (Jasna & Drazen, 2011). Thus, ROS/oxidative stress plays a key role in the pathogenesis of ulcerogenesis, gastric inflammation, *H. pylori* carcinogenesis through cellular oxidation and organ dysfunction (Suzuki, Nishizawa, Tsugawa, Mogami, & Hibi, 2012; Tandon et al., 2004).

3.6 Neurological disorders

Together with genetic disorders or lifestyle factors, reactive oxygen species (ROS) and imbalanced metabolism can lead to various neurodegenerative disorders like Alzheimer's, Parkinson's disease, aging, states of confusion, dementia, and other neuronal complications (Kim, Kim, Rhie, & Yoon, 2015). Collective oxidative stress in homeostasis of neuron cells and abnormal metabolic paths in brain might induce development of dementia and cognitive decline (Gandhi & Abramov, 2012). Protein oxidation, lipid peroxidation and mitochondrial DNA mutations may be components in progressive Parkinson's disease (Uttara, Singh, Zamboni, & Mahajan, 2009). Considering the above discussion, one can conclude that oxidative stress plays major role in the pathogenesis of neurodegenerative disorders.

The collective oxidation patterns and abnormal metabolic paths are specific indicators of oxidative stress, which can lead to arthritis, cancers, cardiovascular problems, diabetes, obesity, non-carcinogenesis inflammatory disease, hypertension, neurodegenerative syndromes, and other age-related disorders. Thus, the prevention/management of oxidative stress mechanisms may be beneficial in the treatment of various chronic diseases or in delaying pathogenesis.

4. Structure–activity of plant metabolites

Currently, the importance of traditional medicine and nutritious foods in the prevention/management of various diseases is well documented (Sen & Chakraborty, 2017). In some countries, people have begun to adopt a daily diet rich in vitamins and antioxidant foods like fruits, vegetables, whole-grains and flavors to overcome the lifestyle disorders, infectious diseases, and promote a healthier life (Biesalski et al., 1997; Rammohan & Bhaskar, 2021). Foods high in natural metabolites are beneficial in maintaining a healthy lifestyle through antioxidant activity (Chun et al., 2005). Natural metabolites, such as ascorbic acid (vitamin C), β-carotene (provitamin A), α-tocopherol (vitamin E), resveratrol, lycopene, quercetin, delphinidin, curcumin, and epigallocatechin gallate (EGCG) have been identified (Rice-Evans, Miller, & Paganga, 1996) as potent antioxidants that are key constituents of fruits and vegetables.

Phenolic metabolites such as, quercetin, resveratrol, delphinidin and epigallocatechin gallate (EGCG) have been found to be antioxidants that eliminate ROS activity (Vaya & Aviram, 2001). Plant phenolics inhibit oxidative enzymes and scavenge/quench reactive oxygen species (Bhaskar et al., 2020; Terao, 2009). Structures of phenolic compounds, such as *O*-dihydroxyl phenyl group (catechol group), diphenyl propionate are responsible for potent radical scavenging activity as described in Fig. 5.

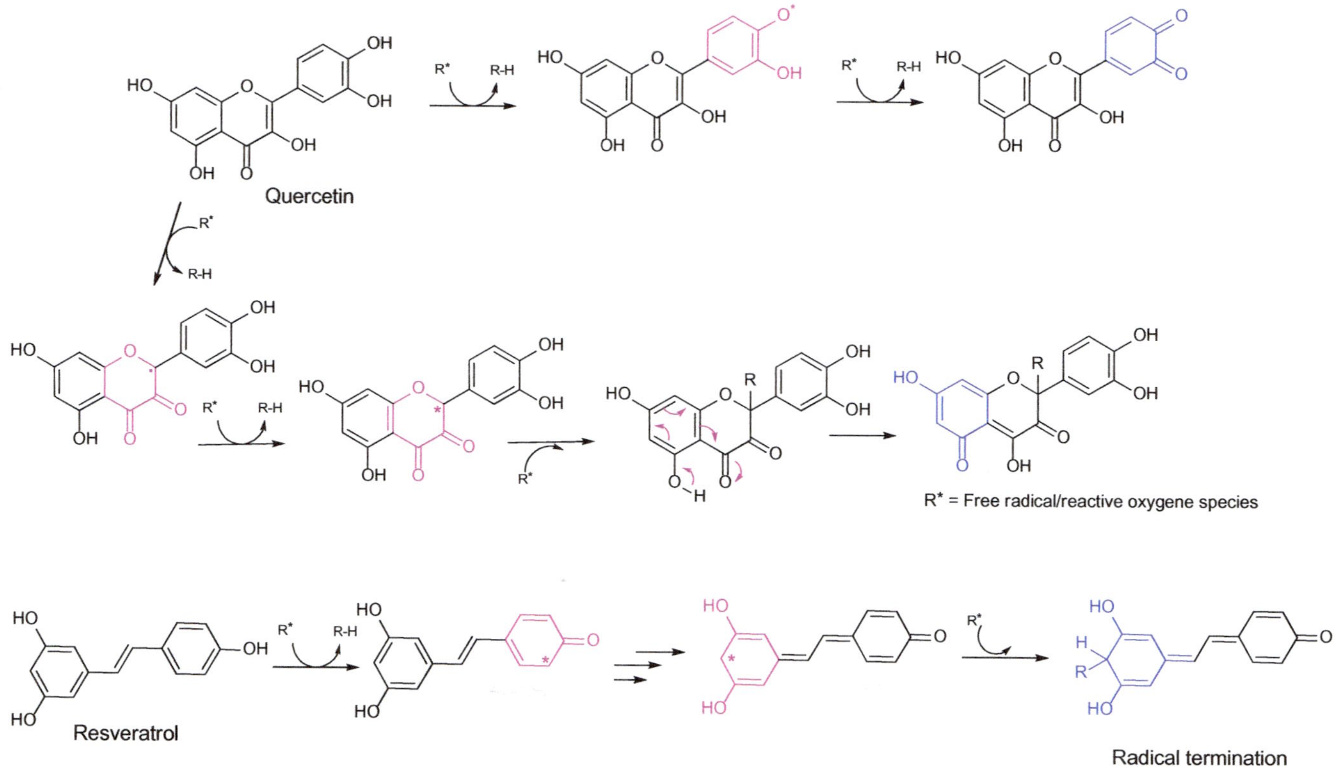

Fig. 5 The conceivable radical scavenging mechanisms of nominated polyphenols.

Our *in vitro* quantified density functional theory (DFT) discloses that antioxidant ability of phenolic metabolite(s) depends on molecular structure and phenolic hydroxyl group polarizability (Rammohan et al., 2020). Phenolic metabolites exert copious mechanisms to reconcile pathogenesis/preventive effects of diseases by suppressing activities of proteins, enzymes, and cytokines (Lobo et al., 2010; Terao, 2009). For instance, epigallocatechin gallate (EGCG), a dietary potent antioxidant tea metabolite, shows a wide-range of pharmacological activities, such as antioxidant, antidiabetic, anti-obesity, anti-inflammatory, anti-viral, cardiovascular- and neuro-protection (Xing, Zhang, Qi, Tsao, & Mine, 2019). Consequently, the foods ironic with natural antioxidants can regulate chronic diseases through reducing the all categories of oxidative damage. Therefore, this comprehensive investigation supports the daily-intake of antioxidant-rich fruits, nuts, vegetables and dietary supplements of natural origin in precise quantities.

5. Conclusions

As a summary, the excessive concentrations of reactive oxygen species (ROS)/free radicals can develop various oxidative damages such as protein oxidation, lipid peroxidation, and DNA mutations. These unbalanced oxidative metabolic reactions can lead to oxidative stress with chronic genomic mutations and transcriptional errors. Therefore, the treatment of these multifaceted oxidative stresses and associated diseases requires consideration. Various studies have revealed that oxidative stress is associated with diseases like atherosclerosis, cancers, cardiovascular diseases, cataracts, diabetes, hypertension, neurological disorders, and age-linked muscular degenerative conditions. Therefore, prevention/management of oxidative stress may be beneficial in treating or delaying the occurrence of numerous chronic diseases.

Polyphenolic containing fruits, vegetables, grains and beverages have been found to have potent antioxidant properties that rectify chronic oxidation patterns. Natural antioxidants are of great importance due to their actions in preventing various pathogenic diseases including angiogenesis and metastasis of cancers, tumors, neurological disorders. In addition, consuming antioxidant rich-foods may be valuable in promoting healthy-life and nutritional benefits. Despite the uncertain myths on the molecular mechanism of antioxidants, the consumption of limited dosages of phenolic rich-foods can heal chronic diseases.

Acknowledgments

The authors acknowledge the Grants Council of the President of the Russian Federation (# НШ-2700.2020.3) and Russian Scientific Foundation (Grant # 21-13-00304).

References

Abbas, M., Saeed, F., Anjum, F. M., Afzaal, M., Tufail, T., Bashir, M. S., et al. (2017). Natural polyphenols: An overview. *International Journal of Food Properties, 20*(8), 1689–1699. https://doi.org/10.1080/10942912.2016.1220393.

Arfin, S., Jha, N. K., Jha, S. K., Kesari, K. K., Ruokolainen, J., Roychoudhury, S., et al. (2021). Oxidative stress in cancer cell metabolism. *Antioxidants, 10*(5), 642. https://doi.org/10.3390/antiox10050642.

Aruoma, O. I. (1998). Free radicals, oxidative stress, and antioxidants in human health and disease. *Journal of the American Oil Chemists' Society, 75*(2), 199–212. https://doi.org/10.1007/s11746-998-0032-9.

Aruoma, O. I. (1999). Antioxidant actions of plant foods: Use of oxidative DNA damage as a tool for studying antioxidant efficacy. *Free Radical Research, 30*(6), 419–427. https://doi.org/10.1080/10715769900300461.

Asmat, U., Abad, K., & Ismail, K. (2016). Diabetes mellitus and oxidative stress—A concise review. *Saudi Pharmaceutical Journal, 24*(5), 547–553. https://doi.org/10.1016/j.jsps.2015.03.013.

Ay, M., Charli, A., Jin, H., Anantharam, V., Kanthasamy, A., & Kanthasamy, A. G. (2016). Quercetin. In *Nutraceuticals* (pp. 447–452). Academic Press. https://doi.org/10.1016/B978-0-12-802147-7.00032-2.

Azad, N., Rojanasakul, Y., & Vallyathan, V. (2008). Inflammation and lung cancer: Roles of reactive oxygen/nitrogen species. *Journal of Toxicology and Environmental Health, Part B, 11*(1), 1–15.

Bandyopadhyay, D., Biswas, K., Bhattacharyya, M., Reiter, R. J., & Banerjee, R. K. (2002). Involvement of reactive oxygen species in gastric ulceration: Protection by melatonin. *Indian Journal of Experimental Biology, 40*, 693–705. http://hdl.handle.net/123456789/23512.

Battino, M., Forbes-Hernández, T. Y., Gasparrini, M., Afrin, S., Cianciosi, D., Zhang, J., et al. (2019). Relevance of functional foods in the Mediterranean diet: The role of olive oil, berries and honey in the prevention of cancer and cardiovascular diseases. *Critical Reviews in Food Science and Nutrition, 59*(6), 893–920. https://doi.org/10.1080/10408398.2018.1526165.

Bhaskar, B. V., Babu, T. M. C., Rammohan, A., Zheng, G. Y., Zyryanov, G. V., & Gu, W. (2020). Structure-based virtual screening of *Pseudomonas aeruginosa* LpxA inhibitors using pharmacophore-based approach. *Biomolecules, 10*(2), 266. https://doi.org/10.3390/biom10020266.

Biesalski, H. K., Böhles, H., Esterbauer, H., Fürst, P., Gey, F., Hundsdörfer, G., et al. (1997). Antioxidant vitamins in prevention. *Clinical Nutrition, 16*(3), 151–155. https://doi.org/10.1016/S0261-5614(97)80245-2.

Biswas, S. K. (2016). Does the interdependence between oxidative stress and inflammation explain the antioxidant paradox? *Oxidative Medicine and Cellular Longevity, 2016*, 1–9. https://doi.org/10.1155/2016/5698931.

Bjørklund, G., & Chirumbolo, S. (2017). Role of oxidative stress and antioxidants in daily nutrition and human health. *Nutrition, 33*, 311–321. https://doi.org/10.1016/j.nut.2016.07.018.

Boyer, J., & Liu, R. H. (2004). Apple phytochemicals and their health benefits. *Nutrition Journal, 3*(1), 5. https://doi.org/10.1186/1475-2891-3-5.

Chatterjee, S. (2016). Chapter two-oxidative stress, inflammation, and disease. *Oxidative Stress and Biomaterials, 35–58*. https://doi.org/10.1016/B978-0-12-803269-5.00002-4.

Chun, O. K., Kim, D. O., Smith, N., Schroeder, D., Han, J. T., & Lee, C. Y. (2005). Daily consumption of phenolics and total antioxidant capacity from fruit and vegetables in the American diet. *Journal of the Science of Food and Agriculture, 85*(10), 1715–1724. https://doi.org/10.1002/jsfa.2176.

Crozier, A., Del Rio, D., & Clifford, M. N. (2010). Bioavailability of dietary flavonoids and phenolic compounds. *Molecular Aspects of Medicine, 31*(6), 446–467. https://doi.org/10.1016/j.mam.2010.09.007.

David, A. V. A., Arulmoli, R., & Parasuraman, S. (2016). Overviews of biological importance of quercetin: A bioactive flavonoid. *Pharmacognosy Reviews, 10*(20), 84. https://doi.org/10.4103/0973-7847.194044.

Dewick, P. M. (2002). *Medicinal natural products: A biosynthetic approach* (2nd edition, pp. 7–30). England, UK: John Wiley & Sons.

Dolinsky, V. W., & Dyck, J. R. (2011). Calorie restriction and resveratrol in cardiovascular health and disease. *Biochimica et Biophysica Acta (BBA) - Molecular Basis of Disease, 1812*(11), 1477–1489. https://doi.org/10.1016/j.bbadis.2011.06.010.

Działo, M., Mierziak, J., Korzun, U., Preisner, M., Szopa, J., & Kulma, A. (2016). The potential of plant phenolics in prevention and therapy of skin disorders. *International Journal of Molecular Sciences, 17*(2), 160. https://dx.doi.org/10.3390%2Fijms17020160.

Ferlay, J., Colombet, M., Soerjomataram, I., Mathers, C., Parkin, D. M., Piñeros, M., et al. (2019). Estimating the global cancer incidence and mortality in 2018: GLOBOCAN sources and methods. *International Journal of Cancer, 144*(8), 1941–1953. https://doi.org/10.1002/ijc.31937.

Gandhi, S., & Abramov, A. Y. (2012). Mechanism of oxidative stress in neurodegeneration. *Oxidative Medicine and Cellular Longevity, 2012*, 1–11. https://doi.org/10.1155/2012/428010.

Giacco, F., & Brownlee, M. (2010). Oxidative stress and diabetic complications. *Circulation Research, 107*(9), 1058–1070. https://doi.org/10.1161/CIRCRESAHA.110.223545.

Gupta-Elera, G., Garrett, A. R., Robison, R. A., & O'Neill, K. L. (2012). The role of oxidative stress in prostate cancer. *European Journal of Cancer Prevention, 21*(2), 155–162. https://doi.10.1097/CEJ.0b013e32834a8002.

Han, X., Shen, T., & Lou, H. (2007). Dietary polyphenols and their biological significance. *International Journal of Molecular Sciences, 8*(9), 950–988. https://doi.org/10.3390/i8090950.

Hayes, J. D., Dinkova-Kostova, A. T., & Tew, K. D. (2020). Oxidative stress in cancer. *Cancer Cell, 38*(2), 167–197. https://doi.org/10.1016/j.ccell.2020.06.001.

He, X., & Liu, R. H. (2008). Phytochemicals of apple peels: Isolation, structure elucidation, and their antiproliferative and antioxidant activities. *Journal of Agricultural and Food Chemistry, 56*(21), 9905–9910. https://doi.org/10.1021/jf8015255.

Hertog, M. G., Hollman, P. C., & Van de Putte, B. (1993). Content of potentially anticarcinogenic flavonoids of tea infusions, wines, and fruit juices. *Journal of Agricultural and Food Chemistry, 41*(8), 1242–1246. https://doi.org/10.1021/jf00032a015.

Hitchon, C. A., & El-Gabalawy, H. S. (2004). Oxidation in rheumatoid arthritis. *Arthritis Research & Therapy, 6*(6), 1–14. https://doi.org/10.1186/ar1447.

Hollman, P. H., & Katan, M. B. (1999). Dietary flavonoids: Intake, health effects and bioavailability. *Food and Chemical Toxicology, 37*(9–10), 937–942. https://doi.org/10.1016/S0278-6915(99)00079-4.

Imran, A., Arshad, M. U., Mehmood, S., Ahmed, R. S., Butt, M. S., Ahmed, A., et al. (2018). Oxidative stress diminishing perspectives of green and black tea polyphenols: A mechanistic approach. In J. Wong (Ed.), *Polyphenols* (pp. 25–50). https://doi.org/10.5772/intechopen.75933.

Jasna, D., & Drazen, S. (2011). Oxidative stress pathway driven by inflammation in gastric mucosa. In P. Tonino (Ed.), *Gastritis and gastric cancer: New insights in Gastroprotection, Diagnosis and Treatments* IntechOpen. https://doi.org/10.5772/23875.

Kennedy, D. O. (2014). Polyphenols and the human brain: Plant "secondary metabolite" ecologic roles and endogenous signaling functions drive benefits. *Advances in Nutrition*, *5*(5), 515–533. https://doi.org/10.3945/an.114.006320.

Key, T. J., Schatzkin, A., Willett, W. C., Allen, N. E., Spencer, E. A., & Travis, R. C. (2004). Diet, nutrition and the prevention of cancer. *Public Health Nutrition*, 7(1a), 187–200. https://doi.org/10.1079/PHN2003588.

Kim, G. H., Kim, J. E., Rhie, S. J., & Yoon, S. (2015). The role of oxidative stress in neurodegenerative diseases. *Experimental Neurobiology*, *24*(4), 325–340. https://doi.org/10.5607/en.2015.24.4.325.

Kuhnau, J. (1976). Flavonoids. A class of semi-essential food components: Their role in human nutrition. *World Review of Nutrition and Dietetics*, *24*, 117–191. https://doi.org/10.1159/000399407.

Kumar, S., & Pandey, A. K. (2013). Chemistry and biological activities of flavonoids: An overview. *The Scientific World Journal*, *2013*, 1–17. https://doi.org/10.1155/2013/162750.

Lepetsos, P., & Papavassiliou, A. G. (2016). ROS/oxidative stress signaling in osteoarthritis. *Biochimica et Biophysica Acta (BBA) - Molecular Basis of Disease*, *1862*(4), 576–591. https://doi.org/10.1016/j.bbadis.2016.01.003.

Lobo, V., Patil, A., Phatak, A., & Chandra, N. (2010). Free radicals, antioxidants and functional foods: Impact on human health. *Pharmacognosy Reviews*, *4*(8), 118. https://doi.org/10.4103/0973-7847.70902.

Lugrin, J., Rosenblatt-Velin, N., Parapanov, R., & Liaudet, L. (2014). The role of oxidative stress during inflammatory processes. *Biological Chemistry*, *395*(2), 203–230. https://doi.org/10.1515/hsz-2013-0241.

Manach, C., Scalbert, A., Morand, C., Rémésy, C., & Jiménez, L. (2004). Polyphenols: Food sources and bioavailability. *The American Journal of Clinical Nutrition*, *79*(5), 727–747. https://doi.org/10.1093/ajcn/79.5.727.

Mrduljas, N., Kresic, G., & Bilusic, T. (2017). Polyphenols: Food sources and health benefits. In M. C. Hudea (Ed.), *Functional food-improve health through adequate food* (pp. 22–41). UK: Intech Open. https://doi.org/10.5772/intechopen.68862.

Nair, P. M., & Vining, L. C. (1965). Cinnamic acid hydroxylase in spinach. *Phytochemistry*, *4*(1), 161–168. https://doi.org/10.1016/S0031-9422(00)86159-2.

Nita, M., & Grzybowski, A. (2016). The role of the reactive oxygen species and oxidative stress in the pathomechanism of the age-related ocular diseases and other pathologies of the anterior and posterior eye segments in adults. *Oxidative Medicine and Cellular Longevity*, *2016*, 1–23. https://doi.org/10.1155/2016/3164734.

Noda, N., & Wakasugi, H. (2001). Cancer and oxidative stress. *Japan Medical Association Journal*, *44*(12), 535–539.

Nourazarian, A. R., Kangari, P., & Salmaninejad, A. (2014). Roles of oxidative stress in the development and progression of breast cancer. *Asian Pacific Journal of Cancer Prevention*, *15*(12), 4745–4751. https://doi.org/10.7314/APJCP.2014.15.12.4745.

Patel, R. P., Boersma, B. J., Crawford, J. H., Hogg, N., Kirk, M., Kalyanaraman, B., et al. (2001). Antioxidant mechanisms of isoflavones in lipid systems: Paradoxical effects of peroxyl radical scavenging. *Free Radical Biology and Medicine*, *31*(12), 1570–1581. https://doi.org/10.1016/S0891-5849(01)00737-7.

Pietta, P. G. (2000). Flavonoids as antioxidants. *Journal of Natural Products*, *63*(7), 1035–1042. https://doi.org/10.1021/np9904509.

Prasad, S., Gupta, S. C., Pandey, M. K., Tyagi, A. K., & Deb, L. (2016). Oxidative stress and cancer: Advances and challenges. *Oxidative Medicine and Cellular Longevity*, *2016*. https://doi.org/10.1155/2016/5010423.

Pycia, K., Kapusta, I., Jaworska, G., & Jankowska, A. (2019). Antioxidant properties, profile of polyphenolic compounds and tocopherol content in various walnut (*Juglans regia* L.) varieties. *European Food Research and Technology, 245*(3), 607–616. https://doi.org/10.1007/s00217-018-3184-3.

Rammohan, A., & Bhaskar, B. V. (2021). In Atta-ur-Rahman (Ed.), *Frontiers in Clinical Drug Research-Diabetes and Obesity: 6. Flavonoids as prominent anti-diabetic agents* (pp. 31–71). https://doi.org/10.2174/97898114791991200600004.

Rammohan, A., Bhaskar, B. V., Camilo, A., Jr., Gunasekar, D., Gu, W., & Zyryanov, G. V. (2020). In silico, in vitro antioxidant and density functional theory based structure activity relationship studies of plant polyphenolics as prominent natural antioxidants. *Arabian Journal of Chemistry, 13*(2), 3690–3701. https://doi.org/10.1016/j.arabjc.2019.12.017.

Rice-Evans, C. A., Miller, N. J., & Paganga, G. (1996). Structure-antioxidant activity relationships of flavonoids and phenolic acids. *Free Radical Biology and Medicine, 20*(7), 933–956. https://doi.org/10.1016/0891-5849(95)02227-9.

Salehi, B., Azzini, E., Zucca, P., Maria Varoni, E., Anil Kumar, N. V., Dini, L., et al. (2020). Plant-derived bioactives and oxidative stress-related disorders: A key trend towards healthy aging and longevity promotion. *Applied Sciences, 10*(3), 947. https://doi.org/10.3390/app10030947.

Salehi, B., Fokou, P. V. T., Sharifi-Rad, M., Zucca, P., Pezzani, R., Martins, N., et al. (2019). The therapeutic potential of naringenin: A review of clinical trials. *Pharmaceuticals, 12*(1), 11. https://doi.org/10.3390/ph12010011.

Sen, S., & Chakraborty, R. (2017). Revival, modernization and integration of Indian traditional herbal medicine in clinical practice: Importance, challenges and future. *Journal of Traditional and Complementary Medicine*, 7(2), 234–244. https://doi.org/10.1016/j.jtcme.2016.05.006.

Shankar, G. M., Antony, J., & Anto, R. J. (2015). Quercetin and tryptanthrin: Two broad spectrum anticancer agents for future chemotherapeutic interventions. In P. D. Boyer, & E. G. Krebs (Eds.), *Vol. 37. The enzymes* (pp. 43–72). Academic Press. https://doi.org/10.1016/bs.enz.2015.05.001.

Snezhkina, A. V., Kudryavtseva, A. V., Kardymon, O. L., Savvateeva, M. V., Melnikova, N. V., Krasnov, G. S., et al. (2019). ROS generation and antioxidant defense systems in normal and malignant cells. *Oxidative Medicine and Cellular Longevity, 2019*, 1–17. https://doi.org/10.1155/2019/6175804.

Song, W. O., Chun, O. K., Hwang, I., Shin, H. S., Kim, B. G., Kim, K. S., et al. (2007). Soy isoflavones as safe functional ingredients. *Journal of Medicinal Food, 10*(4), 571–580. https://doi.org/10.1089/jmf.2006.0620.

Steelman, L. S., Abrams, S. L., Whelan, J., Bertrand, F. E., Ludwig, D. E., Bäsecke, J., et al. (2008). Contributions of the Raf/MEK/ERK, PI3K/PTEN/Akt/mTOR and Jak/STAT pathways to leukemia. *Leukemia, 22*(4), 686–707. https://doi.org/10.1038/leu.2008.26.

Suzuki, H., Nishizawa, T., Tsugawa, H., Mogami, S., & Hibi, T. (2012). Roles of oxidative stress in stomach disorders. *Journal of Clinical Biochemistry and Nutrition, 50*(1), 35–39. https://doi.org/10.3164/jcbn.11-115SR.

Tan, B. L., Norhaizan, M. E., & Liew, W. P. P. (2018). Nutrients and oxidative stress: Friend or foe? *Oxidative Medicine and Cellular Longevity, 2018*. https://doi.org/10.1155/2018/9719584.

Tan, B. L., Norhaizan, M. E., Liew, W. P. P., & Sulaiman Rahman, H. (2018). Antioxidant and oxidative stress: A mutual interplay in age-related diseases. *Frontiers in Pharmacology, 9*, 1162. https://doi.org/10.3389/fphar.2018.01162.

Tandon, R., Khanna, R. D., Dorababu, M., & Goel, R. K. (2004). Oxidative stress and antioxidants status in peptic ulcer and gastric carcinoma. *Indian Journal of Physiology and Pharmacology, 48*(1), 115–118.

Terao, J. (2009). Dietary flavonoids as antioxidants. In T. Yoshikawa (Ed.), *Vol. 61. Food factors for health promotion* (pp. 87–94). Karger Publishers. https://doi.org/10.1159/000212741.

Tresserra-Rimbau, A., Rimm, E. B., Medina-Remón, A., Martínez-González, M. A., López-Sabater, M. C., Covas, M. I., et al. (2014). Polyphenol intake and mortality risk: A re-analysis of the PREDIMED trial. *BMC Medicine*, *12*(1), 77. https://doi.org/10.1186/1741-7015-12-77.

Tu, S. H., Chen, L. C., & Ho, Y. S. (2017). An apple a day to prevent cancer formation: Reducing cancer risk with flavonoids. *Journal of Food and Drug Analysis*, *25*(1), 119–124. https://doi.org/10.1016/j.jfda.2016.10.016.

Uttara, B., Singh, A. V., Zamboni, P., & Mahajan, R. (2009). Oxidative stress and neurodegenerative diseases: A review of upstream and downstream antioxidant therapeutic options. *Current Neuropharmacology*, 7(1), 65–74.

Vasanthi, H. R., & Parameswari, R. P. (2010). Indian spices for healthy heart-an overview. *Current Cardiology Reviews*, *6*(4), 274–279. https://doi.org/10.2174/157340310793566172.

Vaya, J., & Aviram, M. (2001). Nutritional antioxidants mechanisms of action, analyses of activities and medical applications. *Current Medicinal Chemistry: Immunology, Endocrine & Metabolic Agents*, *1*(1), 99–117. https://doi.org/10.2174/1568013013359168.

Vidavalur, R., Otani, H., Singal, P. K., & Maulik, N. (2006). Significance of wine and resveratrol in cardiovascular disease: French paradox revisited. *Experimental and Clinical Cardiology*, *11*(3), 217.

World Health Organization. (2003). *Diet, nutrition, and the prevention of chronic diseases: Report of a joint WHO/FAO expert consultation. Vol. 916.* Geneva: World Health Organization.

Wrigley, C. W., Corke, H., Seetharaman, K., & Faubion, J. (Eds.). (2015). *Encyclopedia of food grains* Academic Press.

Xing, L., Zhang, H., Qi, R., Tsao, R., & Mine, Y. (2019). Recent advances in the understanding of the health benefits and molecular mechanisms associated with green tea polyphenols. *Journal of Agricultural and Food Chemistry*, *67*(4), 1029–1043. https://doi.org/10.1021/acs.jafc.8b06146.

Yan, Z., Zhong, Y., Duan, Y., Chen, Q., & Li, F. (2020). Antioxidant mechanism of tea polyphenols and its impact on health benefits. *Animal Nutrition*. https://doi.org/10.1016/j.aninu.2020.01.001.

Yoo, S. J., Go, E., Kim, Y. E., Lee, S., & Kwon, J. (2016). Roles of reactive oxygen species in rheumatoid arthritis pathogenesis. *Journal of Rheumatic Diseases*, *23*(6), 340–347. https://doi.org/10.4078/jrd.2016.23.6.340.

Zuo, X., Tian, C., Zhao, N., Ren, W., Meng, Y., Jin, X., et al. (2014). Tea polyphenols alleviate high fat and high glucose-induced endothelial hyperpermeability by attenuating ROS production via NADPH oxidase pathway. *BMC Research Notes*, 7(1), 120. https://doi.org/10.1186/1756-0500-7-120.

CHAPTER FIFTEEN

Protein L-isoAspartyl Methyltransferase (PIMT) and antioxidants in plants

Shraboni Ghosh[†] and Manoj Majee*
National Institute of Plant Genome Research, New Delhi, India
*Corresponding author: e-mail address: manojmajee@nipgr.ac.in

Contents

Abstract

All life forms, including plants, accumulate reactive oxygen species (ROS) as a byproduct of metabolism; however, environmental stresses, including abiotic stresses and pathogen attacks, cause enhanced accumulation of ROS in plants. The increased accumulation of ROS often causes oxidative damage to cells. Organisms are able to maintain levels of ROS below permissible limits by several mechanisms, including efficient antioxidant systems. In addition to antioxidant systems, recent studies suggest that protein L-isoaspartyl methyltransferase (PIMT), a highly conserved protein repair enzyme across evolutionary diverse organisms, plays a critical role in maintaining ROS homeostasis by repairing isoaspartyl-mediated damage to antioxidants in plants. Under stress conditions, antioxidant proteins undergo spontaneous isoaspartyl (isoAsp) modification

[†] Present address: Department of Biosciences, Durham University, Stockton Road, Durham DH1 3LE, United Kingdom.

Vitamins and Hormones, Volume 121
ISSN 0083-6729
https://doi.org/10.1016/bs.vh.2022.10.005

which is often detrimental to protein structure and function. This reduces the catalytic action of antioxidants and disturbs the ROS homeostasis of cells. This chapter focuses on PIMT and its interaction with antioxidants in plants, where PIMT constitutes a secondary level of protection by shielding a primary level of antioxidants from dysfunction and permitting them to guard during unfavorable situations.

1. Introduction

Reactive oxygen species (ROS) are often considered by-products of aerobic metabolism. ROS molecules can be either free radical (superoxide anion, hydroxyl radical) or non-radical molecule (hydrogen peroxide, singlet oxygen). In contrast to their destructive role, ROS are also recognized as second messengers in various biological processes, especially under stressful conditions (Bailly, 2004; Ghosh & Ghosh, 2021a; Kant, Tyagi, Ghosh, & Jha, 2019; Mittler, 2002). The fate of ROS rests on the equilibrium between the generation of ROS and its scavenging. As ROS molecules have a dual role (destructive or signaling), it is important for the cells to regulate ROS levels to dodge any kind of oxidative damage and avoid their complete eradication from the system (Neill, Desikan, & Hancock, 2002). In plants, ROS scavenging is done by a well-organized system of antioxidants comprising enzymatic and non-enzymatic antioxidants. The non enzymatic antioxidants include ascorbate, glutathione, carotenoids, vitamins etc. The antioxidant enzyme system includes superoxide dismutase (SOD), catalase (CAT), guaiacol peroxidase (GPX), enzymes of ascorbate—glutathione pathway such as ascorbate peroxidase, monodehydroascorbate reductase (MDHAR), dehydroascorbate reductase (DHAR), glutathione reductase (GR) etc. (Bailly, 2004; Mittler, 2002).

Antioxidants are susceptible to various modifications that negatively affect their biological activity. For example, recent studies have shown that antioxidants like superoxide dismutase (SOD) and catalase (CAT) undergo isoaspartyl modification under stressful environments that negatively influence their functional competence. Therefore, preservation of a strong antioxidant activity is essential to maintain ROS homeostasis during stressful environments, which is also associated with improved stress tolerance under external stimuli (Chen, Nayak, et al., 2010; Chen, Zhang, & Shen, 2010; Zaefyzadeh, Quliyev, Babayeva, & Abbasov, 2009). Organisms, including plants, have developed specialized protein repair enzyme systems to mitigate protein damages. The deleterious alterations in proteins are recognized and reversed by protein repair mechanisms Among protein repairing systems, peptidyl-prolyl cis-transisomerases

(PPIases) are enzymes that convert the abnormal cis-proline residues to normal trans-proline (Schiene & Fischer, 2000).

Additionally, methionine sulfoxide reductase (MSR) recognizes methionine sulfoxide residues in impaired proteins and reduces them back to normal methionine residues (Grimaud et al., 2001; Ruan et al., 2002). Another highly conserved protein repairing enzyme, PROTEIN L-ISOASPARTYL O-METHYLTRANSFERASE (PIMT, EC 2.1.1.77), recognizes and repairs abnormal isoAsp residues in proteins and peptides. Formation of abnormal L-isoAsp residues is the most common racemization in almost all organisms, and it is found to increase with age (Cloos & Christgau, 2002). This chapter highlights the relationship between antioxidants and PIMT during abiotic stress conditions. We have given an overview of the mechanism of PIMT action in repairing antioxidants.

2. Abiotic stress and reactive oxygen species

Abiotic stress can be defined as the negative effect of inorganic factors on living organisms. High temperature or low temperature, salinity, drought or flooding, nutrient deficiency or metal toxicity are a few of the major abiotic stress conditions a plant faces daily during its life cycle. In the agricultural sector, abiotic stress is one of the main reasons for massive monetary and production losses (Wang, Vinocur, & Altman, 2003; Wania, Kumar, Shriram, & Sah, 2016). Abiotic stress has become a major concern affecting crop productivity due to a changing environment, pollution, and global warming. Unlike animals, plants are sessile; hence, they need to develop highly specialized mechanisms for combating stressful environmental conditions. Thus, plants have developed an array of physiological and metabolic responses through activation of stress-responsive genes to survive in unfavorable conditions. Generation of ROS or reactive nitrogen species (RNS) is one of the initial responses of plants that help them to survive under stressful environments. They further regulate many signaling pathways through activation of secondary messengers, induction of transcription and alterations in enzyme activity (Lamotte et al., 2015; Mengel, Chaki, Shekariesfahlan, & Lidermayr, 2013).

At a low level, ROS thus act as a signaling molecules, but when a stressful environment persists, they are accumulated at higher concentrations which lead to oxidative damage and ultimately cell death (Halliwell & Gutteridge, 1999; Stadtman, 2004; Valko, Rhodes, Moncol, Izakovic, & Mazur, 2006). ROS mainly includes hydrogen peroxide (H_2O_2), superoxide radical ($O_2^{\bullet -}$), hydroxyl radical ($OH^{\bullet}$) and singlet oxygen ($^1O_2$). Generation of

ROS primarily results from excitation or partial reduction of molecular oxygen. They are considered as detrimental by-products of necessary cellular metabolism in oxygen-consuming organisms (Apel & Hirt, 2004; Miller, Suzuki, Ciftci-Yilmaz, & Mittler, 2010). ROS, as signaling molecules, control many important aspects of plant development, abiotic and biotic stress responses (Apel & Hirt, 2004; Mittler, Vanderauwera, Gollery, & Van Breusegem, 2004). To combat abiotic stress tolerance, stress-tolerant plants with boosted plant defense mechanisms should be generated. For improved tolerance to abiotic stress, the conventional approach for engineering plants comprises intensification of the endogenous arrangement by intruding at various levels of the response, starting from sensors and signaling/regulatory elements, e.g., kinases, transcription factors, to direct-action genes or effectors, e.g., antioxidants, Heat Shock Proteins (HSPs), and enzymes for the biosynthesis of osmoprotectants. Gene manipulation at the post-translational level of signaling cascade, regulatory systems based on small RNAs, epigenetic control of gene expression and overlapping effects of several hormones; allow different approaches for achieving stress tolerance.

3. Antioxidants in plants

Changes in environmental conditions lead to ROS formation in plants and animals. Generation of such harmful molecules lead to undesired alterations in proteins, nucleic acid, sugars, lipids, pigment, and other essential biomolecules. During favorable conditions, a basal level of ROS is maintained in a cell. When stress is perceived, ROS levels are elevated at different sites like cell wall, cell membrane, chloroplast, endoplasmic reticulum, peroxisomes and the mitochondria. Under light conditions, chloroplast and peroxisomes are the primary sites for ROS production, whereas, during dark conditions, mitochondria are the major sites of ROS generation (Bailly, 2004; Choudhury, Panda, Sahoo, & Panda, 2013; Das & Roychoudhury, 2014; Mittler, 2002). To protect the cell from such alterations, plants have developed a robust defense system involving antioxidants. Antioxidants can be enzymatic or non-enzymatic. The enzymatic antioxidants include ascorbate peroxidase (APX), CAT, dehydroascorbate reductases (DHAR), guaiacol peroxidase (GOPX), glutathione peroxidase (GPX), glutathione reductases (GR), glutathione-S- transferase (GST), monodehydroascorbate reductases (MDHAR), and SOD. These antioxidants reduce ROS either by breaking down of ROS or removing free radical and non-free radical molecules (Bailly, 2004; Mittler, 2002). The details about these antioxidants have been tabulated in Table 1.

Table 1 List of antioxidants in plants.

S. no	Antioxidative enzymes	Enzyme code	Localization	Function
1.	Ascorbate peroxidase (APX)	EC 1.11.1.11	tAPX is found in thylakoid; gmAPX is located at glyoxisome membrane; sAPX is found in chloroplast stroma; and cAPX is found in cytosol	APX catalyzes detoxification of H_2O_2
2.	Catalase (CAT)	EC 1.11.1.6	CAT1 and CAT2 are found in peroxisomes and cytosol; CAT3 is localized in mitochondria	CAT catalyzes the conversion of hydrogen peroxide into oxygen and water
3.	Dehydroascorbate reductases (DHAR)	EC 1.8.5.1	DHARs are located chiefly in the cytosol	DHAR regulate ascorbic acid redox state
4.	Guaiacol peroxidase (GOPX)	EC 1.11.1.7	GOPX is present in the cytosol, vacuole, cell wall, and apoplast	GOPX oxidizes aromatic electron donors (like guaiacol and pyrogallol) by using hydrogen peroxide
5.	Glutathione peroxidase (GPX)	EC 1.11.1.9	GPX is localized to chloroplast, mitochondria, cytosol, and ER	GPX removes hydrogen peroxides from a cell
6.	Glutathione reductases (GR)	EC 1.6.4.2	GR is found in the chloroplast stroma, mitochondria, cytosol, and peroxisomes	GR catalyzes NADPH-dependent reduction of oxidized glutathione (GSSG) to its reduced form (GSH)
7.	Glutathione-S-transferase (GST)	EC 2.5.1.18	Primarily found in the cytosol. GSTT1, GSTT2, and GSTT3 are located in the peroxisomes. GSTT3L and GSTU12 are found in the nucleus; GSTL2 is present in chloroplast	GST is involved in the detoxification of herbicides, phytohormone homeostasis, and sequestration of anthocyanin into vacuoles
8.	Monodehydro-ascorbate reductases (MDHAR)	EC 1.6.5.4	MDHAR is mainly found in chloroplast and cytosol, it is also present in peroxisomes, glyoxysomes, and mitochondria	MDHAR catalyzes the reduction of the monodehydroascorbate-e radical, also scavange H_2O_2
9.	Superoxide dismutase (SOD)	EC 1.15.1.1	Fe-SOD is found in plastids; Cu/Zn-SOD are found in cytosol and chloroplast; Mn-SOD is found in the mitochondrial matrix and peroxisomes	SOD catalyzes the removal of superoxide anion by dismutating it into oxygen and hydrogen peroxide

Non-enzymatic antioxidants comprise alpha-tocopherol (VitE), ascorbic acid (ASH or VitC), glutathione (GSH), phenolic compounds, some alkaloids, amino acids like proline, carotenoids, and flavonoids. They work by interrupting free radical chain reactions (Ahmad, Umar, & Sharma, 2010; Gill & Tuteja, 2010).

4. Protein L-isoAspartyl Methyltransferases

Under stressful environments and upon aging, proteins become vulnerable and undergo multiple spontaneous covalent modifications (Clarke, 2003). Because of their unstable behavior, proteins can experience a range of critical modifications like chemical alterations, including glycation, oxidation, deamidation, racemization, and isomerization etc. (Stadtman, 1988). These modifications may change their original structure and inherent role.

To prevail over such protein damage, organisms, including plants, have evolved specialized protein repairing enzyme systems. The detrimental modifications in proteins are recognized and reversed by protein repair mechanisms that allow the damaged proteins to recover. PIMT is a protein repairing enzyme which recognizes L-isoaspartate (or D-aspartate) residues in proteins (Aswad, 1984; McFadden & Clarke, 1982). Several other spontaneous protein alterations occur in a cell which have no known repair mechanisms. In that case, such damaged proteins are degraded by proteolytic pathways, which release free amino acids required for synthesizing new proteins.

PIMT recognizes deamidated asparagine or isomerized aspartate residues. With the help of AdoMet, a methyl group is transferred to the carboxyl group of L-isoAsp, leading to formation of methyl esters and release of methanol, as a by-product. Eventually, L-isoAsp are converted back to L-Asp *via* an intermediate structure, Succinimide. Succinimide hydrolyzes to generate a mixture of L-isoAsp and L-Asp in a 3:1 ratio. After many repetitive cycles, a considerable amount of deleterious L-isoAsp residues is converted back to normal L-Asp residues (McFadden & Clarke, 1987).

PIMT was coincidently discovered by Axelrod and Daly in 1965 while investigating the metabolism of AdoMet in bovine pituitary extracts (Axelrod & Daly, 1965). In the beginning, it was named as 'methanol forming enzyme' due to the presence of radioactive methanol in the pituitary extracts. The first plant *PIMT* gene was described from wheat (Mudgett & Clarke, 1993). Unlike animals, there are two genes encoding for PIMT in plants. PIMT1 and PIMT2 are members of the class I family of AdoMet-dependent methyltransferases. They usually have five conserved motifs [Pre region-I: T(I/V)SAPHM(H/V)A, Region-I: LD(V/I)GSG(T/S)G, Region-II: APY (D/N)AIHVG, Region-III: QL(K/A)(P/X)GGR(M/L)(V/I)(I/V), Post region-III: V(R/I)YPLTS] (Kamble & Majee, 2020).

PIMT repairing action generally depends on S-adenosyl methionine (AdoMet), which acts as a cofactor (Fig. 1). When PIMT recognizes an isoAsp residue in any protein, with the help of AdoMet, it catalyzes the transfer of a methyl group from AdoMet to the free α-carboxyl group of L-isoAsp residues to form a methyl ester and S-adenosyl homocysteine (AdoHcy). Next, L-isoAsp α-methyl ester is converted back to L-succinimide by the release of methanol. Finally, L-succinimide hydrolyzes to give rise to L-isoAsp or L-Asp. Like L-isoAsp, PIMT can also identify and repair D-Asp via AdoMet-dependent methyl esterification reaction (Aswad, Paranandi, & Schurter, 2000; Brennan, Anderson, Jia, Waygood, & Clarke, 1994; Lowenson & Clarke, 1992; McFadden & Clarke, 1987).

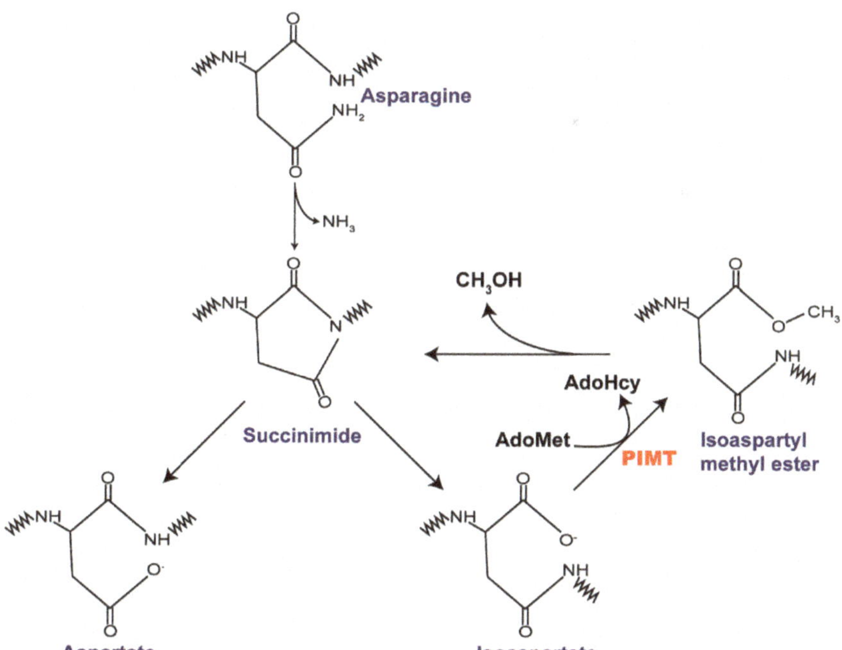

Fig. 1 Mechanism of PIMT repair.

5. Role of PIMT in plants

5.1 PIMT maintains seed longevity and germination vigor

Being sessile, plants cannot move away from unfavorable conditions and thus have evolved tight regulatory mechanism for their survival. In this context, seed occupies a very crucial positions in plant's life cycle, because seeds have a very unique ability to sense environmental cues and accordingly survive in adverse environmental conditions. Seeds posses various mechanisms to remain viable and retain germination ability for prolong period of time. In these mechanisms, protein repair mechanisms appear to play a vital role for the survival of seeds particularly for an extended period. PIMT is one among these protein repairing enzymes which play a crucial role in protecting structure and functions of seed proteins, which are otherwise vulnerable from unfavorable modifications in desiccated seeds. The presence and role of PIMT proteins in seeds has been reported in various plant species (Table 2). The first PIMT coding gene in plants was reported in wheat (Mudgett & Clarke, 1993). Initially, PIMT was linked to exceptionally high seed longevity in sacred lotus (one of the world's longest living seeds) as the enzyme was found to be highly accumulated in their seeds (Shen-Miller, Mudgett, Schopf, Clarke, & Berger, 1995). In tomato, abundant PIMT activity was seen till 48 h post-imbibition; later a decline in activity was observed after germination (Kester, Geneve, & Houtz, 1997). In addition, seventeen-year-old barley (*Hordeum vulgare* cv Himalaya) with low vigor seed lot was revealed to possess greater isoAsp amount due to low PIMT activity (Mudgett, Lowenson, & Clarke, 1997) and subsequently Kumar, Houtz, and Knowles (1999) showed elevation in PIMT activity in aged potato tubers. In Arabidopsis, overexpression of PIMT1 resulted in enhanced germination vigor and longevity by preventing abnormal L-isoaspartyl accumulation in seed proteins, whereas lower levels of PIMT1 elevated sensitivity to aging treatments and loss of seed vigor in Arabidopsis mutant under stressful germination conditions (Oge et al., 2008). In Arabidopsis, *AtPIMT2* was detected at the time of seed maturation, indicating a role in seed development (Xu et al., 2004). Later, participation of CaPIMT2 was also documented in seed vigor and longevity in chickpea (Verma et al., 2013; Verma, Singh, Kaur, & Majee, 2010). Therefore, it was evident that both PIMT1 and PIMT2 (isoforms of PIMT) facilitate improved seed longevity and germination vigor. Both PIMT and PIMT2 was shown to be developmentally regulated in various

Table 2 List of characterized PIMT gene(s) and their physiological roles in various plant species.

Organisms	*PIMT-* Gene(s)	Physiological role	References
Triticum aestivum	*PIMT*	Not reported	Mudgett and Clarke (1993)
Arabidopsis thaliana	*PIMT1*	Seed longevity and germination vigor Abiotic stress tolerance	Oge et al. (2008) Ghosh, Kamble, Verma, et al. (2020), Ghosh, Kamble, and Majee (2020)
	PIMT2	Abiotic stress tolerance	Ghosh, Kamble, Verma, et al. (2020), Ghosh, Kamble, and Majee (2020)
Cicer arietinum	*PIMT1*	Seed vigor and longevity	Verma et al. (2013)
	PIMT2	Seed vigor and longevity	Verma et al. (2013)
Oryza sativa	*PIMT1*	Seed vigor and longevity Seed desiccation tolerance	Petla et al. (2016) Wei et al. (2015) Kamble and Majee (2022)
	PIMT2	Seed vigor and longevity Seed desiccation tolerance	Petla et al. (2016) Kamble and Majee (2022)
Oryza coarctata	*PIMT1*	Seed vigor and longevity Seed trait improvement (upon ectopic expression in other plant species, but not in native system *O. coarctata* due to the lack of PIMT expression in seed)	Kamble and Majee (2022) Kamble, Petla, Ghosh, Achary, and Majee (2022)
	PIMT2	Seed vigor and longevity, Seed trait improvement(upon ectopic expression in other plant species, but not in *O. coarctata* due to the lack of PIMT expression n seed)	Kamble and Majee (2022) Kamble, Petla, et al. (2022)

plant species. In Arabidopsis, AtABI4 was shown to regulate the expression of AtPIMT1 to impart seed vigor and longevity (Kamble, Ghosh, Achary, & Majee, 2022). In rice, different isoforms of PIMT express differentially during seed maturation. PIMT particularly OsPIMT2 was found to be mainly localized in the embryo, but not in endosperms, of rice seeds. Subsequently,

PIMT was shown to maintain ROS homeostasis in seeds. Furthermore, a link between antioxidants and OsPIMT was defined in rice seeds. They demonstrated that OsPIMT1 and OsPIMT2 might protect age-induced damages to antioxidants like APX and CAT in seeds that help to provide improved seed vigor and longevity (Petla et al., 2016). Recently, Kamble & Majee, 2022 revealed that a PIMT-mediated protein repair system is vital for seed desiccation tolerance in rice. Wild rice with a weak repair system showed desiccation intolerance. Studies have hypothesized that PIMT enzymes protect several proteins, which are implicated in seed vigor and longevity, from isoAsp induced damage, and allow them to remain functional for maintaining seed vigor and viability. Several studies have reported that protective proteins like LATE EMBRYOGENESIS ABUNDANT (LEA) proteins, HEAT SHOCK PROTEINS (HSP), etc. which, participate in seed desiccation tolerance are susceptible to isoAsp induced modification that negatively influences their biological role (Chen, Nayak, et al., 2010; Chen, Zhang, et al., 2010; Nayak et al., 2013).

5.2 PIMT promotes stress adaption

Plants have established complex machineries to survive during stressful conditions. During stress conditions, plants activate several enzymes and hormones to survive such conditions. PIMT play pivotal role in regulating many stress tolerance mechanisms in animals and plants. Initial reports in the last decade suggested expression of PIMT is dependent on environmental conditions. Expression and activity of PIMT are induced during stress conditions. In wheat (*Triticum aestivum*) seedlings, dehydration and salinity stress promote PIMT genes' transcription and repairing activity (Mudgett & Clarke, 1994). Kindrachuk et al., 2003 showed that during heat stress conditions, over accumulation of PIMT boosts heat shock survival capabilities in *Escherichia coli*. Furthermore, PIMT from chickpea resulted in increased survival of *E. coli* under oxidative stress (Verma et al., 2010). Elevated PIMT activity was observed when corn seeds were challenged with osmotic or salt stress (Thapar, Kim, & Clarke, 2001). Recently, Ghosh, Kamble, Verma, et al. (2020) demonstrated PIMT-mediated repair as a second tier of protection in the cellular system in response to abiotic stress, where PIMT protects the first tier antioxidants, which, when functional, detoxifies ROS.

6. PIMT and antioxidants

6.1 Superoxide dismutase

Superoxide dismutases (SODs) are one of the most important antioxidants in a cell. They catalyze the dismutation of superoxides into oxygen and hydrogen peroxide. Superoxides are produced at multiple locations within a cell. Therefore, SODs are localized at several sites like mitochondria, chloroplast, glyoxysome, peroxisome, cytosol etc. (Fridovich, 1986).

Generally, superoxide is generated by photosystem I (PSI) (chloroplast), electron transport chains (mitochondria and peroxisomes) and xanthine dehydrogenase (cytosol) (Vaahtera, Brosche, Wrzaczek, & Kangasjarvi, 2014; Zarepour et al., 2010). Abiotic or biotic stresses can perturb electron transportation in such organelles, thus elevating superoxide generation. Environmental changes can also disturb the antioxidant systems involved in ROS scavenging, endorsing stress-responsive accumulation of superoxide ions (You & Chan, 2015). Besides, superoxide ions are also produced in the initial step of stress signaling. For example, during a pathogenic attack, an oxidative burst in the apoplast is the first step of plant defense and depends on plasma membrane-bound NADPH oxidases (Torres, Jones, & Dangl, 2006).

Antioxidants like superoxide dismutase (SOD) and catalase (CAT) become susceptible to isoAsp formation in response to stress. PIMT recognizes and repairs the isoAsp residues and thus restores their catalytic efficacies. The repaired SOD further performs dismutation of superoxide to hydrogen peroxide. CAT then decomposes the hydrogen peroxide into water and oxygen.

Stress-induced buildup of superoxide ions contributes to injuries at a cellular level. Superoxides can permanently inactivate some enzymes through oxidation of their iron-sulfur centers, and activate hydroxyl radical production, which then reacts with other biomolecules (Farmer & Mueller, 2013). Superoxide dismutases catalyze the rapid dismutation of superoxide, consequently diminish the threat of hydroxyl radical accumulation via metal-catalyzed reactions. SOD-mediated dismutation is 10 thousand times quicker than other spontaneous reactions. Depending on the metal cofactor in the active center, there are three categories of SOD enzymes in plants: Cu/Zn-SOD, Mn-SOD, Fe-SOD. Among three, Cu/Zn-SOD is the most abundant isozyme in chloroplast stroma, cytosol, peroxisomes, and apoplast.

Mn-SOD is mainly localized in mitochondria and peroxisomes but has also been detected in both apoplast and cell walls. Fe-SOD expression is relatively low and is constrained to chloroplast stroma of some plant species (Mittler et al., 2004).

SOD encoding genes are highly sensitive to external cues, and their increased activity is usually correlated with improved tolerance against stresses (Sharma, Jha, Dubey, & Pessarakli, 2012). They are encoded by a small multigene family, and the first plant SOD gene was cloned from maize (Cannon, White, & Scandalios, 1987). In Arabidopsis, there are seven isoforms for SOD, one MnSOD (MSD1), three FeSODs (FSD1, FSD2, FSD3), and three CuZnSODs (CSD1, CSD2, CSD3) (Kliebenstein, Monde, & Last, 1998). In response to stresses like heat, oxidative and salinity, SODs become vulnerable to isoAsp formation. Accumulation of isoAsp residues at asparagine sites in SOD enzymes leads to decreased dismutation efficacy. In such conditions, PIMT recognizes these damaged SOD proteins (MSD1 and CSD2) and restores their catalytic efficiency by repairing the isoAsp residues (Fig. 2) (Ghosh, Kamble, & Majee, 2020; Ghosh, Kamble, Verma, et al., 2020). Therefore, PIMT-mediated protein repair is an integral part of the effective functioning of SOD system.

6.2 Catalase

Catalase was the first antioxidant to be discovered. It is present in all the domains of life. According to Loew (1900), 'there seems to exist no plant and no animal which is without that particular enzyme.' In general, a typical catalase protein is composed s of polypeptides of 50–70 kDa in mass that are arranged into tetramers, where each monomer contains a heme prosthetic group (Regelsberger et al., 2001). They have a fast turnover rate and catalyze the decomposition of hydrogen peroxide into water and oxygen. Hydrogen

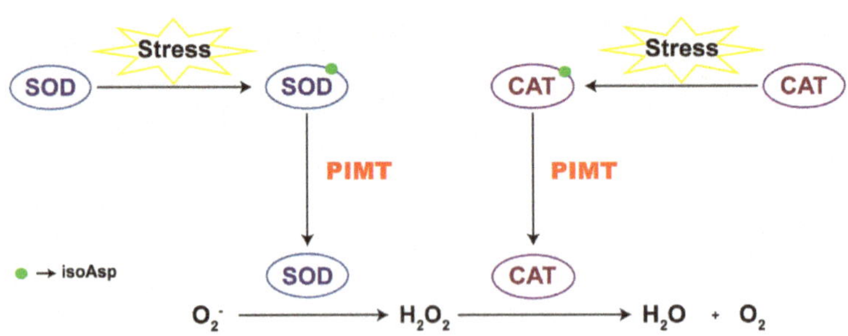

Fig. 2 PIMT-mediated repair of Antioxidants.

peroxide is produced directly by reduction of an oxygen molecule, indirectly through dismutation of superoxide radical or via electron transport chains (Foyer & Noctor, 2000; Mittler et al., 2004).

In angiosperms, there are three genes encoding for catalase. For example, Arabidopsis has three genes encoding for *CAT1*, *CAT2*, and *CAT3* (Frugoli et al., 1996). The expression of CAT2 and CAT3 is controlled by circadian rhythm, with contrasting day-night rhythms in abundance (Michael & McClung, 2002). In addition, expression of these genes is induced in response to different external factors like drought and salinity stress (Xing, Jia, & Zhang, 2007). CAT1 is strongly expressed in photosynthetic tissues, CAT2 is linked with vascular tissues. CAT3 is mainly expressed in seeds and reproductive tissues (Mhamdi et al., 2010). The protein sequence of all the CAT genes contains 492 amino acids, with similarity between sequences, whereas nucleotide sequences are relatively dissimilar. The resemblance between the proteins makes it tough to synthesize isoform-specific antibodies. In-gel antibody analysis can be achieved after electrophoretic separation of CAT2 and CAT3 isoforms (Smykowski, Zimmermann, & Zentgraf, 2010; Zimmermann, Heinlein, Orendi, & Zentgraf, 2006). Catalase enzymes are highly light-sensitive and have a high turnover rate. Overall, abiotic stresses that decrease protein turnover rate also reduce CAT enzymatic efficiency. For example, external stimuli like cold, salt, and high light may lead to a reduction in CAT protein level by accelerated protein damage (Hertwig, Streb, & Feierabend, 1992; Volk & Feierabend, 1989). In response to abiotic stress conditions, catalase enzymes become prone to isoAsp formation. These deleterious residues interfere in the ROS scavenging action of catalase. Out of the three CATs, CAT2, and CAT3 are repaired by PIMT during elevated temperatures (Fig. 2). In short, PIMT facilitates catalases to regain their enzymatic efficacy under stress (Ghosh, Kamble, Verma, et al., 2020). The functional analysis and control mechanism of class III catalase, CAT1 needs to be elucidated. It will be interesting to investigate if PIMT also repairs CAT1 under adverse conditions. Interaction of PIMT with catalase along with resulting impact can be of great significance for stress-responsive pathways.

7. Identification of isoAsp susceptible proteins

Since the discovery of PIMT, generations of isoAsp in proteins and peptides have been detected throughout the animal and plant kingdom. This spontaneous modification is not random in nature. isoAsp residues are generally formed in long-lived proteins and exist in organs with low metabolic activity (Kamble & Majee, 2020). Identification of isoAsp susceptible proteins will offer evidence of involvement of PIMT in various biological processes. Despite that, identification of isoAsp residues in proteins is a challenging task in PIMT research, mainly because of L-isoaspartyl methylester, which is highly unstable in nature and has a low level of methylatable residues in proteins. Hence, not many isoAsp susceptible proteins have been recognized until now. Nevertheless, with time technologies have become advanced, allowing site-specific identification of isoAsp residues in proteins.

For global analysis of isoAsp content in any protein sample, ISOQUANT detection kit (Promega, Madison, WI) can be used. Total protein or any recombinant protein is incubated with PIMT and S-adenosyl L-methionine. S-adenosyl homocysteine (a methylation by-product), is detected by reversed-phase High-Performance Liquid Chromatography (HPLC) and further quantified by comparing the peak area to a standard curve (Johnson & Aswad, 1991; Kharbanda et al., 2007; Schurter & Aswad, 2000). Another approach involves radiolabeled S-adenosyl L-methionine in the reaction. After PIMT repair, radioactive methanol is formed as a by-product. The electrons released are counted using a liquid scintillation counter (Ghosh, Kamble, Verma, et al., 2020; Petla et al., 2016; Verma et al., 2010).

Another proteomics-based technique on blot methylation (OBM) is used by the researchers to dodge high-resolution separation techniques like HPLC. In OBM, total protein is separated by 2D gel electrophoresis and blotted onto the membrane. Proteins with isoAsp residues are methylated on blot by covering it with PIMT and AdoMet. Lastly, PIMT substrates are recognized by peptide mass fingerprinting in 2D gel spots, which run parallel (Vigneswara et al., 2006; Zhu, Doyle, Mamula, & Aswad, 2006).

Chen, Nayak, et al. (2010), Chen, Zhang, et al. (2010) described phage display and bio-panning to recognize isoAsp susceptible proteins in plants. In this technique, the protein extraction step, which can yield unwanted isoAsp formation during protein extraction, and the intervention of the seed storage protein targets that mask other low abundant substrates of PIMT, could be avoided.

Identification of isoAsp residues in proteins like antioxidants can be useful for predicting their fate. By exploiting elegant techniques such as mass spectrometry, we can recognize site-specific Asp and Asn residues susceptible to isoAsp formation in any peptide. isoAsp residues are formed by deamidation of Asn or isomerization of Asp. The identification of deamidation of Asn in proteins or peptides is relatively easier than the identification of isomerization of Asp. As the residue mass of isoAsp is equivalent to the parent Asp, hence, calculation of molecular mass at the peptide level is not enough to detect site-specific isomerization (Zhang, Yip, & Katta, 2011). When combined with chromatographic separation, such analysis may identify the presence of isoAsp in certain peptides. For the identification of isomerization of Asp, MS/MS should comprise an electron capture dissociation (ECD)/electrontransfer dissociation (ETD) as they generate a unique y-ion.

By using MS/MS approach, Ghosh, Kamble, Verma, et al., 2020 identified class I and class III of catalase and two isoforms of superoxide dismutase (MSD1 and CSD2) to be isoAsp susceptible. The formation of isoAsp in catalase during extreme conditions leads to a decline in their catalytic efficacy. However, after PIMT repair, their ROS detoxification capability was restored. Hence by repairing antioxidants, PIMT plays a crucial role in ROS scavenging during abiotic stress. Identification of more antioxidants susceptible to isoAsp formation will lead to conceptual advancement of the antioxidative defense system developed by the plants to overcome stressful environments.

8. Conclusion

PIMT is a well-known protein repair enzyme found in all domains of life, including plants. Although PIMT was discovered many years back, and its activity has been detected in a wide range of plants, PIMT studies are generally restricted to stress biology. The function of PIMT is reported to extended life span and survivability across a range of species.

According to the literature, higher plants possess two PIMT coding genes, which are involved in maintaining seed vigor and longevity. Recently, PIMTs were also shown to be involved in abiotic stress tolerance by restoring the catalytic activities of antioxidants including CAT and SOD. Like other PIMT substrates, antioxidants become vulnerable to isoAsp formation during prolonged stress conditions, which is then recognized and repaired by PIMT.

Thus, the PIMT-mediated repair system acts as a first tier of cell protection in response to abiotic stress. PIMT protects the second-tier enzymes like antioxidants that facilitate in detoxifying ROS. Investigation of the relationship between PIMT and other antioxidants would indeed offer further understanding of PIMT function under stressful situations in plants.

References

Ahmad, P., Umar, S., & Sharma, S. (2010). Mechanism of free radical scavenging and role of Phytohormones in plants under abiotic stresses. In M. Ashraf, M. Ozturk, & M. Ahmad (Eds.), *Plant Adaptation and Phytoremediation*. Dordrecht: Springer.

Apel, K., & Hirt, H. (2004). Reactive oxygen species: Metabolism, oxidative stress, and signal transduction. *Annual Review of Plant Biology*, *55*, 373–399.

Aswad, D. W. (1984). Stoichiometric methylation of porcine adrenocorticotropin by protein carboxyl methyltransferase requires deamidation of asparagine 25. *Journal of Biological Chemistry*, *259*, 10714–10721.

Aswad, D. W., Paranandi, M. V., & Schurter, B. T. (2000). Isoaspartate in peptides and proteins: Formation, significance, and analysis. *Journal of Pharmaceutical and Biomedical Analysis*, *21*, 1129–1136.

Axelrod, J., & Daly, J. (1965). Pituitary gland: Enzymic formation of methanol from S-adenosylmethionine. *Science*, *150*, 892–893.

Bailly, C. (2004). Active oxygen species and antioxidants in seed biology. *Seed Science Research*, *14*, 93–107.

Brennan, T. V., Anderson, J. W., Jia, Z., Waygood, E. B., & Clarke, S. (1994). Repair of spontaneously deamidated HPr phosphocarrier protein catalyzed by the L-isoaspartate-(D-aspartate) O-methyltransferase. *Journal of Biological Chemistry*, *269*, 24586–24595.

Cannon, R. E., White, J. A., & Scandalios, J. G. (1987). Cloning of cDNA for maize superoxide dismutase 2 (SOD2). *Proceedings of the National Academy of Sciences of the United States of America*, *84*(1), 179–183.

Chen, T., Nayak, N., Majee, S. M., Lowenson, J., Schäfermeyer, K. R., Eliopoulos, A. C., et al. (2010). Substrates of the Arabidopsis thaliana protein isoaspartyl methyltransferase 1 identified using phage display and biopanning. *Journal of Biological Chemistry*, *285*, 37281–37292.

Chen, Q., Zhang, M., & Shen, S. (2010). Effect of salt on malondialdehyde and antioxidant enzymes in seedling roots of Jerusalem artichoke (Helianthus tuberosus L.). *Acta Physiologiae Plantarum*, *33*, 273–278.

Choudhury, S., Panda, P., Sahoo, L., & Panda, S. K. (2013). Reactive oxygen species signaling in plants under abiotic stress. *Plant Signaling & Behavior*, *8*, 4.

Clarke, S. (2003). Aging as war between chemical and biochemical processes: Protein methylation and the recognition of age-damaged proteins for repair. *Ageing Research Reviews*, *2*, 263–285.

Cloos, P. A., & Christgau, S. (2002). Non-enzymatic covalent modifications of proteins: Mechanisms, physiological consequences and clinical applications. *Matrix Biology*, *21*, 39–52.

Das, K., & Roychoudhury, A. (2014). Reactive oxygen species (ROS) and response of antioxidants as ROS-scavengers during environmental stress in plants. *Frontiers in Environmental Science*, *2*, 53.

Farmer, E. E., & Mueller, M. J. (2013). ROS-mediated lipid peroxidation and RES-activated signaling. *Annual Review of Plant Biology*, *64*, 429e450.

Foyer, C. H., & Noctor, G. (2000). Oxygen processing in photosynthesis: Regulation and signalling. *New Phytologist*, *146*, 359–388.

Fridovich, I. (1986). Biological effects of the superoxide radical. *Archives of Biochemistry and Biophysics*, *247*(1), 1–11.

Frugoli, J. A., Zhong, H. H., Nuccio, M. L., McCourt, P., McPeek, M. A., Thomas, T. L., et al. (1996). Catalase is encoded by a multigene family in Arabidopsis thaliana (L.) Heynh. *Plant Physiology*, *112*, 327–336.

Ghosh, S., & Ghosh, S. (2021a). Hitting hard times: Effect of abiotic stress on root physiology. In S. Mukherjee, & F. Baluška (Eds.), *Rhizobiology: Molecular Physiology of Plant Roots. Signaling and Communication in Plants*. Cham: Springer.

Ghosh, S., Kamble, N. U., & Majee, M. (2020). A protein repairing enzyme, protein L-isoaspartyl methyltransferase is involved in salinity stress tolerance by increasing efficiency of ROS-scavenging enzymes. *Environmental and Experimental Botany*, *180*, 104266.

Ghosh, S., Kamble, N. U., Verma, P., Salvi, P., Petla, B. P., Roy, S., et al. (2020). Arabidopsis protein L-Isoaspartyl methyltransferase repairs isoaspartyl damage to antioxidant enzymes and increases heat and oxidative stress tolerance. *Journal of Biological Chemistry*, *295*, 783–799.

Gill, S. S., & Tuteja, N. (2010). Reactive oxygen species and antioxidant machinery in abiotic stress tolerance in crop plants. *Plant Physiology and Biochemistry*, *48*, 909–930.

Grimaud, R., Ezraty, B., Mitchell, J. K., Lafitte, D., Briand, C., Derrick, P. J., et al. (2001). Repair of oxidized proteins: Identification of a new methionine sulfoxide reductase. *Journal of Biological Chemistry*, *276*, 48915–48920.

Halliwell, B., & Gutteridge, J. M. C. (1999). *Free radicals in biology and medicine* (3rd ed.). New York: Oxford University Press.

Hertwig, B., Streb, P., & Feierabend, J. (1992). Light dependence of catalase synthesis and degradation in leaves and the influence of interfering stress conditions. *Plant Physiology*, *100*, 1547–1553.

Johnson, B. A., & Aswad, D. W. (1991). Optimal conditions for the use of protein L-isoaspartyl methyltransferase in assessing the isoaspartate content of peptides and proteins. *Analytical Biochemistry*, *192*, 384–391.

Kamble, N. U., Ghosh, S., Achary, R. K., & Majee, M. (2022). Arabidopsis ABSCISIC ACID INSENSITIVE4 targets protein L-isoaspartyl methyltransferase1 in seed. *Planta*, *256*, 30.

Kamble, N. U., & Majee, M. (2020). Protein l-isoaspartyl methyltransferase (PIMT) in plants: Regulations and functions. *Biochemical Journal*, *477*(22), 4453–4471.

Kamble, N. U., & Majee, M. (2022). ABI transcription factors and protein L-isoaspartyl methyltransferase module mediate seed desiccation tolerance and longevity in Oryza sativa. *Development*, *149*(11), dev 200600.

Kamble, N. U., Petla, B. P., Ghosh, S., Achary, R. K., & Majee, M. (2022). Oryza coarctata protein L-isoaspartyl methyltransferase (PIMT) repairs isoaspartyl modification to antioxidative enzymes and is implicated in seed traits in rice. *Environmental and Experimental Botany*, *202*, 105027, ISSN 0098-8472, https://doi.org/10.1016/j.envexpbot.2022.105027.

Kant, R., Tyagi, K., Ghosh, S., & Jha, G. (2019). Host Alternative NADH:Ubiquinone oxidoreductase serves as a susceptibility factor to promote pathogenesis of Rhizoctonia solani in Plants. *Phytopathology*, *109*(10), 1741–1750.

Kester, S. T., Geneve, R. L., & Houtz, R. L. (1997). Priming and accelerated ageing affect L-isoaspartyl methyltransferase activity in tomato (Lycopersicon esculentum Mill.) seed. *Journal of Experimental Botany*, *48*, 943–949.

Kharbanda, K. K., Mailliard, M. E., Baldwin, C. R., Beckenhauer, H. C., Sorrell, M. F., & Tuma, D. J. (2007). Betaine attenuates alcoholic steatosis by restoring phosphatidylcholine generation via the phosphatidylethanolamine methyltransferase pathway. *Journal of Hepatology*, *46*(2), 314–321.

Kindrachuk, J., Parent, J., Davies, G. F., Dinsmore, M., Attah-Poku, S., & Napper, S. (2003). Overexpression of L-Isoaspartate O-methyltransferase in Escherichia coli increases heat shock survival by a mechanism independent of methyltransferase activity. *Journal of Biological Chemistry*, *278*, 50880–50886.

Kliebenstein, D. J., Monde, R. A., & Last, R. L. (1998). Superoxide dismutase in Arabidopsis: An eclectic enzyme family with disparate regulation and protein localization. *Plant Physiology*, *118*, 637–650.

Kumar, G. N. M., Houtz, R. L., & Knowles, N. R. (1999). Age induced protein modifications and increased proteolysis in potato seed tubers. *Plant Physiology*, *119*, 89–99.

Lamotte, O., Bertoldo, J. B., Besson-Bard, A., Rosnoblet, C., Aimé, S., Hichami, S., et al. (2015). Protein S-nitrosylation: Specificity and identification strategies in plants. *Frontiers in Chemistry, 2*, 114.

Loew, O. (1900). A new enzyme of general occurrence in organisms. *Science, 11*, 701–702.

Lowenson, J. D., & Clarke, S. (1992). Recognition of D-asaprtyl residues in polypeptides by the erythrocyte L-isoaspartyl/D-asaprtyl protein methyltransferase. Implications for the repair hypothesis. *Journal of Biological Chemistry, 267*, 5985–5995.

McFadden, P. N., & Clarke, S. (1982). Methylation at D-aspartyl residues in erythrocytes: Possible step in the repair of aged membrane proteins. *Proceedings of the National Academy of Sciences of the United States of America, 79*, 2460–2464.

McFadden, P. N., & Clarke, S. (1987). Conversion of isoaspartyl peptides to normal peptides: Implications for the cellular repair of damaged proteins. *Proceedings of the National Academy of Sciences of the United States of America, 84*, 2595–2599.

Mengel, A., Chaki, M., Shekariesfahlan, A., & Lidermayr, C. (2013). Effect of nitric oxide on gene transcription—S-nitrosylation of nuclear proteins. *Frontiers in Plant Science, 4*, 293.

Mhamdi, A., Queval, G., Chaouch, S., Vanderauwera, S., Van Breusegem, F., & Noctor, G. (2010). Catalase function in plants: A focus on *Arabidopsis* mutants as stress-mimic models. *Journal of Experimental Botany, 61*(15), 4197–4220.

Michael, T. P., & McClung, C. R. (2002). Phase-specific circadian clock regulatory elements in Arabidopsis. *Plant Physiology, 130*, 627–638.

Miller, G., Suzuki, N., Ciftci-Yilmaz, S., & Mittler, R. (2010). Reactive oxygen species homeostasis and signalling during drought and salinity stresses. *Plant, Cell & Environment, 33*, 453–467.

Mittler, R. (2002). Oxidative stress, antioxidants and stress tolerance. *Trends in Plant Science, 7*, 405–410.

Mittler, R., Vanderauwera, S., Gollery, M., & Van Breusegem, F. (2004). Reactive oxygen gene network of plants. *Trends in Plant Science, 9*, 490–498.

Mudgett, M. B., & Clarke, S. (1993). Characterization of plant L-Isoaspartyl methyltransferases that may be involved in seed survival: Purification, cloning, and sequence analysis of the wheat germ enzyme. *Biochemistry, 32*, 11100–11111.

Mudgett, M. B., & Clarke, S. (1994). Hormonal and environmental responsiveness of a developmentally regulated protein repair L-isoaspartyl methyltransferase in wheat. *Journal of Biological Chemistry, 269*, 25605–25612.

Mudgett, M. B., Lowenson, J. D., & Clarke, S. (1997). Protein repair L-isoaspartyl methyltransferases in plants (phylogenetic distribution and the accumulation of substrate proteins in aged barley seeds). *Plant Physiology, 115*, 1481–1489.

Nayak, N. R., Putnam, A. A., Addepalli, B., Lowenson, J. D., Chen, T., Jankowsky, E., et al. (2013). An Arabidopsis ATP-dependent, DEAD-box RNA helicase loses activity upon IsoAsp formation but is restored protein isoaspartyl methyltransferase. *The Plant Cell, 25*, 2573–2586.

Neill, S., Desikan, R., & Hancock, J. (2002). Hydrogen peroxide signalling. *Current Opinion in Plant Biology, 5*, 388–395.

Oge, L., Bourdais, G., Bove, J., Collet, B., Godin, B., Granier, F., et al. (2008). Protein repair L-Isoaspartyl methyltransferase 1 is involved in both seed longevity and germination vigor in Arabidopsis. *The Plant Cell, 20*, 3022–3037.

Petla, B. P., Kamble, N. U., Kumar, M., Verma, P., Ghosh, S., Singh, A., et al. (2016). Rice protein L-isoaspartyl methyltransferase isoforms differentially accumulate during seed maturation to restrict deleterious isoAsp and reactive oxygen species accumulation and are implicated in seed vigor and longevity. *New Phytologist, 211*, 627–645.

Regelsberger, G., Jakopitsch, C., Furtmuller, P. G., Rueker, F., Switala, J., Loewen, P. C., et al. (2001). The role of distal tryptophan in the bifunctional activity of catalase-peroxidases. *Biochemical Society Transactions, 29*, 99–105.

Ruan, H., Tang, X. D., Chen, M. L., Joiner, M. L. A., Sun, G., Brot, N., et al. (2002). High-quality life extension by the enzyme peptide methionine sulfoxide reductase. *Proceedings of the National Academy of Sciences of the United States of America, 99*, 2748–2753.

Schiene, C., & Fischer, G. (2000). Enzymes that catalyze the restructuring of proteins. *Current Opinion in Structural Biology, 10*, 40–45.

Schurter, B. T., & Aswad, D. W. (2000). Analysis of isoaspartate in peptides and proteins without the use of radioisotopes. *Analytical Biochemistry, 282*, 227–231.

Sharma, P., Jha, A. B., Dubey, R. S., & Pessarakli, M. (2012). Reactive oxygen species, oxidative damage, and antioxidant defence mechanism in plants under stressful conditions. *Journal of Botany, 2012*, 217037.

Shen-Miller, J., Mudgett, M. B., Schopf, J. W., Clarke, S., & Berger, R. (1995). Exceptional seed longevity and robust growth: Ancient sacred Lotus from China. *American Journal of Botany, 82*, 1367–1380.

Smykowski, A., Zimmermann, P., & Zentgraf, U. (2010). G-box binding factor 1 reduces CATALASE2 expression and regulates the onset of leaf senescence in Arabidopsis thaliana. *Plant Physiology, 153*, 1321–1331.

Stadtman, E. R. (1988). Protein modification in aging. *The Journals of Gerontology, 43*, B112–B120.

Stadtman, E. R. (2004). Role of oxidant species in aging. *Current Medicinal Chemistry, 11*, 1105–1112.

Thapar, N., Kim, A. K., & Clarke, S. (2001). Distinct patterns of expression but similar biochemical properties of protein L-isoaspartyl methyltransferase in higher plants. *Plant Physiology, 125*, 1023–1035.

Torres, M. A., Jones, J. D. G., & Dangl, J. L. (2006). Reactive oxygen species signaling in response to pathogens. *Plant Physiology, 141*, 373e378.

Vaahtera, L., Brosche, M., Wrzaczek, M., & Kangasjarvi, J. (2014). Specificity in ROS signaling and transcript signatures. *Antioxidant Redox Signals, 21*, 1422e1441.

Valko, M., Rhodes, C. J., Moncol, J., Izakovic, M., & Mazur, M. (2006). Free radicals, metals and antioxidants in oxidative stress-induced cancer. *Chemico-Biological Interactions, 160*, 1–40.

Verma, P., Kaur, H., Petla, B. P., Rao, V., Saxena, S. C., & Majee, M. (2013). Protein L-isoaspartyl methyltransferase 2 is differentially expressed in chickpea and enhances seed vigor and longevity by reducing abnormal isoaspartyl accumulation predominantly in seed nuclear proteins. *Plant Physiology, 161*, 1141–1157.

Verma, P., Singh, A., Kaur, H., & Majee, M. (2010). Protein L-isoaspartyl methyltransferase1 (CaPIMT1) from chickpea mitigates oxidative stress-induced growth inhibition of Escherichia coli. *Planta, 231*, 329–336.

Vigneswara, V., Lowenson, J. D., Powell, C. D., Thakur, M., Bailey, K., Clarke, S., et al. (2006). Proteomic identification of novel substrates of a protein isoaspartyl methyltransferase repair enzyme. *Journal of Biological Chemistry, 281*, 32619–32629.

Volk, S., & Feierabend, J. (1989). Photoinactivation of catalase at low temperature and its relevance to photosynthetic and peroxide metabolism in leaves. *Plant, Cell and Environment, 12*, 701–712.

Wang, W., Vinocur, B., & Altman, A. (2003). Plant responses to drought, salinity and extreme temperatures: Towards genetic engineering for stress tolerance. *Planta, 28*, 1–14.

Wania, S. H., Kumar, V., Shriram, V., & Sah, S. K. (2016). Phytohormones and their metabolic engineering for abiotic stress tolerance in crop plants. *The Crop Journal, 4*, 162–176.

Wei, Y., Huibin, X., Lirong, D., Yongsheng, Z., et al. (2015). Protein repair L-isoaspartyl methyltransferase 1 (PIMT1) in rice improves seed longevity by preserving embryo vigor and viability. *Plant Molecular Biology, 89*, 475–492. https://doi.org/10.1007/s11103-015-0383-1.

Xing, Y., Jia, W., & Zhang, J. (2007). AtMEK1 mediates stress-induced gene expression of CAT1 catalase by triggering H2O2 production in Arabidopsis. *Journal of Experimental Botany*, *58*, 2969–2981.

Xu, Q., Belcastro, M. P., Villa, S. T., Dinkins, R. D., Clarke, S. G., & Downie, A. B. (2004). A second protein L-Isoaspartyl methyltransferase gene in Arabidopsis produces two transcripts whose products are sequestered in the nucleus. *Plant Physiology*, *136*, 2652–2664.

You, J., & Chan, Z. (2015). ROS regulation during abiotic stress responses in crop plants. *Frontiers in Plant Science*, *6*, 1092e1107.

Zaefyzadeh, M., Quliyev, R. A., Babayeva, S. M., & Abbasov, M. A. (2009). The effect of the interaction between genotypes and drought stress on the superoxide dismutase and chlorophyll content in durum wheat landraces. *Turkish Journal of Biology*, *33*, 1–7.

Zarepour, M., Kaspari, K., Stagge, S., Rethmeier, R., Mendel, R., & Bittner, F. (2010). Xanthine dehydrogenase atxdh1 from Arabidopsis thaliana is a potent producer of superoxide anions via its nadh oxidase activity. *Plant Molecular Biology*, *72*, 301e310.

Zhang, J., Yip, H., & Katta, V. (2011). Identification of isomerization and racemization of aspartate in the Asp-Asp motifs of a therapeutic protein. *Analytical Biochemistry*, *410*, 234–243.

Zhu, J. X., Doyle, H. A., Mamula, M. J., & Aswad, D. W. (2006). Protein repair in the brain, proteomic analysis of endogenous substrates for protein L-isoaspartyl methyltransferase in mouse brain. *Journal of Biological Chemistry*, *281*, 33802–33813.

Zimmermann, P., Heinlein, C., Orendi, G., & Zentgraf, U. (2006). Senescence-specific regulation of catalase in Arabidopsis thaliana (L.) Heynh. *Plant, Cell and Environment*, *29*, 1049–1056.

Kolibri 360 GmbH
Ferdinand-Jühlke-Straße 7
99095 Erfurt, Deutschland
produktsicherheit@kolibri360.de